Springer Proceedings in Energy

The series Springer Proceedings in Energy covers a broad range of multidisciplinary subjects in those research fields closely related to present and future forms of energy as a resource for human societies. Typically based on material presented at conferences, workshops and similar scientific meetings, volumes published in this series will constitute comprehensive state-of-the-art references on energy-related science and technology studies. The subjects of these conferences will fall typically within these broad categories:

- Energy Efficiency
- Fossil Fuels
- Nuclear Energy
- Policy, Economics, Management & Transport
- Renewable and Green Energy
- Systems, Storage and Harvesting
- Materials for Energy

eBook Volumes in the Springer Proceedings in Energy will be available online in the world's most extensive eBook collection, as part of the Springer Energy eBook Collection. To submit a proposal or for further inquiries, please contact the Springer Editor in your region:

Kamiya Khatter (India)
Email: kamiya.khatter@springer.com
Loyola D'Silva (All other countries)
Email: loyola.dsilva@springer.com

Diego Rossit · Pedro Moreno-Bernal ·
Carlos Enrique Torres-Aguilar
Editors

Proceedings of the VII Ibero-American Congress of Smart Cities, ICSC-Cities 2024, 12–14 November, San Carlos, Costa Rica

Energy Applications in Smart Cities

CITIES
CIUDADES INTELIGENTES TOTALMENTE INTEGRALES, EFICIENTES Y SOSTENIBLES

Springer

Editors
Diego Rossit
Department of Engineering
Universidad Nacional del Sur
Bahía Blanca, Buenos Aires, Argentina

INMABB
CONICET
Bahía Blanca, Buenos Aires, Argentina

Carlos Enrique Torres-Aguilar
Academic Division of Engineering
and Architecture
Universidad Juárez Autónoma de Tabasco
Cunduacán, Tabasco, Mexico

Pedro Moreno-Bernal
Universidad Autónoma del Estado de
Morelos
Cuernavaca, Morelos, Mexico

ISSN 2352-2534 ISSN 2352-2542 (electronic)
Springer Proceedings in Energy
ISBN 978-981-96-4300-4 ISBN 978-981-96-4301-1 (eBook)
https://doi.org/10.1007/978-981-96-4301-1

This Springer imprint is published by the registered company Springer Nature Singapore Pte Ltd.
The registered company address is: 152 Beach Road, #21-01/04 Gateway East, Singapore 189721, Singapore

Organization

General Chairs

Luis Hernández Callejo, Universidad de Valladolid, Spain
Sergio Nesmachnow, Universidad de la República, Uruguay
Pedro Moreno Bernal, Universidad Autónoma del Estado de Morelos, México

Publication Chairs

Sergio Nesmachnow, Universidad de la República, Uruguay
Luis Hernández Callejo, Universidad de Valladolid, Spain, Spain
Pedro Moreno Bernal, Universidad Autónoma del Estado de Morelos, México
Diego Rossit, Universidad Nacional del Sur, Argentina
Carlos Torres Aguilar, Universidad Juárez Autónoma de Tabasco, México
Andrei Tchernykh, CICESE Research Center, México

Submission and Conference Management Chair

Diego Rossit, Universidad Nacional del Sur, Argentina

Organization Committee

Luis Hernandez Callejo, Universidad de Valladolid, Spain
Sergio Nesmachnow, Universidad de la República, Uruguay
Pedro Moreno Bernal, Universidad Autónoma del Estado de Morelos, México

Leonardo Cardinale Villalobos, Tecnológico de Costa Rica, Costa Rica
Jesús de la Casa Hernández, Universidad de Jaén, Spain

Local Organization Chair

Leonardo Cardinale Villalobos, Tecnológico de Costa Rica

Local Organization Committee

Paola Vega Castillo, Tecnológico de Costa Rica
Patricia Rivera Figueroa, Tecnológico de Costa Rica
Lady Fernández Mora, Tecnológico de Costa Rica
Cinthia Carvajal Alfaro, Tecnológico de Costa Rica
Carolina Bermúdez Quesada, Tecnológico de Costa Rica
Eduardo Valverde Masis, Tecnológico de Costa Rica

Program Committee Chairs

Luis Hernández Callejo, Universidad de Valladolid, Spain, Spain
Sergio Nesmachnow, Universidad de la República, Uruguay
Pedro Moreno Bernal, Universidad Autónoma del Estado de Morelos, México
Diego Rossit, Universidad Nacional del Sur, Argentina
Carlos Torres Aguilar, Universidad Juárez Autónoma de Tabasco, México

Program Committee

Adriana Correa, Universidad de Valladolid, Spain
Adrian Lara, Universidad de Costa Rica, Costa Rica
Adrián Toncovich, Universidad Nacional del Sur, Argentina
Agustín Barrera-Sánchez, Centro Nacional de Investigación y Desarrollo Tecnológico, México
Alberto Redondo Plaza, Universidad de Valladolid, Spain
Alfonso García Alvaro, Universidad de Valladolid, Spain
Alicia Martinez, Centro Nacional de Investigación y Desarrollo Tecnológico, México
Ana Carolina Olivera, Universidad Nacional de Cuyo—CONICET, Argentina
Ana Sánchez, Centro Nacional de Investigación y Desarrollo Tecnológico, México
Andrés Adolfo Navarro Newball, Pontificia Universidad Javeriana, Colombia

Andres Felipe Fuentes Vasquez, Pontificia Universidad Javeriana, Colombia
Angela Ferreira, Instituto Politécnico do Bragança, Portugal
Ángel Hernández Jiménez, Instituto Politécnico Nacional, México
Ángel Zorita Lamadrid, Universidad de Valladolid, Spain
Antonella Cavallin, Universidad Nacional del Sur, Argentina
Antonio Cano Ortega, University de Jáen, Spain
Antonio Mauttone, Universidad de la República, Uruguay
Antonio Muñoz, Universidad de Málaga, Spain
Araceli Ávila, Centro Nacional de Investigación y Desarrollo Tecnológico, México
Armando Huicochea, Universidad Autónoma del Estado de Morelos, México
Aura Dolores Zambrano Rendón, Escuela Superior Politécnica Agropecuaria de Manabí, Ecuador
Beatriz Martínez Bahena, Universidad Autónoma del Estado de Morelos, México
Begoña González Landín, Universidad de Las Palmas de Gran Canaria, Spain
Belén Carro, Universidad de Valladolid, Spain
Belisario Morales Morales, Instituto Tecnológico Superior de Centla, México
Benjamin Alvarez Alor, Centro Nacional de Investigación y Desarrollo Tecnológico, México
Bernardo Pulido-Gaytan, National College, Ireland
Bouras Abdelkarim, Badji Mokhtar University, Algeria
Bruno Canizes, Instituto Politécnico do Porto, Portugal
Bruno Dias, Federal University of Juiz de Fora, Brazil
Bruno França, Fluminense Federal University, Brazil
Alberto Ochoa-Zezatti, Universidad Autónoma de Ciudad Juárez, México
Carlos Baier, Universidad de Talca, Chile
Carlos Grande, Universidad Centroamericana José Simeón Cañas, El Salvador
Carlos Kornuta, Universidad Gastón Dachary, Argentina
Carlos Torres-Aguilar, Universidad Juárez Autónoma de Tabasco, México
Carmen Vasquez, Universidad Nacional Experimental Politécnica Antonio José de Sucre, Venezuela
César Varela, Centro Nacional de Investigación y Desarrollo Tecnológico, México
Christian Cintrano, Universidad de Málaga, Spain
Claudio Paz, Universidad Tecnológica Nacional, Argentina
Claudio Risso, Universidad de la República, Uruguay
Cristina Sáez Blázquez, Universidad de Salamanca, Spain
Daniel Morinigo-Sotelo, Universidad de Valladolid, Spain
Daniel Rossit, Universidad Nacional del Sur—CONICET, Argentina
David Balladares de la Cruz, Universidad Juárez Autónoma de Tabasco, México
David Peña Morales, Universidad de Cádiz, Spain
Deyslen Mariano, Instituto Tecnológico de Santo Domingo, República Dominicana
Diego Alberto Godoy, Universidad Gastón Dachary, Argentina
Diego Arcos-Aviles, Universidad de las Fuerzas Armadas ESPE, Ecuador
Diego Rossit, Universidad Nacional del Sur—CONICET, Argentina
Edgar Vazquez Beltran, Universidad de Sonora, México
Edgar Vicente Macias Melo, Universidad Juárez Autónoma de Tabasco, México

Eduardo Espinosa, Universidad Católica de la Santísima Concepción, Chile
Eduardo Fernández, Universidad de la República, Uruguay
Efrain Simá-Moo, Centro Nacional de Investigación y Desarrollo Tecnológico, México
Elmer Iván Sánchez-Rabadán, Universidad Autónoma del Estado de Morelos, México
Emanuela Giancola, Instituto Politécnico Nacional, Spain
Emmanuel Millan, Universidad Nacional de Cuyo, Argentina
Enrique González, Universidad de Salamanca, Spain
Esteban Mocskos, Universidad de Buenos Aires, Argentina
Fábio Castro, Instituto Politécnico do Porto, Portugal
Federico Alonso Pecina, Universidad Autónoma del Estado de Morelos, México
Federico Serra, Universidad Nacional de San Luis, Argentina
Fernando Lezama, Instituto Politécnico do Porto, Portugal
Fernando Mezzano, Universidad Nacional de San Luis, Argentina
Florencia Caro, Universidad de la República, Uruguay
Franciele Weschenfelder, Universidad Tecnológica del Uruguay, Uruguay
Francisco Beltrán-Carbajal, Universidad Autónoma Metropolitana, México
Francisco Esteban, Universidad Nacional de San Luis, Argentina
Francisco José Sánchez Sutil, University de Jáen, Spain
Francisco Magaña-Hernández, Universidad Juárez Autónoma de Tabasco, México
Francisco Valbuena, Universidad de Valladolid, Spain
Gabriel Villalobos, Universidad de Costa Rica, Costa Rica
Gaston Frias, Universidad Nacional de San Luis—CONICET, Argentina
Gilberto Martinez, Instituto Politécnico Nacional, México
Gina Paola Maestre Gongora, Universidad de Antioquia, Colombia
Giovanni Mochi, Università degli Studi di Perugia, Italia
Guillermo Magaldi, Universidad Nacional de San Luis, Argentina
Guillermo Santamaría Bonfil, Universidad Nacional Autónoma de México, México
Gustavo Medina Angel, Universidad Autónoma del Estado de Morelos, México
Héctor Chicchiarini, Universidad Nacional del Sur—CONICET, Argentina
Hector Felipe Mateo Romero, Universidad de Valladolid, Spain
Hortensia Amaris, Universidad Carlos III de Madrid, Spain
Ignacio Turias, Universidad de Cádiz, Spain
Irene Lebrusán, Universidad Autónoma de Madrid, Spain
Iván Hernández-Pérez, Universidad Juárez Autónoma de Tabasco, México
Ivonne Yazmín Arce García, Universidad Autónoma del Estado de Morelos, México
Jamal Toutouh, Massachusetts Institute of Technology, Spain
Javier Rocher, Universitat Politècnica de València, Spain
Javier Uriarte, Universidad de Sonora, México
Jesús Armando Aguilar, Universidad Autónoma de Baja California, México
Jesús Del Carmen Peralta Abarca, Universidad Autónoma del Estado de Morelos, México
Jesús Hernandez, University de Jáen, Spain
Jesús Vegas, Universidad de Valladolid, Spain

Joao Coelho, Instituto Politécnico do Bragança, Portugal
Joao Soares, Instituto Politécnico do Porto, Portugal
Jonathan Muraña, Universidad de la República, Uruguay
Jorge De La Cruz, Aalborg University, Dinamarca
Jorge López-Rebollo, Universidad de Salamanca, Spain
José Alberto Hernández, Universidad Autónoma del Estado de Morelos, México
José Almeida, Instituto Politécnico do Porto, Portugal
José Barbosa, Instituto Politécnico do Bragança, Portugal
Jose Ignacio Morales Aragonés, Zigurat Institute Technology, Spain
José Ramón Aira, Universidad Politécnica de Madrid, Spain
Juan Chavat, Universidad de la República, Uruguay
Juan Espinoza, Universidad de Cuenca, Ecuador
Juan Humberto Sossa Azuela, Instituto Politécnico Nacional, México
Juan Manuel Ramírez Alcaraz, Universidad de Colima, México
Juan-Pablo Peñaloza, Universidad Nacional de San Luis, Argentina
Juan Pavón, Universidad Complutense de Madrid, Spain
Juan Coca, Universidad de Valladolid, Spain
Juan Serrano-Arellano, Instituto Tecnológico de Pachuca, México
Karla María Aguilar Castro, Universidad Juárez Autónoma de Tabasco, México
Laene Oliveira Soares, Centro Federal de Educação Tecnológica Celso Suckow da Fonseca, Brazil
Lenin Ramos, Instituto Politécnico Nacional, México
Leticia Neria Monroy, Universidad Autónoma del Estado de Morelos, México
Lilian Obregon, Universidad de Valladolid, Spain
Lorena Parra, Universitat Politècnica de València, Spain
Lucas Iacono, Know-Center GmbH, Austria
Lucas Martín Fernandez, Universidad Nacional de San Luis, Argentina
Luis Aarón Martínez Figueroa, Universidad Centroamericana José Simeón Cañas, El Salvador
Luis Angeles, Universidad Juárez Autónoma de Tabasco, México
Luis Montané Jiménez, Universidad Veracruzana, México
Luis Hernández-Callejo, Universidad de Valladolid, Spain
Luis Manuel Navas Gracia, Universidad de Valladolid, Spain
Luis Marrone, Universidad Nacional de La Plata, Argentina
Luis Torres, Universidad Nacional de Río Cuarto, Argentina
Magdalena Irurtia, Administración Nacional de Usinas y Trasmisiones Eléctricas, Uruguay
Marcin Seredynski, E-Bus Competence Center, Francia
Margherita Valle Pilia, Universidad Latinoamericana de Ciencia y Tecnología, Costa Rica
Mariana Coccola, Universidad Tecnológica Nacional—CONICET, Argentina
Mariano Frutos, Universidad Nacional del Sur—CONICET, Argentina
María Sánchez-Aparicio, Universidad Politécnica de Madrid, Spain
Maria Teresa Cepero, Universidad Veracruzana, México
Marina Codina Escolar, Universitat Politècnica de Catalunya, Spain

Mario Andrés Paredes Valverde, Tecnológico Nacional de México, México
Mario Carbono de la Rosa, Universidad Nacional Autónoma de México, México
Martin Rodríguez Vázquez, Tecnológico Nacional de México, México
Matias Tello, Universidad Nacional de San Luis, Argentina
Miguel Angel Che Pan, Centro Nacional de Investigación y Desarrollo Tecnológico, México
Miguel Ángel Juárez Merino, Universidad Nacional Autónoma de México, México
Miguel-Ángel Muñoz-García, Universidad Politécnica de Madrid, Spain
Miguel Aybar, Instituto Tecnológico de Santo Domingo, República Dominicana
Miguel Davila, Universidad Politécnica Salesiana, Ecuador
Miguel Jimenez, Universidad Autónoma de Campeche, México
Monica Alonso, University Carlos III de Madrid, Spain
Ney Balderramo, Universidad Técnica de Manabí, Ecuador
Nicolás Plaza-Gómez, Universidad de Valladolid, Spain
Noelia Soledad Pinto, Universidad Tecnológica Nacional, Argentina
Olalla García Pérez, Universidad Europea de Madrid, Spain
Oscar Danilo Montoya Giraldo, Universidad Distrital Francisco José de Caldas, Colombia
Oscar Duque-Perez, Universidad de Valladolid, Spain
Oscar Izquierdo, Centro de Investigaciones Energéticas Medioambientales y Tecnológicas, Spain
Osvaldo González, Universidad Nacional de Asunción, Paraguay
Outmane Oubram, Universidad Autónoma del Estado de Morelos, México
Pablo Daniel Godoy, Universidad Nacional de Cuyo, Argentina
Paula de Andrés Anaya, Universidad de Salamanca, Spain
Paula Peña-Carro, Centro de Investigaciones Energéticas Medioambientales y Tecnológicas, Spain
Pedro Moreno-Bernal, Universidad Autónoma del Estado de Morelos, México
Pedro Piñeyro, Universidad de la República, Uruguay
Ponciano Jorge Escamilla-Ambrosio, Instituto Politécnico Nacional, México
Raúl Alberto López Meraz, Universidad Veracruzana, México
Renzo Massobrio, University of Antwerp, Bélgica
Ricardo Beltran-Chacon, Centro de Investigación en Materiales Avanzados, México
Roberto Flores, Universidad Autónoma del Estado de Morelos, México
Roberto Lemaitre, Universidad de Costa Rica, Costa Rica
Roberto Quintal Palomo, Universidad Autónoma de Yucatan, México
Roberto Ramírez Basterrechea, Independent Researcher, Spain
Roberto Villafafila, Universitat Politècnica de Catalunya, Spain
Rodrigo Alonso-Suárez, Universidad de la República, Uruguay
Rodrigo Porteiro, Universidad de la República, Uruguay
Rodrigo Ramirez-Pisco, Universitat Carlemany, Andorra
Rogelio Vargas Lopez, Universidad Autónoma de Guadalajara, México
Roger Castillo Palomera, Universidad Juárez Autónoma de Tabasco, México
Rosenberg Romero, Universidad Autónoma del Estado de Morelos, México
Rubén-Iván Bolaños, Instituto Tecnológico Metropolitano, República Dominicana

Preface

This SPE volume presents selected articles from the 7th edition of the Iberoamerican Congress on Smart Cities (ICSC-CITIES 2024), held from November 12 to November 14, 2024 in San Carlos, Costa Rica, in a mixed modality, with face-to-face and on-line talks and article presentations. This event continues the successful six previous editions of the congress, held in Soria, Spain, in 2018 and 2019, Costa Rica in 2020, Cancún, México, in 2021, Cuenca, Ecuador, 2022, and Mexico City and Cuernavaca in 2023.

The main goal of ICSC-CITIES 2024 was to provide a forum for researchers, scientists, teachers, decision-makers, postgraduate students and practitioners from different countries in IberoAmerica and worldwide to share their current initiatives related to Smart Cities. Articles in this volume address five relevant topics (Energy approaches for smart cities, Energy efficiency and sustainability, Electric systems for smart cities, Control strategies for smart grid, and Other innovative developments for smart cities) covering several areas of research and applications.

The main program consisted of three discussion panels on Energy Transition of Smart Cities, AI applied in business, and Cybersecurity for Smart Cities, along with 10 talks from experts, 19 oral presentations sessions and a poster session from international authors, highlighting recent developments in areas related to smart cities. Over three hundred distinguished participants from 42 countries gathered face-to-face or virtually for the congress. The Program Committee of ICSC-CITIES 2024 received 129 manuscripts. 93 submissions were accepted for oral presentation and the best 30, whose contents are within Energy-related areas, were selected to be published in this SPE volume. All articles have undergone a careful single-blind peer-review process, by at least three subject-matter experts before being selected for publication.

We would like to express our deep gratitude to all the contributors of ICSC-CITIES 2024, the congress organizers, and to the authors and reviewers for their endeavors that made the paper reviewing process efficiently available. We also thank the participants of the congress, our institutions, and all readers of this SPE volume.

Bahía Blanca, Argentina — Diego Rossit
Cuernavaca, Mexico — Pedro Moreno-Bernal
Cunduacán, Mexico — Carlos Enrique Torres-Aguilar
December 2024

Contents

that it learns how to adjust its parameters through reinforcement learning. So, in the initial stage, it is not necessary to set the fixed parameter (it sets lower and upper values to RS, Rsh, Iph, Io, and m based on the literature), since the algorithm can find the most suitable one in the process of trial and error.

Liu et al. [20], by its turn, propose an advanced shuffled frog-leaping algorithm (SFLBS) method based on genetic algorithms. SFLBS algorithm aims at finding the minimum value of RMSE related to the PV parameters (considering the single, double, and three diode models) and compares the results with other algorithms found in the literature. It also defines lower and upper values of Rs, RSh, Iph, Io, and m based on the literature and uses these parameters to run the proposed SFLBS algorithm. To validate the performance of the algorithm, the RMSE was also used, and the results were compared with other state-of-the-art algorithms and with the experimental data under different environmental conditions.

Cervellini et al. [6] propose a method based on genetic algorithms to obtain the parameters of the one-diode model of PV cells. The method combines two approaches: (i) use of genetic algorithm (GA) to minimize the error for the entire set of data; and (ii) seek for the error minimization in selected I–V points (open-circuit, short-circuit, and MPP). GA is applied to extract the parameters not provided by the manufacturer's datasheet. The results are compared to the manufacturers' data and the quadratic error between the currents is calculated. This technique also requires the establishment of boundaries to the Rs, Rsh, m, and Is values.

Cortés et al. [5] propose a method to estimate the parameters of a one-diode model of a PV cell based on artificial neural networks (ANN) and the multilayer perceptron concept. The steps of the ANN are applied, considering the different inputs relates to current, voltage, and resistance. The Symmetric Mean Absolute Percentage Error (SMAPE) is used as an error metric to compare the results to the literature and the experimental data.

Li et al. [21] propose a dynamic self-adaptive and mutual-comparison teaching–learning-based optimization (DMTLBO) to extract the parameters of PV cells (considering one, two, and three diode models). RMSE is used to evaluate the performance of the algorithm, comparing them to other algorithms found in the literature. As the previous studies that apply algorithms, it also requires the definition of lower and upper boundaries to the Rs, RSh, Iph, Io, and m values to run the proposed DMTLBO algorithm and estimate the parameters of the single diode model, double diode model, and PV module models.

Regarding the application, different types of PV and cells are considered as Kyocera KC200GT, RTC France, and Shell SM55. Besides testing the proposed methods under the standard conditions, most of the paper evaluate different scenarios, varying the irradiance and/or the temperature values.

Although, there are different methods (analytic and numerical) and algorithms dedicated to extracting the parameters of the one diode model, one of the problems related to these approaches is that most of the time, an initial value of, at least, one parameter must be determined to define the others. For example, Orioli [11] define an initial value of the shunt resistor (Rsh) and diode ideally factor (m) to calculate the other parameters based on the proposed equations, while Álvarez et al. [12] adopts an

Part I
Energy Approaches for Smart Cities

Chapter 1
Passive Electroluminescence and Photoluminescence in Photovoltaic Modules: Synchronous and Asynchronous Lock-In Approaches

Alberto Redondo-Plaza, Halleluyah Kupolati, Victor Ndeti Ngungu, José Ignacio Morales-Aragones, Víctor Alonso-Gómez, Michael Njoroge Gitau, and Luis Hernández-Callejo

Abstract The integration of renewable energy for electricity generation has become imperative to meet environmental goals and curb global warming. Photovoltaic energy is one of the technologies with the highest potential and growth prospects. In photovoltaic power plants, periodic inspection techniques for the modules may be necessary. Among various inspection techniques, electroluminescence and photoluminescence imaging are of significant interest as they provide information about module performance and detect various types of failures. This article presents a technique named passive electroluminescence and photoluminescence that enables luminescence acquisition during daylight conditions while the photovoltaic power plant is in normal operation. The technique is based on an electronic board that is connected within the photovoltaic string and modulates the signal. A lock-in

A. Redondo-Plaza (✉) · L. Hernández-Callejo
Department of Agricultural and Forestry Engineering, University of Valladolid, Soria, Spain
e-mail: albertogregorio.redondo@uva.es

L. Hernández-Callejo
e-mail: luis.hernandez.callejo@uva.es

H. Kupolati · V. N. Ngungu · M. N. Gitau
Department of Electrical, Electronic & Computer Engineering, University of Pretoria, Pretoria, South Africa
e-mail: u15292747@tuks.co.za

M. N. Gitau
e-mail: njoroge.gitau@up.ac.za

J. I. Morales-Aragones · V. Alonso-Gómez
Department of Applied Physics, University of Valladolid, Valladolid, Spain
e-mail: joseignacio.morales@uva.es

V. Alonso-Gómez
e-mail: victor.alonso.gomez@uva.es

D. Rossit et al. (eds.), *Proceedings of the VII Ibero-American Congress of Smart Cities, ICSC-Cities 2024, 12-14 November, San Carlos, Costa Rica*, Springer Proceedings in Energy, https://doi.org/10.1007/978-981-96-4301-1_1

approach is also implemented, which involves obtaining several pictures under low and high emission states, which are then processed to remove the background signal generated by sunlight. The study focuses on two lock-in approaches: synchronous and asynchronous, which differ in whether the camera and the electronic board are coordinated. Both approaches have been successfully implemented, yielding similar results.

Keywords Solar energy · Photovoltaic · Inspection techniques · Luminescence · Electroluminescence · Photoluminescence and lock-in

1.1 Introduction

Global warming has become one of the main issues that requires addressing by modern society [1]. Global warming is mainly caused by carbon emissions, which globally reached 35.6 Gt CO_2 in 2023 [2]. The power sector remains the main contributor to carbon emissions, accounting for 38.4% of global CO_2 emissions [2]. In addition, 70% of total global CO_2 emissions can be associated with cities, while two-thirds of energy consumption occurs in urban areas [3]. These facts highlight the need for efforts to limit carbon emissions in both cities and the power sector.

City-integrated renewable energy systems have become a feasible solution for enhancing the sustainability in urban areas, curbing carbon emissions [4]. Among the several technologies, photovoltaic energy presents a huge potential due to its easy installation, high integration, and vast potential in urban areas. In fact, several studies have shown that photovoltaics could satisfy more than 50% of the electrical needs of several cities [4]. Photovoltaic energy development is not only expected in cities, but also in large utility photovoltaic farms, where forecasts indicate that photovoltaic energy will be the electrical technology with the highest installed capacity in the world by 2027, reaching a global power of 2359 GWp [5].

Inspections techniques are required during the operation and maintenance of photovoltaic power plants to maximize profitability and prevent safety issues [6, 7]. Photovoltaic solar modules are the critical part of photovoltaic power plant, as they are dealing with several environmental loads such as hail, thermal changes, snow and wind load, moisture or ultraviolet radiation. These facts led to a degradation in photovoltaic modules [8], and eventually the creation of faults [9, 10]. Inspection techniques in photovoltaic power plants such as visual inspection, infrared imaging, current–voltage measurements and luminescence imaging are able to detect these failures [11].

Luminescence techniques have recently generated significant interest in photovoltaic power plant inspections [12]. This type of imaging technique can provide more useful information than other methods, such as infrared inspections or current–voltage measurements. Luminescence images are based on the capture of the emission generated by the crystalline structures that make up the solar cells. In silicon solar cell technology, this emission corresponds to a wavelength of 1150 nm [13].

Moreover, luminescence emission can be generated by current injection into the photovoltaic module, termed Electroluminescence (EL), or by optical excitation with a suitable light source, termed Photoluminescence (PL). Both EL and PL images can detect a wide range of faults, such as cracks, finger failures, bypass diode failures, potential-induced degradation (PID), corrosion, series or shunt resistance issues, and mechanical damages, among others [9].

Traditionally, luminescence images have been obtained using a power supply to inject current into the photovoltaic module to produce the EL effect. The EL signal is captured with an InGaAs or silicon camera during nighttime, as the luminescence emission is much lower than sunlight intensity [12]. Since it is also desirable to acquire images during the day, new approaches have been developed, not only to enable luminescence imaging during daylight but also to avoid the use of an external energy source such as a power supply.

Several outdoor luminescence techniques have been developed to acquire EL or PL images during high irradiance conditions. Most of these techniques are based on a lock-in approach [14–19]. The lock-in approach involves modulating an EL or PL signal in the photovoltaic module, meaning the EL/PL signal alternates between a high and low state periodically. The imaging acquisition setup for daylight approaches includes an InGaAs camera and a bandpass filter. InGaAs cameras are highly sensitive within the luminescence emission wavelength, while bandpass filters block different luminescence wavelengths. Both factors help to increase the ratio between luminescence intensity and the background intensity generated by sunlight. During the measurement process, a set of images is captured while the EL/PL signal is being modulated. Thus, some of the images capture the EL/PL signal and the background, while others capture only the background signal. Table 1.1 shows all the techniques that can modulate an EL or PL signal in a photovoltaic module and therefore allow luminescence imaging through a lock-in approach.

As previously explained, the lock-in technique results in a stack of images where the EL/PL signal is either high or null. It becomes imperative to separate this set of images into two groups: those corresponding to high EL/PL emission and those corresponding to null EL/PL emission. This separation allows for processing all the images through various operations to obtain a final EL or PL image. At this point, two approaches can be implemented: synchronous or asynchronous. The synchronous approach involves a system that synchronizes the camera with the device that modulates the signal, making it possible to know the emission state of each image during the measurement process. Conversely, the asynchronous approach does not require synchronization; the camera acquires images at its maximum frequency while the EL/PL signal is being modulated. Therefore, it is necessary to implement an algorithm capable of separating the images in the stack into the two mentioned groups.

This article evaluates these two methods using the passive EL and PL technique, where an electronic device has been developed to modulate EL and PL signal without the need of a power supply. The second section of the manuscript presents the electronic device and acquisition setup, as well as a description of the synchronous and asynchronous approaches. The third section shows the obtained results, and finally, the discussion and conclusion can be found in the last section.

Table 1.1 Techniques for modulating luminescence emission in photovoltaic modules

Technique	Reference	Description
EL bias switching method	[14]	EL signal is modulated using a programmable power supply connected to the module which changes the state of the module between current injection and open circuit
PL switching method	[14]	PL signal is modulated using an electronic device connected to the module which changes the state of the module between open circuit and short circuit
Control cell method (PL)	[16]	PL signal is modulated through a control solar cell which is shaded or not to change the state of all the solar cells under the same substring between open circuit and maximum power point
Inverter control method (PL)	[17]	PL signal is modulated through the solar inverter within a photovoltaic string. Solar inverter can change the string state between open circuit and maximum power point
Self-source EL	[18]	EL signal is modulated through an electronic device which collect the current generated by the module and inject it then in the same module, modulating EL signal
Passive EL and PL	[19]	The EL signal is modulated through an electronic device that collects the current generated by the module and then injects it back into the same module, modulating the EL signal

1.2 Materials and Methods

1.2.1 Lock-In Technique

The lock-in technique can be used to obtain EL and PL images during high irradiance conditions. This technique is based on capturing a set of images where the luminescence emission is periodically high and low. To achieve this modulation of the signal, it is necessary to change the working point of the module. Figure 1.1 shows a current–voltage curve of an illuminated photovoltaic module, as well as the luminescence intensity as a function of the operational voltage of the module.

When the module is under short circuit (SC), luminescence emission is null. From SC to open circuit (OC), the current–voltage curve is in the first quadrant, which means the PV module works as a generator. Here, PL emission can be found, being maximum at OC conditions and almost null at maximum power point (MPP). Working at higher voltages than OC means current injection (CI) into the module, and it starts behaving as a load. In this situation, EL emission is added to the PL signal.

Considering the above, it is mandatory to change the working point of the module between two states where luminescence emission is high and null/low. For EL imaging, it is necessary to change the working point between CI (high EL signal) and OC (null EL signal), while PL imaging requires a change between OC (high PL signal) and MPP (low PL signal) or SC (null PL signal).

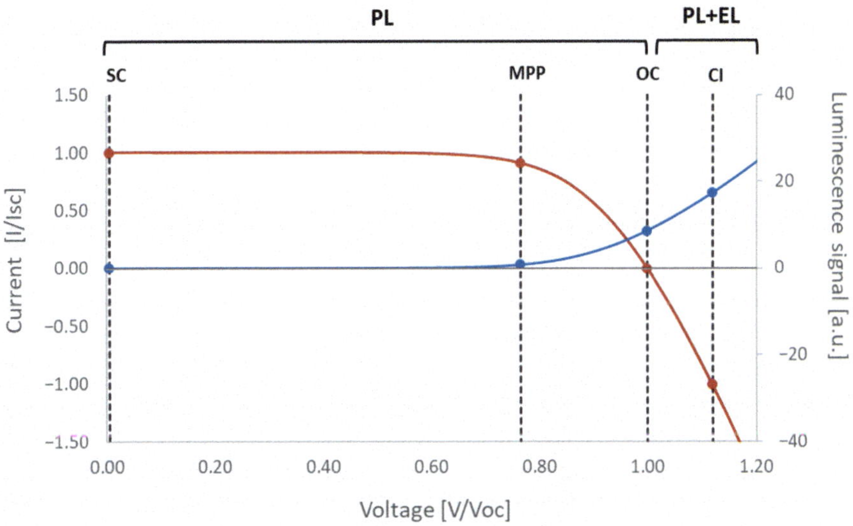

Fig. 1.1 Illuminated current–voltage curve of a PV module and luminescence signal intensity

The set of images must be processed. All the images under high and low luminescence emission are averaged. Therefore, a subtraction of both averages allows for the elimination of the background signal generated by the sunlight. Since it is necessary to know which images have been taken under high and low emission states, two strategies can be implemented: synchronous or asynchronous approach. It should be noted that it is necessary to obtain several images (typically more than 100 images) to achieve a result of good quality, as the background signal is much higher than the luminescence signal intensity.

Synchronous lock-in approach

The synchronous approach involves a coordination system between the camera and the device that modulates the EL/PL signal. When the control system sets a new working point in the module, a trigger to capture a picture is generated in the camera. This synchronization requires continuous communication between the camera, the device that modulates the signal, and the control system, allowing for labeling all the pictures during the acquisition process.

Asynchronous lock-in approach

The asynchronous approach does not require continuous communication and synchronization between the camera and the modulating device. Instead, the camera captures frames at its maximum frequency, while the device modulates the luminescence signal at a frequency higher than twice the camera's acquisition frequency. As the frames cannot be labeled during the acquisition process, postprocessing of the image set must involve an algorithm that classifies the frames into high and low/null luminescence emission.

In the present study, a basic algorithm has been developed to automatically classify all the images obtained during the acquisition process. It is based on comparing the average pixel intensity of the image that is to be classified with the average pixel intensity of the images that have been immediately taken before and after. Therefore, if the image's average intensity is higher than the average of the surrounding images, the image is classified as a high EL/PL signal picture; otherwise, it is labelled as a low/null EL/PL signal picture.

1.2.2 Passive Electroluminescence Set-Up

For a passive EL technique an electronic board have been developed. The power topology of the board is displayed in Fig. 2a. It is based on three MOSFET transistors and two diodes. The electronic board must be connected within the photovoltaic string and through the control of the transistor it is able to set four different states (SC, OC, MPP and CI) in the photovoltaic module which is connected. The truth table shown in Fig. 2a indicates which transistors must be conducting to achieve the mentioned working states of the module. The mentioned topology has been integrated in a printed circuit board as Fig. 2b shows. This printed circuit board include a voltage regulator which use the voltage module to produce a 5 V source which is required in the electronic board. Therefore, the device does not require an external battery to work. Moreover, a Bluetooth module (HC-06) is connected to the board, so it can be controlled by an external device such as a laptop or smartphone.

The complete acquisition set-up is displayed in Fig. 1.3. A control system has been developed in python. This python program is running in a laptop which in connected to the InGaAs camera (Hamamatsu C12741-03) through a USB cable and is also connected to the electronic board through Bluetooth port. The control system uses the camera API to control it and synchronous and asynchronous approaches have been properly programed. After the acquisition process the software can process the images stack according to the selected approach. Note that the software is run on a laptop with an intel core i7-11800H.

The synchronous approach involves sending a command to the board and waiting until the board changes the module working point and sends back a response to the laptop. The InGaAs camera is configured in trigger capture mode, and when the response is received on the laptop, a trigger is generated in the camera to capture a picture. Once the picture has been obtained, the program sends the following command to the board, repeating the previous process until all the images have been collected. Then, the images are processed to obtain a final EL or PL signal.

The asynchronous approach involves sending a command and argument to the electronic board. When the board receives them, it starts modulating EL or PL at a frequency determined by the received argument. Here, the camera is configured in a continuous capture mode, where images are taken at the maximum speed of the camera. Once all the images have been obtained, another command is sent to

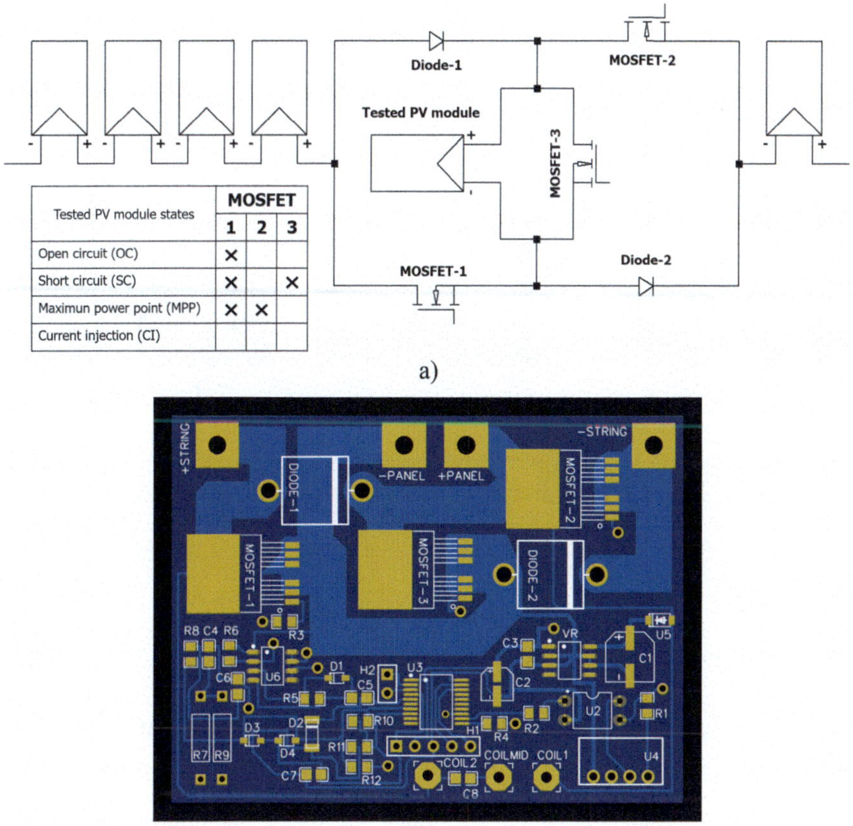

Tested PV module states	MOSFET 1	MOSFET 2	MOSFET 3
Open circuit (OC)	×		
Short circuit (SC)	×		×
Maximun power point (MPP)	×	×	
Current injection (CI)			

Fig. 1.2 Developed electronic device for EL modulation

the board to restore the normal performance of the module. Then, the developed algorithm is used to classify and process the images.

In the experimental setup, the board has been connected within a PV string made of eleven 235 Wp PV polycrystalline modules (TSM-235-PC05A). The string is connected to a PV string inverter, and measurements have been performed during normal operation of the PV facility under high irradiance conditions (800–1000 W/m^2).

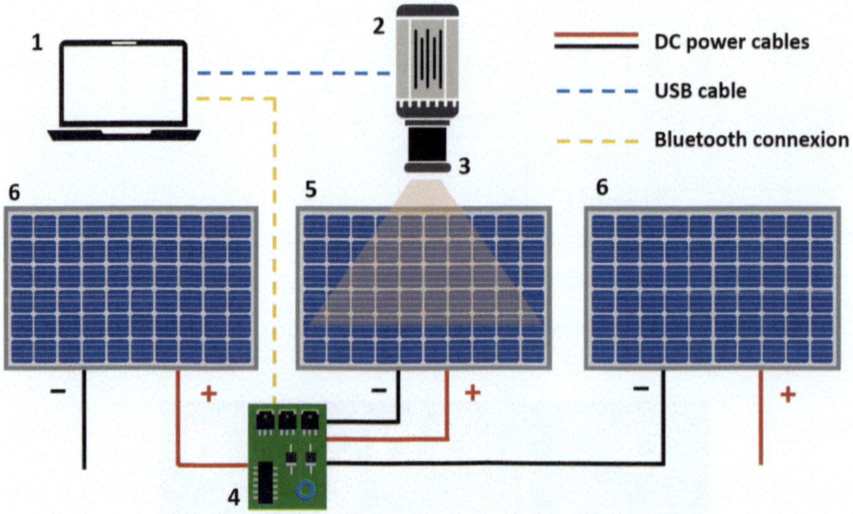

Fig. 1.3 Passive EL set-up for imaging acquisition. Laptop (1), InGaAs camera (2), bandpass filter (3), electronic board (4), tested PV module (5) and other modules of the PV string (6)

1.3 Results

During the measurement process, EL and PL pictures have been successfully obtained using both asynchronous and synchronous approaches, modulating the luminescence signal with the developed electronic board. This allows for luminescence measurement during the normal operation of a PV plant in daylight conditions. The obtained images, as well as the main parameters of the measurement process, can be observed in Fig. 1.4. All the image stacks were taken on the same day under similar irradiance conditions (800–1000 W/m2). The exposure time was set to 2 ms, and 1000 pictures were obtained in both measurement approaches. These images are complemented with their Signal to Noise Ratio (SNR), a metric that has been previously developed to evaluate the quality of luminescence pictures during daylight conditions [20].

It is possible to see that both synchronous and asynchronous approaches offer similar results, providing almost the same final images. Moreover, this qualitative analysis is also complemented and supported by the calculation of the Signal to Noise Ratio, which is very similar across all images, resulting in a comparable quality in the final pictures for both approaches. However, slightly higher quality can be observed in the synchronous strategy. This could be explained by the error in the development algorithm for the automatic classification of the images, combined with the fact that some of the pictures taken in the asynchronous approach could be captured during the transition between one state and another.

Nevertheless, the asynchronous approach could be more advantageous in some aspects. First, as synchronization is not required, the integration of the camera and board is much easier. Moreover, this approach also allows maximizing the acquisition

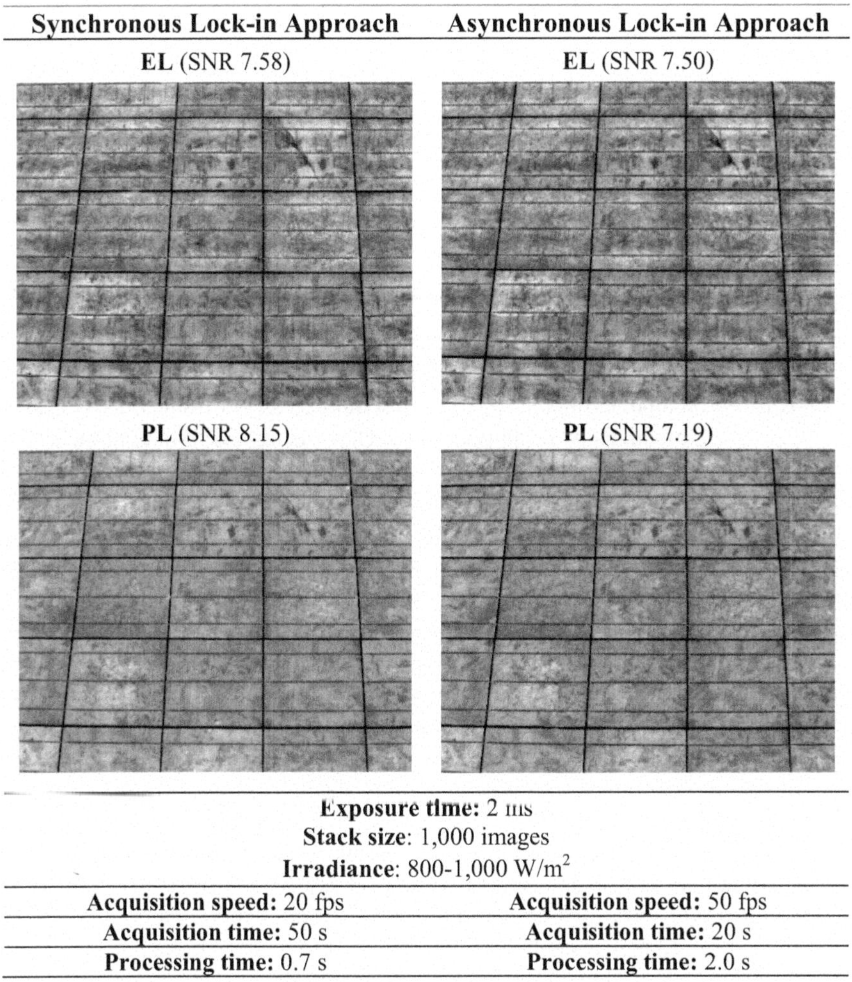

Fig. 1.4 EL and PL images obtained by synchronous and asynchronous approaches

speed of the camera, working at 50 fps compared to 20 fps in the synchronous approach. For an acquisition of 1000 images, the synchronous approach requires 50 s, while the asynchronous approach just requires 20 s. Conversely, the asynchronous approach needs a higher processing time, as the classification algorithm must be implemented, leading to a processing time for 1000 images of around 2.0 s, compared to 0.7 s in the synchronous approach. Note that processing time will depend on the computer processing unit characteristics while the acquisition time is limited by the camera speed.

Finally, it is worth mentioning that different patterns can be observed in EL and PL images. In EL images, a higher signal is observed near the metal contacts of the cells, as the luminescence signal is induced by electrical excitation. However, PL images display a more homogeneous distribution of the signal, as the excitation source is sunlight. Therefore, both EL and PL images can provide similar but complementary information, with EL offering a better understanding of the electrical performance of the module, while PL enhances the evaluation of the quality and condition of the crystalline structures of the PV cells.

1.4 Discussion and Conclusion

The present article has demonstrated the feasibility of obtaining both EL and PL pictures under high irradiance conditions while a PV plant is in operation. This technique is based on modulating the EL/PL signal through an electronic board capable of changing the module's working point, as well as using a lock-in approach. The lock-in approach involves obtaining a stack of images taken during the modulation of the EL or PL signal in the tested PV device. Two lock-in approaches have been evaluated: synchronous and asynchronous. While the synchronous approach can label the pictures during the measurement process for further processing, the asynchronous approach requires an algorithm to classify the images.

The study has shown that both approaches can be successfully implemented, yielding similar results. A lower quality (approximately 2–10% lower) is expected in the asynchronous approach, which can vary depending on the used algorithm and external factors such as the luminescence signal intensity of the PV module or the irradiance. However, the asynchronous approach offers a shorter acquisition time, as the camera can operate at its maximum speed. Additionally, this approach facilitates easier integration with the device that modulates the luminescence signal, as coordination is not required. It could be highly desirable in some applications, such as integration in unmanned aerial vehicles (UAVs).

To summarize, the synchronous approach is recommended due to its higher quality, if integration is feasible and acquisition time is not critical, which could have a significant impact on commercial applications.

Acknowledgements This study was supported by University of Valladolid with 2024 grants for attendance at courses, conferences and workshops relevant for the development of doctoral theses.

Funding Information This work has been funded by the Spanish Ministry of Education through the National Program FPU (grant number FPU21/04288); the Spanish Ministry of Science, Innovation, and Universities through the "Plan Estatal de Investigación Científica, Técnica y de Innovación" (project ID: PID2023-148369OB-C43); and the University of Valladolid with ERASMUS + program.

Disclosure of Interests The authors have no competing interests to declare that are relevant to the content of this article.

References

1. J. Houghton, Global warming. Rep. Prog. Phys. **68**, 1343–1403 (2005). https://doi.org/10.1088/0034-4885/68/6/R02
2. Z. Liu, Z. Deng, S.J. Davis, P. Ciais, Global carbon emissions in 2023. Nat. Rev. Earth Environ. **5**, 253–254 (2024). https://doi.org/10.1038/s43017-024-00532-2
3. M.G.M. Almihat, M.T.E. Kahn, K. Aboalez, A.M. Almaktoof, Energy and sustainable development in smart cities: an overview. Smart Cities **5**, 1389–1408 (2022). https://doi.org/10.3390/smartcities5040071
4. D.M. Kammen, D.A. Sunter, City-integrated renewable energy for urban sustainability. Science **2016**(352), 922–928 (1979). https://doi.org/10.1126/science.aad9302
5. International Energy Agency (IEA), *Renewables 2022: Analysis and Forecast to 2027* (2022)
6. L. Hernández-Callejo, S. Gallardo-Saavedra, V. Alonso-Gómez, A review of photovoltaic systems: design operation and maintenance. Solar Energy **188**, 426–440 (2019). https://doi.org/10.1016/j.solener.2019.06.017
7. A. Peinado Gonzalo, A. Pliego Marugán, F.P. García Márquez, Survey of maintenance management for photovoltaic power systems. Renew. Sustain. Energy Rev.**134**, 110347 (2020). https://doi.org/10.1016/j.rser.2020.110347
8. D.C. Jordan, S.R. Kurtz, Photovoltaic degradation rates—an analytical review. Prog. Photovoltaics Res. Appl. **21**, 12–29 (2013). https://doi.org/10.1002/pip.1182
9. M. Köntges, U. Jahn, K.A. Berger, T. Friesen, *Review of Failures of Photovoltaic Modules* (2014)
10. Y.-Y. Hong, R.A. Pula, Methods of photovoltaic fault detection and classification: a review. Energy Rep. **8**, 5898–5929 (2022). https://doi.org/10.1016/j.egyr.2022.04.043
11. M. Waqar Akram, G. Li, Y. Jin, X. Chen, Failures of photovoltaic modules and their detection: a review. Appl. Energy **313**, 118822 (2022). https://doi.org/10.1016/j.apenergy.2022.118822
12. O. Kunz, J. Schlipf, A. Fladung, Y.S. Khoo, K. Bedrich, T. Trupke, Z. Hameiri, Outdoor luminescence imaging of field-deployed PV modules. Prog. Energy **4**, 042014 (2022). https://doi.org/10.1088/2516-1083/ac9a33
13. M.H. Ulrike Jahn, M. Köntges, D. Parlevliet, M. Pagg, I. Tsanakas, J.S. Stein, K.A. Berger, S. Ranta, R.H. French, M. Richter et al, *Review on Infrared and Electroluminescence Imaging for PV Field Applications* (2018)
14. M. Guada, Á. Moretón, S. Rodríguez-Conde, L.A. Sánchez, M. Martínez, M.Á. González, J. Jiménez, L. Pérez, V. Parra, O. Martínez, Daylight luminescence system for silicon solar panels based on a bias switching method. Energy Sci. Eng. **8**, 3839–3853 (2020). https://doi.org/10.1002/ese3.781
15. L. Stoicescu, M. Reuter, J.H. Werner, DaySy: luminiscence imaging of PV modules in daylight, in *Proceedings of the 29th European Photovoltaic Solar Energy Conference Exhibition* (2014)
16. R. Bhoopathy, O. Kunz, M. Juhl, T. Trupke, Z. Hameiri, Outdoor photoluminescence imaging of photovoltaic modules with sunlight excitation. Prog. Photovolt. Res. Appl. **26**, 69–73 (2018). https://doi.org/10.1002/pip.2946
17. M. Vuković, I. Eriksdatter Høiaas, M. Jakovljević, A. Svarstad Flø, E. Olsen, I. Burud, Photoluminescence imaging of silicon modules in a string. Prog. Photovolt. Res. Appl. **30**, 436–446 (2022). https://doi.org/10.1002/pip.3525
18. T. Kropp, M. Berner, L. Stoicescu, J.H. Werner, Self-sourced daylight electroluminescence from photovoltaic modules. IEEE J. Photovolt. **7**, 1184–1189 (2017). https://doi.org/10.1109/JPHOTOV.2017.2714188

19. A. Redondo-Plaza, J.I. Morales-Aragonés, S. Gallardo-Saavedra, H.F. Mateo-Romero, S. Araujo-Rendón, Á.L. Zorita-Lamadrid, V. Alonso-Gómez, L. Hernández-Callejo, Passive electroluminescence and photoluminescence imaging acquisition of photovoltaic modules. Sensors **24**, 1539 (2024). https://doi.org/10.3390/s24051539
20. C. Mantel, G.A. Dos Reis Benatto, N. Riedel, S. Thorsteinsson, P.B. Poulsen, H. Parikh, S. Spataru, D. Sera, S. Forchhammer, SNR study of outdoor electroluminescence images under high sun irradiation, in *Proceedings of the 2018 IEEE 7th World Conference on Photovoltaic Energy Conversion (WCPEC) (A Joint Conference of 45th IEEE PVSC, 28th PVSEC & 34th EU PVSEC)*, IEEE, June 2018, pp. 3285–3289

Chapter 2
One Diode Model Intrinsic Parameter Extraction from Solar Module Dark I/V Curves with Low Computational Cost

José Ignacio Morales Aragonés, Luis Hernández-Callejo, Víctor Alonso Gómez, Alberto Redondo Plaza, Vanessa de Almeida Guimaraes, and Santiago Araujo Rendón

Abstract A double mathematical transformation is proposed in this work for the ease of solar modules parameter extraction. This transformation allows to define a distance between I–V measured data and the theoretical values corresponding to the one diode model of solar modules/cells, suitable for fitting measured curves with a low computational cost. Moreover, this transformation makes the iteration process inherent to one diode model theoretical points calculation to result always convergent. For validation of this method, a real laboratory circuit suited to the one diode model has been built, allowing the variation of series and shunt resistances, and a 25 I–V curves dataset has been created from 5 different values of series and shunt resistances.

Keywords Module dark I–V curves · Parameter extraction · Photovoltaic

---

J. I. Morales Aragonés · L. Hernández-Callejo (✉) · A. Redondo Plaza
Universidad de Valladolid, Campus Universitario Duques de Soria, Soria, Spain
e-mail: luis.hernandez.callejo@uva.es

V. Alonso Gómez
Universidad de Valladolid, Campus Miguel Delibes, Valladolid, Spain

V. de Almeida Guimaraes
COPPEAD Graduate School of Business—Federal University of Rio de Janeiro, Rio de Janeiro, Brazil

S. Araujo Rendón
Universidad Rey Juan Carlos, Campus Miguel Delibes, Móstoles (Madrid), Spain

D. Rossit et al. (eds.), *Proceedings of the VII Ibero-American Congress of Smart Cities, ICSC-Cities 2024, 12-14 November, San Carlos, Costa Rica*, Springer Proceedings in Energy, https://doi.org/10.1007/978-981-96-4301-1_2

2.1 Introduction

Countries around the world have been seeking alternative sources of energy to help in the achievement of greenhouse gas (GHG) emission targets of the environmental agreements, such as the Paris Agreement. According to World Resource Institute—WRI [1], after the United Nations Climate Change Conference 2021—COP26, 74 countries committed to a net-zero target of GHG emissions: most of them by 2050, as Brazil, Spain, Italy, Japan and Portugal; some of them in early stages as Finland (by 2035) and Germany (by 2045); and others in later stages, as China by 2060.

In this context, solar, wind, hydro, biofuels, among others are considered the key to a less carbon-intensive and more sustainable energy system [2]. Nevertheless, the non-renewable sources of energy represented almost 70% of the electricity generated in 2021 and coal, alone, was responsible for 56.12% of the total generated energy [3]. Even considering that the share of renewable sources has increased 6% from 2020 to 2021, the International Energy Agency (IEA) report on the Electricity Market shows that the current energy matrix is not enough to reduce the emissions to the level of the 2050 net-zero emission targets [3]. Therefore, it points out the need of increasing energy efficiency and promoting low carbon sources of energy.

Among the renewable energy source, solar photovoltaic—PV is pointed to as one with a potential of growing. The solar PV alone accounted for 60% of all renewable power expansion projected by IEA to 2021 and it should increase in the next years due to its lower costs when compared to other renewable (as hydropower, bioenergy, geothermal) and even to non-renewable (as coal and natural gas) [4].

Among the challenges related to PV solar, one may highlight the management of the system. Therefore, many papers are dedicated to extracting the parameters of the one diode model to predict the behavior of an entire PV system and manage it [5]. The preference for the one diode model is justified by its relation between accuracy and simplicity [6]. Sudhakar Babu et al. [7] explains that the double diode model presents a higher accuracy, but the complexity of extracting its unknown parameters leads to the more frequent adoption of the one diode model.

In the literature, there are papers focused on evaluating the I–V characteristics of PV cells [8], especially under different illumination conditions [9, 10], while others are proposing a new one-diode model to represent the I–V curves [11]. Nevertheless, most of the papers are dedicated to proposing methods or algorithms to extract the five unknown parameters of the one diode model.

Álvarez et al. [12] propose an analytical technique to model the behavior of PV cells based on two approaches: (i) define the value of the diode ideally factor (m) in order to extract the other parameters (resistance serie—Rs, shunt resistance—Rsh, photocurrent—Iph, and saturation current—Io) by the application of analytical models, and; (ii) develop a simplified approach to the Lambert W-function, to obtain the value of Rs and to solve the implicit equation of 1-diode/2-resistance equivalent circuit model in order to derive the output current in relation to the output voltage. A case study was carried out comparing the fitting of I–V curves from literature and experimental PV cells and its accuracy was evaluated based on the root-mean-square error—RMSE analysis.

Reddy and Yammani [13] apply the forward–backward linear least square error (FB-LLSE) curve fitting method to estimate the parameters of PV cells. The proposed method searches for the best fit of a set of data points by minimizing the sum of the squares errors from the plotted I–V curve. Its performance is evaluated by applying four techniques: RMSE, mean absolute error (MAE), normalized root mean square error (NRMSE), and normalized mean absolute error (NMAE). The obtained results are compared with other methods in the literature.

Chaibi et al. [14] propose a technique that combines an iterative process to determine the value of Rsh with a numerical study based on the manufacturer's datasheet information. To each type of PV cell, the minimum and maximum values of Rsh are determined based on the literature. For each iteration on Rsh, the other parameters are calculated by using the datasheet equations at standard conditions (i.e. temperature equal to 25 °C and irradiance equal to 1000 W/m^2). The criteria to extract the parameters is to find the lowest error between the datasheet and simulated powers at the maximum power point (MPP). Also, the I–V characteristics provided by manufacturers and literature-based techniques are compared to the simulated ones.

Wei et al. [15] propose an analytical method to obtain the five parameters of PV cell based on the relationship between the slope of the I–V curve and the parameters of the one-diode model. The proposed analytical solution method is based on two steps: (i) establish five individual algebraic equations to each parameter (Iph, m, Io, Rs, and Rsh); and (ii) sampling some key points, avoiding the open-circuit voltage point. To compare the accuracy with other analytical methods, RMSE is applied.

Based on an extension of the Cheung method, Victor-Tapio Rangel-Kuoppa et al. [16] propose a method for the extraction of the Rsh and Isat of an I–V measurement of the one-diode model of a solar cell. The proposed procedure is applied to two simulated I–V curves (under darkness and illumination) to determine the parameters and their accuracy is tested by evaluating the integral percentage error (%). The results and the comparison with the literature are presented and discussed in [17].

Ali et al. [18] presents a numerical method to estimate the parameters of a PV cell and compare their performances under different solar irradiances. It uses data of current and voltage of MPP and open circuit provided by the manufacturer to calculate m, Rs, Rsh. Based on them, they deduce the saturation current and photocurrent. To extract the parameters from real measurements, the authors apply the Fractional-order Darwinian particle swarm optimization (FODPSO) method through experimental data collected from a PV platform. To validate the proposed approach, the estimated data is compared to the experimental data considering the I–V curves collected from three different day hours with different irradiances.

Regarding the algorithms, we found different approaches in the literature. Hu et al. [19] propose a differential evolution algorithm with reinforcement learning—RLDE to simulate the I–V data and the parameters of the single diode, double diode, and PV module models. The algorithm search for the minimum RMSE to find the optimal values of the parameters. To validate the performance of the algorithm, the RMSE was used, and the results were compared with other state-of-the-art algorithms and with the experimental data from the manufacturers' datasheet under different environmental conditions. The authors claim that the benefit of their algorithm is that it learns how to adjust its parameters through reinforcement learning. So, in the initial stage, it is not necessary to set the fixed parameter (it sets lower and upper values to RS, Rsh, Iph, Io, and m based on the literature), since the algorithm can find the most suitable one in the process of trial and error.

Liu et al. [20], by its turn, propose an advanced shuffled frog-leaping algorithm (SFLBS) method based on genetic algorithms. SFLBS algorithm aims at finding the minimum value of RMSE related to the PV parameters (considering the single, double, and three diode models) and compares the results with other algorithms found in the literature. It also defines lower and upper values of Rs, RSh, Iph, Io, and m based on the literature and uses these parameters to run the proposed SFLBS algorithm. To validate the performance of the algorithm, the RMSE was also used, and the results were compared with other state-of-the-art algorithms and with the experimental data under different environmental conditions.

Cervellini et al. [6] propose a method based on genetic algorithms to obtain the parameters of the one-diode model of PV cells. The method combines two approaches: (i) use of genetic algorithm (GA) to minimize the error for the entire set of data; and (ii) seek for the error minimization in selected I–V points (open-circuit, short-circuit, and MPP). GA is applied to extract the parameters not provided by the manufacturer's datasheet. The results are compared to the manufacturers' data and the quadratic error between the currents is calculated. This technique also requires the establishment of boundaries to the Rs, Rsh, m, and Is values.

Cortés et al. [5] propose a method to estimate the parameters of a one-diode model of a PV cell based on artificial neural networks (ANN) and the multilayer perceptron concept. The steps of the ANN are applied, considering the different inputs relates to current, voltage, and resistance. The Symmetric Mean Absolute Percentage Error (SMAPE) is used as an error metric to compare the results to the literature and the experimental data.

Li et al. [21] propose a dynamic self-adaptive and mutual-comparison teaching–learning-based optimization (DMTLBO) to extract the parameters of PV cells (considering one, two, and three diode models). RMSE is used to evaluate the performance of the algorithm, comparing them to other algorithms found in the literature. As the previous studies that apply algorithms, it also requires the definition of lower and upper boundaries to the Rs, RSh, Iph, Io, and m values to run the proposed DMTLBO algorithm and estimate the parameters of the single diode model, double diode model, and PV module models.

Regarding the application, different types of PV and cells are considered as Kyocera KC200GT, RTC France, and Shell SM55. Besides testing the proposed methods under the standard conditions, most of the paper evaluate different scenarios, varying the irradiance and/or the temperature values.

Although, there are different methods (analytic and numerical) and algorithms dedicated to extracting the parameters of the one diode model, one of the problems related to these approaches is that most of the time, an initial value of, at least, one parameter must be determined to define the others. For example, Orioli [11] define an initial value of the shunt resistor (Rsh) and diode ideally factor (m) to calculate the other parameters based on the proposed equations, while Álvarez et al. [12] adopts an initial value of m to determine the other parameters. Other approaches set boundaries for the parameters in order to find the ideal value of one or more parameters (as in [6, 19]).

In this context, Reddy and Yammani [13] emphasize that a method that does not require any initial information about the classical parameters (m, Rsh, Iph, Io, and Rs) nor use the datasheet information (as open-circuit voltage, current and voltage at MPP) is an advance to the literature. In fact, different papers are trying to propose methods that eliminate the need for sampling the slope of the I–V curve at the MPP and open circuit point (that is required in most of the analytical methods), as seen in [15]. Others are trying to extract the parameters of PV cells using only the manufacturer's data, without any measurement or prior knowledge [5].

Therefore, the objective of this article is to present a new method for parameter extraction in PV modules, from I–V curves in darkness. In order to validate this method a laboratory device has been built corresponding to the 1-diode model circuit. This way it is possible to take I–V measurements over this device knowing previously the values of the intrinsic parameters: diode saturation current, diode ideality factor, series resistance and shunt resistance. The extrinsic parameter (photogeneration current) that depends on the external irradiation is not considered in this method because within the one diode model it is simply an additive factor and does not modify the I–V curve shape. The method is suitable for parameter extraction from measured dark I–V curves, but also through a proper data transformation for clean illumination I–V curves where no bypass diodes effects are observed. For parameter extraction method validation, an I–V curves dataset has been created from measurements over the laboratory device, where the shunt and series resistances has been varied in a reasonable range, labelling each curve with both actual values. The current saturation and ideality factor (that depends on the diodes choice) cannot be varied but are determined via precise measurements and labels the whole dataset. After this introduction, the article continues as explained below: Sect. 2.2 with materials and methods; Sect. 2.3 with the results; and Sect. 2.4 with the conclusions.

2.2 Materials and Methods

In this section, the used materials and the methods will be presented. As explained in the introduction, a 1 diode model real circuit has been built and allowed us for a dark I–V curves dataset to be created. The fitting of a real solar module to the one diode model is out of the scope of this work and the measurements made has been used to validate the performance of the parameter extraction method proposed only over this particular model. From this dark I–V curves dataset, the new method for parameter extraction will be applied. The parameter extraction method will be explained in the methods section.

2.2.1 Materials

The shape of a PV module I–V curve can be explained according to the 1-diode model (a classic model in scientific literature). Therefore, for this scientific work a laboratory device based on the topology of the 1-diode model has been built, simulating a PV module of N_S photovoltaic cells in series.

For a later analysis of the measurements over this device the intrinsic parameters of the chosen diodes such as the saturation current, ideality factor and intrinsic series and parallel resistances must be determined. Characterization of these parameters is crucial for labelling the dataset measured from the electronic device that emulates the electrical behavior of a photovoltaic solar module.

From now on, the ideality factor will be integrated with temperature and the constants on the exponent of the diode equation in a unique parameter called junction voltage (Te), so the diode equation will state like this Eq. (2.1):

$$I = I_0\left[\mathrm{e}\frac{Vd}{Te} - 1\right] \tag{2.1}$$

Two different characterization setups have been implemented. The first one is optimized to obtain the series resistance and junction voltage and the second one is suitable to obtain the shunt resistance and the saturation current.

The intrinsic series resistance can be calculated from the asymptotic slope of the curve in the conduction region of the diode (first quadrant). The expected series resistance is a very low value, so for a precise characterization a setup with fifty diodes in series has been built, and this way the series resistance is multiplied fifty times. Also, the equivalent junction voltage of the setup result fifty times greater than the individual diodes one. The first experimental setup is shown in Fig. 2.1. Since temperature has been integrated in the parameter "Te" it is quite important to keep its value constant during the I–V curve tracing. For this, a fast sweep was performed via a capacitor charge. First the switch in the figure is closed and a constant current is stablished through it. The measurement starts opening the switch and the subsequent capacitor charge perform the sweep while the V and I values are captured by an oscilloscope.

The fitting of the curve obtained with this first setup allows the determination of series resistance and the parameter "Te".

The expected parasitic shunt resistance of a diode is very high, and the saturation current is very low, so for a precise measurement of this parameters 100 diodes were soldered in parallel, and this way the shunt resistance of the association is 100 times lower, and the saturation current is 100 times higher. The intrinsic parallel resistance can be obtained from the slope of the curve in the reverse current region of the diode (third quadrant) and the saturation current correspond to the intersection of the asymptotic line with the current axis. See schematic design and experimental setup in Fig. 2.2.

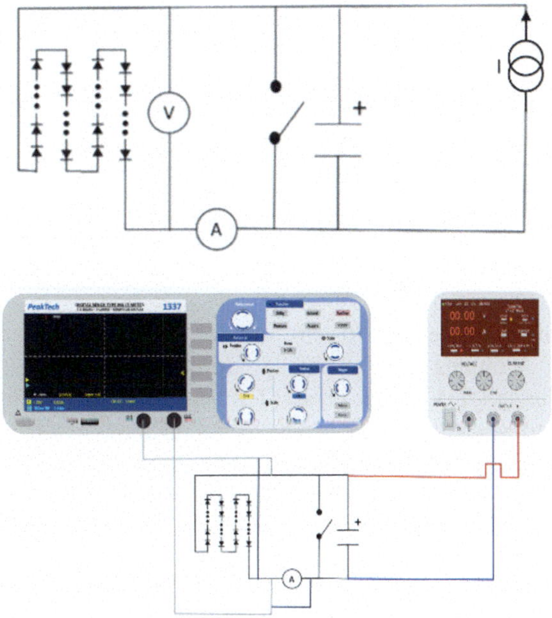

Fig. 2.1 PV module according to the 1-diode model. Experiment to obtain the parasitic serial resistance and junction voltage of the diodes

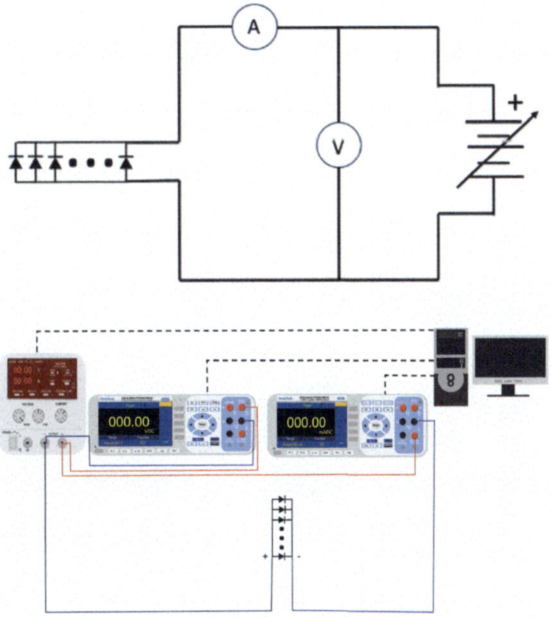

Fig. 2.2 PV module according to the 1-diode model. Experiment to obtain the saturation current and shunt resistance of the diodes

The results of the measurements at 25 °C for the four intrinsic parameters of the diodes from both setups are shown in Table 2.1.

After the determination of this fixed parameters, we were ready to build the laboratory device for the I–V dataset creation.

The typical series resistance of a discrete diode (Table 2.1) is on the order of 10 times greater than the typical value of a photovoltaic cell (units of mOhm), and the typical shunt resistance of diodes is much greater than the solar cell's [22], so to proper simulate individual solar cells it is convenient to solder 10 diodes in parallel in order to adjust the resistances to more realistic values. These groups of parallel diodes simulating each single cell can be soldered in series to simulate a photovoltaic solar module with 10 cells in series (see Fig. 2.3).

With this setup and adding different values of resistors in series and in parallel the dark I–V curves dataset has been created, labelling each of them with the four corresponding device parameters (total Series resistance -Rs-, total Shunt Resistance -Rp-, equivalent Saturation Current -Io = 10 × I_{0d} - and equivalent Junction Voltage -Te = 10 × T_{ed} -). This dataset was intended for our proposed extraction method, but it is suitable for validation of any other extraction algorithm.

Table 2.1 Diode intrinsic parameters obtained from characterization measurements series and parallel at 25 °C

Parameters			
Series measurements		Shunt measurements	
T_{eS} (association, mV)	2200	I_{0P} (association, A)	1.52E−07
R_{SS} (association, Ω)	2.04	R_{PP} (association, Ω)	1.05E+09
T_{ed} (diode, mV)	44	I_{0d} (diode, A)	1.52E−09
R_{Sd} (diode, Ω)	4.08E−2	R_{Pd} (diode, Ω)	1.05E+11

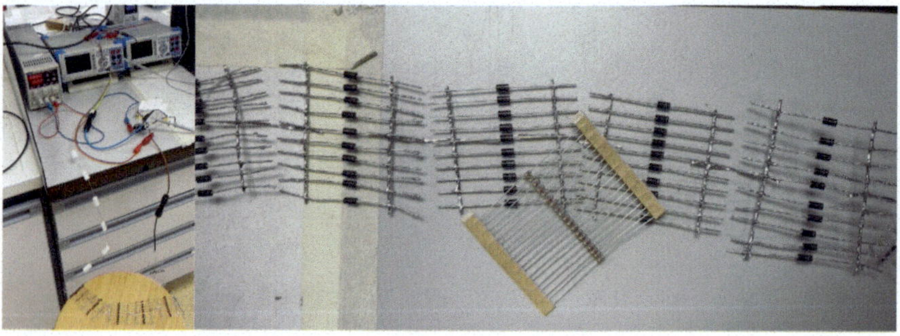

Fig. 2.3 Assembly of the emulated PV module based on the 1-diode model

2.2.2 Methods

In this section the proposed distance definition for fitting and parameter extraction will be explained As is well known, a photovoltaic cell (photovoltaic module) can be modelled by the one diode model, described by the equation shown in (2.2), where the photocurrent term has been omitted because we work from dark curves.

$$I = I_0\left[\mathrm{e}^{\frac{v - IR_s}{Te}} - 1\right] + \frac{V - IR_s}{R_P} \tag{2.2}$$

The proposed approach consists in making the following transformations first to prepare the equation for the fitting procedure:

- A normalization that leads to a curve with $I_{\max} = 1,\ V_{\max} - 1$.
- An inversion and translation that leads to a pseudo-illumination curve with $I_{SC} = 1, V_{OC} = 1$.

Therefore, these transformations can be seen in Eq. (2.3).

$$I' = 1 - \frac{I}{k_i},\ V' = \frac{V}{k_v} \tag{2.3}$$

where $k_i = I_{\max}, k_v = V_{\max}$

Now, these transformations are implemented in Eq. (2.4).

$$\begin{aligned} k_i(1 - I') &= I_0\left[e^{\frac{V'k_v - k_i(1 - I')R_s}{Te}} - 1\right] + \frac{V'k_v - k_i(1 - I')R_s}{R_P}; \\ (1 - I') &= \frac{I_0}{k_i}\left[e^{\frac{V' - \frac{k_i}{k_v}(1 - I')R_s}{\frac{Te}{k_v}}} - 1\right] + \frac{V' - \frac{k_i}{k_v}(1 - I')R_s}{\frac{k_i}{k_v}R_P} \end{aligned} \tag{2.4}$$

Next, the following parameters are renamed as follows (Eq. (2.5)):

$$I_0' = \frac{I_0}{k_i},\ R_S' = \frac{k_i}{k_v}R_S,\ R_P' = \frac{k_i}{k_v}R_P,\ T_e' = \frac{T_e}{k_v} \tag{2.5}$$

So, Eq. (2.3) becomes Eq. (2.6).

$$(1 - I') = I_0'\left[e^{\frac{V' - (1 - I')R_S'}{T_e'}} - 1\right] + \frac{V' - (1 - I')R_S'}{R_P'} \tag{2.6}$$

Now, this curve is defined in the first quadrant intersecting the axes in I = 1 and V = 1 and due to the particular shape of I–V curves, it is close to the first quadrant circumference with radius = 1. When measurements errors are present in both axes magnitudes the optimal distance between the measured data and the analytic curve to be minimized is the orthogonal distance, so for our transformed curves is convenient to take this distance over the radius, that is not so far from the orthogonal distance (for a unity circumference is exactly the orthogonal distance).

This suggests a new transformation to polar coordinates. Then a distance from a point (I_0, V_0) to the curve can be defined with the squared difference along the radius. This new approach leads to a distance quite like the orthogonal distance for curves with this kind of shape and it is far more accurate than the least square distance with cartesian coordinates, that for measurement errors present in both I and V can lead to great errors in the fitting process.

Therefore, a new transformation to polar coordinates is introduced, as shown in Eq. (2.7).

$$I' = r\,sen\theta, \quad V' = r\cos\theta, \quad r = \sqrt{(I')^2 + (V')^2}, \quad \theta = arctg\left(\frac{I'}{V'}\right) \tag{2.7}$$

The above changes are introduced in Eq. (2.6) and we obtain Eq. (2.8).

$$\begin{gathered}(1 - r\,sen\theta) = I_0'\left[e\frac{r(\cos\theta + R_S'\,sen\theta) - R_S'}{T_e'} - 1\right] + \frac{r(\cos\theta + R_S'\,sen\theta) - R_S'}{R_P'}; \\ r = \frac{1 + \frac{R_S'}{R_P'} - I_0'\left[e^{\frac{r(\cos\theta + R_S'\,sen\theta) - R_S'}{T_e'}} - 1\right]}{sen\theta\left(1 + \frac{R_S'}{R_P'}\right) + \cos\theta/R_P'}\end{gathered} \tag{2.8}$$

This transformation is specific for the original curve it was applied to, because the values of I_{max}, V_{max} were used in the transformation. Therefore, the transformed curve will intersect the axes at $[1, 1]$ only for the transformed values of the parameters $\left[R_S', R_P', I_0', T_e'\right]$ that come from the original values $[R_S, R_P, I_0, R_S]$. However, the axis I' is always intersected by the curve at $I' = 1$ because of how the first transformation was perform. The intersection with I' is the transformed point $(I = 0, V = 0)$ in the original curve and then $I' = 1 - \frac{I}{k_i} = 1$. Indeed, if we substitute $\theta = \Pi/2\ rad$ which is the angle that is associated with the intersection of the curve with the axis in the equation for the radius, Eq. (2.9) is obtained.

$$r = \frac{1 + \frac{R_S'}{R_P'} - I_0'\left[e^{\frac{rR_S' - R_S'}{T_e'}} - 1\right]}{1 + \frac{R_S'}{R_P'}} \tag{2.9}$$

This equation is always right for $r = 1$ regardless of the values of the parameters $\left[R_S', R_P', I_0', T_e'\right]$. On the other hand, the intersection with the $V'axis$ at $V' = 1$ only happens for the transformed parameters $\left[R_S', R_P', I_0', T_e'\right]$. It is possible use this characteristic of our transformed curve to reduce the number of parameters to evaluate to fit experimental data. The transformed parameters $\left[R_S', R_P', I_0', T_e'\right]$ must generate a curve that intersects with the axis at $V' = 1$. By implementing this fact in the equation of the radius, it is possible to obtain a new equation (Eq. (2.10)) that involves the four parameters and allow us to eliminate one of them for the fitting procedure. For the intersection of the curve at $V' = 1$, the result is $r = 1,\ \theta = 0\,rad$.

$$1 = \frac{1 + \frac{R'_S}{R'_P} - I'_0\left[e^{\frac{1-R'_S}{T'_e}} - 1\right]}{\frac{1}{R'_P}};$$

$$1 = R'_P + R'_S - I'_0 R'_P\left[e^{\frac{1 - R'_S}{T'_e}} - 1\right] \tag{2.10}$$

It is also known that in the area were the curve intersects the V' axis, the curve is more sensible to changes in the following parameters: I'_0, T'_e. Therefore, and to reduce the number of parameters, it is possible to express I'_0 as a function of the rest of the parameters, as shown in Eq. (2.11).

$$I'_0 = \frac{R'_P + R'_S - 1}{R'_P\left[e^{\frac{1-R'_S}{T'_e}} - 1\right]} \tag{2.11}$$

Finally, it must realise that the accuracy of this whole procedure relays on the exactitude of the measurement of $[I_{\max}, V_{\max}]$. It is fundamental to ensure a high accuracy of this measurement, for example taking the average of several measurements over this final point.

2.3 Results

In this section, the research results will be presented. In the first part of the section, the results of the precise measurements of the starting parameters over the laboratory device are shown. In the second part of the section the parameters values obtained applying the extraction method proposed are presented and compared with the device's starting values.

2.3.1 Results of Dataset Creation

To obtain the I–V curves dataset, a matrix of measurements has been made over the one diode model laboratory device, adding 5 different values for the series resistances and 5 different values for the shunt resistances. This implies that a total of 25 different combinations (see Table 2.2). The different resistance values of the resistors used within the device have been carefully measured, since they show some tolerance over the standard value. These measurements have been performed with a linear regression fitting of an I–V set of points. The values shown in Table 2.2 for series resistance includes the parasitic resistance of the diodes.

Table 2.2 Resistance values obtained for series and shunt resistors within the laboratory device

Series resistance (ohms, Ω)	Shunt resistance (ohms, Ω)
Rs1 = 0.143	Rp1 = 216
Rs2 = 0.247	Rp2 = 454
Rs3 = 0.348	Rp3 = 549
Rs4 = 0.451	Rp4 = 669
Rs5 = 0.552	Rp5 = infinite

The I–V one diode model curves dataset was created from all the possible combinations of series and shunt resistances, and each curve has been labelled with the following code: "Rsx_Rpy.csv", where x is the ordinal number of series resistance enumeration (from 1 to 5), and y is the ordinal number of shunt resistance (from 1 to 4 or "X" for infinite).

During all the measurements the temperature has been maintained constant at 25 °C to be sure that the intrinsic parameters of the diodes (Te and Io) keep on the same value determined for them in the previous characterization of the diodes, performed also at 25 °C.

2.3.2 *Results of Parameter Extraction Based on the New Proposed Method*

Once all the curves already mentioned in the previous section have been obtained, each one of them is subjected to the new double mathematical transformation and fitted to the theoretical equation by means of a simple least squares method over the radius as described before.

Since the transformed theoretical Eq. (2.8) is an implicit one, calculation of its points for each set of parameters require using a non-analytical method, and for a low computational cost we used a simple iteration algorithm starting form a guess value of the radius on the right side of (2.8) to obtain a new radius on the left side, and reinserting this new value on the right side again for each iteration.

This simple algorithm could show convergence problems depending on the equation, but we found it to be always convergent for the transformed Eq. (2.8), because the proximity of the radius to the value 1 along all the angle range. This is one of the remarkable advantages of our proposed transformation.

As an example, Fig. 2.4 shows a curve fitted with our method (blue) and the measured points (red) after the proposed transformation on polar coordinates.

Table 2.3 shows all the values extracted with the new proposed distance definition and subsequent fitting for the 25 measured curves of the dataset along with the real values corresponding to each curve in order to know the error of the parameter estimation.

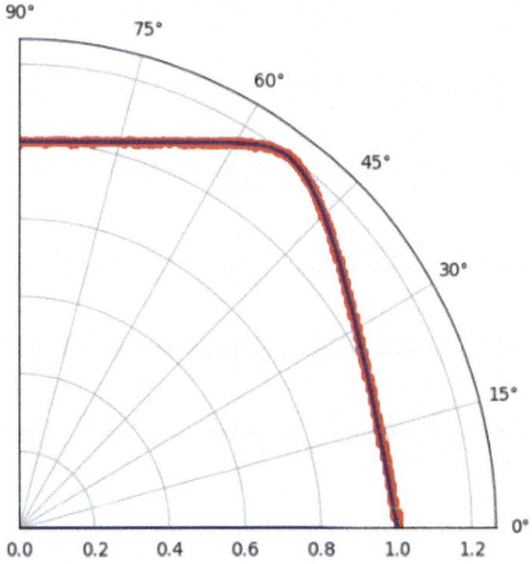

Fig. 2.4 Fitted curve according to new method (blue) versus actual measured curve (red) (Polar coordinates)

As can be seen in Table 2.3, the values extracted with the new model for the series resistance show an error under 10% in all cases and the average error is 6.9%.

With respect to the shunt resistance, the values are not so accurate as usually happens in literature. However, for the two lower values labelled in the Dataset (216 and 454 ohms) error is always under 100%, and values under 50% are more frequent. As the main purpose of parameter extraction is the diagnostics of degradation in solar modules this allow at least for detecting low shunt resistance condition.

For the shunt resistance value of 454 ohms on a module as the one simulated with our laboratory device (Voc around 7.5 V) the power loss will be around 0.13 W. Considering a solar module power generation of 29 W (with a typical photocurrent of 3.3 A), the power loss due to shunt resistance will be 0.4% for 454 ohms, so the extraction process described will always alert for power losses bigger than this value.

2.4 Conclusions

In this paper a new method based on a double mathematical transformation has been proposed for the extraction of the main parameters of a PV module from I–V curves in darkness. For the validation of the method an electronic circuit with the one diode model topology has been built. On this circuit the values of serial and shunt resistances can be varied to create an I–V curves dataset that can be useful not only for our method validation but also to validate any other methods based on the one diode model. This way, a dataset of 25 curves has been created and used for our method evaluation.

The curves on this dataset have been passed to the new parameter extraction method, obtaining the estimated values of series resistance and shunt resistance. Since these resistance values are labelled on each curve, it is possible to evaluate the accuracy of the extraction performed.

Table 2.3 Results of the extraction of the main parameters from the 25 measurements carried out

Dataset	Rs dataset (Ω)	Rs (adjusted, Ω)	Rs error (%)	Rsh dataset (Ω)	Rsh (adjusted, Ω)	Rp error (%)
1	0.170	0.154	9.47	216	179.9	16.7
2	0.170	0.185	8.72	454	365.8	19.4
3	0.170	0.171	0.55	549	444.0	19.1
4	0.170	0.166	2.37	669	> 1000	> 100
5	0.170	0.175	2.90	> 1000	> 1000	> 100
6	0.278	0.253	9.13	216	166.2	23.1
7	0.278	0.268	3.49	454	818.2	80.2
8	0.278	0.258	7.28	549	> 1000	> 100
9	0.278	0.273	1.79	669	> 1000	> 100
10	0.278	0.258	7.20	> 1000	> 1000	> 100
11	0.380	0.356	6.34	216	225.0	4.2
12	0.380	0.355	6.65	454	886.3	95.2
13	0.380	0.356	6.35	549	704.9	28.4
14	0.380	0.376	1.06	669	416.5	37.7
15	0.380	0.415	9.21	> 1000	> 1000	> 100
16	0.476	0.471	1.03	216	258.3	19.6
17	0.476	0.463	2.80	454	476.8	5.0
18	0.476	0.468	1.69	549	633.1	15.3
19	0.476	0.479	0.75	669	480.5	28.2
20	0.476	0.476	0.02	> 1000	> 1000	> 100
21	0.589	0.604	2.58	216	981.0	> 100
22	0.589	0.569	3.39	454	182.3	59.8
23	0.589	0.571	3.03	549	214.5	60.9
24	0.589	0.573	2.76	669	> 1000	> 100
25	0.589	0.608	3.26	> 1000	> 1000	> 100

The values obtained showed an average error of 6.9% for the series resistance and for the two lower values of shunt resistance (worst case) most of the errors are under 50%, that will allow for detecting low shunt resistance conditions.

Anyway, one of the most remarkable advantages of the mathematical transformation proposed for parameter extraction is that with a very simple iteration algorithm the calculation of the theoretical values of the I–V curve results to be always convergent, and at least for the tests performed in this work, the parameters results and processing time are essentially independent of the starting values assigned at the beginning of the process.

Regarding future work, it will be interesting analyze the error evolution with the resistance values variation and to test some kind of weight functions that will allow for minimizing the errors.

References

1. World Resource Institute—WRI, How National Net-Zero Targets Stack Up After the COP26 Climate Summit. https://www.wri.org/insights/how-countries-net-zero-targets-stack-up-cop26#:~:text=1.,net-zeroemissionsbymidcentury.&text=Aftertherecent26thUN,aimfor2050orearlier (2021)
2. International Energy Agency—IEA, Renewables 2020—Analysis and Forecast to 2025 (2020)
3. International Energy Agency—IEA, Electricity Market Report (2022)
4. International Energy Agency—IEA, Renewables 2021—Analysis and Forecast to 2026 (2021)
5. B. Cortés, R.T. Sánchez, J.J. Flores, Characterization of a polycrystalline photovoltaic cell using artificial neural networks. Sol. Energy **196**, 157–167 (2020). https://doi.org/10.1016/j.solener.2019.12.012
6. M.P. Cervellini, N.I. Echeverria, P.D. Antoszczuk et al., Optimized parameter extraction method for photovoltaic devices model. IEEE Lat. Am. Trans. **14**, 1959–1965 (2016). https://doi.org/10.1109/TLA.2016.7483540
7. T. Sudhakar Babu, J. Prasanth Ram, K. Sangeetha et al., Parameter extraction of two diode solar PV model using Fireworks algorithm. Sol. Energy **140**, 265–276 (2016). https://doi.org/10.1016/j.solener.2016.10.044
8. A Sabadus, M Paulescu, On the nature of the one-diode solar cell model parameters. Energies (Basel) **14** (2021). https://doi.org/10.3390/en14133974
9. Y. Zhang, Q. Gao, Y. Yu, Z. Liu, Comparison of double-side and equivalent single-side illumination methods for measuring the I–V characteristics of bifacial photovoltaic devices. IEEE J. Photovolt **8**, 397–403 (2018). https://doi.org/10.1109/JPHOTOV.2017.2778226
10. D.R. Ahmed, I.R. Mohammed, H.M. Abdullah et al., The correlation of device parameters with illumination energy to explore the performance of a monocrystalline silicon solar module. Silicon (2021). https://doi.org/10.1007/s12633-021-00966-z
11. A. Orioli, An accurate one-diode model suited to represent the current-voltage characteristics of crystalline and thin-film photovoltaic modules. Renew. Energy **145**, 725–743 (2020). https://doi.org/10.1016/j.renene.2019.06.074
12. J.M. Álvarez, D. Alfonso-Corcuera, E. Roibás-Millán et al, Analytical modeling of current voltage photovoltaic performance: an easy approach to solar panel behavior. Appl. Sci. (Switzerland) **11** (2021). https://doi.org/10.3390/app11094250
13. S. Sreekantha Reddy, C. Yammani, Parameter extraction of single-diode photovoltaic module using experimental current–voltage data. Int. J. Cir. Theory Appl. **1–19** (2021). https://doi.org/10.1002/cta.3133
14. Y. Chaibi, A. Allouhi, M. Salhi, A simple iterative method to determine the electrical parameters of photovoltaic cell. J. Clean. Prod. **269**, 122363 (2020). https://doi.org/10.1016/j.jclepro.2020.122363
15. D. Wei, M. Wei, H. Cai et al., Parameters extraction method of PV model based on key points of I–V curve. Energy Convers. Manag. **209**, 112656 (2020). https://doi.org/10.1016/j.enconman.2020.112656
16. V.T. Rangel-Kuoppa, M.D.L. Albor-Aguilera, C. Hérnandez-Vásquez et al, Shunt resistance and saturation current determination in CdTe and CIGS solar cells. Part 1: A new theoretical procedure and comparison with other methodologies. Semicond. Sci. Technol. **33** (2018). https://doi.org/10.1088/1361-6641/aab017
17. V.T. Rangel-Kuoppa, M.D.L. Albor-Aguilera, C. Hérnandez-Vásquez et al, Shunt resistance and saturation current determination in CdTe and CIGS solar cells. Part 2: Application to experimental IV measurements and comparison with other methods. Semicond. Sci. Technol. **33** (2018). https://doi.org/10.1088/1361-6641/aab018
18. M.H. Ali, A. Rabhi, S. Haddad, A. El Hajjaji, Real-time determination of solar cell parameters. J. Electron. Mater. **46**, 6535–6543 (2017). https://doi.org/10.1007/s11664-017-5697-0
19. Z. Hu, W. Gong, S. Li, Reinforcement learning-based differential evolution for parameters extraction of photovoltaic models. Energy Rep. **7**, 916–928 (2021). https://doi.org/10.1016/j.egyr.2021.01.096

20. Y. Liu, A.A. Heidari, X. Ye et al., Evolutionary shuffled frog leaping with memory pool for parameter optimization. Energy Rep. **7**, 584–606 (2021). https://doi.org/10.1016/j.egyr.2021.01.001
21. L. Li, G. Xiong, X. Yuan et al., Parameter extraction of photovoltaic models using a dynamic self-adaptive and mutual—comparison teaching-learning-based optimization. IEEE Access **9**, 52425–52441 (2021). https://doi.org/10.1109/ACCESS.2021.3069748
22. H.A. Koffi, A.A. Yankson, A.F. Hughes et al., Determination of the series resistance of a solar cell through its maximum power point. Afr. J. Sci. Technol. Innov. Dev. **12**, 699–702 (2020). https://doi.org/10.1080/20421338.2020.1731073

Chapter 3
Renewable Communities Combining Anaerobic Digestion and Solar Energy: A Case Study in the Province of Soria (Spain)

Alfonso García Álvaro, Víctor Alonso Gómez, Jesús Armando Aguilar-Jiménez, and Ignacio de Godos Crespo

Abstract The increase in energy consumption, primarily reliant on fossil fuels, has led to a continuous rise in carbon dioxide (CO_2) emissions over many years, particularly in the last two decades, which has intensified the effects of climate change. Urban areas, which concentrate both population and industry, exhibit the highest emission levels, making their transition to a system based on renewable energy sources crucial for reducing environmental impact. Renewable energy communities are proposed as a dual-solution approach: (i) an innovative and sustainable model for local energy management and production, and (ii) a strategy that generates fewer greenhouse gas emissions. This study presents a model of an energy community that utilizes organic waste generated in three urban centers of varying sizes (500, 5000, and 40,000 inhabitants), supplemented by solar energy, with the aim of meeting the energy needs of municipal buildings. The results demonstrate that, while the system proves more efficient in smaller urban centers, the percentage of municipal energy consumption that can be covered by this model exceeds 70% in larger urban areas.

Keywords Anaerobic digestion · Biomethane · Energies hybridization · Solar energy · Waste to energy

---

A. García Álvaro (✉) · I. de Godos Crespo
Institute of Sustainable Process, University of Valladolid, Valladolid, Spain
e-mail: agarcia@uva.es

Department of Chemical Engineering and Environmental Technology, University of Valladolid (UVA), Campus Universitario Duques de Soria, Soria, Spain

V. Alonso Gómez
Department of Applied Physics, University of Valladolid (UVA), Campus Universitario Duques de Soria, Soria, Spain

J. A. Aguilar-Jiménez
Centro de Estudios de las Energías Renovables, Instituto de Ingeniería, Universidad Autónoma de Baja California, Mexicali, Baja California, México

D. Rossit et al. (eds.), *Proceedings of the VII Ibero-American Congress of Smart Cities, ICSC-Cities 2024, 12-14 November, San Carlos, Costa Rica*, Springer Proceedings in Energy, https://doi.org/10.1007/978-981-96-4301-1_3

3.1 Introduction

The increase in energy consumption, largely based on fossil fuels, has led to a rise in carbon dioxide (CO_2) emissions over the years, particularly in the last 20 years, contributing to the effects of climate change. Urban environments, where population and industry are concentrated, are the main sources of higher emission rates. Therefore, their transition to systems based on renewable energy sources is essential to mitigate this environmental impact [1]. Renewable energy communities are proposed as a dual-solution approach: (i) an innovative and sustainable model for local energy management and production, and (ii) a model that generates fewer greenhouse gas emissions.

Provinces such as Soria in Spain, despite enjoying good environmental health, are increasingly recognizing the importance of maintaining this status. In Soria's capital, the main urban and industrial center, a project called "Soria 2030" is underway, as part of the 2030 Agenda, aiming to promote a modern, carbon-free city, accelerate the transition to a circular economy, and support the development of a bioeconomy in the region [2, 3].

This research aims to conduct a study in three representative localities of the Soria province with different population sizes, focusing on the implementation of an energy community model utilizing their local renewable resources. On one hand, urban organic waste would be separated and transformed into bioenergy through anaerobic digestion, while on the other, solar energy, an abundant resource in the province, would be converted into electricity and heat at the same urban organic waste management facility (Fig. 3.1). The two proposed technologies exhibit a high level of technological development [4, 5]; however, their combined use has not been thoroughly studied, which could potentially present a viable energy model based on local resources.

In this study, the unique contribution lies in the establishment of an energy community model that utilises organic waste from diverse urban centres with populations ranging from 500 to 40,000 inhabitants, supplemented by solar energy. This contrasts with other studies that focus on isolated livestock farms, residential buildings, or specific waste types like cattle manure, swine manure, or food waste as can be seen in Table 3.1. By targeting urban solid waste within anaerobic digestion systems, this study demonstrates an innovative waste-to-energy pathway, adaptable to urban settings with varying population densities. Unlike studies utilising concentrated solar plants or hybrid power plants, this research employs hybrid solar integration in mesophilic conditions (35 °C), promoting scalability and resilience in urban environments. The findings contribute a replicable framework for sustainable energy generation and waste management in mid-sized urban areas, especially pertinent for decentralised energy systems in regions lacking extensive energy infrastructure.

This initiative also aligns directly with the Sustainable Development Goals (SDGs), particularly SDG 7 and SDG 11: "Affordable and Clean Energy" and "Sustainable Cities and Communities." These energy communities promote access to

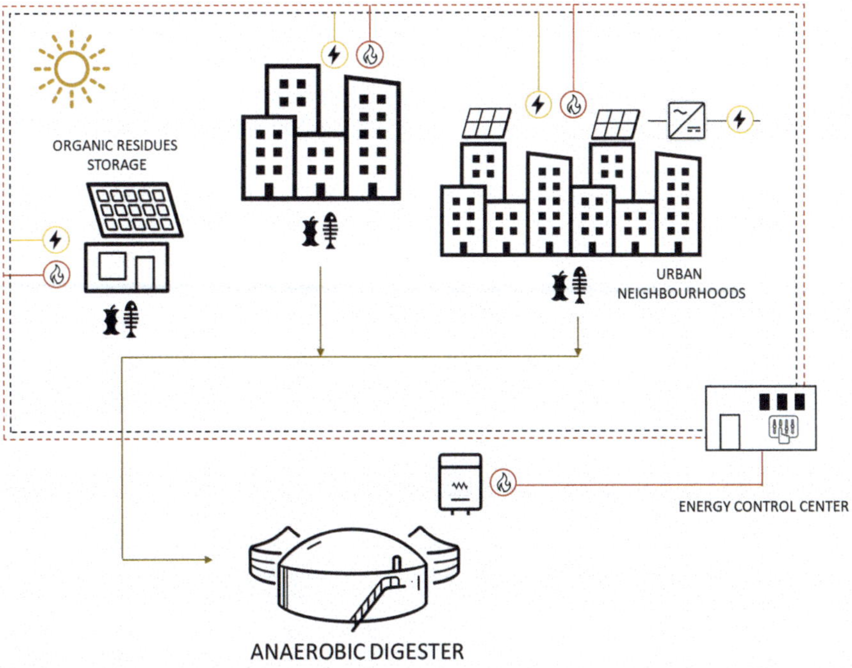

Fig. 3.1 Schematic diagram of the AD/S system studied

sustainable, decentralized, and renewable energy sources, fostering the development of less dependent and more environmentally friendly towns and cities [13, 14]. Additionally, it supports SDG 12, SDG 13, and SDG 15: "Climate Action," "Responsible Consumption and Production," and "Life on Land." The proposal aims to reduce greenhouse gas emissions, minimize environmental impact, preserve local ecosystems, and encourage more responsible and efficient resource consumption [15, 16].

3.2 Materials and Methods

3.2.1 Anaerobic Digestion/Solar (AD/S) System Description

The scheme of the proposed system can be seen in Fig. 3.1. The system consists of a biogas plant with an anaerobic reactor working in mesophilic conditions, at 35 °C, and a set of hybrid solar modules (Photovoltaic/thermal) located in the organic residues storage and connected to an energy control center.

Table 3.1 Review of the anaerobic digestion and solar energy hybridization system researches

Reference	System proposed	Substrate	AD details	Study location
[6]	A solar greenhouse to optimize the available radiation to increase the medium temperature in the digester in Rural areas in China	Slurry	Ambiente conditions	Xuzhou (China)
[7]	Integration of solar thermal energy to reduce the amount of biogas used (Comparison of different collectors)	Organic fraction of municipal solid waste	Mesophilic/ thermophilic conditions	Different cities in Italy
[8]	A model study of a MicroGrids with photovoltaic array and an anaerobic digester providing the energy to dairy farmers' association system	Cattle manure	No information	Albarrancho (Bolivia)
[9]	Use of hybrid solar panels tu increase the biomethane production in isolated livestock farms	Swine manure	Mesophilic conditions (35 °C)	Soria (Spain), Iowa (United States), Odense (Denmark), Santa Catarina (Brazil) and Laixi (China)
[10]	solar-heated digesters are sized to meet the methane needs for two large-scale applications: household (pure electricity)	Food waste from students' restaurants	Thermophilic temperature (55 °C)	South Korea
[11]	hybrid solar-anaerobic digestion system for a cleaner energy in residential buildings	Kitchen waste (KW) and blackwater	Mesophilic conditions (37 °C)	Different cities in Morocco
[12]	Hybrid renewable power plant composed of a Concentrated Solar Plant with molten salt storage sub-system, a Rankine steam cycle power plant and a biomass anaerobic digester	Solid urban waste (120,000 inhabitants)	No information	Messaad (Algeria)
This study	Energy community that utilizes organic waste generated in three urban centers of varying sizes (500, 5000, and 40,000 inhabitants), supplemented by solar energy	Solid urban waste	Mesophilic conditions (35 °C)	Different urban areas in Soria (Spain)

The anaerobic digester is the tank where the biological reactions of transformation of the organic matter of the substrate will be carried out. As a by-product, biogas, with a high energy content, is obtained [17].

The thermal energy captured by the solar module field is transferred to a thermal energy storage tank and a heat exchanger is used to heat the digester and the organic substrate. On the other hand, the electricity produced is managed from the energy control center.

3.2.2 Design System

The substrate used in the anaerobic digestion is the urban organic residues and the quantities of this organic waste produced in each location were estimated, considering the population and the waste produced per person. Simultaneously, the biogas and solar plant were sized to calculate both the energy production and demand in each proposed system (scenarios 1, 2, and 3). Finally, an energy balance was performed for each location.

Based on the final results, an estimate was made of the percentage of municipal energy consumption that could be covered if a biogas plant were implemented to manage the organic waste generated in each of the study locations.

The organic waste management of the locations is proposed. For this purpose, an anaerobic cylindrical reactor has been designed for each case depending on the waste's quantity. The operating parameters of a hydraulic residence time (HRT) of 45 days and a working temperature of 35 °C have been selected considering previous works [18–20]. The system characteristics can be seen in Table 3.2.

Urban organic waste, as a substrate, can be very changing in composition and organic content. Representative average characteristics have been measured for each location based on taken samples. Biochemical biogas production experiments were performed to estimate the biogas production potential of the three substrates, following the methodology developed by Holliger et al. [21], under mesophilic temperature conditions (35 °C). The analytical characterization needed was carried out using laboratory equipment to determine the following parameters: Total Solids, Volatile Solids, and pH and following the American Public Health Association standard methods [22].

The solar installation will be based on the installation of hybrid solar panels on the roof of the organic residue's storage with different surface areas in each location. The thermal efficiency of 60% and electrical efficiency of 17% [9].

3.2.3 Energy Balance

The energy balance took into account the consumption and production of energy throughout the process in the form of heat and electricity, considering Eq. 3.1–3.6

Table 3.2 Characteristics of the proposed system

	Quintana Redonda (Spain)	El Burgo de Osma (Spain)	Soria (Spain)
Anaerobic digester			
Work volume (m^3)	10.5	83.6	444.9
Work temperature (°C)	35	35	35
HRT: Hydraulic retention time (days)	45	45	45
Substrate: Organic urban residues			
Managed volume (t/year)	56.9	452.1	2405.7
Volatile Solids (kg/m3)	250	250	250
Biogas yield (m^3/t VS)	601.0	625.0	552.0
η_{bo} is the boiler efficiency (%)	90	90	90
Solar system: Hybrid technology			
Surface area (m^2)	10	80	400
η_e thermal conversion efficiency (%)	60	60	60
η_{pv} photoelectric conversion efficiency (%)	17	17	17

$$E_{Balance}\left(\frac{W \cdot h}{d}\right) = E_{prod} - E_{Demand} \tag{3.1}$$

$$E_{Electricity\ Balance}\left(\frac{W \cdot h}{d}\right) = E_{E\ prod} - E_{E\ demand} \tag{3.2}$$

$$E_{Heat\ Balance}\left(\frac{W \cdot h}{d}\right) = E_{H\ prod} - \left[E_{H\ digester\ demand} + E_{H\ substrate\ demand}\right] \tag{3.3}$$

where E is energy and H is heat.

$$E_{Electricity\ Balance}\left(\frac{W \cdot h}{d}\right) = E_{E\ prod} - E_{E\ demand} \tag{3.4}$$

$$E_{Heat\ prod}\left(\frac{W \cdot h}{d}\right) = \left[Irr\left(\frac{kWh}{m^2}\right) * solar\ panel_{area}\left(m^2\right) * \eta_e(\%) * \eta_{ts}(\%)\right] + \left[biogas_{boiler} * \eta_{bo}\right] \tag{3.5}$$

where η_e is the solar panel thermal efficiency (%), η_{ts} is the thermal system efficiency taking account different possible thermal losses (operational issues…) and η_{bo} is the boiler efficiency.

$$E_{electricity\ prod}\left(\frac{W \cdot h}{d}\right) = Irr\left(\frac{kWh}{m^2}\right) * Solar\ panel_{area}\left(m^2\right) * \eta_{pv}(\%) * \eta_{es}(\%) \tag{3.6}$$

where η_{pv} photoelectric conversion efficiency (%), η_{es} is the electricity system efficiency taking account different possible electricity losses.

3.2.4 Study Localizations

In order to analyze the effect of the hybrid AD/S system under different population size scenarios, three towns/cities were selected for the study: Quintana Redonda (Spain) with 506 inhabitants, El Burgo de Osma (Spain) with a population of 5166 inhabitants and Soria (Spain) with 40,096 inhabitants.

The meteorological information for each location corresponds to an average year considering the data from 2005. Table 3.3 shows the main characteristics of the regions analyzed. Since the three localities studied have different ambient temperatures and solar radiation, the thermal energy generation needs of the AD/S system to satisfy the demand of the anaerobic reactor will be different.

The annual thermal energy required for the anaerobic reactor was calculated from the thermal energy that would be required to maintain the reactor and the input substrate at 35°C throughout the year (Eq. 3.7–3.9). Similarly, the electricity required for the system takes into account the reactor mixing and pumping needs.

$$Digester\ volume\left(m^3\right) = \frac{substrate\ V}{day}\left(\frac{m^3}{day}\right) * HRT(days) * 1.3(biogas\ factor) \tag{3.7}$$

$$E_{Heat\ digester}\left(\frac{W \cdot h}{d}\right) = K\left(\frac{W}{m^2 \cdot {}^\circ C}\right) * A\left(m^2\right) * \Delta(T_{ext} - T_{int})({}^\circ C) * \frac{24h}{day} \tag{3.8}$$

$$E_{Heat\ substrate}\left(\frac{W \cdot h}{d}\right) = Q\left(\frac{subst\ m^3}{day}\right) * \left(\frac{4.2MJ}{{}^\circ C \cdot m^3}\right) * \Delta(T_{ext} - T_{subst})({}^\circ C) \tag{3.9}$$

Table 3.3 Cities selected for the study

	Quintana Redonda (Spain)	El Burgo de Osma (Spain)	Soria (Spain)
Latitude and longitude	41.6, – 2.6	41.6, – 91.7	41.7, – 2.5
Annual irradiation (MWh/m^2)	1620.5	1642.1	1649.1
Average annual temperature (°C)	11.1	11.5	10.8

$$E_{electricity\ demand}\left(\frac{W \cdot h}{d}\right) = \sum EP(W) * WT\left(\frac{h}{d}\right) \tag{3.10}$$

In the last stage, the energy consumption of municipal buildings in an urban center in Spain will be considered to calculate the % of this energy requirement that could be covered with the energy produced in the AD/S system for each locality. Considering information sources from IDEA and EUROSTAT [23, 24], energy consumption values has been considered and shown in Table 3.4.

3.3 Results and Discussion

3.3.1 AD Experiments

Once the Biochemical biogas production experiments were completed and the data processed, the biogas production potential from organic waste in each location was defined and presented in Table 3.5. The three substrates were similar in composition and the biological degradation was also analogous with biogas productions between 552 in Soria and 625 in Quintana Redonda m^3/t VS. This small difference could be attributed to the composition of organic waste from rural areas might have a higher proportion of biodegradable materials, such as agricultural residues or untreated food waste, which are more efficient for biogas generation. These results align with the literature, which discusses differences in municipal solid waste between small towns and larger cities [25].

3.3.2 Energy Balance

Counting all the inputs and outputs of the proposed system at each location, the following energy balances per month have been obtained and are shown in Fig. 3.2.

The energy balance is similar across the three locations, with a consistent year-round biogas production of 4276 kWh per month in Quintana Redonda, 3531 kWh in El Burgo de Osma, and 165,997 kWh in Soria, as organic waste is available throughout the year. On the other hand, both the energy production from solar modules and the energy consumption of the anaerobic reactor are more seasonal. There is higher production during the summer months, when solar irradiance is at its peak, and greater heat losses during the winter months due to low temperatures, as reported in the work of García Álvaro et al. [9].

Considering the population of each locality and the energy balance, in Fig. 3a, the energy produced per inhabitant was calculated to range from 58 to 107 kWh annually, with the highest value obtained for the smallest municipality (Quintana Redonda) and the lowest for the bigger one (Soria). On the contrary, It can be observed in

Table 3.4 Municipal energy consumption in each study location

		Quintana Redonda (Spain)			El Burgo de Osma (Spain)			Soria (Spain)		
	Energy consumption (kWh/year)	Facilities	Size factor (%)	Consumption (kWh)	Facilities	Size factor (%)	Consumption (kWh)	Facilities	Size factor (%)	Consumption (kWh)
City council	100,000	1	20	20,000	1	50	50,000	1	100	100,000
Schools	150,000	1	20	30,000	3	50	225,000	10	100	1,500,000
Cultural centres	50,000	1	20	10,000	2	100	100,000	3	100	150,000
Sports facilities	200,000	1	20	40,000	3	100	600,000	5	100	1,000,000
Health centres	150,000	1	15	22,500	1	100	150,000	3	100	450,000

Table 3.5 Potential energy transformable from anaerobic digestion of organic waste in the three study locations

	Quintana Redonda (Spain)	El Burgo de Osma (Spain)	Soria (Spain)
Inhabitants	506	5166	40,096
Organic residues produced (t/inhab year)	0.150	0.175	0.200
Managed volume (%)	75%	50%	30%
Managed volume (t/year)	56.9	452.0	2405.8
Biogas yield (m^3/t VS)	601	625	552
Biogas (m^3/year)	8553	70,629	331,995

Fig. 3b, that with the approach to utilize the energy for the municipal needs, the larger the municipality is, the more efficient the proposed system becomes. In the case of Soria, the system can cover more than 70% of the energy needs while the smaller localities, Quintana Redonda or El Burgo de Osma, only arrives up to 45% of the needs.

3.4 Conclusions

The management of municipal solid waste through the biological process of anaerobic digestion in a biogas plant is proposed as an alternative for three municipalities of different sizes in the province of Soria.

This approach offers the advantage of energy recovery from waste management, and when combined with solar energy from the waste storage facility, it can meet more than 70% of the municipal energy needs in an urban center like Soria, with over 40,000 inhabitants. The proposed system is also efficient in smaller municipalities, providing higher energy production per capita, though covering a smaller percentage of municipal energy needs—around 45% for both El Burgo de Osma, representing a town of 5000 inhabitants, and Quintana Redonda, a village of 500 inhabitants.

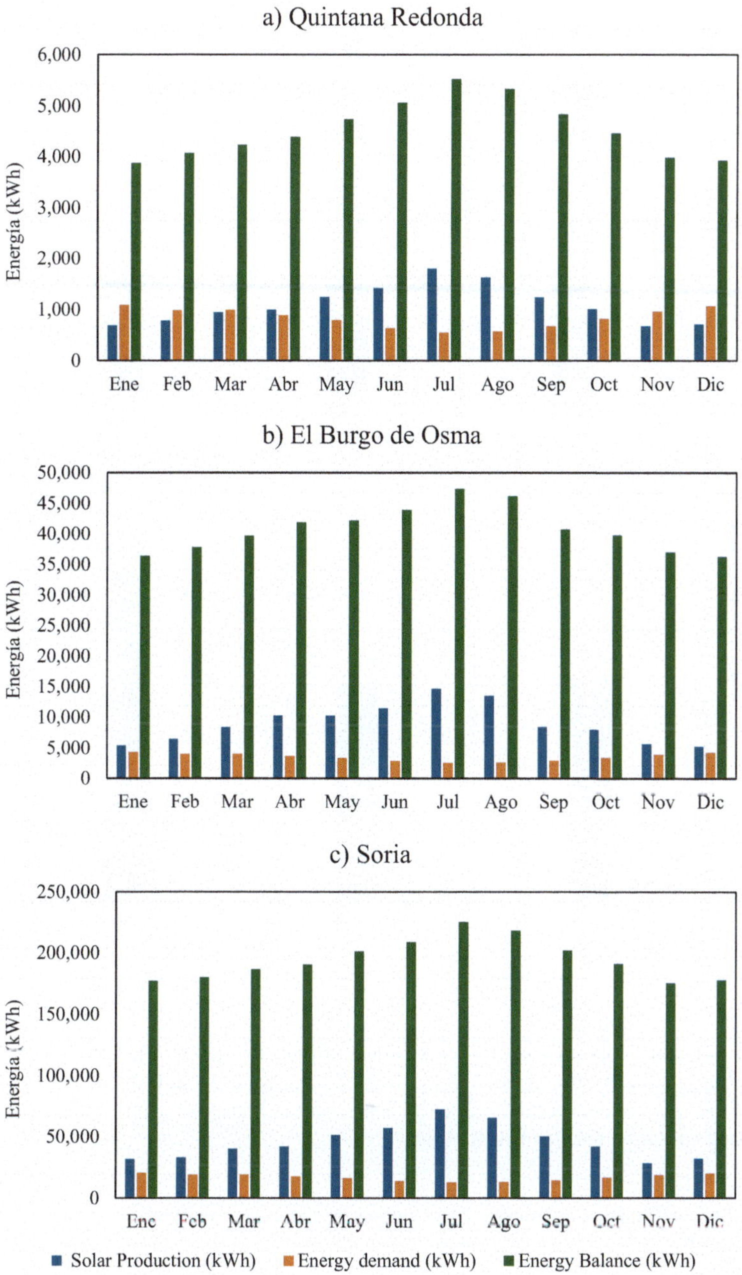

Fig. 3.2 Energy consumed and produced each month in **a** Quintana Redonda (Spain), **b** El Burgo de Osma (Spain) and **c** Soria (Spain)

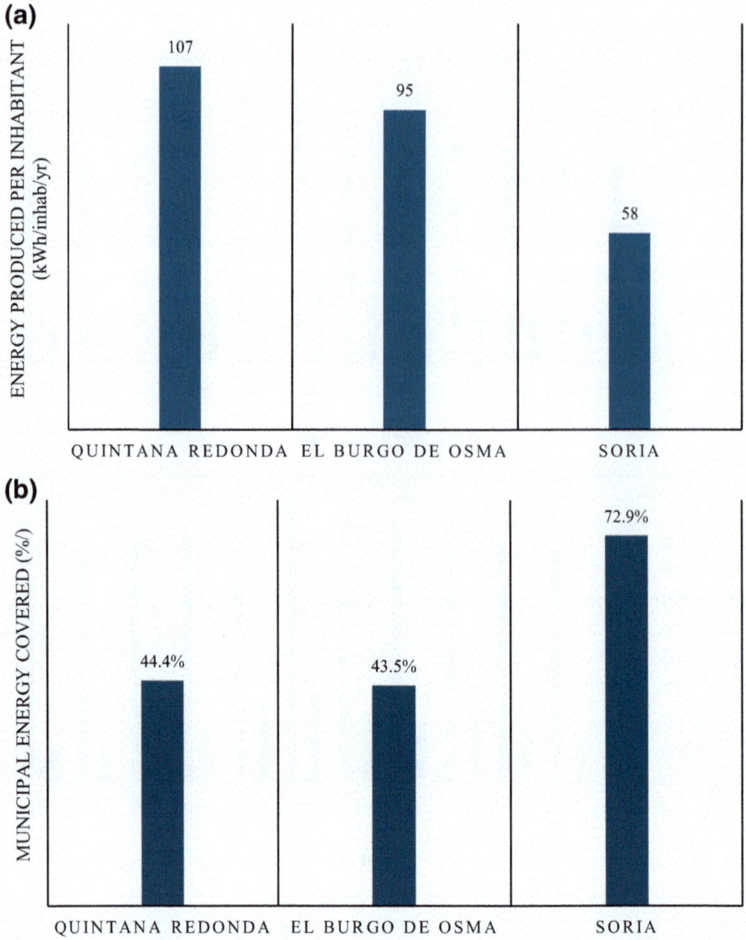

Fig. 3.3 **a** Energy generated per inhabitant per year in each study municipality and **b** Percentage of municipal energy needs covered in each study location

Acknowledgements This work has been written in collaboration with the "*Cátedra Agenda Urbana 2030 para el desarrollo local*" of the University of Valladolid and the municipalities of Soria, Burgo de Osma and Quintana Redonda.

Declaration of Competing Interest The authors declare that they have no known competing financial interests or personal relationships that could have appeared to influence the work reported in this paper.

References

1. M.J. Burke, J.C. Stephens, Political power and renewable energy futures: a critical review. Energy Res. Soc. Sci. **35**, 78–93 (2018). https://doi.org/10.1016/j.erss.2017.10.018
2. Ayuntamiento de Soria, Hoja de Ruta para alcanzar la neutralidad en emisiones a 2030 (2020)
3. A.G. Álvaro, C.R. Palomar, V. de Almeida Guimarães, E.B. Hedo, R.M. Torre, I. de Godos Crespo, Developmental perspectives of the biofuel-based economy, in *Biofuels in Circular Economy*, ed. by S.A. Bandh, F.A. Malla (Springer, Singapore, 2022), pp. 133–156. https://doi.org/10.1007/978-981-19-5837-3_9
4. J.S. Sravan, A. Tharak, S.V. Mohan, Chapter 1—Status of biogas production and biogas upgrading: a global scenario, in *Emerging Technologies and Biological Systems for Biogas Upgrading*, ed. by N. Aryal, L.D. Mørck Ottosen, M.V. Wegener Kofoed, D. Pant (Academic Press, USA, 2021), pp. 3–26. https://doi.org/10.1016/B978-0-12-822808-1.00002-7
5. L. Eicke, A. Eicke, M. Hafner, Solar power generation, in *The Palgrave Handbook of International Energy Economics*,ed. by M. Hafner, G. Luciani (Springer, Cham, 2022), pp. 157–169. https://doi.org/10.1007/978-3-030-86884-0_9
6. X. Su, X. Shao, Y. Geng, S. Tian, Y. Huang, Optimization of feedstock and insulating strategies to enhance biogas production of solar-assisted biodigester system. Renew. Energy **197**, 59–68 (2022). https://doi.org/10.1016/j.renene.2022.07.090
7. L. Lombardi, B. Mendecka, S. Fabrizi, Solar integrated anaerobic digester: energy savings and economics. Energies (Basel) **13**(17) (2020). https://doi.org/10.3390/en13174292
8. J. Villarroel-Schneider, S. Balderrama, C. Sánchez, E. Cardozo, A. Malmquist, A. Martin, Open-source model applied for techno-economic optimization of a hybrid solar PV biogas-based polygeneration plant: The case of a dairy farmers' association in central Bolivia. Energy Convers. Manag. **291**, 117223 (2023). https://doi.org/10.1016/j.enconman.2023.117223
9. A. García Álvaro, C.R. Palomar, R.M. Torre, D.H. Redondo, I. de Godos Crespo, Hybridization of anaerobic digestion with solar energy: a solution for isolated livestock farms, *Energy Conversion and Management: X*, p. 100488 (2023). https://doi.org/10.1016/j.ecmx.2023.100488
10. M. El Ibrahimi, I. Khay, A. El Maakoul, A. Berrada, M. Bakhouya, Chapter 9—Hybrid solar-heated anaerobic digestion system for electricity and heat cogeneration, in *Hybrid Energy System Models*, ed. by A. Berrada, R. El Mrabet (Academic Press, USA, 2021), pp. 297–342. https://doi.org/10.1016/B978-0-12-821403-9.00009-3
11. M. Mahir, A. El Maakoul, I. Khay, M. Bakhouya, A hybrid building-integrated solar-powered anaerobic digestion system: Energy, economic, and environmental analysis in Morocco. J. Build. Eng. **95**, 110280 (2024). https://doi.org/10.1016/j.jobe.2024.110280
12. S. Pedrazzi, F. Masetti, G. Allesina, P. Tartarini, Hybridization of solar power plants with biogas from anaerobic digestion: a modeled case study. AIP Conf. Proc. **2191**(1), 020126 (2019). https://doi.org/10.1063/1.5138859
13. United Nations, Sustainable Development Goals : SDG 7—Ensure access to affordable, reliable, sustainable and modern energy for all (2015)
14. United Nations, Summary Progress Update 2021 : SDG 11—Sustainable cities and communities (2021)
15. United Nations, Summary Progress Update 2021 : SDG 13—Climate action (2021)
16. United Nations, Summary Progress Update 2021 : SDG 15—Life on land (2021)
17. P.G. Kougias, I. Angelidaki, Biogas and its opportunities—a review. Front. Environ. Sci. Eng. **12**(3), 14 (2018). https://doi.org/10.1007/s11783-018-1037-8
18. D. Deublein, A. Steinhauser, Biogas from waste and renewable resources. Focus Catal. **2011**(9), 8 (2011). https://doi.org/10.1016/S1351-4180(11)70392-0
19. A.G. Alvaro, C.R. Palomar, D.H. Redondo, R.M. Torre, I. de Godos Crespo, Simultaneous production of biogas and volatile fatty acids through anaerobic digestion using cereal straw as substrate. Environ. Technol. Innov. 03215 (2023). https://doi.org/10.1016/J.ETI.2023.103215

20. A. García Álvaro, I. Mediavilla, C. Ruiz Palomar, L.S. Esteban, I. de Godos Crespo, Energy valorization of solid residue from steam distillation of aromatic shrubs. Ind. Crops Prod. **222**, 119485 (2024). https://doi.org/10.1016/j.indcrop.2024.119485
21. C. Holliger et al., Towards a standardization of biomethane potential tests. Water Sci. Technol. **74**(11), 2515–2522 (2016). https://doi.org/10.2166/wst.2016.336
22. A.W. APHA, Standard Methods for the Examination of Water & Wastewater, 22nd edn (2012)
23. EUROSTAT, Energy Performance of Buildings Directive (2024)
24. Institute for Diversification and Energy Saving (IDAE), Energy rating for buildings (2018)
25. L. Breitenmoser et al., Methane potential from municipal biowaste: insights from six communities in Maharashtra, India. Bioresour. Technol. **254**, 224–230 (2018). https://doi.org/10.1016/j.biortech.2018.01.074

Chapter 4
Direct Demand Side Management of Residential Electric Water Heaters in Uruguay's Electric Grid

Magdalena Irurtia, Rodrigo Porteiro, and Mario Ibarburu

Abstract Balancing electricity supply and demand is a critical challenge in power system operation, particularly with the growing reliance on inherently variable renewable energy sources. Direct Demand Side Management (DDSM) offers a promising solution by directly controlling energy consumption to stabilize the grid. Uruguay, a regional leader in renewable energy, faces increasing challenges due to its high share of wind and solar power. This study explores the implementation of a DDSM program using residential electric water heaters (EWHs) in Uruguay. By leveraging the thermal storage capacity of these devices, the program aims to reduce generation, transmission and distribution costs while minimizing disruptions to users. The study proposes a DDSM program and evaluates its impact by modeling the interruption capacity of EWHs using physical dynamics models, historical consumption data, and performing simulations with MOP, a specialized software for simulating and optimizing the operation of power systems developed by UTE, the state owned power utility of Uruguay. The findings show that the DDSM strategy can significantly lower peak demand, reduce operational costs, and improve system flexibility, positioning it as a sustainable solution for future energy management. Specifically, the results show a 2% reduction in peak demand, 4 million USD in savings from generation costs, and 18.5 million USD in grid costs, representing up to 1.5% of the total annual system expenses.

Keywords Demand management · Electric water heaters · Smart grids

M. Irurtia (✉) · R. Porteiro · M. Ibarburu
UTE, Administración Nacional de Usinas y Trasmisiones Eléctricas, Montevideo, Uruguay
e-mail: mirurtia@ute.com.uy

R. Porteiro
e-mail: rporteiro@ute.com.uy

M. Ibarburu
e-mail: mibarburu@ute.com.uy

D. Rossit et al. (eds.), *Proceedings of the VII Ibero-American Congress of Smart Cities, ICSC-Cities 2024, 12-14 November, San Carlos, Costa Rica*, Springer Proceedings in Energy, https://doi.org/10.1007/978-981-96-4301-1_4

4.1 Introduction

The growing integration of renewable energy sources, such as wind and solar, into power grids globally has introduced substantial variability in power generation. This variability poses new challenges for maintaining the stability of the supply-demand balance, a task traditionally handled by conventional generation methods. As a result, the increasing reliance on renewables demands innovative approaches to enhance grid flexibility and ensure the continued reliability of power systems.

Direct Demand Side Management (DDSM) is one such promising solution. Unlike traditional demand-side management, which relies on price signals to encourage consumer behavior changes, DDSM enables the direct, centralized control of energy-consuming devices [1]. By dynamically adjusting demand in real-time based on grid conditions [2], DDSM can significantly enhance system performance, reduce operational costs, and facilitate the integration of a larger share of renewable energy [3]. Recent advancements in communication technologies, coupled with declining costs of control systems, have made large-scale implementation of DDSM increasingly feasible [4].

Uruguay offers a compelling case study for DDSM due to its leadership in renewable energy adoption within Latin America. Since 2007, the country has undergone a profound energy transition, resulting in a generation mix that heavily features wind and solar power. According to data from UTE [5], in 2023, 92% of Uruguay's electricity production was sourced from renewable energy, with wind power contributing 37% and solar photovoltaic 4%. While this transition has significantly reduced the country's reliance on fossil fuels, it has also increased the unpredictability of power generation, challenging grid management [6].

Residential EWHs, present in 87% of Uruguayan households and accounting for over a third of residential electricity consumption, offer an ideal opportunity for implementing DDSM. These devices, with their thermal storage capability, can be remotely managed to shift consumption away from peak periods, enhancing grid stability without compromising user comfort.

This paper investigates the implementation and outcomes of a DDSM program utilizing residential EWHs in Uruguay. Through comprehensive modeling and simulation, the study examines the program's impact on peak demand reduction, system costs, and user comfort. The findings highlight DDSM's potential as an effective strategy for advancing the sustainability and resilience of power systems, and provide valuable insights for further research and practical applications in energy management.

4.2 Direct Demand Side Management Using Electric Water Heaters

In this section, a description of the problem is presented, followed by an overview of the operation and control in DDSM. Finally, related work is reviewed.

4.2.1 Problem Description

The rapid integration of renewable energy sources into power grids worldwide has introduced new challenges, primarily due to the intermittent and variable nature of resources like wind and solar power. This variability complicates the management of power systems, especially in balancing supply and demand. In Uruguay, where renewable generation accounts for a large portion of the generation mix, this challenge is particularly pronounced. The unpredictability of wind and solar power has led to higher operational complexity, making it necessary to find flexible demand-side solutions.

Residential EWHs, which are a common household appliance in Uruguay, present a valuable opportunity for implementing Direct Demand Side Management (DDSM). These devices are highly flexible because their operation can be shifted away from peak demand periods without severely affecting user comfort [7]. Effective DDSM strategies can help optimize energy usage, reduce operational costs, and improve overall system efficiency by aligning EWH operation with periods of lower grid demand or higher renewable energy availability.

4.2.2 Electric Water Heater Operation and Control in DDSM

EWHs are well-suited for DDSM programs due to their predictable consumption patterns and thermal storage capacity. These devices operate via a thermostat that activates the heating element when the water temperature drops below a preset threshold T_{LOW} and deactivates it upon reaching T_{HIGH}. The predictable interaction between thermostat settings and hot water demand enables the identification of optimal periods for interrupting operation.

In a DDSM framework, EWHs can be remotely controlled to pause heating during peak demand and resume it during off-peak hours with minimal disruption. Successful implementation balances user comfort, thermal storage, and recovery timing [8], supported by robust communication technologies [2]. To achieve this, smart controllers would enable real-time remote control, adjusting EWHs' operation based on grid conditions. This approach, with an estimated cost of 10 USD per unit, offers a cost-effective alternative to large-scale storage solutions like batteries or pumped hydro [9], and while the investment would be substantial, it would be justified by long-term savings.

4.2.3 Related Work

Demand Side Management (DSM) has evolved significantly over the past decades, with Direct Load Control (DLC) emerging as a pivotal strategy for enhancing grid reliability and integrating renewable energy [10]. Early implementations of DLC in the mid-20th century demonstrated the potential for centralized management of household appliances, such as heating and cooling systems, to mitigate peak demand and improve system efficiency [4].

In the context of residential EWHs, some research has highlighted their effectiveness in demand response programs. For instance, studies conducted in the United States during the 1960s and 1970s demonstrated that DLC of EWHs could significantly reduce peak loads and operational costs. More recent advancements in smart grid technologies have enabled real-time control of these devices, allowing utilities to optimize their operation based on grid conditions [4].

Internationally, several countries have successfully implemented DDSM programs. In Australia, direct control of residential EWHs has been used to manage peak demand during heatwaves, while Switzerland has integrated EWHs into virtual power plants to provide ancillary grid services, such as frequency regulation. These international programs underscore the flexibility and reliability of residential EWHs as a resource for demand-side management [4].

However, Uruguay presents a unique case for DDSM implementation. The high penetration of renewables introduces significant variability and complexity in grid management [11], making solutions like DDSM particularly relevant. Unlike countries with more diversified energy infrastructures, Uruguay's smaller, centralized grid offers both challenges and opportunities for scaling DDSM.

A key initiative in Uruguay is *UTE's Smart Water Heater Pilot Program*, launched in 2017, which introduced smart timers for residential EWHs. This program demonstrated the feasibility of reducing peak demand and achieving energy savings through direct load control, laying the groundwork for future large-scale DDSM implementations.

While Uruguay can learn from these global experiences, its unique energy landscape requires tailored solutions. This study builds on both international and local experiences, providing a comprehensive analysis of the technical, economic, and social factors that influence the success of DDSM programs in Uruguay. By focusing on the country's specific context, this research can provide insights for policymakers, utility companies, and other stakeholders involved in sustainable energy management.

4.3 The Proposed Approach

This section outlines the methodological framework developed to evaluate the DDSM program applied to residential EWHs in Uruguay. By combining comprehensive data analysis, system modeling and simulations, this approach not only assesses the feasibility of the program but also quantifies its potential impact on grid performance and stability in a renewable-energy-dominant context.

4.3.1 Initial Data

This study utilizes consumption data from 224 residential EWHs in Uruguay from ECD-UY dataset [12], spanning various time periods and covering a range of household settings. Each device is associated with a CSV file containing detailed power measurements recorded at minute-level granularity, including the following key attributes:

- *Customer ID*: A unique identifier for each EWH.
- *Timer IMEI*: A unique identifier for the measuring device.
- *Timestamp*: The recorded measurement time in Unix format.
- *Power Measurement*: The recorded power consumption in Watts (W).

Additionally, a supplementary dataset provides detailed technical specifications for a broad range of EWHs configurations available in Uruguay, including water storage volume and electric power rating.

The datasets are essential for understanding the demand-side dynamics of residential EWHs, enabling the simulation of their interruption potential and recovery behavior under various operational scenarios. By linking consumption patterns with technical specifications, the study aims to assess the feasibility of integrating these devices into a DDSM program and quantify their impact on grid performance in Uruguay.

4.3.2 Characterization of EWH Operation

States of operation. Understanding the operational behavior of EWH is essential for their integration into a DDSM program. These devices operate in five distinct states, which are determined by water temperature and hot water demand, as illustrated in Fig. 4.1:

- **State 1**: The unit is off, with no water inflow or outflow. The water temperature is below T_{LOW}, and there is no hot water demand.

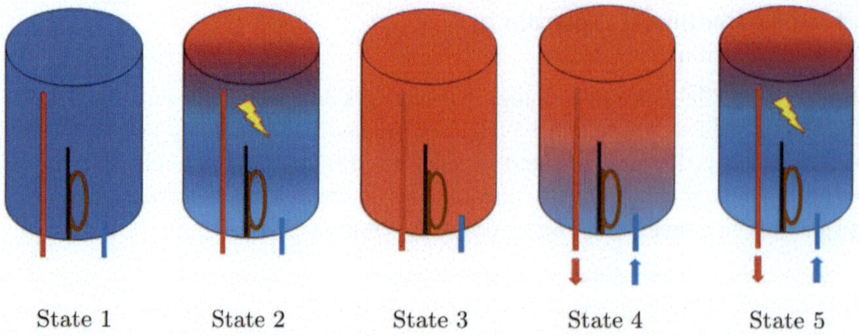

Fig. 4.1 States of operation for EWHs

- **State 2**: The unit is on and heating the water, but there is no water inflow or outflow. This state occurs when the EWH has been idle or when heat loss has reduced the water temperature below T_{LOW}.
- **State 3**: The unit is off, with no water inflow or outflow, and the water temperature is between T_{HIGH} and T_{LOW}.
- **State 4**: The unit is off, but there is both water inflow and outflow (hot water demand). The water temperature is between T_{HIGH} and T_{LOW}.
- **State 5**: The unit is on, and both water inflow and outflow are present (hot water demand). The temperature has dropped below T_{LOW}, causing the EWH to activate.

Of the five operational states, States 2 and 5 are the most relevant for the DDSM program, as they correspond to periods when the EWH is actively consuming electricity and can be managed to reduce peak demand.

Types of heating cycles. EWHs have two types of heating cycles:

- **Recovery cycles**: These occur when heat is lost to the environment through the tank walls, and the EWH activates to maintain water temperature.
- **Usage cycles**: These are triggered by hot water consumption, requiring more energy to heat incoming cold water.

Only usage cycles are considered interruptible within the DDSM framework, as they involve longer heating periods and provide more flexibility for load shifting. Although recovery cycles help maintain water temperature, they are short and have a negligible impact on system-level demand management. Thus, distinguishing between these two types of cycles is useful for effective implementation of the DDSM program.

Operational modeling. To accurately simulate the behavior of EWHs within a DDSM framework, their thermal dynamics across the predefined operational states are modeled. The approach accounts for both the EWH's energy input and hot water consumption dynamics, incorporating key variables such as tank water volume, EWH power, and overall water temperature. This modeling is based on several simplifying assumptions, including negligible heat loss to the environment and uniform water temperature within the tank.

The equations for each operational state are solved to estimate the maximum interruption time that preserves user comfort, as detailed in Sect. 4.3.3. The detailed equations are omitted for brevity, as the focus is on the broader system behavior rather than the granular thermal dynamics.

4.3.3 *Maximum Interruption Periods*

The maximum interruption period for each EWH represents the longest duration the device can remain powered off without compromising hot water availability. Its calculation is grounded in several key assumptions:

1. The EWH must be actively powered to be controllable.
2. Interruptions are only performed in the last part of usage cycles.
3. Energy recovery must be completed before the next usage cycle begins.

Calculating the maximum interruption period involves analyzing the EWHs's power consumption and applying physical models to simulate its thermal dynamics. The goal is to determine the longest interruption time for each device while maintaining the water temperature above the established comfort threshold $T_{LIM} = 40°$ C. Since direct temperature measurements inside the EWHs are unavailable, only power consumption data is used to estimate the operational states described in Sect. 4.3.2.

In the general case of a usage cycle (see Fig. 4.2a), the EWH would begin in State 5 (EWH on with hot water usage), with the water at temperature T_2 . After a period $t_{X1} - t_{X2}$, the hot water demand would cease, and the EWH would transition to the State 2 (restoration state), with the water initially at temperature T_1 ($T_1 > 40°$ C), and remaining in this state for a period t_{X2} (until the water temperature reaches T_{HIGH}). However, if the EWH were turned off at temperature T_2 (see Fig. 4.2b), it would transition to State 4 (hot water usage with the EWH off). During this phase, the water temperature would drop further to T_3 ($T_3 < T_1$ due to the lack of heating) and after a period $t_{X1} - t_{X2}$, the EWH would reach State 1, stabilizing at temperature T_3.

For each EWH configuration, the maximum interruption time t_{X1} is calculated by determining the largest value for which T_3 remains greater than $T_{LIM} = 40°$ C. This preserves user comfort while optimizes the load shift potential in the DDSM program.

4.3.4 *Recovery Cycles*

Recovery cycles occur when the EWH is actively raising the water temperature from T_{LOW} to T_{HIGH}, without any water consumption. The duration of these cycles, referred to as the *recovery time*, is calculated based on a thermal energy balance that

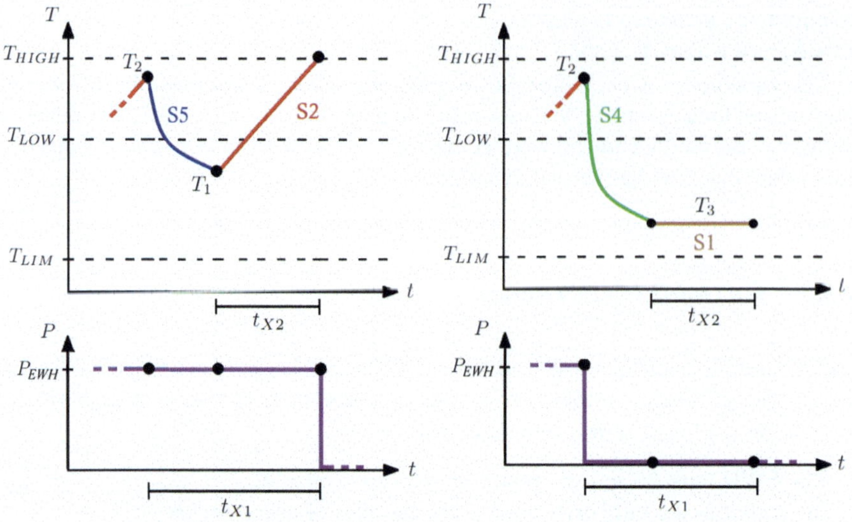

(a) Without interruption. (b) With interruption.

Fig. 4.2 Water temperature and power consumption of EWH during and after a heating cycle. Panel **a** shows normal operation, while panel **b** shows impact of interrupting the heating cycle

incorporates the physical characteristics of the EWH, including tank volume, water density, specific heat capacity, and the heater's power rating.

The recovery time is crucial in calculating demand reduction for each EWH, as it directly relates to the maximum interruption period. The detailed analysis is further explored in Sect. 4.4.3.

4.3.5 Interruptible Cycles

Interruptions are only performed during usage cycles. However, to maintain user comfort, the interruptible cycles for each EWH are constrained. The underlying hypothesis is that if multiple usage cycles occur in close succession, interrupting the first cycle may not provide sufficient time to recover the unsupplied energy before the next cycle starts. To address this, two key parameters are introduced:

- **Interruption window**: The period of the day during which interruptions can occur for the fleet of EWHs.
- **Time between interruptible cycles**: The time elapsed between the end of one interruptible cycle and the beginning of the next. This parameter is specific to each EWH configuration.

Defining the interruption window is critical for maximizing the system's benefits, while the time between interruptible cycles ensures that the recovery process can be completed before the next usage cycle begins, maintaining user comfort.

Once both settings are established for an EWH, the interruptible cycles are determined, directly impacting the device's peak demand reduction capacity.

4.3.6 Energy Interruption Potential

Once selected the variables presented in Sect. 4.3.5 and defined the maximum interruption period according to what was explained in Sect. 4.3.3, the peak demand reduction for each monitored EWH can be calculated by processing the minute-level historical consumption data from the device. These results are then generalized to the entire fleet of k residential EWH units participating in the program. The detailed calculation procedure, outlined in Sect. 4.4.4, provides the interruptible demand on an hourly basis throughout the study period.

4.3.7 Interruption Modes and Demand Management Case Selection

The pair of variables defined in Sect. 4.3.5 determines an *interruption mode*, which is associated with a specific demand reduction for each hour of the study period, resulting in potential benefits from demand interruptions based on that mode.

An interruption mode, combined with an specific *recovery window*-during which deferred energy is restored while ensuring system stability and user comfort-defines a *demand management case*. Throughout this study, different demand management cases will be considered in order to maximize the potential of the DDSM program, taking into account that the state of the system continuously changes from one day to the next in terms of demand and marginal costs, meaning the most favorable strategy may vary daily. Section 4.4.4 explores the selection process among different interruption modes and demand management cases, with the goal of maximizing the economic benefits of the DDSM program.

The methodological framework in Sect. 4.3, especially the operational modeling of EWH dynamics, forms the basis for the experimental analysis in Sect. 4.4. Simulations using historical data quantify the DDSM model's impact on peak demand and system costs, providing an assessment of its potential benefits under different conditions.

4.4 Experimental Analysis

In the following section, the experimental analysis is described.

4.4.1 Computational Platform and Software

The simulation of the Uruguayan electrical system was performed using MOP [13], a Java-based code developed by UTE for the optimization and simulation of power system generation. The data processing required for calculating the demand reduction and assessing the benefits of implementing the DDSM program was performed using Python. All simulations and computations were conducted on system equipped with an Intel vPro i7 processor and 16 GB of RAM.

4.4.2 Simulating the System Operation

A *reference simulation* is conducted using the MOP model to represent the operation of the Uruguayan power system in the year 2025, which determines the study period. This simulation assumes no implementation of the DDSM program and no expansion in generation capacity, reflecting typical system operating conditions based on historical data.

The operation of the Uruguayan system was simulated using MOP in each of the 114 hydraulic inflow scenarios registered since 1909. The outputs generated by MOP, which include profiles of marginal cost of generation and system demand for each hour of 2025, are used to evaluate the benefits from demand-side management under different strategies, assessing the impact of the DDSM program on peak demand reduction and overall cost savings, as will be explained throughout this section.

4.4.3 Maximum Interruption Period and Recovery Time

By calculating the maximum interruption period and the recovery time for each EWH in the dataset of configurations available in Uruguay, a direct relationship between these two parameters is established. For each EWH, the ratio of recovery time to maximum interruption period is approximately 0.34. This empirically derived ratio, $m = 0.34$, forms the basis for estimating the energy interruption potential, as detailed in Sect. 4.4.4.

4.4.4 Energy Interruption Potential Calculation

For a defined interruption mode (see Sect. 4.3.7), the following steps are applied to the consumption data for each EWH:

1. *Data alignment*: The data is processed to ensure minute-level granularity, removing gaps or duplicate entries. Timestamps are adjusted to the local time zone (Uruguay).
2. *Identifying recovery cycles*: Heating cycles are clustered by duration, with shorter cycles identified as recovery cycles. The mode of the shorter cluster is used to estimate recovery times.
3. *Filtering recovery cycles*: Recovery cycles are excluded from the analysis, leaving only usage cycles for evaluation.
4. *Identifying eligible usage cycles*: Only the last usage cycle within the defined interruption window is selected for each day.
5. *Calculating maximum interruption time*: The maximum interruption time is calculated by dividing the recovery time by $m = 0.34$ (see Sect. 4.4.3).
6. *Trimming usage cycles*: The final portion of each selected usage cycle is reduced by the maximum interruption time.
7. *Determining interruptible power*: The interruptible power for each EWH is calculated as the difference between power consumption before and after trimming the usage cycle.

This process generates a new dataset containing minute-level interruptible power data for each of the 224 EWHs monitored. This results are averaged across the 224 units to produce an hourly interruptible power profile representative of the entire fleet. Assuming the DDSM program covers $k = 1000000$ residential EWH units (the estimated total fleet in Uruguay), the maximum interruptible demand for each hour over the study period is calculated by multiplying the average hourly interruptible power by k, for the defined interruption mode.

4.4.5 Demand Management Case Selection

As mentioned in Sect. 4.3.7, to establish the potential benefits of implementing the DDSM program, the most favorable demand management cases must be identified choosing the interruption mode and the recovery window.

To select the interruption mode, three features were considered:

1. *Energy interruption potential*: This represents the total energy that can be interrupted during an specific period.
2. *Unserved energy*: The energy that cannot be supplied to consumers due to lack of generation, leading to unmet demand. In Uruguay, unserved energy costs are stratified into four levels, with increasing unit costs as the deficit deepens. These

costs reflect both the loss of consumer comfort and the broader economic impact of energy shortfalls.

3. *Surplus energy*: Refers to the energy that could be produced, but is not generated because is not enough demand. Within the DDSM program, surplus energy serves as a valuable resource for recovering energy deferred during EWH interruptions. By leveraging these energy surpluses, the program ensures that deferred energy is recovered without destabilizing the grid.

The Energy interruption potential is obtained as explained in Sect. 4.4.4, while the unserved energy and surplus energy are derived from the reference simulation.

The most suitable interruption windows for the different demand management cases are selected based on a high amount of energy interruption potential, along with the hours of the day exhibiting the maximum expected unserved energy.

In each case, the time between interruptible cycles is set equal to the duration of the interruption window, ensuring that only the final usage cycle within the interruption window is considered for interruption.

Finally, the optimal recovery window, during which deferred energy is restored, is defined identically for all cases. The recovery energy is distributed proportionally to the system's surplus energy curve during the early morning hours (00:00 to 07:00).

4.4.6 Costs Reduction Estimations

This section outlines the process used to estimate the economic benefits achieved through the implementation of the DDSM program.

Grid fixed cost reduction. Lowering peak demand offers substantial economic grid benefits. Peak demand plays a critical role in determining the capacity and expansion needs of transmission and distribution networks. According to Government Decree N$^{\underline{o}}$ 17/024 [14], the savings from reducing the transmission peak are 288,000 USD/(MW·year), and according to the Uruguayan regulator [15], the costs of reducing the distribution peak are 132,000 USD/(MW·year). Therefore, the total savings for the entire network amount to $\xi = 420{,}000$ USD/(MW·year). This value includes capital recovery, operational and maintenance costs for the distribution and transmission network, commercial expenses, and energy losses associated with distribution activities. Revenues generated in this scheme must be sufficient to maintain service quality, support network expansion, and ensure a reasonable return for the utility.

To calculate grid fixed cost reduction, let P_{MAX} be the maximum demand peak for the whole system in the study period without DDSM. For each day of the study period, the hourly maximum interrumpible demand of EWHs is subtracted from the total demand in order to reduce system peak demand of the day. The new peak demand of the year, P'_{MAX}, is the maximum of the resulting daily peaks. This strategy prioritizes the reduction of grid fixed over reduction of generation cost. The reduction of grid fixed is equal to $\left(P_{\text{MAX}} - P'_{\text{MAX}}\right)\xi$. Grid critical days refer to those in which the demand exceeds the new peak P'_{MAX}. During these days, interruptions are performed with the sole purpose to reduce peak system demand.

During periods where demand exceeds P'_{MAX} and falls outside the defined interruption window, special interventions are required to manage the system. In these critical periods, EWHs consumption is shifted to off-peak hours, effectively deferring demand. The Value of Interrupted Energy VEI and the Value of Recovered Energy VER, as defined by Eqs. 4.4 and 4.3, are applied to quantify the benefits of these manual shifts, ensuring consistency in the overall economic valuation.

Generation cost reduction. Using the results from the reference simulation, the analysis proceeds as follows:

1. *Defining operational day*: To fully analyze the impact of the DDSM, an *operational day* is defined, starting at 10:00 am on one day and ending at 9:00 am the following day. This framework allows for both the interruption of EWHs and the recovery of energy to occur within the same period, creating a structured approach for energy interruption and recovery.
2. *Quantifying potential savings*: For each operational day $r \in [1, 2, \ldots, 365]$ and for each hydrological scenario $w \in [1, 2, \ldots, 114]$, the system is evaluated across $i \in [1, 2, \ldots, N]$ demand management cases. The DDSM program is applied only if an economic benefit is identified for that particular day and scenario.
The daily benefit, $DB(r, w)$, represents the economic savings generated each day by the DDSM program. It is calculated by determining the amount of energy interrupted and the amount of energy recovered during off-peak hours. For each operational day r and each hydrological scenario w, the daily benefit is given by Eq. 4.1:

$$DB(r, w) = \max_{i=1:N} (B_i(r, w), 0) \tag{4.1}$$

where $B_i(r, w)$ is the benefit for a given demand management case i, calculated as the difference between the VEI and the VER, according to Eq. 4.2:

$$B_i(r, w) = VEI_i(r, w) - VER_i(r, w) \tag{4.2}$$

The Value of Interrupted Energy $VEI_i(r, w)$ is calculated as the sum of the interrupted energy (IE) across all hours within the interruption window, weighted by the marginal cost of electricity at each hour, as shown in Eq. 4.3:

$$VEI_i(r, w) = \sum_{h \in \text{interruption window}} cmg(h, w) \times IE_i(h, w) \tag{4.3}$$

Similarly, the Value of Recovered Energy $VER_i(r, w)$ is determined by summing the recovered energy (RE) during the recovery window, weighted by the marginal cost of electricity during recovery hours, as shown in Eq. 4.4.

$$VER_i(r, w) = \sum_{h \in \text{recovery window}} cmg(h, w) \times RE_i(h, w) \tag{4.4}$$

Here, h refers to the hours within the operational day, and $cmg(h, w)$ represents the marginal cost of electricity, which varies by hour and scenario. This ensures the model accurately accounts for fluctuations in energy prices and grid conditions. The total annual benefit $AB(w)$ for each hydrological scenario is calculated by summing the daily benefits across all days of the year (Eq. 4.5).

$$AB(w) = \sum_{r=1}^{365} DB(r, w) \tag{4.5}$$

Finally, the expected annual benefit EAB is obtained by averaging the annual benefits across all 114 hydrological scenarios (Eq. 4.6):

$$EAB = \frac{1}{114} \sum_{w=1}^{114} AB(w) \tag{4.6}$$

This provides a comprehensive, marginal-cost-based valuation of the economic benefits associated with the DDSM program under a variety of hydrological conditions.

4.5 Results and Discussion

This section outlines the results of the simulations assessing the impact of the proposed strategy. Key metrics such as peak demand reduction, cost savings, and system reliability are analyzed based on historical consumption data and modeled scenarios. These findings demonstrate the potential of DDSM to enhance grid performance while maintaining user comfort.

4.5.1 Selected Demand Management Cases

A total of 11 demand management cases were analyzed, each focusing on different interruption windows within the 18:00 to 24:00 range, all using the same recovery window during the early morning hours, as described in Sect. 4.4.5. While the specifics of each case are not detailed here, Fig. 4.3 illustrates the interruption and recovery power profiles for one of the demand management cases.

The results indicate that extending the interruption window reduces the capacity to interrupt during earlier hours, where system failures are more critical. However,

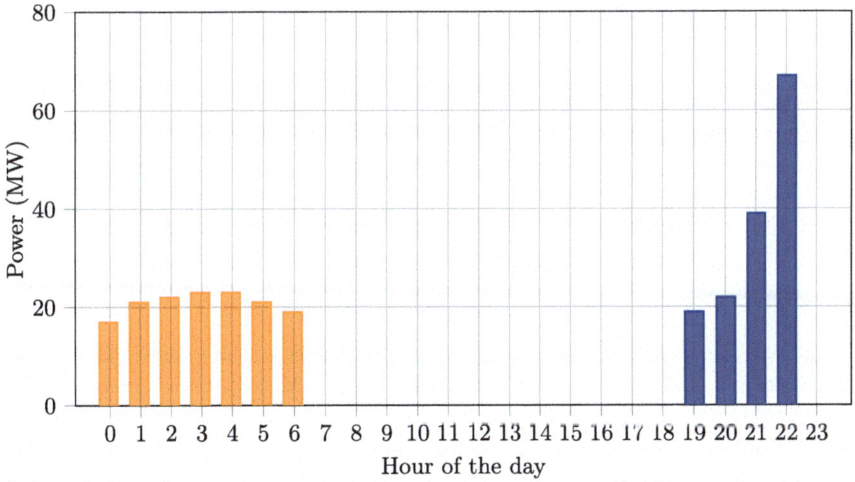

Fig. 4.3 Interruption power (blue) and recovery power (orange) profiles for a demand management case.

while failure power in the late hours of the day is not as significant, the maximum interruption power during this time is considerable, competing with interruptions in earlier hours.

4.5.2 Peak Demand Reduction

The implementation of the DDSM strategy can reduce the system's peak demand by approximately 2%, lowering the maximum demand from 2404 to 2360 MW (44 MW of reduction). This reduction is relevant given the high penetration of renewable energy sources, which increase the variability and unpredictability of the grid.

4.5.3 Cost Reduction

The proposed DDSM program can achieve annual savings of approximately 4 million USD in generation costs and 18.5 million USD in savings from reducing the annual peak demand for the distribution and transmission systems, as outlined in Sect. 4.4.6.

Therefore, the total average savings of implementing the program, including generation, transmission, and distribution, is 22.5 million USD. These savings represent around 1.5% of the total annual expenses of Uruguay's system, emphasizing the financial viability of the DDSM program.

4.6 Conclusions and Future Work

The implementation of DDSM using residential EWHs in Uruguay's electrical grid is a transformative strategy for optimizing energy management in renewable energy dominated systems. This study demonstrates that DDSM can significantly reduce peak demand and deliver substantial economic savings by leveraging the inherent flexibility of EWH operation to reduce consumption during peak hours, while maintaining user comfort. The findings underscore DDSM's potential to enhance grid sustainability and resilience, particularly in the face of fluctuating renewable generation.

The results presented here are a conservative estimate of the system's true potential, offering a valuable benchmark, because the DDSM policy was determined by heuristics. However, industrial implementations may vary due to unmodeled factors such as infrastructure costs and environmental impacts, which were not fully addressed in this analysis. Further research should expand upon these limitations to provide a more comprehensive evaluation of DDSM's feasibility and long-term benefits. A line of future work includes the implementation of a model that optimizes the DDSM policy.

Looking forward, the integration of DDSM strategies has global implications for energy management. As renewable energy sources continue to proliferate, grid stability challenges will become more common across countries worldwide. The lessons learned from Uruguay's implementation could serve as a model for other nations facing similar challenges. Moreover, the scalability of this program to larger, more complex grids and its adaptability to other flexible loads offer promising avenues for future research. Ultimately, DDSM could play a fundamental role in helping countries achieve their renewable energy targets, reduce carbon emissions, and ensure the long-term sustainability of power systems globally.

References

1. S.N. Bragagnolo, J.C. Vaschetti, F. Magnago, J.C. Gomez-Targarona, Gestión de la demanda en las redes inteligentes. perspectiva y control desde el usuario y la distribuidora. Información tecnológica **31**(3), 159–170 (2020)
2. J.I.P. Arriaga, L.J. Sánchez, M.P. de Tembleque. La gestión de la demanda de electricidad vol. i. Documento de trabajo **65** (2005)
3. A.R.S. Vidal, L.A.A. Jacobs, L.S. Batista, An evolutionary approach for the demand side management optimization in smart grid, in *2014 IEEE Symposium on Computational Intelligence Applications in Smart Grid (CIASG)* (IEEE, 2014), pp. 1–7
4. V.Z. Gjorgievski, N. Markovska, A. Abazi, N. Duić, The potential of power-to-heat demand response to improve the flexibility of the energy system: an empirical review. Renew. Sustain. Energy Rev. **138**, 110489 (2021)
5. UTE, Ute en cifras 2023. gerencia planificación, normas corporativas (2023). Accessed 2024
6. CAF, Uruguay, líder en fuentes renovables (2021). Accessed 2024-09-21
7. R. Porteiro, J. Chavat, S. Nesmachnow, A thermal discomfort index for demand response control in residential water heaters. Appl. Sci. **11**(21), 10048 (2021)

8. T. Ericson, Direct load control of residential water heaters. Energy Policy **37**(9), 3502–3512 (2009)
9. P. Armstrong, D. Ager, I. Thompson, M. McCulloch, Improving the energy storage capability of hot water tanks through wall material specification. Energy **78**, 128–140 (2014)
10. F. García Martínez, D. Trebolle, M. Goudó Navarro, J.M. Galán, P. Linares, A. Conchado, Gestión de la demanda eléctrica. Anales de Mecánica y Electricidad **XC**(IV), 26–32 (2013)
11. R. Bertoni, P. Messina, The "wind revolution" in Uruguay and the role of the public sector in guiding energy transitions, in *Energy Transitions in Latin America: The Tough Route to Sustainable Development* (Springer, 2023), pp. 195–216
12. J. Chavat, S. Nesmachnow, J. Graneri, G. Alvez, ECD-UY, detailed household electricity consumption dataset of Uruguay. Sci. Data **9**(1) (2022)
13. MOP, Modelo de operación desarrollado por ute (2024). Accessed 2024-09-21
14. IMPO, Decreto 17/2024 (2024). Accessed 2024-09-21
15. URSEA, Consulta pública: cargos de distribución (2024). Accessed 2024-09-21

Part II
Energy Efficiency and Sustainability

Chapter 5
Wind Power Generation Forecasting: A Real-World Nation-Wide Case Study in Uruguay

Sergio Nesmachnow and Claudio Risso

Abstract This article discusses the utilization of Artificial Neural Networks for predicting wind power generation in Uruguay. The precise estimation of power output plays a vital role in designing a dependable wind power generation infrastructure. Accurate predictions enable the implementation of efficient planning, management, and distribution strategies for the produced power, thereby enhancing the performance and the efficacy of the system. The research incorporates actual wind power generation data from Uruguay spanning the period between 2018 and 2022. The forecasting is performed using a Long Short-Term Memory artificial neural network. The primary findings suggest that the proposed methodology has a good prediction accuracy, with an average root mean square error of 0.12. The computed forecasts can be effectively employed for planning and scheduling to enhance service quality.

Keywords Renewable energy · Wind power · Sustainability · Artificial neural networks · Forecasting · Real-wold case study

5.1 Introduction

The significance of renewable energy sources like wind and solar power has grown substantially in efforts to diminish greenhouse gas emissions and address climate change. Nevertheless, the fluctuating and intermittent nature of renewable energy sources presents notable challenges for their seamless integration into the power grid. Precise and dependable forecasts of renewable energy generation are critical to offer effective strategies for their integration into smart grid operational frameworks.

This research was developed within a joint project with the Uruguayan electricity company, Administración Nacional de Usinas y Trasmisiones Eléctricas (UTE).

S. Nesmachnow (✉) · C. Risso
Universidad de la República, Montevideo, Uruguay
e-mail: sergion@fing.edu.uy

C. Risso
e-mail: crisso@fing.edu.uy

D. Rossit et al. (eds.), *Proceedings of the VII Ibero-American Congress of Smart Cities, ICSC-Cities 2024, 12-14 November, San Carlos, Costa Rica*, Springer Proceedings in Energy, https://doi.org/10.1007/978-981-96-4301-1_5

These predictions play a pivotal role in optimizing energy administration, as well as in guaranteeing the stability and reliability of the electrical grid.

Recent years have witnessed notable advancements in the realm of forecasting models for renewable energy generation. The proposed forecasting models leverage a variety of data sources, encompassing historical energy production records, meteorological projections, and other pertinent information. Approaches applying computational intelligence and deep learning models have been proposed to forecast the output of renewable energy sources with remarkable precision. The advantages stemming from precise forecasts of renewable energy generation are substantial, encompassing enhanced grid reliability, diminished energy expenses, and heightened integration of renewable energy sources. Specifically, precise prediction of power output assumes critical importance in the configuration of a dependable wind power generation infrastructure. This capability facilitates efficient planning, administration, and distribution of the produced wind power, ensuring optimal functionality and efficacy of the system.

Following this line of research, this manuscript studies the use of computational intelligence models, namely Artificial Neural Networks (ANN) to forecast renewable (wind) power generation in Uruguay. Several repositories of raeal data are taken into account in the analysis, spanning the period between January 2018 to August 2022. The reported research has been developed in a collaboration project between Universidad de la República, Uruguay, and the National Electricity Company UTE.

The significance of the reported research is based on the relevance of renewable energy sources for the country. In the last 15 years, Uruguay has significantly advanced in the shift towards renewable energy sources, which accounts for more than 90% of the total generation. Wind power has a 40% of the total share. Thus, the availability of accurate prediction models and tools for renewable energy generation is crucial, especially considering the high variability of wind in the country. When properly integrated in automated management tools withing the paradigm of smart electric grid, forecasting tools contribute to optimize and provide robustness and reliability to the service [1].

The study applies a Long Short Term Memory (LSTM) ANN for forecasting, a model that computed accurate predictions in our previous articles about forecasting solar photovoltaic power generation [13, 14]. The ANN-based approach is trained and evaluated using real time series data from wind power generation in Uruguay. The forecasting capabilities of the computational intelligence model were notable for a highly variable resource such as wind. The accuracy results, evaluated by the Root Mean Square Error (RMSE) metric, indicated a high precision of the generated forecasted time series. In turn, these outcomes are very useful to be applied in management, optimization, and decision-making models for the novel smart electrical grids.

The manuscript is organized as described next. Section 5.2 describes the addressed problem of forecasting wind power generation and presents a review of relevant related works, including those developed in Uruguay. The applied ANN approach is described in Sect. 5.3, along with the implementation details of the LSTM model.

The experimental evaluation is reported in Sect. 5.4, including a discussion of the main results and their relevance. Section 5.5 present the conclusions and comments some important lines for future work.

5.2 The Wind Power Generation Forecasting Problem and Related Works

This section introduces the wind power generation forecasting problem and comments articles from the literature review.

5.2.1 Wind Power Generation Forecasting

The problem at hand involves predicting the time series of wind power generation in Uruguay, based on authentic data supplied by the National electricity utility UTE (Administración Nacional de Usinas y Trasmisiones Eléctricas). Precise forecasting of renewable energy generation is of major importance for Uruguay to progress towards a more sustainable and dependable energy infrastructure. The country's efforts in this area, led by UTE, serve as a model for other countries seeking to integrate renewable energy sources into their electrical grids.

In the considered problem, the predictions are calculated hourly over a span of three days (72 h). This time horizon is defined for the need of predictions to effectively aid in determining the most suitable scheduling and distribution of existing electricity generation resources to satisfy the electricity demand within a power system. The related scheduling and control tasks entail deciding on the operation of power plants, the optimal quantity of power each plant should produce, and the timing for dispatching the generated electricity to fulfill the load demands. This task is critical to ensure the dependable and effective functioning of the power system, while simultaneously reducing expenses, enhancing grid reliability, and fostering the integration of renewable energy sources.

Instead of a single prediction, multiple values are determined for the next forecasting horizon. The challenge of forecasting over multiple hours presents a notably more complex task compared to single-hour power generation forecasting (usually, within a forecast horizon of one or a few hours), which has been addressed in prior studies. The forecasting method is required to anticipate a series of successive values within the wind power generation time series in Uruguay. The data used for training and evaluation of the applied learning model consists of historical observations of wind power generation and estimations derived from numerical weather prediction models.

Taking into account the defined forecasting horizon, the problem is under the medium-to-long-term forecasting classification category [2]. While some researchers have examined long-term forecasting problems spanning from one month to one year, aiming to leverage seasonal fluctuations and specific climate or generation trends, the typical delimitation often tag forecasts beyond 24 h as long-term [19]. The task of multi-hour forecasting poses greater complexity compared to short-term (up to six hours) and medium-term (six hours to a day) forecasting challenges. As the forecasting horizon extends, the margin of error typically amplifies for predictions further down the timeline. This means that the forecast precision diminishes as the time frame lengthens, making it more difficult to accurately project electricity generation or other pertinent variables for subsequent hours within the forecast horizon.

5.2.2 Literature Review

Many academic articles have addressed different flavors of the wind power generation forecasting problem, considering its relevance for management and operation of electricity systems.

Early studies about forecasting models for wind power generation revealed that prediction errors notably escalate when considering longer forecasting horizons [23]. Notably, the Wind Power Management System (WPMS), developed by the German Solar Energy Research Institute, demonstrated outstanding accuracy, with RMSE values within the range of 7–19% for immediate and short-term predictions [7].

The Armines Wind Power Prediction System (AWPPS), developed by Ecole des Mines de Paris, France, showcased accuracies ranging from 5% of the nominal power for a single farm when forecasting one hour ahead to 15% for predictions spanning 48 h [9]. Comparable accuracy levels were observed with other numerical predictors like the eWind power prediction Tool, developed by the Truewind company, USA, showcasing an accuracy of 11.7% for one-day ahead forecasting [24], and 16–19% for WPFS, from the Chinese Electric Power Science Institute [5].

Computational intelligence methods have also shown great prediction accuracy, even improving over standard numerical models. Demolli et al. [6] analyzed supervised and non-supervised methods for wind power generation forecasting. The analysis showed that the approach applying Support Vector Regression computed the best results, using mean wind speed as covariate input. However, the dispersion of RMSE values tended to be high, due to unexpected fluctuations on wind speed.

Feature extraction methods based on Convolutional Neural Network (CNN) architectures have been frequently applied in the analysis of wind speed patterns. Wang et al. [22] applied a CNN using downsampling to extract useful hidden features from wind power time series. Then, ensemble methods were applied to obtain both deterministic and non-deterministic forecasts. Competitive and robust results were reported.

Zhu et al. [25] applied a CNN model for wind power generation prediction up to sixteen time steps ahead. A two-dimensional CNN structure was used to extract useful features to deal with uncertainty on wind speed and direction. The model was evaluated using data from a real wind farm in Belgium, showing accurate predictive capabilities.

The use of weather information as covariant to improve the learning process was studied by Li et al. [8]. The approach applied a two-dimensional CNN combined with a LSTM and a multi-layer perception for multi-dimensional meteorological feature extraction. The proposed model showed good results in the evaluation over a case study involving a wind farm in Gansu, China.

In Uruguay, Zucatelli et al. [26] used ANNs for short-to-medium-term wind power and wind power ramp forecastings in Soriano. Precise results were computed for one hour ahead forecastings, but the accuracy of results diminished when extending the forecasting horizon to six hours.

Overall, ANN architectures including recurrent and convolutional layers, Long Short-Term Memory, and hybrid models, have been the most chosen models for wind power generation forecasting. Ancillary techniques, such as data decomposition, feature selection, and using different external data as covariant in the learning process, have often contributed to improve the forecasting accuracy [3].

Accurate wind power generation forecasts are crucial for assisting national grid managers and operators. Among other relevant applications related to the development of the Uruguayan the smart grid, our research group has applied computational intelligence methods for stochastic optimization and management of renewable energy resources [20], demand forecasting and planning of residential [4, 11, 18], large [10, 15, 16, 21], and industrial customers [17].

5.3 Forecasting Wind Power Generation Using Artificial Neural Networks

This section reports the development of predictors for wind power generation applying neural networks.

5.3.1 Description of the Proposed Approach

The proposed approach is based on the implementation of forecasting methods using the LSTM architecture for neural networks. LSTM models have the capabilities for properly modeling time series and relations. In turn, the proposed LSTM model showed good prediction accuracy when applied to solar photovoltaic power generation forecasting in our previous article [14].

The forecasting approach is classified as a multi-step, direct, and multi-variate schema. On the one hand, regarding to the number of predictions to be made, the multi-step approach is followed, where many steps are considered to be predicted within the forecasting horizon. Taking into account the features of the related smart grid management problem, 72 h ahead are considered, and blocks of 72 predictions are made, one prediction for each hour in the future. Then, regarding the inputs considered to make the prediction of each block, different models are created to be able to receive multiple inputs and create multiple outputs. This procedure is known as a sequence-to-sequence direct model, which is (computationally) more efficient than models that perform the predictions step-by-step (indirect or recursive models). On the other hand, regarding the data used as input, the approach is multivariate since it receives multiple inputs: the time series of previous observations in a given configurable window and (optionally) a number of additional information to be used as covariant in the learning process. In the reported research, the estimations estimations from numerical weather prediction models are used as covariant.

Pre-processing and formatting procedures are applied to the input data to align with the format expected by the input data layers of the LSTM architecture, according to the applied multi-dimensional learning approach.

5.3.2 *Implementation Details of the LSTM Approach*

Two configurations were studied for the architecture of the predictive LSTM model. Both architectures follows the sequential model, organized as a stack of layers, and have one tensor as input layer. Then, the two configurations differ in the number and dimension of hidden layers:

- Model LSTM 1: The hidden layers include one LSTM layer with dimension 100 cells with tanh activation function, and one dense layer with dimension 100 with ReLU activation function.
- Model LSTM 1: The hidden layers include one LSTM layer with dimension 200 cells with tanh activation function, one dense layer with dimension 100 with ReLU activation function, and one dense layer with dimension 100 with ReLU activation function.

The stacked dense layer(s) after the LSTM layer are applied to take advantage of their capability of molding the output of the LSTM layer into a shape and distribution compatible with the problem at hand. Finally, Both configurations uses the same output layer: a dense layer with dimension 72, according to the defined goal for the multi-step forecasting problem.

In the approach that do not use the estimations as covariant, the input layer is a tensor of dimension $1 \times H \times 1$, where H is the number of previous hourly observations taken as input. In the approach that uses the estimations as covariant for learning, the input layer is a tensor of dimension $1 \times H \times 1$.

The studied configurations for the LSTM model are presented in Fig. 5.1 (model LSTM 1) and Fig. 5.2 (model LSTM 2). The proposed models have 54,424 parameters (model 1) and 189,772 parameters (model 2). All of the parameters are trainable.

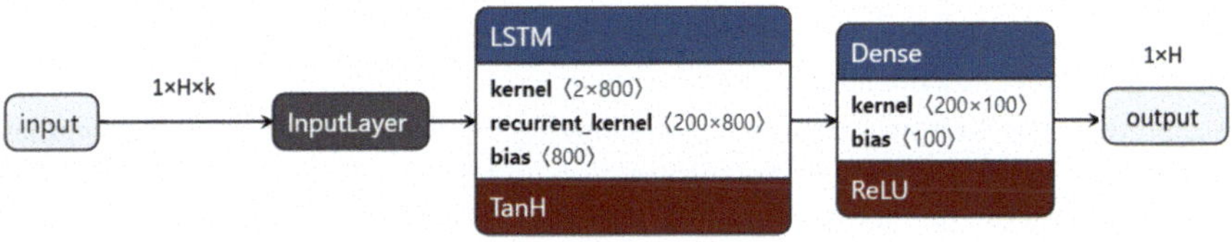

Fig. 5.1 LSTM neural network model, configuration 1

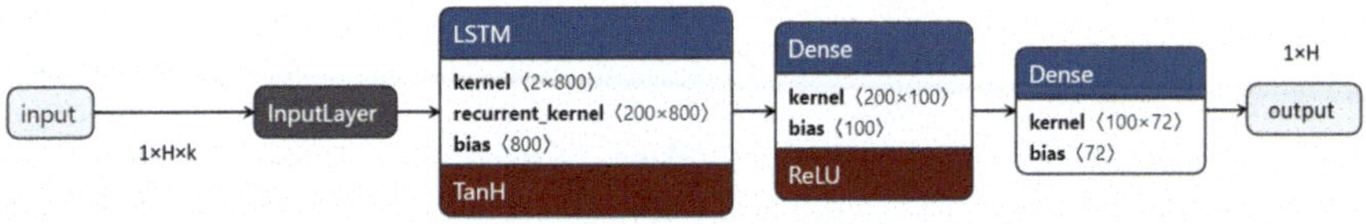

Fig. 5.2 LSTM neural network model, configuration 2

5.3.3 Methodology for Training and Validation

The model training process employs a sliding window approach as a data augmentation technique for increase the number of samples in the training dataset. Applying data augmentation is imperative because following the conventional method of splitting the dataset into a part for training and another for validation (e.g., 80 and 20%, as usual) would yield a small training dataset comprising disjoint training blocks of 72 h each. The sliding window method enables a significant expansion in the number of training blocks compared to the original blocks in the input database. The optimization criterion used to train the neural network involves minimizing the RMSE for each period to be forecasted. RMSE serves as the loss function of the proposed LSTM, due to its ability to offer a robust assessment of prediction accuracy within the training dataset. It is favored over an absolute metric like Mean Absolute Error (MAE) because RMSE imposes heavier penalties on forecast inaccuracies.

The forecasting evaluation process applies a walk-forward strategy. Walk-forward is a cross-validation technique used to determine the accuracy of predictive models. The technique is specially useful for data with a temporal sequence order and applies a rationale based on a realistic evaluation about how the trained model will behave and generalize to deal with new, unseen data. Assuming as known the data for a previous (training) period, it is used to forecast the series of data for the next period. The accuracy of the model is evaluated using the RMSE metric. The training and validation periods are then moved forward in time, by adding new data to the training dataset and removing the oldest data. The process is repeated until covering the whole validation dataset. This way, the model is continuously evaluated by assessing metrics collected from each test period.

Although the walk-forward strategy demands larger volumes of data and also more computation time for validation than other techniques, it provides important advantages, including: (i) the temporal order of samples and observations is maintained, allowing for a more precise detection of temporal patterns; (ii) the continuous evaluation makes it more easily adapt to pattern changes and specific events; and (iii) avoids using future data to train the model, thus reducing the data leaking phenomenon that arises when training data includes information about the target data, which will not be available when the model is used for real forecasting.

The quantity of past observations considered for forecasting is a parameter within the evaluation process. During the evaluation of errors over the validation dataset, both the overall RMSE value and the RMSE values for each predicted hour are computed, aiming to assess prediction quality as the forecast horizon lengthens. The training optimizer employed was the Adam variant of the Stochastic Gradient Descent algorithm. A batch training method was utilized, considering training blocks processed through the LSTM neural network in each batch. Preliminary experiments demonstrated that a batch size of 24 blocks provided a correct training behavior.

5.4 Training and Forecasting Results

This section describes the data considered for training and validation and reports the experimental evaluation of the multi-step, multivariate LSTM model for wind power generation forecasting.

5.4.1 Development and Execution Platform

The LSTM learning model was developed using the Tensorflow and Keras libraries of the Python programming language. Additional libraries (NumPy and Pandas) were used for data processing. The experimental evaluation was performed using the high-performance computing platform of National Supercomputing Center, Uruguay (Cluster-UY) [12].

The computational infrastructure used for the execution of experiments consisted of HPE DL380 Gen10 servers with Intel Xeon-Gold 6138 at 2.00 GHz. Parallel computing strategies were applied for the LSTM training, using Tesla P100 graphic processing units with 12 GB RAM.

5.4.2 Data Repositories

The main source of data is the information about wind power generation in Uruguay, gathered by UTE. Data records correspond to hourly measures of wind power generation and the global demand for electric power in the country.

The historical time series of wind power generation data used in the study spans from January 1st, 2018 to August 28th, 2022. Relevant information stored in the dataset include, among others:

- Timestamp: a correlative identifier of the measurement.
- Date and hour: month, day, year, hour, and minutes of the measurement.
- Delivered power (in MWh), i.e., the effective power delivered to the distribution network after the eventual reductions to the generation to match the demand.
- Curtails (in MWh), i.e., the deliberate reduction in order to balance energy supply and demand or due to transmission constraints.
- Total generated power (in MWh), i.e., the total generated power delivered by the wind power infrastructure.
- Total installed power (in MWh), the maximum power generation of the installed wind turbines.
- Plant Load Factor, i.e., the percentage of power delivered by the wind plants corresponding to its installed capacity, which is a measure of the capacity utilization of the plants.

Exhaustive pre-processing tasks were needed to sanitize the raw data logs containing the information. Outliers detection, imputation, correction, and interpolation techniques were applied to that end. Additional repositories containing meteorological data were also consulted and processed to solve discrepancies and ambiguities.

An example of the historical time series of wind power generation data is presented in Listing 1, resulting after the sensitization process. Data correspond to the first ten measurements for June 9th, 2017.

```
TimeStamp   Date     Hour Delivered  Curtails Total  Installed   Plant Load Factor
3817        9/6/17   000        677         0   677       1403            0.482537
3818        9/6/17   100        578        40   618       1403            0.440485
3819        9/6/17   200        487        98   585       1403            0.416964
3820        9/6/17   300        431       169   600       1403            0.427655
3821        9/6/17   400        420        75   495       1403            0.352815
3822        9/6/17   500        444       126   570       1403            0.406272
3823        9/6/17   600        462         5   467       1403            0.332858
3824        9/6/17   700        549         0   549       1403            0.391304
3825        9/6/17   800        620         0   620       1403            0.441910
3826        9/6/17   900        545         0   545       1403            0.388453
```

The data sample in Listing 1 shows interesting facts of the processed data. The non-zero curtailment values demonstrate the needed management to match generation and demand, especially considering that wind power is not stored (due to economic reasons, it is more convenient to simply discard it). Although it is a fixed value for the sample records showed in the listing, the total installed power field is relevant, considering that in the period accounted in the study new wind farms were built and started operation in the country. For the reported day, values of the Plant Load Factor field show that the wind farms were operating at a moderate rate, generating between 33 and 48% of the installed capacity in the country.

Figure 5.3 presents samples of the available historical data of wind power generation (Plant Load Factor values) used for the research (provided by UTE). The

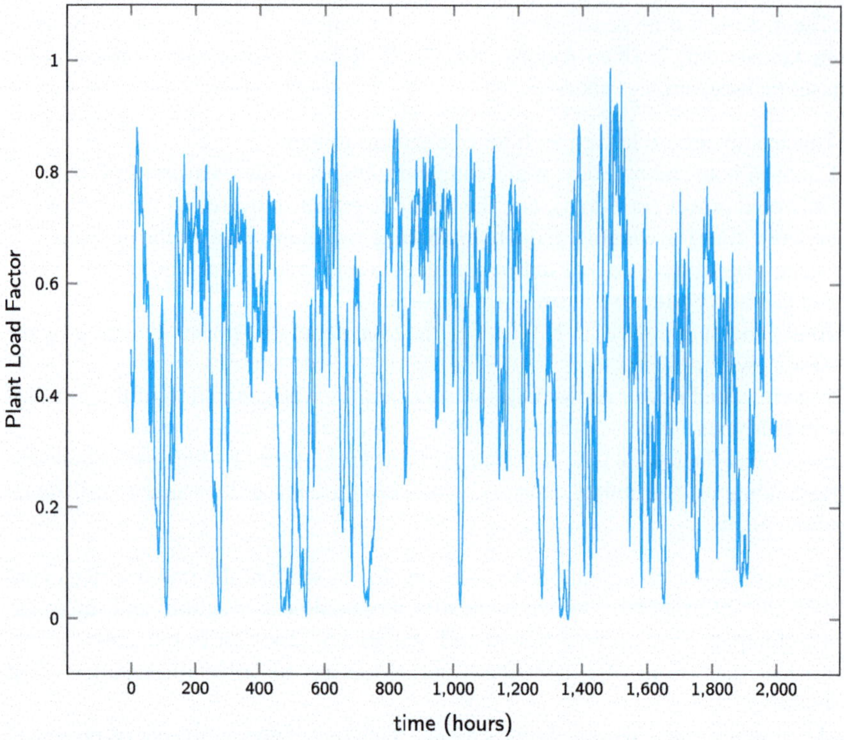

(a) Data sample: 3000 timesteps

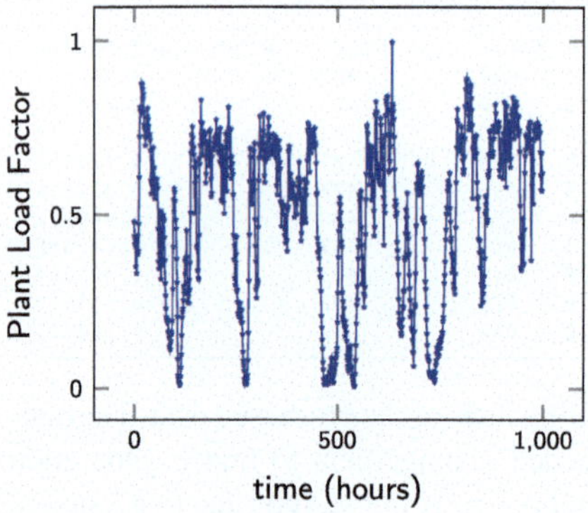

(b) Data sample: 1000 timesteps

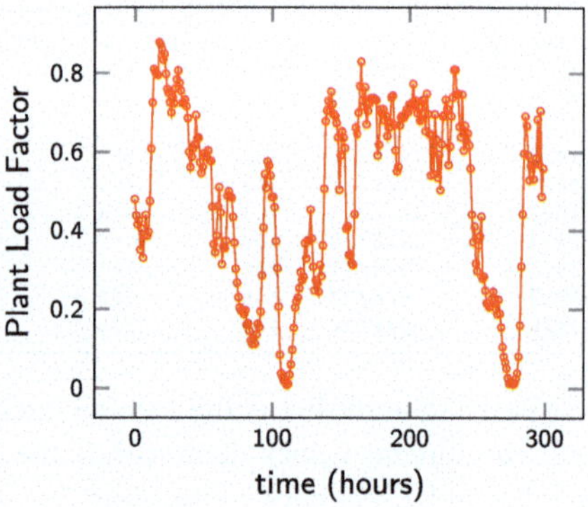

(c) Data sample: 300 timesteps

Fig. 5.3 Samples of wind power generation data from the information provided by UTE (Plant Load Factor values) samples show the great variability of wind power generation, as a consequence of the very variable availability of the win resource. Values closer to the maximum installed power generation capacity are reported for hours 635 (Plant Load Factor over 0.98) and 1506 (Plant Load Factor over 0.92), whereas almost null generation are reported when no wind is available, such as for hours 110 to 112 (Plant Load Factor less than 0.015), 275, 469 to 474, and even values of Plant Load Factor below to 0.01 for hours 543, 1333 to 1337, and 1354 to 1360 for very calm periods of time, with no wind.

5.4.3 Results of the Training Phase

The proposed LSTM model was trained with the Plant Load Factor values. The values are already normalized, as Plant Load Factor is the quotient of the delivered power by the wind plants and the overall installed capacity. Working with the normalized values instead of real delivered power has three main advantages: (i) normalized values are less prone to training pathologies arising due to the dominance of large values; (ii) it allows a simpler design of the LSTM architecture, removing the need for including an additional layer for the normalization of input values; and (iii) makes the approach more robust and extensible to increments in the installed wind power generation capacity in the country.

The training was performed using the augmented dataset, built applying the described sliding window technique. From an initial number of 635 blocks in the original input file, a total number of 40,752 training blocks were obtained using data augmentation.

The number of epochs for training was set to 100, yet both variants of the LSTM neural network achieved highly reliable RMSE results for each predicted period in about 40 epochs (i.e., RMSE below 0.015). These outcomes signify a suitable level of accuracy for the forecasts conducted on the training dataset. The training procedure demanded approximately 10 min to complete on a conventional Cluster-UY node equipped with a Tesla P100 GPU.

The typical evolution behavior of the loss function (RMSE) during training for both studied LSTM configurations are reported in Fig. 5.4.

The graphics in Fig. 5.4 shows that model LSTM 2, using a larger LSTM layer and one additional dense layer is able to compute lower values of the loss function faster than model LSTM 1. In any case, both models achieved good values of RMSE in training experiments.

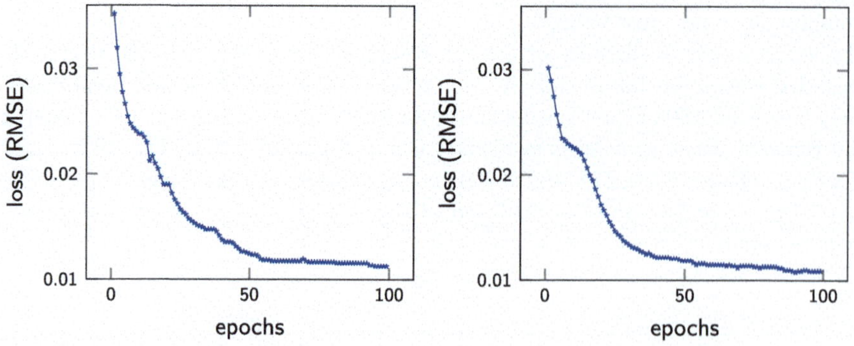

(a) Model LSTM 1 (b) Model LSTM 2

Fig. 5.4 Loss function (RMSE) evolution during training for the two studied configurations of the LSTM model for wind power generation forecasting

Table 5.1 Summary of validation results

model	min. RMSE	hour	max. RMSE	hour	avg. RMSE
Model LSTM 1	0.07	1	0.180	72	0.14
Model LSTM 2	0.06	1	0.165	72	0.12

5.4.4 Results of the Validation Phase

After training, the studied models were applied to the validation dataset. Table 5.1 summarizes the RMSE results in validation.

Overall, model LSTM 2 performed better than model LSTM 1. The RMSE values ranged between 0.07 and 0.18 for model LSTM 1 and between 0.06 and 0.165 for model LSTM 2. When considering the full validation dataset, the average RMSE computed by model LSTM 2 (0.12) was 16.5% better than the average RMSE computed by model LSTM 1 (0.14).

The best forecasting results were computed for the first hour in the future. The computed values were highly competitive with the ones proposed in the related literature, especially considering the high variability of the wind resource in Uruguay and that simple LSTM architectures were applied and the analysis. The forecasting errors increased for subsequent hours in the future. Sample RMSE values for the studied LSTM models for each predicted hour in the forecasting period are reported in Fig. 5.5.

The comparison of RMSE values in Fig. 5.5 shows that model LSTM 2 has a more smooth and robust pattern of results, and also computed better overall values. Both models computed significantly worse results after hour 64, suggesting that the accumulation of errors makes more difficult the forecast. Specific modifications

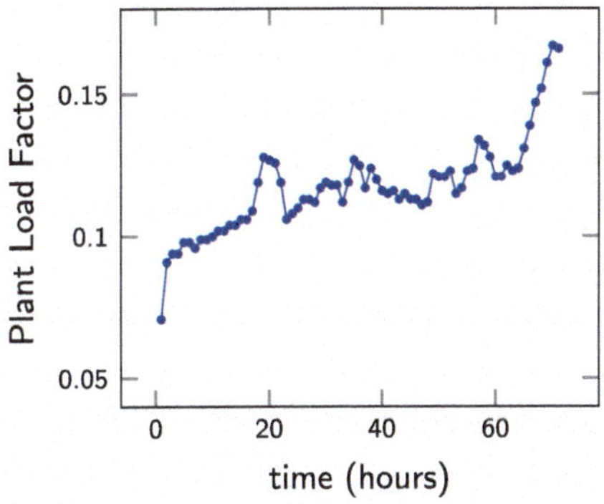

(a) Model LSTM 1

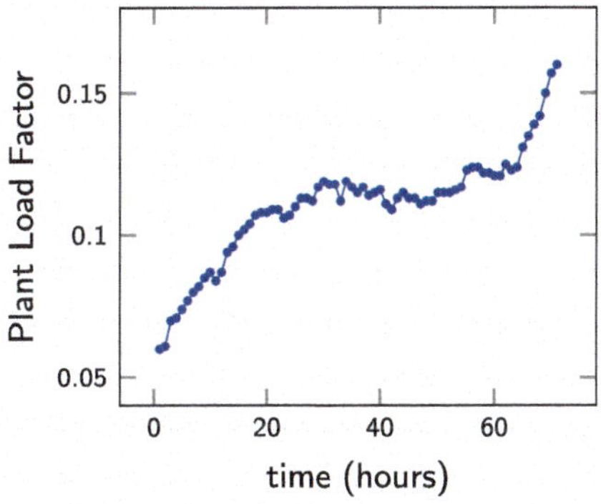

(b) Model LSTM 2

Fig. 5.5 Sample RMSE values for the studied LSTM models for each predicted hour in the forecasting period

Table 5.2 Average RMSE values for each hour for model LSTM 2

H	RMSE	Hour	RMSE	Hour	RMSE	Hour	RMSE	Hour	RMSE	Hour	RMSE
1	0.056	2	0.098	3	0.112	4	0.107	5	0.108	6	0.104
7	0.100	8	0.100	9	0.095	10	0.086	11	0.080	12	0.079
13	0.086	14	0.092	15	0.101	16	0.104	17	0.108	18	0.105
19	0.112	20	0.122	21	0.123	22	0.126	23	0.125	24	0.111
25	0.104	26	0.106	27	0.107	28	0.104	29	0.109	30	0.119
31	0.126	32	0.123	33	0.118	34	0.115	35	0.124	36	0.133
37	0.131	38	0.126	39	0.128	40	0.116	41	0.115	42	0.111
43	0.109	44	0.111	45	0.112	46	0.113	47	0.119	48	0.120
49	0.120	50	0.132	51	0.131	52	0.122	53	0.122	54	0.112
55	0.113	56	0.122	57	0.116	58	0.119	59	0.123	60	0.125
61	0.121	62	0.121	63	0.125	64	0.123	65	0.119	66	0.121
67	0.129	68	0.138	69	0.144	70	0.157	71	0.165	72	0.165

should be applied to the forecasting model to address the loss of accuracy for the last hours in the forecasting horizon. In any case, this is not a significant drawback of the proposed approach, mainly because the LSTM models computed accurate results and because the forecast might be applied within the forecasting horizon (e.g., at hour 40) and eventually update the model, to mitigate the loss of accuracy in advanced hours.

Table 5.2 reports the average RMSE values for the best model (LSTM 2) for each hour in the forecasting horizon.

Two examples of the wind power generation predictions using model LSTM 2 are presented in Figs. 5.6 and 5.7.

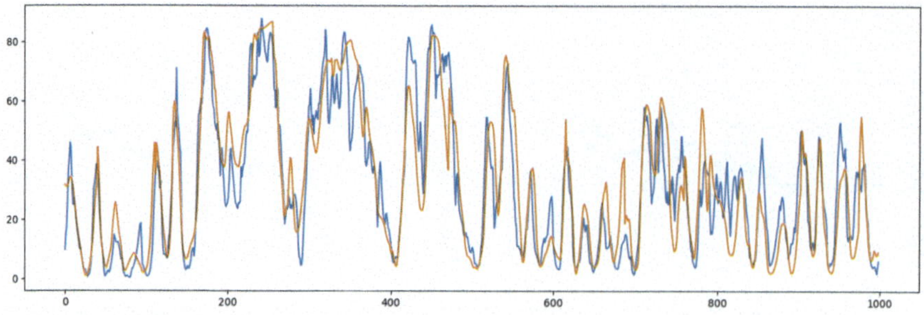

Fig. 5.6 Example of forecasts by model LSTM 2 (orange) and actual (blue) wind power generation values

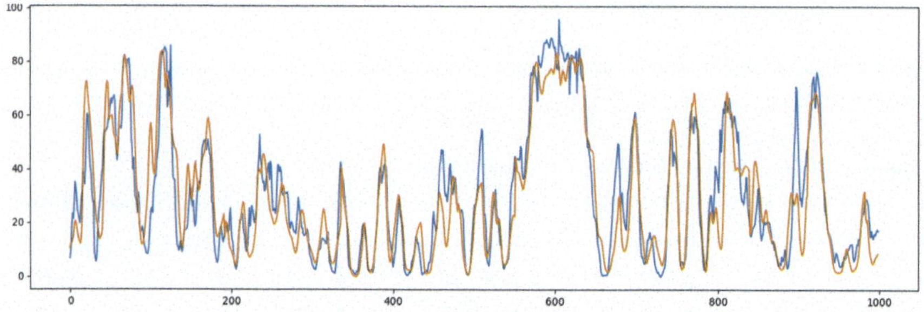

Fig. 5.7 Example of forecasts by model LSTM 2 (orange) and actual (blue) wind power generation values

5.5 Conclusions and Future Work

This article studied the application of LSTM neural network models for the problem of wind power generation forecasting. The approach was applied to a real case study of forecasting wind power generation in a country level.

Two configurations were studied for the LSTM architecture. The training experiments were performed using real wind power generation data from the Uruguayan electricity company, UTE. The validations results showed that the LSTM configuration using a larger LSTM layer stacked with two Dense layers. The accuracy of the model was correct, with RSME values between 0.06 and 0.165 for the 72 h forecasting horizon considered. The forecasts are useful for management and optimization of the electric grid.

The main lines for future work are related to consider different ANN models and extend the forecasting problem to other prediction horizons.

References

1. J. Aguiar, M. Pérez, An insight of deep learning based demand forecasting in smart grids. Sensors **23**(3), 1467 (2023)
2. R. Ahmed, V. Sreeram, Y. Mishra, M. Arif, A review and evaluation of the state-of-the-art in PV solar power forecasting: techniques and optimization. Renew. Sustain. Energy Rev. **124**, 109792 (2020)
3. G. Alkhayat, R. Mehmood, A review and taxonomy of wind and solar energy forecasting methods based on deep learning. Energy AI **4**, 100060 (2021)
4. J. Chavat, S. Nesmachnow, J. Graneri, G. Alvez, ECD-UY, detailed household electricity consumption dataset of Uruguay. Sci. Data **9**(1) (2022)
5. X. Chen, Y. Jiang, K. Yu, Y. Liao, J. Xie, Q. Wu, Combined time-varying forecast based on the proper scoring approach for wind power generation. J. Eng. **2017**(14), 2655–2659 (2017)
6. H. Demolli, A. Dokuz, A. Ecemis, M. Gokcek, Wind power forecasting based on daily wind speed data using machine learning algorithms. Energy Convers. Manag. **198**, 111823 (2019)
7. G. Giebel, L. Landberg, G. Kariniotakis, R. Brownsword, State-of-the-art on methods and software tools for short-term prediction of wind energy production, in *European Wind Energy Conference & Exhibition* (2023)
8. G. Li, T. Wang, F. Hu, T. Liu, The application of deep neural network algorithm considering multi-dimensional meteorological feature extraction in short-term wind speed prediction, in *3rd Information Technology, Networking, Electronic and Automation Control Conference*. IEEE (2019)
9. C. Monteiro, R. Bessa, V. Miranda, A. Botterud, J. Wang, G. Conzelmann, *Wind Power Forecasting: State of the Art 2009*. Tech. Rep. ANL/DIS-10-1 (Argonne National Laboratory, Illinois, USA, 2009)
10. J. Muraña, S. Nesmachnow, Simulation and evaluation of multicriteria planning heuristics for demand response in datacenters. Simulation **99**(3), 291–310 (2021)
11. S. Nesmachnow, G. Colacurcio, D. Rossit, J. Toutouh, F. Luna, Optimizing household energy planning in smart cities: a multiobjective approach. Rev. Fac. Ing. Univ. Antioquia **101** (2021)
12. S. Nesmachnow, S. Iturriaga, Cluster-UY: collaborative scientific high performance computing in Uruguay. Supercomput., Commun. Comput. Inf. Sci. **1151**, 188–202 (2019)
13. S. Nesmachnow, C. Risso, Forecasting solar photovoltaic power generation: a real-world case study in Uruguay. In: *VI Iberoamerican Congress on Smart Cities*, pp. 188–202 (2023)
14. S. Nesmachnow, C. Risso, A computational intelligence approach for solar photovoltaic power generation forecasting. Renew. Energies **2**(1) (2024)
15. S. Nesmachnow, D. Rossit, J. Toutouh, F. Luna, An explicit evolutionary approach for multiobjective energy consumption planning considering user preferences in smart homes. Int. J. Ind. Eng. Comput. **12**(4), 365–380 (2021)
16. R. Porteiro, J. Chavat, S. Nesmachnow, A thermal discomfort index for demand response control in residential water heaters. Appl. Sci. **11**(21), 10048 (2021)
17. R. Porteiro, L. Hernández-Callejo, S. Nesmachnow, Electricity demand forecasting in industrial and residential facilities using ensemble machine learning. Rev. Fac. Ing. Univ. Antioquia **102**, 9–25 (2020)
18. R. Porteiro, S. Nesmachnow, P. Moreno, C.E. Torres, Computational intelligence for residential electricity consumption assessment: detecting air conditioner use in households. Sustain. Energy Technol. Assessments **58**, 103319 (2023)
19. Y. Ren, P. Suganthan, N. Srikanth, Ensemble methods for wind and solar power forecasting—a state-of-the-art review. Renew. Sustain. Energy Rev. **50**, 82–91 (2015)
20. C. Risso, G. Guerberoff, A learning-based methodology to optimally fit short-term wind-energy bands. Appl. Sci. **11**(11), 5137 (2021)
21. D. Rossit, S. Nesmachnow, J. Toutouh, F. Luna, Scheduling deferrable electric appliances in smart homes: a bi-objective stochastic optimization approach. Math. Biosci. Eng. **19**(1), 34–65 (2022)

22. H. Wang, G. Li, G. Wang, J. Peng, H. Jiang, Y. Liu, Deep learning based ensemble approach for probabilistic wind power forecasting. Appl. Energy **188**, 56–70 (2017)
23. X. Wang, P. Guo, X. Huang, A review of wind power forecasting models. Energy Procedia **12**, 770–778 (2011)
24. Y. Wang, R. Zou, F. Liu, L. Zhang, Q. Liu, A review of wind speed and wind power forecasting with deep neural networks. Appl. Energy **304**, 117766 (2021)
25. A. Zhu, X. Li, Z. Mo, R. Wu, Wind power prediction based on a convolutional neural network, in *International Conference on Circuits, Devices and Systems*. IEEE (2017)
26. P. Zucatelli, E. Nascimento, A. Santos, A. Arce, D. Moreira, An investigation on deep learning and wavelet transform to nowcast wind power and wind power ramp: a case study in Brazil and Uruguay. Energy **230**, 120842 (2021)

Chapter 6
Spectral Analysis for Anticipating Critical Failures in Lithium-Ion Batteries: A Wavelet Approach

Mario Carbonó dela Rosa, J. Gómez, José Miguel Sánchez, Adalberto Ospino C., Eduard Antonio Mantilla Torres, and Víctor Alonso-Gómez

Abstract The increasing demand of energy density and charging rate of lithium- ion batteries rise concerns about safety hazard. Critical battery failures, such as thermal runaway, can lead to explosions and pose serious safety risks for users. In this work, we developed a novel method for early fault detection in lithium-ion batteries based on the spectral analysis of the cycling test signals. We analyzed voltage and current of cycling tests in lithium-ion batteries with silicon base anodes. The spectral analysis is performed by the Fourier and wavelet transforms (continuous, cross and coherence). Results show that prior to an irregular event or critical failure, new frequencies appear superimposed on the main cycling one, signaling the onset of degradation or failure. We concluded that spectral analysis of the cycling signal enables the early detection of

M. C. dela Rosa (✉) · E. A. Mantilla Torres
Escuela de Ciencias Básicas Ingeniería y Tecnología, Universidad Nacional Abierta y a Distancia, Bogotá, Cundinamarca, Colombia
e-mail: mario.carbono@unad.edu.co

E. A. Mantilla Torres
e-mail: eduard.mantilla@unad.edu.co

J. Gómez
Instituto de Energías Renovables, Universidad Nacional Autónoma de México, Temixco, Morelos, México
e-mail: jagoo@ier.unam.mx

J. M. Sánchez · A. Ospino C.
Departamento de Energía, Universidad de La Costa, CUC, Barranquilla, Atlántico, Colombia
e-mail: jsanchez80@cuc.edu.co

A. Ospino C.
e-mail: aospino8@cuc.edu.co

V. Alonso-Gómez
Universidad de Valladolid, Valladolid, Spain
e-mail: victor.alonso.gomez@uva.es

D. Rossit et al. (eds.), *Proceedings of the VII Ibero-American Congress of Smart Cities, ICSC-Cities 2024, 12-14 November, San Carlos, Costa Rica*, Springer Proceedings in Energy, https://doi.org/10.1007/978-981-96-4301-1_6

battery failures. This method could be implemented in battery management systems to reduce the risk of critical failures.

Keywords Lithium-ion battery · Failure detection · Climate change

6.1 Introduction

To reduce automotive emissions, improve air quality, and mitigate global temperature rise, many countries have adopted increasingly strict regulatory laws, the United Nations' International Panel of Climate Change (IPCC), emphasizes the many indicators on climate change and recommends that the world society respond to the serious problems [1, 2]. In this sense, the development of clean, safe, and efficient technologies has been promoted, these include in the field of transportation: electric vehicles (EVs), hybrid electric vehicles (HEVs) and plug-in hybrid electric vehicles (PHEVs) [3].

Electric vehicles are not a new technology, the first ones were introduced in the late 19th and early twentieth century. At some point between 1900 and 1930 they were more popular and numerous than internal combustion vehicles. They had a range of up to 50 km, and their lead-acid batteries could be recharged in 8 h. Gradually these vehicles fell into disuse because no electrical network was built to allow them to be charged and compared to internal combustion vehicles, the latter had a longer autonomy, could be refueled in a few minutes and its fuel began to be cheap and available everywhere [4]. Electric vehicles (EVs) have emerged as a viable solution to environmental challenges due to their superior efficiency and reduced emissions compared to internal combustion engine vehicles. Among the main ones are: No harmful tailpipe emissions such as carbon dioxide, sulfur oxides, nitrogen oxides or solid particles are produced during their operation. In addition, the electric motor is more efficient than the internal combustion engine, most of the electric power supplied is converted into rotary motion over a very wide range of speeds, this means that the engine torque can be transmitted to the wheels without voluminous gearboxes or drive shafts. Also, the electric motor is very quiet, which contributes to reduce noise pollution in cities [5–8].

Decades of research and development in materials science and electronics enabled its resurgence using first nickel metal hydride (NiMH) and now lithium-ion batteries, because batteries play a key factor in the transition to the mass use of electric vehicles and their safety. In this regard, lithium-ion (Li-ion) batteries have dominated the market over other chemistries in EVs [9]. According to historical data the energy density of lithium-ion batteries has increased by 2.5 times, while the cost has decreased by about 81% since 2008 [10]. However, this increase in energy density and reduction in charging times is accompanied by frequent safety issues. Specifically, the complexity of early fault detection, as discussed in recent work, remains a key challenge, as failures like overheating and over-discharge can lead to catastrophic events if not prevented in time [11]. Battery accidents such as spontaneous fires of electric vehicles and the fire of Boeing 787 have been a barrier to its massification and have caused significant economic losses to manufacturers. One of the reasons for this problem is that battery management systems (BMS) do not prevent and diagnose some critical failures such as overheating, over-discharge or short circuits in a timely manner [12–14].

This study presents a novel method based on a spectral of battery voltage over time. Time–frequency analysis method makes it possible to identify the low frequency components of the voltage signal that are related to failures or degradation. We analyzed a battery with a silicon and coal (Si/C) based anode, as this material is considered the most promising because of its high theoretical capacity. Results show that, before an irregular event or critical failure in the battery, arise new frequency components that combine with main during the cycling.

6.2 Methodology

We implemented a hybrid time–frequency analysis in four stages (see Fig. 6.1). First, we performed charge/discharge tests on half cell batteries using a standard protocol. In the second stage, the electrical signal data from the cycling tests were filtered and adapted for subsequent analysis. In the third stage, data processing was carried out using Fourier and wavelet transforms to the battery voltage data obtained from the charge/discharge test, the aim is the identification of the low frequency harmonic components and the symptoms or signals that precede the irregular events failures and degradation of the battery, in the final stage the proposed method for early detection of battery failures was carried out.

We analyzed cycling test data from half-cells that have a silicon mesh working electrode and a lithium metal counter electrode. These tests were performed under a standard charge/discharge protocol (see Table 6.2), cycling the batteries started with constant current (DC) charging until the cut-off voltage was reached. The same procedure was then applied to discharge the cell, thus completing the charge/discharge cycle, and this was repeated for multiple cycles. The database information from the tests performed is summarized in Table 6.1.

6.3 Results and Discussion

The half-cell voltage data from the cycling tests were analyzed using the fast Fourier transform (FFT), as well as continuous, cross, and coherence wavelet transforms. We employed the wavelet toolbox, signal analyzer application libraries, FFT and WT functions, along with other MATLAB tools to perform the data transformations. The sampling frequency (SF) for the FFT was set according to the Nyquist-Shannon sampling theorem [15] such that the SF at least doubled the natural frequency of the signal, in this case the highest frequency of the voltage cycle.

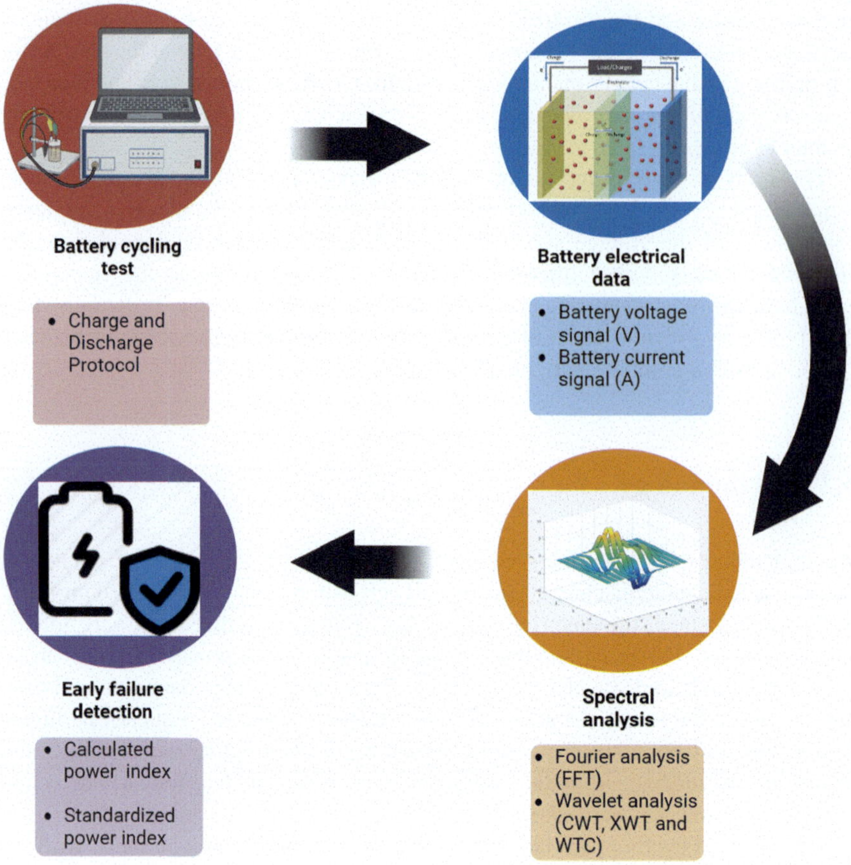

Fig. 6.1 Pathway of the proposed time–frequency analysis method for early battery failure detection

Table 6.1 Parameters of the lithium-ion cell evaluated

Research group	Cell type	Cathode	Anode	Capacity (mAh/g)
IER-UNAM	Half el-cell	Si/C	Li	300

Table 6.2 Charging/discharging protocol applied

	Charge		Discharge			T (°C)
Cell	CC (mA/g)	Cut-off voltage (V)	CC (mA/g)	CV (V)	Cycles	
SiC_HC1	100	1.3	100	0.1	100	25
SiC_HC2	100	1.3	100	0.1	74	25
SiC_HC3	100	1.3	100	0.1	42	25

6.3.1 Frequency Spectrum of Cycling Voltage

The Fourier spectra range was narrowed to highlight the main frequency components and identify those frequencies and amplitude spectra associated with irregularities or failure events in the batteries. The Fourier spectrum for the voltage signal of the SiC_HC1 battery evaluated by IER-UNAM was evaluated at 25 °C and is shown in Fig. 6.2d (The spectra for SiC_HC2 and SiC_HC3 cells are shown in Figs. 6.5d to 6.6d of the supplemental information). In this case, the charge/discharge voltage period is constant during the three cycling tests, the main frequency is clearly identified at 0.509 Hz (with a period of 1.96 h). Low amplitude frequency components below 4.24×10^{-2} Hz are difficult to distinguish in the Fourier spectrum of Fig. 6.2d and between 0.58 and 1.56 Hz is observed an irregular spectrum, this event cannot be related yet to any evident failure or degradation of the cell voltage.

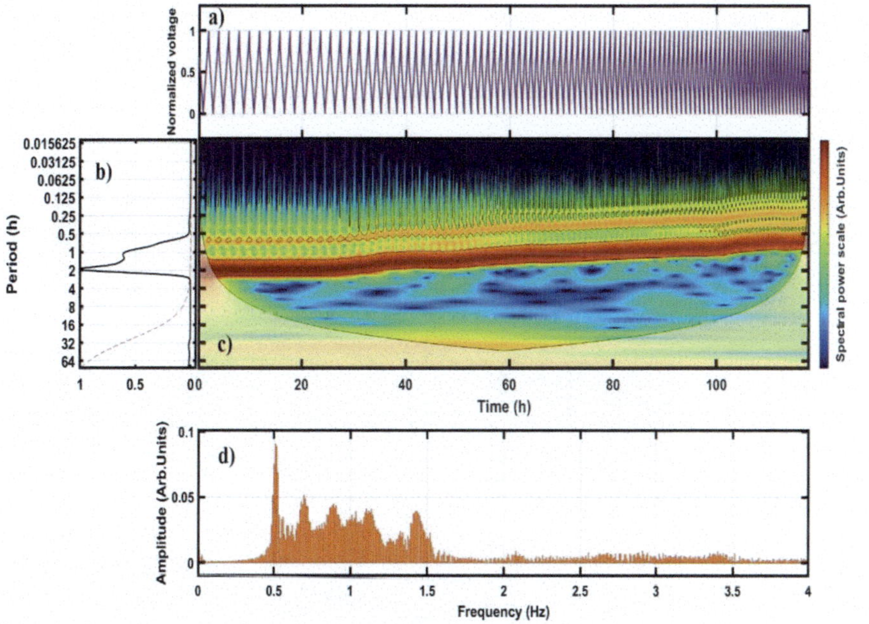

Fig. 6.2 Time series for battery voltage (**a**), The global wavelet is shown as the black curve, and the red dashed line is the 95% red-noise confidence level (**b**), The CWT spectral power and the curved black line that marks the cone of influence (COI) zone outside of which identification or detection of the periods will not be trustable (**c**) and Fourier spectra of the cycling test performed by IER-UNAM to SiC_HC1 cell

6.3.2 Time–Frequency Analysis by the Continuous Wavelet Transforms (CWT)

The battery voltage signal analysis over time obtained during half-cell (SiC_HC1) cycling tests was extended to the time–frequency domain by means of the continuous wavelet transform (CWT). The wavelet transform expansion was performed in terms of a family of wavelets constructed by dilation and translation of the mother wavelet. We selected the Morlet wavelet as the mother wavelet to analyze local spectral power variations across multiple periodicities, aiming to assess power fluctuations within a single non-stationary time series (battery voltage) exhibiting several periodic cycles [16]. The Morlet mother wavelet provides high resolution of the periodicity (frequency) scale, and is complex, allowing us to filter the series into bandwidths [17]. The color bar shown on the right side of Figs. 6.2b, 6.3b and 6.4b indicates the spectral power scale of the wavelet in arbitrary units, deep red is the highest spectral power and blue the lowest.

Figure 6.2c shows the continuous wavelet transform of the time dependent voltage for the SiC_HC1 evaluated by the IER-UNAM. As can be observed, in the middle of

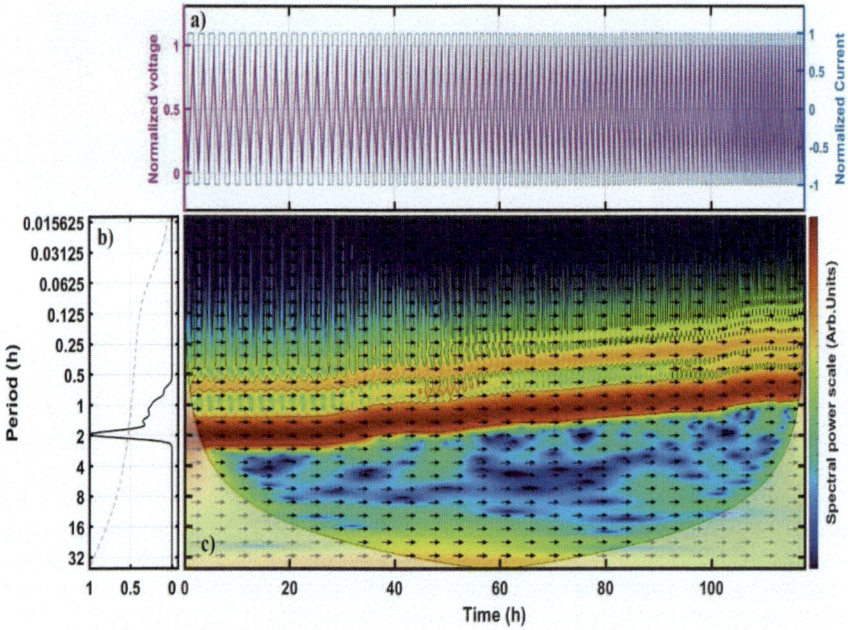

Fig. 6.3 The time series of current and battery voltage (**a**), The global wavelet is shown as the black curve, and the red dashed line is the 95% red-noise confidence level (**b**) and the XWT spectral power and the curved black line that marks the cone of influence (COI) zone outside of which identification or detection of the periods will not be trustable (**c**), of the cycling test performed by IER-UNAM to SiC_HC1 cell

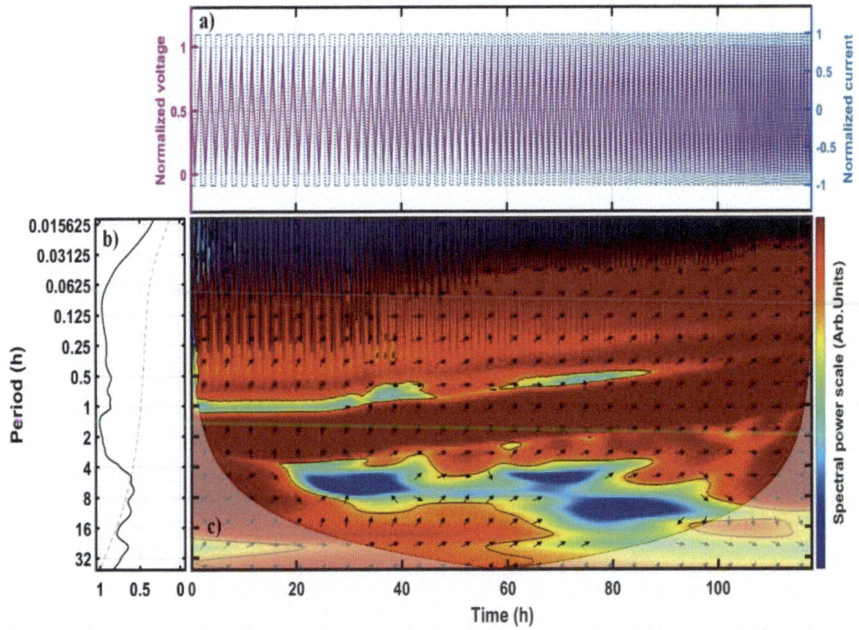

Fig. 6.4 The time series of current and battery voltage (**a**), The global wavelet is shown as the black curve, and the red dashed line is the 95% red-noise confidence level (**b**) and the WTC spectral power and the curved black line that marks the cone of influence (COI) zone outside of which identification or detection of the periods will not be trustable (**c**), of the cycling test performed by IER-UNAM to SiC_HC1 cell

the wavelet spectrum there is a deep red band with the main periodicity of approximately (2 h) as identified in the global wavelet, see Fig. 6.2b. When there are irregular variations in the cycling period, this band becomes warped. For example, at about 30 h, a change in the C rate of battery voltage is detected, obviously these cycles have different frequencies that originate the irregularity in that band. Moreover, an elongated zone of lower spectral intensity (in light green) appears below and extends around the time when the volt- age cycle changed.

A revision to the data base indicates that during this event, the cell capacity dropped, suggesting that these periodicities indicate an internal short duration failure that affected the battery. Similar irregular events are clearly seen in the spectral power of the wavelet at 47 h, when another drop in capacity occurred, however those irregular events are hardly distinguished in the time dependent voltage signal. After 50 h, the spectral power of the wavelet shows that the main periodicity decreases slowly with respect to the period scale, indicating that the C rate of voltage cycling is increasing as can be seen in Fig. 6.2a, and this in turn indicates that a progressive loss of battery energy storage is occurring.

Again, as can be seen in the global wavelet, shortly before 100 h on the time scale, there is a range of periods above the main periodicity located around 1 h. These periods with high spectral power correspond to the second highest amplitude peak identified in the irregular spectrum of Fig. 6.2d. A review of the database indicates that before the appearance of these new periodicities, the battery had lost more than 55% of its capacity. In Figs. 6.5c to 6.6c of the supplemental information are shown the wavelet spectral power of SiC_HC2 and SiC_HC3 cells voltage tested without any critical failure detected.

6.3.3 Time Frequency Analysis by the Cross Wavelet Transform (XWT)

The cross-wavelet spectrum introduced by Hudgins et al. [18], is computed to study the co-varying nature and measures the synchronization in phase and/or frequency of two phenomena/variables X and Y a [19]. The cross-wavelet function (W_{XY}) is defined by Eq. 6.1, as follows:

$$\mathrm{W_{xy}} = \mathrm{W_x} \circledast \mathrm{W}_Y^* \tag{6.1}$$

where $\circledast$ is the Hadamard product, (*) defines a complex conjugation, W_x and W_y are the wavelet transforms of two-time series X and Y which correspond to the voltage and current intensity in normalized units of the SiC_HC1 battery cycle test, as shown in Fig. 6.3a. On the other hand. Figure 6.3c shows the cross wavelet transform results of the interrelation between time series, as the wavelet power spectrum, in Figure 6.3c, where the back arrows indicate the relative phase of the synchronization between the voltage and the current intensity in time–frequency space. In this sense, from left to right and from right to left arrows ($\rightarrow$0° or $\leftarrow$ 180°, respectively) indicate a linear synchronization, in—phase or antiphase, respectively, at a certain frequency between these two phenomena. Any other orientation means that there is a complex, non-linear synchronization [20]. Therefore, we can conclude that the two-time series are in phase along the period and time scale. Moreover, other significant periods below the main one are observed in the global wavelet in Fig. 6.3b. In the supplemental information are shown the XWT spectral power in the Figs. 6.7c to 6.8c of SiC_HC2 and SiC_HC3 cells voltage evaluated without any critical failure detected.

6.3.4 Time–Frequency Analysis by the Wavelet Transforms Coherence (WTC)

The wavelet transform coherence (WTC) is developed by smoothing both the time and the scale of the cross-wavelet spectrum and aims to identify both frequency bands and time intervals in which the time series were related [21], it can be used to analyze the localized correlation coefficients and phase relationships of two time series that contain nonstationary powers at many different frequencies [22]. The WTC in [23] is defined as:

$$R_{xy}(\tau, s) = \frac{\left|S\left(W_{xy}(\tau, s)\right)\right|}{\sqrt{S\left(|W_x(\tau, s)|^2\right).S\left(\left|W_y(\tau, s)\right|^2\right)}} \tag{6.2}$$

R_{xy} takes values between 0 and 1, while S is a smoothing operator applied to both time and frequency (without S, the wavelet coherence would always be equal to 1 and scales). Values closer to 1 provide evidence of a stronger correlation, and values closer to 0 indicate a weaker correlation [24]. Figure 6.4a shows the two-time series in normalized units that were analyzed with the WTC, the first one corresponds to the voltage signal (magenta solid line) and the second one is the battery charge/discharge current (blue dashed line).

As can be seen, Fig. 6.4c shows the results of the coherence analysis, which allows to identify the correlation between the voltage and the current of the battery, that is, the local covariance between these two-time series at each time and frequency. The global wavelet indicates that between 4 and 0.05 h, the region with the highest spectral power density is located, this area (dark red) includes a region between 3.5 and 1.5 h associated with the cycling main frequencies (see Fig. 6.4b, c). As is also observed in Figs. 6.2c and 6.3c, as result of a progressive loss of capacity, the main periodicity presents some discontinuities and a decreasing tendency. Moreover, above the main periodicity and between the time intervals of 0 to 46 h and 56 to 87 h, periodicities of about 1 h appear with irregular spectral power in a medium–low scale.

In the XWT analysis, it is established that the orientation of the arrows and their phase angle represents the correlation between two time series at different scales. Complementarily, if the arrows are pointing up or down, they have phase differences $\pi/2$ and $(-\pi)/2$, respectively, and indicate that the second variable leads the first or that the first variable leads the second, respectively. Otherwise, if the arrow is pointing to the right and has a phase angle $\phi\_xy \in (0, \pi/2)$ or $\phi\_xy \in (-\pi, -\pi/2)$ indicates that the second variable is ahead of the first, finally, whether $\phi\_xy \in (-\pi/2, 0)$ or $\phi\_xy \in (\pi/2, \pi)$ the first variable leading the second [25].

An interesting observation on Fig. 6.4c, in contrast to Fig. 6.3c, is that from the beginning of cycling until 100 h, the arrows point to the right, at an angle of approximately $\pi/4$ in almost all the scales of the period, indicating a time lag between the time series and that the current is ahead of the voltage. After 100 h, both time series tend to be in phase. In Figs. 6.9c to 6.10c of the supplemental information are shown the WTC spectral power of SiC_HC2 and SiC_HC3 cells voltage, tested without any critical failure detected, and the angle of the arrows on the main periodicities indicating a time lag of $\pi/2$ between the two-time series, bringing current ahead of voltage.

6.4 Conclusions

A spectral analysis was conducted on the charge/discharge test cycle voltage of lithium-ion batteries with a silicon-carbon anode. The Fourier transform was employed to identify the main frequencies of the cycles, and the harmonic components present in the voltage signal. A complementary analysis with the continuous, crossed and coherence wavelet transforms showed interesting patterns in the battery cycling tests. First, before an irregular event, critical failure or battery degradation, the continuous and crossed wavelet transform figures show frequency components appearing after a change in the main periodicity of the voltage signal. Second, the coherence wavelet transforms show discontinuous periodicity intervals with medium–low spectral power, a time lag between the time series where the current is ahead of the voltage and a decreasing trend in the main periodicity associated with a loss of battery capacity.

This study is among the first to demonstrate the potential of spectral analysis methods for anticipating battery failures. In conclusion, these methods demonstrate potential for early detection of battery failures and could be integrated into battery management systems (BMS) to identify early signs of degradation, capacity loss, and failure, significantly reducing the risk of critical failures in electric vehicle battery systems. In addition to the potential applications in electric vehicle battery systems, the proposed spectral analysis methodology could be extended to other energy storage technologies where early fault detection is critical. By improving the safety and reliability of these systems, the method supports the broader adoption of energy storage in renewable energy grids, where the stability and longevity of storage solutions are paramount. Future work could explore adapting this methodology to new battery chemistries, such as solid-state or sodium-ion batteries, to further enhance its applicability across the energy sector.

Acknowledgements The author would like to acknowledge to Universidad Nacional Abierta y a Distancia de Colombia, Universidad Nacional Autónoma de México, Universidad de la Costa and Prenteck Ingeniería for their support in the development of this research.

Disclosure of Interests The authors have no competing interests to declare that are relevant to the content of this article.

Supplemental Information

See Figs. 6.5, 6.6, 6.7, 6.8, 6.9 and 6.10

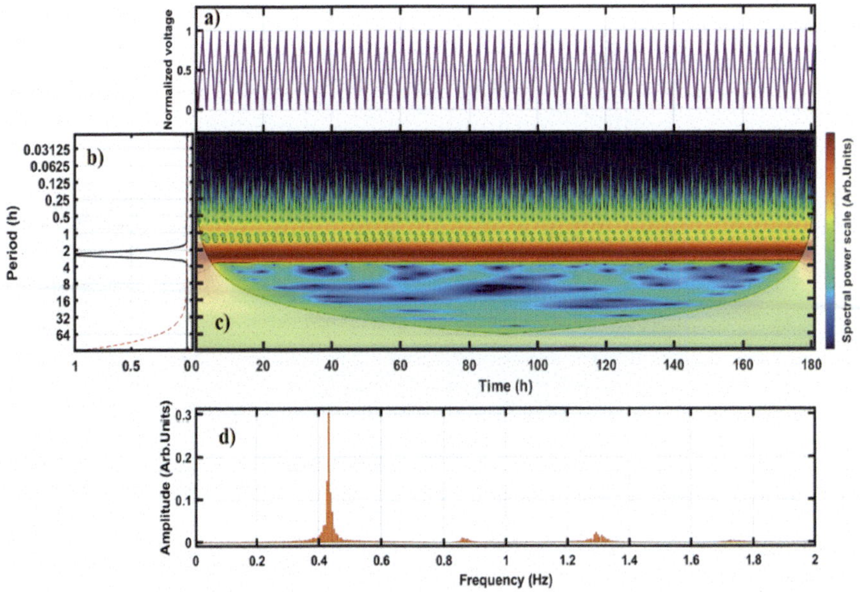

Fig. 6.5 The time series of battery voltage (**a**), global wavelet (**b**), CWT spectral power (**c**) and Fourier spectra of the cycling test performed by IER-UNAM to SiC_HC2 cell

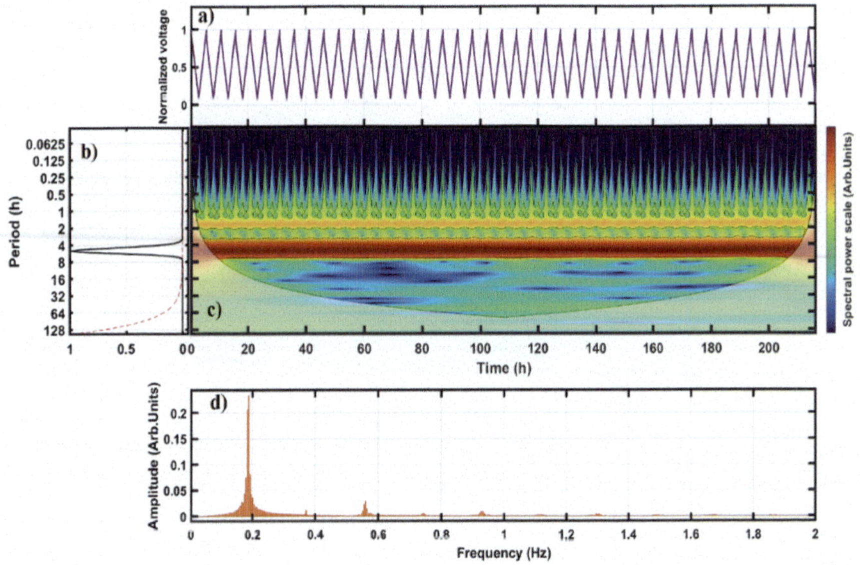

Fig. 6.6 The time series of battery voltage (**a**), global wavelet (**b**), CWT spectral power (**c**) and Fourier spectra of the cycling test performed by IER-UNAM to SiC_HC3 cell

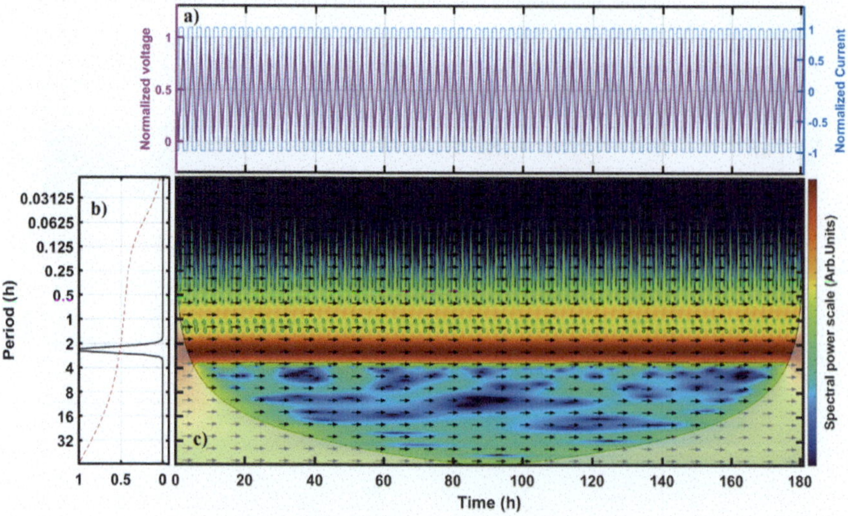

Fig. 6.7 The time series of current and battery voltage (**a**), global wavelet (**b**) and XWT spectral power (**c**), of the cycling test performed by IER-UNAM to SiC_HC2 cell

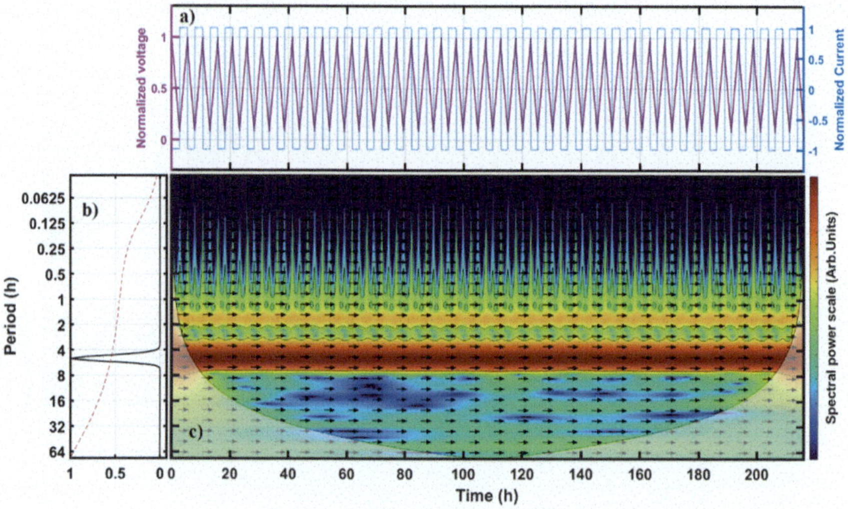

Fig. 6.8 The time series of current and battery voltage (**a**), global wavelet (**b**) and XWT spectral power (**c**), of the cycling test performed by IER-UNAM to SiC_HC3 cell

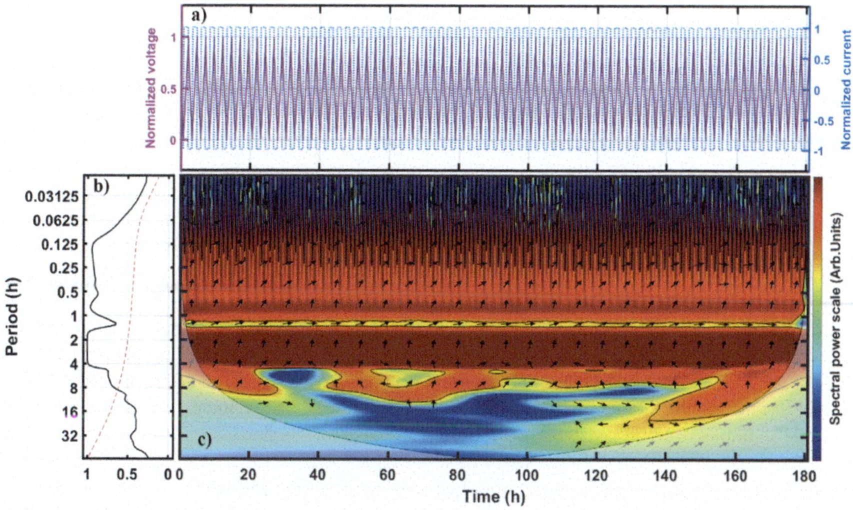

Fig. 6.9 The time series of current and battery voltage (**a**), global wavelet (**b**) and WTC spectral power (**c**), of the cycling test performed by IER-UNAM to SiC_HC2 cell

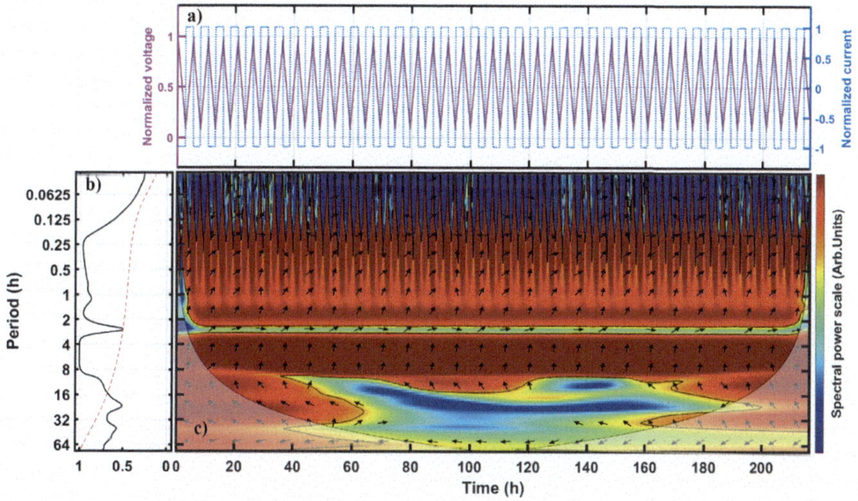

Fig. 6.10 The time series of current and battery voltage (**a**), global wavelet (**b**) and WTC spectral power (**c**), of the cycling test performed by IER-UNAM to SiC_HC3 cell

References

1. A.C.R. Teixeira, J.R. Sodré, Simulation of the impacts on carbon dioxide emissions from replacement of a conventional Brazilian taxi fleet by electric vehicles. Energy **115** (2016). https://doi.org/10.1016/j.energy.2016.07.095
2. H. Lund, B.V. Mathiesen, Energy system analysis of 100% renewable energy systems-the case of Denmark in years 2030 and 2050. Energy **34** (2009). https://doi.org/10.1016/j.energy.2008.04.003
3. Y. Wu, Z. Yang, B. Lin, H. Liu, R. Wang, B. Zhou et al., Energy consumption and CO_2 emission impacts of vehicle electrification in three developed regions of China. Energy Policy **48** (2012). https://doi.org/10.1016/j.enpol.2012.05.060
4. C.C. Chan, The rise & Fall of electric vehicles in 1828–1930: lessons learned. Proc. IEEE **101** (2013). https://doi.org/10.1109/JPROC.2012.2228370
5. J. Axsen, C. Orlebar, S. Skippon, Social influence and consumer preference formation for pro-environmental technology: the case of a U.K. workplace electric-vehicle study. Ecol. Econ. **95** (2013). https://doi.org/10.1016/j.ecolecon.2013.08.009
6. K.G. Høyer, The history of alternative fuels in transportation: the case of electric and hybrid cars. Util. Policy **16** (2008). https://doi.org/10.1016/j.jup.2007.11.001
7. C. Weiller, A. Neely, Using electric vehicles for energy services: industry perspectives. Energy **77** (2014). https://doi.org/10.1016/j.energy.2014.06.066
8. D.V. Lalith, *Electric and Hybrid Electric Vehicles Electric and Hybrid Electric Vehicles* (2021)
9. L.J. Aaldering, C.H. Song, Tracing the technological development trajectory in post-lithium-ion battery technologies: a patent-based approach. J. Clean. Prod. **241** (2019). https://doi.org/10.1016/j.jclepro.2019.118343
10. D. Zhili, L. Boqiang, G. Chunxu, Development path of electric vehicles in China under environmental and energy security constraints. Resour. Conserv. Recycl. **143** (2019). https://doi.org/10.1016/j.resconrec.2018.12.007
11. M.E. Carbonó dela Rosa, G. Velasco Herrera, R. Nava, E. Quiroga González, R. Sosa Echeverría, P. Sánchez Álvarez P et al., A new methodology for early detection of failures in lithium-ion batteries. Energies (Basel) **16**, 1073 (2023). https://doi.org/10.3390/en16031073
12. N. Williard, W. He, C. Hendricks, M. Pecht, Lessons learned from the 787 dreamliner issue on lithium-ion battery reliability. Energies (Basel) **6** (2013). https://doi.org/10.3390/en6094682
13. D. Liu, Y. Song, L. Li, H. Liao, Y. Peng, On-line life cycle health assessment for lithium-ion battery in electric vehicles. J. Clean. Prod. **199** (2018). https://doi.org/10.1016/j.jclepro.2018.06.182
14. R.D. Jilte, R. Kumar, M.H. Ahmadi, Cooling performance of nanofluid submerged vs. nanofluid circulated battery thermal management systems. J. Clean. Prod. **240** (2019). https://doi.org/10.1016/j.jclepro.2019.118131
15. B.P. Lathi, R.A. Green, *Essentials of Digital Signal Processing* (2014). https://doi.org/10.1017/cbo9781107444454
16. C. Torrence, G.P. Compo, A Practical Guide to Wavelet Analysis. Bull. Am. Meteorol. Soc. **79**, 61–78 (1998). https://doi.org/10.1175/1520-0477(1998)079%3c0061:APGTWA%3e2.0.CO;2
17. A. Grinsted, J.C. Moore, S. Jevrejeva, Application of the cross wavelet transform and wavelet coherence to geophysical time series. Nonlinear Process Geophys. **11** (2004). https://doi.org/10.5194/npg-11-561-2004
18. L. Hudgins, C.A. Friehe, M.E. Mayer, Wavelet transforms and atmopsheric turbulence. Phys. Rev. Lett. **71** (1993). https://doi.org/10.1103/PhysRevLett.71.3279
19. W. Soon, K. Dutta, D.R. Legates, V. Velasco, W.J. Zhang, Variation in surface air temperature of China during the 20th century. J. Atmos. Sol. Terr. Phys. **73** (2011). https://doi.org/10.1016/j.jastp.2011.07.007
20. V.M. Velasco Herrera, W. Soon, G. Velasco Herrera, R. Traversi, K. Horiuchi, Generalization of the cross-wavelet function. New Astron **56** (2017). https://doi.org/10.1016/j.newast.2017.04.012

21. P.C. Liu, Wavelet spectrum analysis and ocean wind waves. Wavelet Anal. Appl. **4** (1994). https://doi.org/10.1016/B978-0-08-052087-2.50012-8
22. V. Nourani, M.T. Alami, F.D. Vousoughi, Hybrid of SOM-clustering method and wavelet-ANFIS approach to model and infill missing groundwater level data. J. Hydrol. Eng. **21** (2016). https://doi.org/10.1061/(asce)he.1943-5584.0001398
23. C. Torrence, P.J. Webster, Interdecadal changes in the ENSO-monsoon system. J. Clim. **12** (1999). https://doi.org/10.1175/1520-0442(1999)012<2679:icitem>2.0.co;2
24. I. Akoum, M. Graham, J. Kivihaho, J. Nikkinen, M. Omran, Co-movement of oil and stock prices in the GCC region: a wavelet analysis. Q. Rev. Econ. Finan. **52** (2012). https://doi.org/10.1016/j.qref.2012.07.005
25. A. Fattouh, An emotional model based on wavelet coherence analysis of EEG recordings. BVICAM's Int. J. Inf. Technol. **8** (2016)

Chapter 7
Performance Evaluation of an Earth-Water Heat Exchanger Integrated to a Solar Absorption Cooling System

J. Ríos-Arriola, N. Velázquez-Limón, J. A. Aguilar-Jiménez, S. Islas, G. E. Bejarano-Aguilar, and M. A. Rojas-Casas

Abstract Absorption cooling systems offer a viable alternative for reducing the high electrical energy consumption associated with air conditioning (A/C) in buildings. Their ability to be powered by various thermal energy sources makes them particularly attractive for implementation in areas without access to the electrical grid. However, several factors have hindered the widespread adoption of these systems, including high initial investment costs, low efficiency, and operational limitations tied to site-specific conditions. This study investigates the integration of an earth-to-water heat exchanger to an absorption cooling system considering as case of study the remote community of Puertecitos, BC, Mexico, aimed at reducing freshwater consumption. The TRNSYS software was used to conduct a parametric study of the proposed system's performance. The earth-water heat exchanger saves 8094 kg of water per year, which represents 23.8% of total annual water consumption. Based on the results, it can be concluded that the integration of an earth-water heat exchanger to an absorption cooling system is a technically viable option, because it saves significant amounts of freshwater and allows the system to operate within its nominal conditions for a longer period during high-demand days in the hot season.

Keywords Absorption cooling systems · Earth-to-water heat exchanger · Solar thermal cooling

---

J. Ríos-Arriola · N. Velázquez-Limón (✉) · J. A. Aguilar-Jiménez · S. Islas · G. E. Bejarano-Aguilar · M. A. Rojas-Casas
Centro de Estudio de las Energías Renovables, Instituto de Ingeniería, Universidad Autónoma de Baja California, Mexicali, Baja California, México
e-mail: nicolas.velazquez@uabc.edu.mx

D. Rossit et al. (eds.), *Proceedings of the VII Ibero-American Congress of Smart Cities, ICSC-Cities 2024, 12-14 November, San Carlos, Costa Rica*, Springer Proceedings in Energy, https://doi.org/10.1007/978-981-96-4301-1_7

7.1 Introduction

The use of air conditioning in buildings represents 20% of the world's electrical energy consumption [8]. The increase in global has caused some areas to become warmer and/or desertic [18], increasing the requirement of A/C in buildings. This problem is accentuated in communities isolated from the electrical grid, therefore saving and efficient use of energy is essential [1, 7, 12].

Alternative cooling technologies with low power consumption are commercially available, such as solar thermal absorption cooling. However, their widespread deployment has been constrained by operational challenges, primarily related to the initial capital investment and operational costs associated with the solar collector field. Bellos and Tzivanidis [3] investigated the techno-economic feasibility of single-effect absorption cooling systems driven by evacuated tube collectors (ETC) in various climates. They employed TRNSYS software to assess different configurations of solar collection areas and thermal storage volumes, aiming to minimize the levelized cost of cooling (LCOC). The study concludes that the feasibility of solar-thermal absorption cooling systems is higher in scenarios with high cooling demand and high solar radiation levels.

High levels of solar radiation results in a reduced solar collector field area compared to the area required in regions with lower solar radiation levels to supply a specific power output, decreasing the initial investment and space requirements. Additionally, the operational costs decreases because the solar collectors reach the required temperature to activate the absorption cooling system more rapidly and maintains it for a longer period. Properly sized thermal storage allows the system to operate in the absence of solar radiation, avoiding auxiliary heat consumption and an increase in operational costs.

Shirazi et al. [17] conducted a review on solar-thermal absorption cooling technology and highlighted that one of the main challenges and key areas for improving the efficiency of these systems is the heat rejection medium. They emphasized that in warm and humid climates with high wet-bulb temperatures, cooling towers become limited in their performance. Consequently, they recommend implementing adaptations and/or utilizing alternative heat rejection technologies.

Despite the viability of absorption cooling in regions with high cooling demand and high solar radiation levels due to its low electrical energy consumption, attention must be focused on the cooling subsystem, which typically employs a cooling tower to dissipate residual heat. Cooling tower devices (direct or indirect) limits the system operation when the wet-bulb temperature is high, moreover operating on a regular basis it consumes considerable amounts of water.

The Northwest region of Mexico is predominantly desert characterized by an arid climate according to the Köppen classification [5]. The yearly per capita freshwater availability is 1040 m^3, categorized as regular scarcity according to the Water Stress Indicator (WSI) [11]. Most buildings are equipped with A/C systems, however, the extensive area also includes communities without access to the electrical grid where absorption cooling has been implemented, such as the school of Puertecitos, Baja

California, Mexico. Despite high solar radiation levels in this region, the cooling tower requires approximately 750 kg/day of freshwater [1].

It is well established that shallow soil (0–5 m) exhibits minimal temperature variation compared to the ambient environment [19]. At a certain depth the soil temperature remains relatively constant throughout the year, close to the average ambient temperature onsite. This enables heat exchange (both rejection and extraction) between thermal systems and the soil by means of buried pipes, named according to the fluid they use. The most common types are earth-air heat exchangers (EAHE) and earth-water heat exchangers (EWHE).

Eicker et al. [4] studied a 15 kW absorption chiller activated by solar thermal energy, the authors report that replacing the open wet cooling tower with geothermal heat exchangers for dissipating residual heat into the ground could reduce electrical energy consumption by up to 30%. They highlighted that this heat rejection system only requires a circulation pump to move the fluid through the ground, and pressure drops can be minimized by installing vertical geothermal heat exchangers connected in parallel. The authors do not consider the water savings generated by the EWHE.

Rouag et al. [16] evaluated the replacement of a dry cooling tower (DCT) with an Earth-Water Heat Exchanger (EWHE) to reject residual heat from a solar thermal adsorption cooling system in the semi-arid region of Biskra, Algeria. The EWHE was modeled and sized to ensure that the design maintains the same pressure drop as the DCT. The authors report that with a EWHE diameter of 0.033 m, the minimum required EWHE length must be 100 m to achieve a water outlet temperature of 34 °C. The heat transfer area of the EWHE was calculated to be 10.37 m^2, compared to 270.6 m^2 for the DCT. Their results demonstrated that the EWHE system is a more efficient alternative to the DCT under high ambient temperature conditions.

Al-Qazzaza et al. [2] conducted a comparative analysis between a conventional open wet cooling tower and an Earth-Water Heat Exchanger (EWHE) for reject residual heat in a mechanical vapor compression cooling system. TRNSYS software was used to simulate both heat rejection methods and compare their water and energy consumption. The authors report that the use of the EWHE reduced electrical consumption by 14% and completely eliminated water consumption, equivalent to 9539 m^3/year. The study concluded that the implementation of the EWHE is economically viable, with a payback period calculated at 1.73 years.

Based on the above, it is possible to conclude that the implementation of Earth-to-Water Heat Exchangers (EWHE) for heat rejection in cooling systems is advantageous, as it provides a heat sink with a more stable temperature compared to ambient conditions (dry or wet bulb temperature). Proper sizing of the EWHE can lead to energy savings, as it only requires a pump for operation, unlike cooling towers (both direct and indirect) which require both a pump and a fan. Additionally, the water savings provided by an EWHE, amplified by the growing scarcity of water, significantly enhances its techno-economic feasibility. The present work aims to study the potential of coupling an EWHE in the heat rejection subsystem of a solar thermal absorption cooling systems to reduce their water and electrical energy consumption. A simulation based on measured soil temperature profile will be conducted to quantify the savings, considering the absorption cooling system of the school in Puertecitos as a case study.

7.2 System Description

The proposed EWHE uses a centrifugal pump to circulate the cooling water through its array of vertically "U" pipes connected in parallel. The system incorporates two main headers responsible for extracting and injecting the cooling water. The cooling water carrying residual heat from the absorption cooling process, flows through the EWHE to reject heat by thermal convection to the soil, decreasing its temperature before being directed to the cooling tower or to an absorption chiller, provided its temperature has decreased sufficiently.

Figure 7.1 shows the proposed configuration for the heat rejection subsystem, which includes two strategically placed three-way valves. The first valve is located before the inlet of the Earth-Water Heat Exchanger (EWHE), and its purpose is to redirect the cooling water flow back to the absorption chiller if the fluid temperature is below 30 °C. The second valve is positioned at the outlet of the EWHE, allowing the cooling water flow to be directed to the absorption chiller when the water temperature at the outlet of the EWHE is also below 30 °C.

The design of the Earth-Water Heat Exchanger (EWHE) aims to minimize initial investment and maximize system longevity. Considering the negligible positive impact of using high thermal conductivity materials, as noted by Jakhar et al. [9], it is proposed that the EWHE be constructed from high-density polyethylene (HDPE) a corrosion-resistant material. The pipe diameter should fit within commercially available dimensions, ranging from 4 to 10 inches, while the depth (length) should be between 10 and 100 m. The spacing between geothermal boreholes is expected to range from 0.5 m to 2 m, depending on the available space. With a total cooling water flow rate of 15,000 kg/h, it is essential to consider an adequate number of boreholes in parallel to ensure that each EWHE provides sufficient residence time of the fluid in the subsurface, thus optimizing heat transfer.

7.3 Study Methodology

7.3.1 Simulation Software

Most direct uses of geothermal energy (or low enthalpy geothermal energy) require buried pipes to perform heat exchange with the soil, commonly referred to as geothermal heat exchangers. However, there are few specialized tools for the study

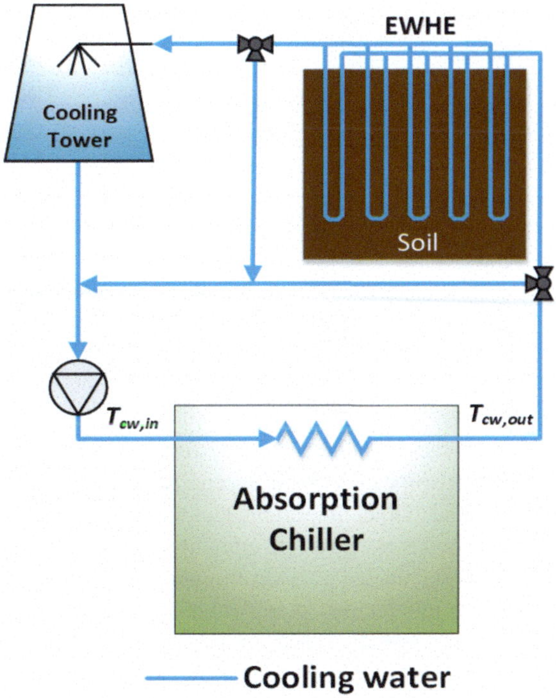

Fig. 7.1 EWHE integrated to the cooling water subsystem

of low-temperature geothermal technologies, and even fewer that allow the joint evaluation of their integration with thermal systems and/or buildings. TRNSYS is a versatile tool for the dynamic simulation of geothermal heat exchangers integrated with cooling systems, particularly in renewable energy applications thus enables the comprehensive modeling of geothermal technologies and their interactions with solar-driven thermal processes. Its modular framework captures key transient behaviors, such as time-dependent solar irradiance, ground thermal response, and ambient conditions, allowing for precise system performance evaluation. Its libraries contain modules that enable the simulation of various geothermal technologies such as: geothermal heat pumps, geothermal foundations, earth-air heat exchangers, earth-water heat exchangers, radiant floors, and underground thermal energy storage tanks.

The Type 557 module of TRNSYS was used to model the behavior of the proposed EWHE. This model assumes a uniform distribution of vertical "U" shaped geothermal heat exchangers, through which a working fluid circulates, exchanging heat by thermal convection with the surrounding soil. Input data for the model includes the specification of one (or more) layers of soil surrounding the geothermal heat exchanger, with known properties such as thermal conductivity (W/m °C) and volumetric heat capacity (J/m^3 °C) [15].

Table 7.1 Modules used for simulation in TRNSYS 17

Component	TRNSYS type	Description
Pump	3	Drives the water trough buried pipes
Time-dependent forcing function	14	Controllers chillers on/off
Weather data processor	15	Climatological database
Day-dependent forcing function sequencer	41	Indicates the day of the week
Absorption chiller	107	Model the absorption chiller
Temperature probe	77	Calculate the soil temperature
Earth to water heat exchanger	557	Model the buried pipes at uniform depth

Several factors influence the performance of a geothermal heat exchanger. Some are related to pipe manufacturing characteristics, such as the diameter, wall thickness, and material. Other important factors include the soil installation depth and the working fluid mass flow rate circulating within the system. Therefore, a parametric study was conducted varying the number of boreholes, boreholes depth, spacing between boreholes and pipe diameter to determine the optimal configuration for the proposed EWHE based on the amount of heat rejected during the hot period (May–October) (Table 7.1).

7.3.2 Soil Temperature Profile

The soil temperature profile is of interest in fields such as agriculture, architecture, and engineering due to its impact on thermal processes in these areas. Given the technical and economic challenges involved in experimental temperature measurement in boreholes or excavations, various mathematical models have been proposed to describe the annual variation of soil temperature at different depths [10, 13, 14]. Consequently, it is possible to make estimations of the soil temperature profile using Eq. (7.1) proposed by Hillel [6], which represents the periodic variation of soil temperature at a given Depth.

$$T_{(z,t)} = T_m + A_0 e^{(-z/d)} sen\left[2\pi(t - t_0) - \frac{z}{d}\right] \tag{7.1}$$

In the Eq. (7.1), T_m represents the annual average of ambient temperature (°C), z is the depth (m), t corresponds to time (h), A_0 is the amplitude of the ambient temperature wave (°C), d refers to the damping depth, and t_0 is the time when the lowest annual temperature occurs (h). The damping depth can be calculated using Eq. (7.2), which allows for the evaluation of the thermal variation attenuation as soil depth increases.

$$d = \sqrt{\frac{2\alpha}{\omega}} \tag{7.2}$$

where α represents the thermal diffusivity of the soil (m^2/s) and is calculated as stated in the Eq. (7.3).

$$\alpha = \frac{k}{\rho C_p} \tag{7.3}$$

where k represents the thermal conductivity (W/m °C), ρ the density (kg/m^3), and C_p the specific heat capacity (J/kg °C) corresponding to the soil at the study site. The variable ω denotes the temperature wave frequency and is calculated as stated in the Eq. (7.4).

$$\omega = \frac{2\pi}{P} \tag{7.4}$$

where P represents the period, which corresponds to the total duration of the temperature wave projection (h). The values T_m, A_0, and t_0 must be obtained from the annual ambient temperature records.

7.4 Results

7.4.1 Soil Temperature Profile

Figure 7.2 presents both the calculated and experimentally measured soil temperature profiles at a depth of 4 m. The linear fit value, R2 = 0.9893, demonstrates that the analytical model employed in TRNSYS to estimate soil temperature accurately represents the observed phenomen. The maximum temperature deviation between the model and experimental data occurs at June 25 (4204 h) with a discrepancy of 0.426 °C.

7.4.2 Parametric Study of EWHE

In order to determine the best configuration of the proposed EWHE, a parametric study was conducted by varying the diameter, vertical length (depth), number of EWHEs in parallel, and spacing between each parallel EWHE. Figure 7.3 shows that the impact of pipe diameter on the heat rejection rate is less significant compared to the spacing between boreholes. The spacing between each borehole significantly affects the heat rejection rate due to the high thermal diffusivity of the sandy soil.

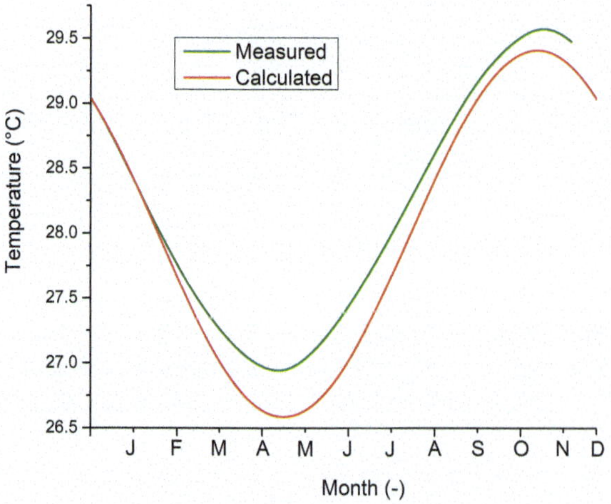

Fig. 7.2 Experimental validation of soil temperature variation at 4 m depth

This property leads to the heat rejected by one EWHE increasing the temperature of the surrounding soil, thereby limiting the heat rejection capacity of adjacent EWHEs. Given the available space at the Puertecitos school, the maximum spacing between boreholes (2 m) was selected as the optimal configuration.

Table 7.2 determined that the optimal diameter is 6 inches (0.1524 m), as reducing from 10 inches (0.254 m) to 8 inches (0.2032 m) increases the heat rejection rate by 1.67%, while reducing from 8 inches (0.2032 m) to 6 inches (0.1524 m) increases

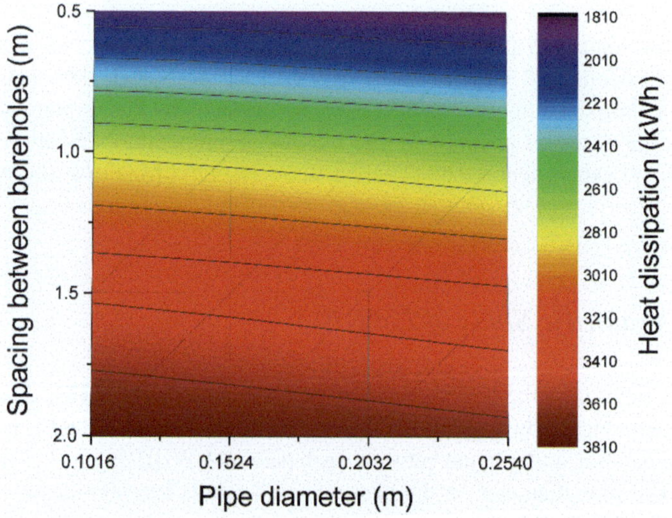

Fig. 7.3 Heat rejection varying EWHE diameter and borehole separation

the rate by 1.63%. However, further decreasing the pipe diameter from 6 inches (0.1524 m) to 4 inches (0.1016 m) results in a heat rejection rate increase of less than 1.2%. Additionally, a smaller pipe diameter reduces the environmental and economic impact associated with land use changes and initial investment. For this first parametric variation, a borehole depth of 30 m and a number of boreholes of 10 were considered.

Based on Table 7.2, it is also possible to corroborate and quantify the qualitative conclusion that a spacing of 2 m is optimal. Increasing the well spacing from 0.5 m to 1 m results in a 45% increase in heat rejection. Further increasing the spacing from 1 m to 1.5 m increases heat rejection by 22%. Finally, increasing the spacing from 1.5 m to 2 m results in a 12% increase in heat rejection. Therefore, 2 m was chosen as the optimal spacing, considering that increasing the spacing from 2 m to 2.5 m yield less than a 10% increase in heat rejection and that greater spacing between boreholes would require more piping to configure the parallel arrangement.

Figure 7.4 shows that both the number of boreholes and their depth significantly influence the heat rejection rate of the proposed EWHE. Selecting the drilling of 20 boreholes at a depth of 100 m appears to be the best option; however, increasing the drilling depth increases both the economic cost and the risk of borehole collapse, and especially because the soil in the study site is predominantly sandy.

Table 7.3 indicates that beyond a borehole depth of 50 m, the increase in heat rejection falls below 10% when extending the depth by an additional 10 m (from 50 to 60 m). Consequently, a borehole depth of 50 m was identified as the optimal configuration. Furthermore, Table 7.3 demonstrates that the optimal number of boreholes is 10. Expanding the number of boreholes from 5 to 10 enhances the heat rejection of the EWHE by 40%, while further increments from 10 to 15 and from 15 to 20 boreholes yield progressively smaller increases in heat rejection, at 16% and 9%, respectively.

In summary, based on the above results was determined that the optimal configuration of the earth-water heat exchanger (EWHE) integrated with a solar absorption cooling are 10 boreholes with 0.1524 m diameter pipes buried at 50 m depth and separated 2 m between each boreholes.

Table 7.2 Heat rejection (kWh) varying EWHE diameter and borehole separation

	Diameter			
Separation	0.01016 m	0.1524 m	0.2032 m	0.254 m
0.5 m	1915.9	1897.4	1850.7	1811.1
1 m	2781.8	2743.2	2695.5	2643.8
1.5 m	3378.7	3338.5	3290.7	3240
2 m	3804.7	3762.7	3715.4	3666.8

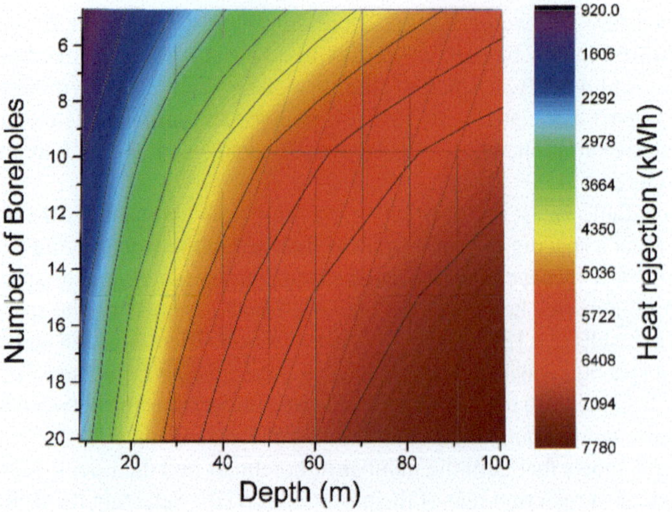

Fig. 7.4 Heat rejection (kWh) varying EWHE diameter and borehole separation

Table 7.3 Heat rejection (kWh) varying EWHE depth and boreholes number

	z									
BN	10 m	20 m	30 m	40 m	50 m	60 m	70 m	80 m	90 m	100 m
5	920.6	1662.4	2335.4	2940.6	3482.4	3968.9	4403.9	4788.7	5132.8	5437.2
10	1561.3	2721.7	3679.2	4460.4	5092.6	5598.3	6013.3	6347.6	6626.6	6854.6
15	2116.3	3548.5	4623.4	5418.4	6008.7	6446.6	6789	7052.8	7269.9	7443.6
20	2611.1	4215.6	5310.3	6058.1	6579.2	6951	7238.1	7455.4	7635.4	7778.6

7.4.3 Operative Study of EWHE

Figure 7.5 shows the performance of the EWHE during the hottest week of the first hot period (4441–4560 h). On average, the operation of the EWHE lowers the cooling water temperature at the cooling tower inlet by 0.97 °C compared to the conventional operation of the absorption cooling system. This is reflected in a reduction of 489 kg of evaporated water in the cooling tower, which represents 14.84% of the total water consumption during that week. The lower temperature of the water exiting the EWHE (compared to the conventional operation) delays the operation of the cooling tower by one hour. Utilizing the EWHE throughout the entire first warm period (May and June) avoids the consumption of 4799 L of water, representing 29.9% of the total water consumption during this period.

Figure 7.6 shows the performance of the EWHE during the critical week of the second hot period (5665–7153 h). On average, the water temperature at the cooling tower inlet is 1.3 °C lower with the EWHE, which results in a reduction of 788 kg

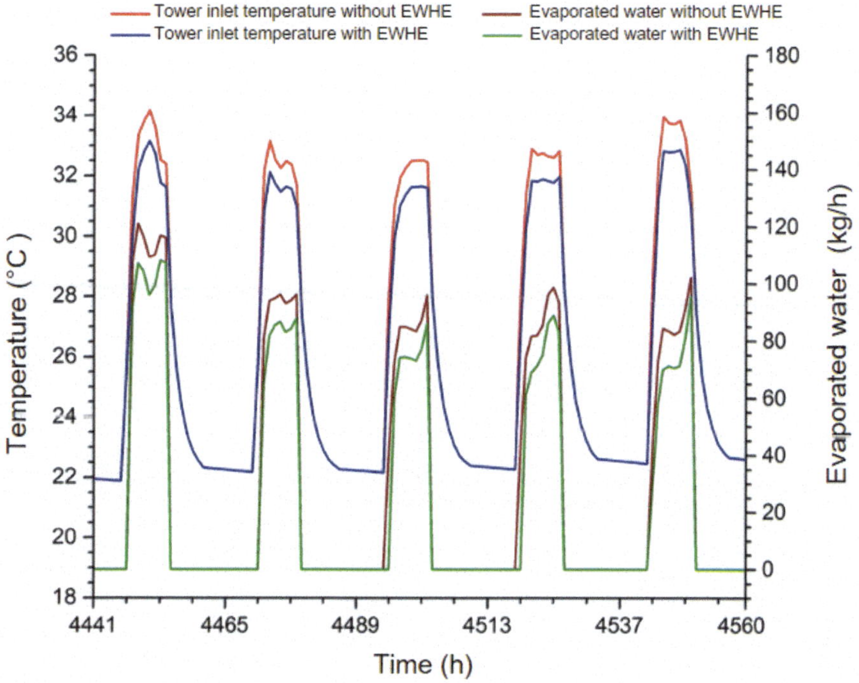

Fig. 7.5 Behavior of the cooling water subsystem during the critical week of the first warm period

of evaporated water, representing 17.9% of the total water consumption during this week. The EWHE performed better on Wednesday, reducing water consumption by an average of 23.9 kg/h. Despite the EWHE delays the cooling tower operation by one hour on Monday and Friday, the daily water saving averages for these days were 22.3 kg/h and 21.5 kg/h, respectively. Employing the EWHE throughout the second warm period (5665–7153 h) results in a reduction of 3295 L of water consumption, equivalent to 18% of the total water usage during this period.

The water savings achieved through the EWHE integration are detailed in Table 7.4, which outlines the water savings for each hot period as well as the total annual water savings. The EWHE demonstrates superior performance during the May–June period, where it decreases water consumption by 4799 kg or 29.94% of the consumption on this period. In comparison, during the second warm period, it reduces water consumption by 3295 kg representing 18.37% of the period total.

The discrepancy in performance between the two hot periods can be attributed to the lower ambient temperature, building temperature, and thermal load experienced during the May–June period relative to the second. This results in the EWHE rejecting less thermal energy into the soil, thereby reducing or mitigating the rise in soil temperature due to heat accumulation. Furthermore, the soil temperature at a depth of 4 m is lower during the first hot period, averaging 26.5 °C, compared to 29.2 °C during the second hot period.

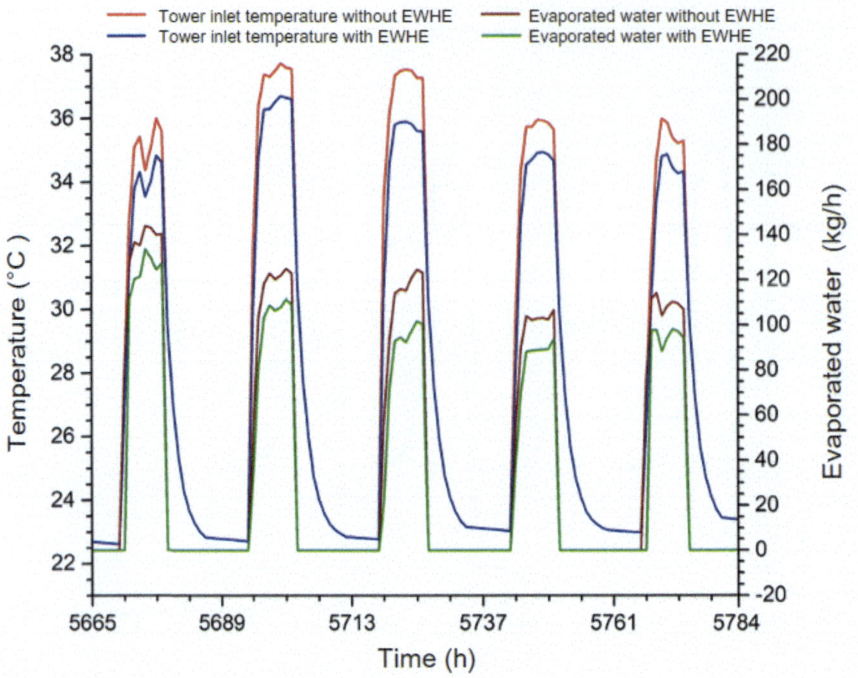

Fig. 7.6 Behavior of the cooling water subsystem during the critical week of the second warm period

Table 7.4 Evaporated water savings generated by the EWHE

Evaporated water (kg/h)	Without EWHE	With EWHE	Savings %
May–June	16,026.2	11,227.1	29.94
August–October	17,939.3	14,644.1	18.37
Total	33,965.5	25,871.2	23.83

7.5 Conclusions

This work technically evaluated the integration of an earth-water heat exchanger (EWHE) to an absorption cooling system for air conditioning a school located in a remote coastal community. TRNSYS simulation software was used in conjunction with experimental weather and soil data obtained on site to evaluate the potential of the EWHE during school hot periods. The results obtained indicate that the EWHE has greater potential during the May–June period, when the ambient, soil and building temperatures are lower than during the August–October season.

During the month of August, ambient temperature and relative humidity reach their highest averages, at 33 °C and 62%, respectively. This poses a critical challenge for the conventional operation of the absorption cooling system, as its cooling tower heat rejection capacity is limited by the high wet bulb temperature when the building's thermal load reaches its peak. Additionally, the average solar radiation (477 W/m^2) is 12.2% lower compared to the average radiation during the May–June period (543 W/m^2), resulting in less useful heat for activating the absorption chiller. This shortage requires auxiliary heat for system operation. In conclusion, although water savings are less compared with May–June season, the operation of the EWHE becomes more critical during the August–October period, as it helps maintain the system's COP close to its nominal value (0.72) for a longer period of time.

The integration of the EWHE to the absorption cooling system reduces (23.83%) approximately a quarter of the annual water consumption in the cooling tower. In addition, on some days it delays the operation of the cooling tower during the first hour of class, avoiding the operation of the cooling water pump and the tower fan, which leads to additional energy savings. Utilizing the EWHE throughout the May–June period (May and June) avoids the consumption of 4799 L of water, representing 29.9% of the total water consumption during this period. During the second hot period (5665–7153 h) results in a reduction of 3295 kg of water consumption, equivalent to 18% of the total water usage during this period.

Acknowledgements The authors would like to express their gratitude to the Engineering Institute of the Autonomous University of Baja California for their invaluable support in carrying out this work, as well as to the "Consejo Nacional de Humanidades, Ciencias y Tecnologías (CONAHCYT)" for granting a scholarship to Juan Ríos-Arriola (CVU 984076) to pursue his PhD studies. This research was funded by "23a Convocatoria Interna UABC" grant number (111/6/C/19/23).

Disclosure of Interests The authors declare that they have no conflict of interest.

References

1. J.A. Aguilar-Jiménez, N. Velázquez-Limón, R. López-Zavala, L.A. González-Uribe, S. Islas, E. Gonzáles, L. Ramírez, R. Beltrán, Optimum operational strategies for a solar absorption cooling system in an isolated school of Mexico. Int. J. Refrig. **112**, 1–13 (2020). https://doi.org/10.1016/j.irefrig.2019.12.010
2. A.H.S. Al-qazzaza, M. Farzaneh-Gorda, H. Niazmand, Systematic analysis with comparison of a chiller plant with horizontal underground heat exchangers and cooling tower. J. Buil. Eng. **92**, 109665 (2024). https://doi.org/10.1016/j.jobe.2024.109665
3. E. Bellos, C. Tzivanidis, C. Symeou, K.A. Antonopoulos, Energetic, exergetic and financial evaluation of a solar driven absorption chiller—a dynamic approach. Energy Convers. Manage. **137**, 34–48 (2017). https://doi.org/10.1016/j.enconman.2017.01.041
4. U. Eicker, D. Pietruschka, R. Pesch, Heat rejection and primary energy efficiency of solar driven absorption cooling systems. Int. J. Refrig. **35**(3), 729–738 (2012). https://doi.org/10.1016/j.ijrefrig.2012.01.012
5. E. García, Modificaciones al Sistema Climático Köppen. In *Instituto de Geografía* (2004). http://www.igeograf.unam.mx/sigg/utilidades/docs/pdfs/publicaciones/geo_siglo21/serie_lib/modific_al_sis.pdf
6. D. Hillel, *Intoduction to Soil Physics* (Academic Press, USA, 1982). https://doi.org/10.1016/C2009-0-03052-9

7. C. Horvath, Y. Hwang, R. Radermacher, W. Gerstler, C.-J. Tang, Waste heat and electrically driven hybrid cooling systems for a high ambient temperature, off-grid application. Energy **66**, 711–721 (2014). https://doi.org/10.1016/j.energy.2013.11.058
8. IEA, The future of cooling, in *Opportunities for energy-efficient air conditioning* (2018). https://doi.org/10.1787/9789264301993-en
9. S. Jakhar, M.S. Soni, N. Gakkhar, Parametric modeling and simulation of photovoltaic panels with earth water heat exchanger cooling. Geoth. Energy **4** (2016). https://doi.org/10.1186/s40517-016-0054-8
10. M. Krarti, D.E. Claridge, J.F. Kreider, Analytical model to predict nonhomogeneous soil temperature variation. J. Solar Energy Eng. Trans. ASME **117**(2), 100–107 (1995). https://doi.org/10.1115/1.2870823
11. T. Manungufala, *Water Scarcity: Classification, Measurement and Management* (2021), pp. 1–14. https://doi.org/10.1007/978-3-319-70061-8_7-1
12. O. Marc, F. Lucas, F. Sinama, E. Monceyron, Experimental investigation of a solar cooling absorption system operating without any backup system under tropical climate. Energy Build. **42**(6), 774–782 (2010). https://doi.org/10.1016/j.enbuild.2009.12.006
13. G. Mihalakakou, M. Santamouris, J.O. Lewis, D.N. Asimakopoulos, On the application of the energy balance equation to predict ground temperature profiles. Sol. Energy **60**(3–4), 181–190 (1997). https://doi.org/10.1016/S0038-092X(97)00012-1
14. O. Ozgener, L. Ozgener, J.W. Tester, A practical approach to predict soil temperature variations for geothermal (ground) heat exchangers applications. Int. J. Heat Mass Transf. **62**, 473–480 (2013). https://doi.org/10.1016/j.ijheatmasstransfer.2013.03.031
15. D. Pahud, G. Hellström, Duct ground heat storage model for TRNSYS (TRNVDST). Proc. Eurotherm Semin. **49**, 127–136 (1996)
16. A. Rouag, A. Benchabane, A. Labed, K. Belhadj, N. Boultif, Applicability of a solar adsorption cooling machine in semiarid regions: proposal of supplementary cooler using earth-water heat exchanger. Int. J. Heat Technol. **34**(2), 281–286 (2016). https://doi.org/10.18280/ijht.340218
17. A. Shirazi, R.A. Taylor, G.L. Morrison, S.D. White, Solar-powered absorption chillers: a comprehensive and critical review. Energy Convers. Manage. **171**(May), 59–81 (2018). https://doi.org/10.1016/j.enconman.2018.05.091
18. F. Stenzel, P. Greve, W. Lucht, S. Tramberend, Y. Wada, D. Gerten, Irrigation of biomass plantations may globally increase water stress more than climate change. Nat. Commun. **12**(1), 1–9 (2021). https://doi.org/10.1038/s41467-021-21640-3
19. R. Vidhi, *Organic Fluids and Passive Cooling in a Supercritical Rankine Cycle for Power Generation from Low Grade Heat Sources* (2014)

Chapter 8
Hybridization of Renewable Energy Sources with Pumped Hydroelectric for Energy System Stability

Alexis V. Terrero, Victor Ocaña, Félix Santos, and Miguel Aybar-Mejía

Abstract The increasing integration of renewable energy sources into energy systems has driven the development of storage technologies, with pumped hydroelectric storage (PHS) as a prominent solution. This document reviews the need for hybridization between renewable sources such as photovoltaic solar energy and wind energy with pumped hydroelectric power to guarantee the energy system's stability and reliability. Solar and wind energy's technical advantages and disadvantages are discussed, as well as the challenges associated with their variability and dependence on external factors such as climate. Likewise, optimization models that combine these renewable sources with storage technologies such as PHS are explored, allowing better management of intermittent energy sources. In addition, relevant case studies are presented, highlighting hybridization's economic and environmental potential. The review emphasizes the importance of developing energy policies that favor investment in hybrid systems and research into emerging technologies. Finally, it suggests the need to continue research on integrating advanced storage technologies to maximize efficiency and improve the resilience of energy systems in the context of decarbonization and climate change.

Keywords Renewable energy · Pumped hydroelectric storage · Photovoltaic solar energy · Wind energy · Energy hybridization · System optimization · Energy stability

A. V. Terrero · V. Ocaña
Area of Basic Sciences, Instituto Tecnológico de Santo Domingo, Santo Domingo, Dominican Republic

F. Santos
Smart Tech and Sustainability (STECHS INTERFAZ SRL), Santo Domingo, Dominican Republic

M. Aybar-Mejía (✉)
Engineering Area, Instituto Tecnológico de Santo Domingo, Santo Domingo, Dominican Republic
e-mail: miguel.aybar@intec.edu.do

D. Rossit et al. (eds.), *Proceedings of the VII Ibero-American Congress of Smart Cities, ICSC-Cities 2024, 12-14 November, San Carlos, Costa Rica*, Springer Proceedings in Energy, https://doi.org/10.1007/978-981-96-4301-1_8

8.1 Introduction

The search for renewable energy sources has become a global priority in recent years. The need to reduce greenhouse gas emissions and decrease our dependence on fossil fuels has driven the development and implementation of different sustainable energy sources.

One of the most prominent renewable energy sources is solar photovoltaic energy. This technology takes advantage of solar radiation to generate electricity. One of its main advantages is that it is an abundant, renewable, and distributed energy source, allowing it to be deployed globally [1]. In addition, during electricity generation, solar PV emits low levels of greenhouse gases, thus reducing the environmental impact [2]. However, the reliance on sunlight availability is a significant disadvantage of this technology, as electricity generation is affected by weather conditions and geographical location [3, 4]. In addition, the high initial investment costs, although with a clear tendency to decrease, and the need for space to install solar panels can also be obstacles to their mass implementation [5].

Wind energy is another renewable source that has gained ground in the energy landscape. Wind power has notable advantages when harnessing the power of the wind to generate electricity. Like solar energy, it is an abundant and renewable source, and its generation has a low emission of greenhouse gases [6]. In addition, wind technology is widely available and mature, making it easy to implement. However, a significant disadvantage is the need for consistent and adequate winds for optimal generation. This implies that not all locations are suitable for installing wind farms, limiting their use to specific windy areas. In addition, the potential impact on avian wildlife and the visual effects on the landscape are essential considerations to consider [7].

To address the intermittent renewable sources, pumped hydroelectric energy (PHS) storage has emerged as a promising solution. This technology allows energy to be stored in water at times of low demand and released when additional electricity is required. Its advantages include large-scale storage capacity, helping to compensate for the variability of other renewable sources, and efficient regulation of the load and frequency of the electricity grid [8, 9].

Pumped hydroelectric storage can respond quickly to fluctuations in demand, providing ancillary services essential to the stability of the electricity system [10]. However, high upfront investment costs, dependence on water availability, and adequate topography are significant disadvantages [11, 12]. Building infrastructure for PHS can be costly and limited by geographical conditions. In addition, the environmental and social impact associated with the construction of reservoirs must also be considered [13].

Considering the above, the main objective of this Bibliographic study is to evaluate the requirements to determine the technical and economic feasibility of hybridizing renewable energy sources with pumped hydroelectric storage, its sizing, and operation through the optimization of the energy system's stability and efficiency.

8.2 Methodology

The literature was collected in articles and reviews (mainly between 2020 and 2025, in the SCOPUS database). The search strategy was based on the concepts of "Renewable energy hybrid system," "pumped hydroelectric storage," "pumped hydro storage," and "Mathematical model optimization. "The bibliographic database obtained from 529 papers was filtrated, receiving the 56 most relevant for the present documentary research. The study identified various methods and criteria for optimizing Hybrid Renewable Energy Systems (HRES) and their challenges. Economic, environmental, and reliability requirements are identified, which must be included in the mathematical design to control (HRES).

8.3 Discussion

8.3.1 Integration and Hybridization of Renewable Energy Sources

The hybridization of various renewable energy sources, such as solar, wind, hydro, and geothermal, maximizes the stability and reliability of the energy grid [14]. Each source has individual limitations—such as intermittency in solar and wind—but hybridization allows their characteristics to complement each other [15]. For example, solar energy is most productive during the day, while wind power can be produced anytime. Adding storage, such as pumped hydro (PHS) and advanced batteries, helps compensate for variability and provides a more consistent supply [16].

In contexts of climate change and growing energy demand, hybridization is essential. Using advanced technologies and optimization through artificial intelligence (such as Gaussian process regression GPR and adaptive algorithms) allows forecasting and managing resource variability [17]. This strategy reduces dependence on fossil fuels and improves the economic viability of projects in decentralized and hard-to-reach communities [18]. Hybrid systems that integrate PTC (Parabolic Trough Collector) and hydrogen systems, in addition to SDS (Solar Dish Stirling) technology, increase efficiency and reduce operating costs, making it easier to apply in low-carbon buildings [19].

Figure 8.1 shows a schematic of a hybrid system that integrates solar, wind, pumped hydroelectric storage, and conventional generation. Predictive control and energy flows make optimizing these sources possible, achieving a more resilient energy system adaptable to climatic and demand variations. Implementing these hybrid and storage technologies contributes significantly to sustainability and development goals in the energy sector.

One example of the hybridization mode used is the one based on hydrogen storage [20]. Allouhi et al. [13] propose a hybrid renewable energy system (HRES) optimized

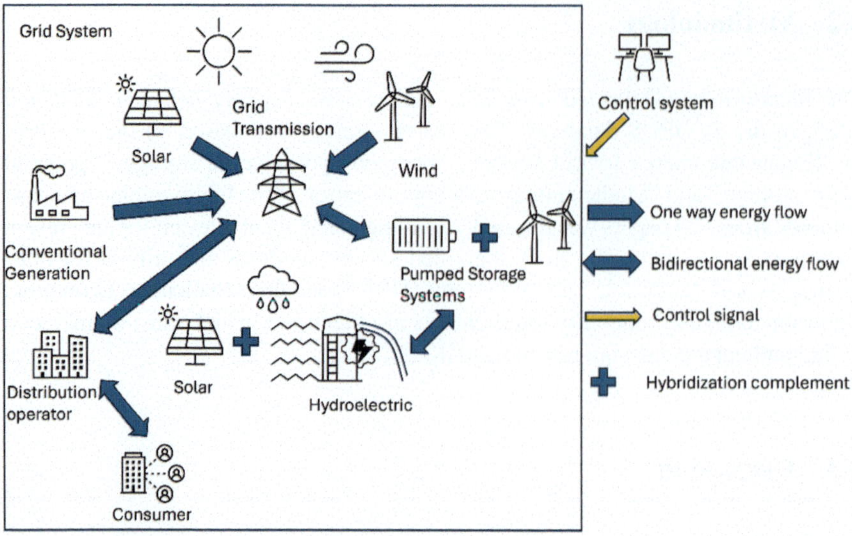

Fig. 8.1 Hybridization of renewable energy sources with HPS

to meet the electricity demand of sustainable residential buildings, with the possibility of generating clean hydrogen [21]. Zhang et al. [22] address the integration of nuclear energy and renewable sources into a hybrid nuclear-renewable energy (N-R-HES) system. Another mode integrates small-scale solar-biohybrid energy using Rankine (steam turbine) and Brayton (gas turbine) cycles based on biogas. Thermodynamic models were developed to characterize the condition of the working fluid and select the most suitable solar harvesting technology for each power generation system [23].

Hybridization based on hydraulic pumping has been widely studied, especially hybrid solar-wind power supply systems based on pumped hydroelectric storage (PHS) [24]. Studies show that the energy storage capacity of hydropower plants can be harnessed to increase renewable energy generation [25]. In this case, the floating PV plant uses the water stored during the day as a free "virtual battery," which increases electricity production by 50% without significantly changing the hydroelectric plant's daily generation.

Implementing a floating PV plant demonstrates how the storage capacity of hydropower plants can be harnessed to increase renewable energy generation by 50%. The plant functions as a 'virtual battery' without significantly altering the hydropower plant's operation, as evidenced in recent studies [26]. Table 8.1 presents examples of hybridizing renewable energy systems with pumped hydroelectric storage (PHS).

Table 8.1 Hybridization of renewable energy systems with pumped hydroelectric storage (PHS)

Feature	References	Detail
Adaptation of PHS at low height	[9]	Use of axial flow pumps and generators
Network stability through PHS	[8]	Adaptation of the PHS for stability in networks with high ER
PHS hybridization with batteries	[10]	Improved reliability of isolated systems
Policies for investment in PHS-BES	[11]	Need for environmental policies and studies for PHS-BES
Floating PV Plant & PHS	[12]	50% increase in grid connection capacity
Economic and social impact of RE in Nigeria	[13]	RD in Nigeria improves employment, GDP, and education
Design and performance of hybrid ER and PHS systems	[14]	8.6% improvements in ER and 3% in energy loss
Underground storage technologies	[15]	Use of UTES and UHS in geological formations
Integer mixed linear programming (PLME) model for PHS	[25, 26]	PLME for investment and optimization decisions in ER systems

Acronyms: Battery Energy storage (BES), Underground Thermal Energy Storage (UTES) and Underground Hydrogen Storage (UHS), gross domestic product (GDP), Renewable energy (ER)

8.3.2 Models Applied for Hybridization (PHS)

In the hybrid renewable energy systems field, in addition to the optimization techniques mentioned above, the technical literature highlights other promising strategies. Hybrid techniques, which combine multiple algorithms, are presented as an effective alternative to overcome the limitations of solutions based on a single algorithm. Some of these techniques include the use of SA-Tabu search, Monte Carlo simulation-PSO, and particle swarm optimization (PSO) with differential evolution (DE) PSO/DE [27], iterative hybrid/genetic algorithm (GA) [28], Artificial Neural Network Fuzzy Interface System (ANFIS), and evolutionary algorithms [29], and simulation–optimization using Monte Carlo Simulation (MCS).

Several models have been studied considering optimal sizing and energy efficiency [30]. Finally, several authors suggest multi-objective optimization, evolutionary algorithms, and particle swarms as effective optimization models [31].

Optimization techniques based on hybrid methods represent a range of promising approaches to solving challenges in hybrid renewable energy systems. The combination of algorithms makes it possible to find optimal configurations that effectively balance economic and technical aspects, thus contributing to the efficiency and sustainability of energy solutions.

Table 8.2 provides a comparative view of models developed for studying hybrid electric power supply systems. Different RES and energy backup/storage systems are combined with the objectives above and models or optimization tools in these models. A diversity of approaches is observed in terms of application, objectives, and models used in the studies analyzed, which suggests the need to generate customized models according to the particularities of the case to be developed.

Although there are established tools for implementing the models computationally, in the case of short-term analyses, it is necessary to explore the use of open-source support in real-time control systems.

Table 8.2 shows interest in the behavior of hybrid systems in rural and remote areas, suggesting the need to continue developing and evaluating these systems at small spatial scales. The timescales are variable, but they show a clear need to verify the behavior of the systems in the short term. However, national and long-term studies suggest the need for the models to be flexible and integrated into more general studies involving various spatial and temporal scales. The assessment of the economic aspects and reliability of these hybrid power supply systems (SHSE) stand out in the objectives of the works analyzed. Although to a lesser extent, social and environmental aspects are also considered, showing a more comprehensive interest in the phenomenon studied.

The literature shows the use of simulation, with models such as HOMER, iHOGA, and others, to evaluate performance and design efficient energy systems. This method could be vital for validating the models developed in different operating situations and be fundamental for performing case studies that strengthen the security of the models' results.

Although there are established tools for the computational implementation of the models, in the case of short-term analyses, it is necessary to explore the use of open-source support in real-time control systems. Table 8.2 shows interest in the behavior of hybrid systems in rural and remote areas, suggesting the need to continue developing and evaluating these systems at small spatial scales. The time scales are variable but show a clear need to verify the behavior of the systems in the short term. However, national and long-term studies suggest the need for the models to be flexible and integrated into more general studies involving various spatial and temporal scales. The assessment of the economic aspects and reliability of these hybrid power supply systems (SHSE) stand out in the objectives of the works analyzed. Although to a lesser extent, social and environmental aspects are also considered, showing a more comprehensive interest in the phenomenon studied.

The literature shows the use of simulation, with models such as HOMER, iHOGA, and others, to evaluate performance and design efficient energy systems. This method could be vital for validating the models developed in different operating situations and be essential for performing case studies that strengthen the security of the models' results.

Table 8.2 Summary of hybrid power supply systems (SHSE) models

Energy technology						Backup/energy storage systems					SD	TD	Objective	Mentioned models	References
SVP	W	H	Bi	Bf	Bd	DG	BS	FC	PHES	HT					
X	X	X	X		X	X	X	X		X	Rural areas	LT	Economic, reliability, and environmental	HOGA	[28]
X			X				X				Rural areas	LT	Economic and environmental	HOMER	[32]
X	X	X	X	X		X	X				Rural areas	ST	Economic and reliability	HOMER, H2RES/ EnergyPlan	[33]
X								X			Remote areas	LT	Economic	HOMER	[34]
X		X	X		X					X	Remote areas	MT	Economic	HOMER, LINDO, and PSO	[35]
X	X				X						Remote areas	ST	Life cycle cost	MATLAB code	[36]
X	X			X					X	X	National	LT	Economic, social, and environmental	LSPS, LOLP, LOLR, or LOLD,	[37]
X		X					X		X		National	LT	Economic, design, and expansion	HOMER	[38]
X	X	X	X						X		National	LT	Economic	–	[39]
X	X	X	X						X		National	ST	Economic	–	[40]
X			X				X			X	Rural areas	ST & MT	Economic, design, and expansion	HOMER and iHOGA	[41]

(continued)

Table 8.2 (continued)

Energy technology						Backup/energy storage systems					SD	TD	Objective	Mentioned models	References
SVP	W	H	Bi	Bf	Bd	DG	BS	FC	PHES	HT					
X	X	X							X		Micro Grid	ST	Economic, design, and expansion	–	[42]
X			X				X		X		Micro Grid	ST	Economic	WOA and EWOA	[30]

Acronyms: SD—Spacial dimension; TD—Time dimension; LT—long term; ST—Short term; MT—Medium term; SPV—Solar photovoltaic, W—Wind power H—Hydropower, Bi—Biomass, Ti/Wa-Tidal/Wwave, Bf/Bd—Biofuel/biodiesel, FC—Fuel cel, PHES—Pumped hydroelectric energy storage, BS—Battery storage; HT—Hydrogen tank storage, DG-Diesel generator. EWOA—Enhanced whale optimization algorithm, WOA—Whale optimization algorithm

8.3.3 Stochastic Analysis and Dynamic Optimization

The reviewed studies show the significant potential of Multi-Objective Optimization (MOO) in hybrid energy systems, where optimization based on economic criteria costs and environmental impacts, in contexts where intermittent solar PV generation is essential in interconnected microgrid-scale energy supply [31]. Potential benefits of MOO are identified, such as improved intermittent resource utilization, reduced energy costs, minimized environmental impacts, and increased efficiency and resilience of energy systems.

Gaps and areas for future work are pointed out, such as the availability of accurate data, the complexity of optimization, the integration of uncertainty, the interoperability of systems, the scalability and replicability of initiatives, long-term performance, the health and well-being of occupants, social and behavioral factors, among others. The need to do more research on accurate data collection, accurate prediction, interoperability, scalability, long-term performance, and impact on residents' health and well-being is highlighted, as well as to explore other metaheuristic optimization algorithms and methods to validate their efficacy and performance compared to conventional approaches such as genetic algorithm (GA).

8.3.4 Uncertainty Analysis Needs Robustness Control Models

The robustness of a mathematical model's solutions is essential to guarantee their validity and reliability under various conditions and slight variations of the scenarios. Robustness analysis seeks to evaluate how a model responds to parameter variations, initial conditions, or assumptions and how these uncertainty-related variations affect the accuracy and stability of the model's predictions.

In modeling energy systems, especially hybrids with renewable energies, there are several events whose sequence and degree of occurrence are difficult to predict. These events cause uncertainties that are often difficult to quantify. These factors relate to economic, political, technical, regulatory, environmental, resource availability, social, and climatic (regional and local) aspects [38]. This affects both expansion plans (long-term) and microgrid control models. Including these uncertainties, having correctly selected the intervals of the uncertainties [43], entails complexities in mathematical modeling and the solution algorithm in terms of calculation [41]. According to information in the literature, uncertainty analysis can be carried out by scenario analysis, probabilistic analysis, or sensitivity analysis [44, 45]. These uncertainty management techniques ensure that the proposed model is robust and flexible. However, the literature consulted shows that few studies perform uncertainty analysis of their models since most of them are based on specific case studies for their validation [46].

Battery energy storage is the predominant option for community and residential HRES. However, their limited lifespan compared to technologies such as pumped hydropower means costly periodic replacements.

The selection of an optimal HRES involves considering conflicting criteria. Based on the literature review, economic and technical aspects are the most prominent criteria. However, structuring and operating these systems based on these criteria alone can fail under specific conditions. Other social, environmental, and political criteria should be considered so that the multi-criteria decision-making framework can be used to structure and operate HRES alternatives [47]. Until now, little attention has been paid to specific methodologies for obtaining estimates of daily load profiles for local consumers [48]. To close this research gap, the development, implementation, and application of models that can be used to predict the daily energy consumption of consumers associated with the microgrid and uncertainty analysis are needed.

8.3.5 Renewable Resource Availability Forecasts

Forecast accuracy is critical in effectively controlling energy generation and demand in renewable resource management. Since factors as diverse as climatic conditions and human behavior significantly influence the generation and consumption of renewable energy, the uncertainty associated with these forecasts is considerable. Therefore, further research to refine the accuracy of these predictions becomes imperative.

Climatic conditions and human behavior strongly influence both renewable energy generation and energy demand, introducing a significant level of uncertainty into forecasts. Therefore, further research to improve prediction accuracy is vital.

In microgrids, research methodologies based on artificial intelligence control distributed energy resources (DERs), loads, weather forecasts, energy markets, and the availability of variable renewable energy resources. Although classic hierarchical control models have been implemented to manage aspects such as maximum power point tracking (MPPT), voltage and frequency regulation, power distribution, protection, fault recovery, and rapid communication, their results would be improved by using prediction models of the parameters that feed the models [49]. A tertiary-level control system could monitor and improve the system's resilience by managing energy, power flow, prosumer market share, customer segmentation, load and generation forecasts, and price prediction. Most control approaches employ neural network algorithms and reinforcement learning applications for decisions about energy distribution and trading, highlighting the importance of reducing uncertainties in predicting data behavior.

In this sense, regardless of the levels of real-time control systems used for hybrid power supply systems, prediction models are vital to help prepare the system to reduce adjustment times and potential system failures.

To address PV and wind generation, the power output between the lower and upper limit of the unit is calculated, constrained by the available forecast [50]. Long-term PV and wind generation forecasts are obtained from reliable national sources; however, due to the high uncertainties, prediction models could improve such predictions, including data compiled in real time close to the forecast point.

In the quest to optimize control in hybrid energy systems, selecting the most suitable architecture for the top level of the control system emerges as a central challenge. This level should optimally coordinate mixed energy resources, load demands, and interactions with grids to improve the overall operation of the energy system. Therefore, it is crucial to have robust prediction models that contribute to system readiness, thereby reducing set-up times and mitigating potential system failures in the dynamic environment of hybrid power supply systems.

8.3.6 Risk and Uncertainty Analysis

Risk analyses play a crucial role in assessing the feasibility and efficiency of integrated energy systems. Various studies have addressed this issue from different perspectives, highlighting the importance of considering the uncertainties and risks associated with integrating renewable energy sources.

Chakraborty et al. [49] examined the improvement of the hybrid thermal-wind-CAES system and its impact on the electricity market, considering the cost of the imbalance [49]. Their analysis revealed that considering the cost of the imbalance in the integration of wind farms significantly affects the system's profitability. In addition, economic risk analysis tools, such as Value-at-Risk (VaR) and Conditional Value-at-Risk (CVaR), were used to assess the system's viability.

Jian et al. [51] conducted a risk analysis of an integrated biomass and concentrated solar energy system using a downward risk restriction approach. This study showed that considering uncertainties in system operation is critical to maximizing system cost-effectiveness and security [51]. Lo et al. [52] reviewed the biomass supply chain, highlighting the need to assess uncertainties related to biomass availability, variations in demand, and changes in material prices to accurately evaluate the economic viability of biomass-based industries [52]. On the other hand, Petkov and Gabrielli [53] analyzed the implementation of Power-to-Hydrogen (PtH2) as seasonal energy storage in low-carbon multi-energy energy systems. Their study underscored the importance of considering uncertainty in the optimal design of energy systems, especially in selecting key technologies and parameters to minimize costs and CO2 emissions [53].

Rossi et al. [54] highlight that properly managing different energy vectors is crucial to achieving the best operational conditions for these communities in energy and economic terms. They also highlight the importance of energy planning tools to evaluate the performance and optimization of these systems, which is fundamental in the transition to cleaner and more sustainable systems [54].

On the other hand, Singh et al. [50] and Singh et al. [55] use optimization algorithms to optimize bidding strategies to maximize social welfare in energy systems integrated with wind energy. Economic risk analysis in hybrid wind power and hydro storage energy systems highlights the importance of minimizing economic and system security risks through metaheuristic algorithms [55].

In addition, Zhang et al. [56] explore how ports can reduce their carbon emissions holistically, highlighting the importance of alternative fuels, innovative technology, and optimization strategies to achieve low-carbon patterns. This study emphasizes the need to consider the characteristics of ports and quantify the properties of renewable energy sources to select the most suitable alternative energy sources, as well as the importance of coordinated optimization of multiple objectives to achieve sustainable development [56].

8.4 Conclusion

Integrating variable renewable energy sources into electricity systems demands sophisticated energy planning and management. Accurate prediction models and the implementation of robust control systems are critical to ensuring the stability and efficiency of these hybrid systems.

Considering uncertainties and risks inherent in renewable energy's variability is crucial to assessing projects' viability and profitability. Hybridization with storage technologies, especially pumped hydropower, emerges as a promising strategy to mitigate renewable sources' intermittency and improve energy systems' resilience. Optimizing hybrid systems requires comprehensively evaluating technical, economic, and environmental factors. Tools such as VaR and CVaR are valuable for quantifying the financial risks of investing in renewable energy.

Further research is needed to develop new storage technologies and improve simulation models for a more sustainable energy future. In addition, formulating clear and flexible energy policies will be essential to incentivize investment in hybridization and storage projects, thus accelerating the transition to cleaner and more resilient energy systems.

Acknowledgements The authors acknowledge the support provided by the Thematic Network 723RT0150, "Network for the large-scale integration of renewable energies in electricity systems (RIBIERSE-CYTED)." Fondocyt Grant funded this research No. (FONDOCYT-2023-1-3C1-0547) in the Dominican Republic.

References

1. J. Yang, J. Han, T. Wu, H. Zhang, L. Shang, Sizing and optimization of on-grid hybrid renewable energy systems considering hydroelectric energy storage. J. Intell. Fuzzy Syst. **40**(1), 1521–1536 (2021)
2. S. Emmanouil, E.I. Nikolopoulos, B. François, C. Brown, E.N. Anagnostou, Evaluating existing water supply reservoirs as small-scale pumped hydroelectric storage options—a case study in Connecticut. Energy **226**, 120354 (2021)
3. IRENA, Renewable power generation costs in 2022, International Renewable Energy Agency (IRENA), Abu Dhabi, (2023)
4. A. Chinnici, G.J. Nathan, B.B. Dally, Hybrid solar-MILD combustion for renewable energy generation. Front. Mech. Eng. **5**, 61 (2019)
5. G. Yang, H. Zhang, W. Wang, B. Liu, C. Lyu, D. Yang, Capacity optimization and economic analysis of PV–hydrogen hybrid systems with physical solar power curve modeling. Energy Convers. Manag. **288**, 117128 (2023)
6. J.D. Ogland-Hand et al., The value of CO2-bulk energy storage with wind in transmission-constrained electric power systems. Energy Convers. Manag. **228**, 1135485 (2021)
7. T.W. Brown, T. Bischof-Niemz, K. Blok, C. Breyer, H. Lund, B.V. Mathiesen, Response to burden of proof: a comprehensive review of the feasibility of 100% renewable-electricity systems. Renew. Sustain. Energy Rev. **92**, 834–847 (2018)
8. F. Piancó et al., Hydroelectric operation for hybridization with a floating photovoltaic plant: a case of study. Renew. Energy **201**, 85–95 (2022)
9. J.P. Hoffstaedt et al., Low-head pumped hydro storage: a review of applicable technologies for design, grid integration, control and modelling. Renew. Sustain. Energy Rev. 158. (2022)
10. B. Alqahtani, et al. Design and performance assessment of a pumped hydro power energy storage connected to a hybrid system of photovoltaics and wind turbines. Energy Convers. Manag. 293. (2023)
11. T.M. Adeyemi-Kayode et al., Impact analysis of renewable energy based generation in West Africa—a case study of Nigeria. Probl. Ekorozwoju **16**(1), 67–78 (2021)
12. A. Behzadi, S.M. Alirahmi, H. Yu, S. Sadrizadeh, An efficient renewable hybridization based on hydrogen storage for peak demand reduction: a rule-based energy control and optimization using machine learning techniques. J. Energy Storage **57**, 106168 (2023)
13. H. Allouhi, A. Allouhi, K.M. Almohammadi, A. Hamrani, A. Jamil, Hybrid renewable energy system for sustainable residential buildings based on solar dish stirling and wind turbine with hydrogen production. Energy Convers. Manag. **270**, 116261 (2022)
14. M.S. Javed, T. Ma, J. Jurasz, M.Y. Amin, Solar and wind power generation systems with pumped hydro storage: review and future perspectives. Renew. Energy **148**, 176–192 (2020)
15. M.R. Elkadeem, S. Wang, A.M. Azmy, E.G. Atiya, Z. Ullah, S.W. Sharshir, A systematic decision-making approach for planning and assessment of hybrid renewable energy-based microgrid with techno-economic optimization: a case study on an urban community in Egypt. Sustain. Cities Soc. **54**, 102013 (2020)
16. R.J. Mahfoud, N.F. Alkayem, Y. Zhang, Y. Zheng, Y. Sun, H.H. Alhelou, Optimal operation of pumped hydro storage-based energy systems: a compendium of current challenges and future perspectives. Renew. Sustain. Energy Rev. **178**, 113267 (2023)
17. Sandia.gov, DOE Global Energy Storage Database, (2022)
18. J.I. Pérez-Díaz, M. Chazarra, J. García-González, G. Cavazzini, A. Stoppato, Trends and challenges in the operation of pumped-storage hydropower plants. Renew. Sustain. Energy Rev. **44**, 767–784 (2015)
19. P. Denholm, E. Ela, B. Kirby, M. Milligan, Role of energy storage with renewable electricity generation, National Renewable Energy Lab. (NREL), Golden, CO, (2010)
20. EIA, Annual energy review 2008; Energy Information Administration, (2008)
21. R. Carapellucci, L. Giordano, A comparison between hydrogen and oxygen depolarized cathode for green hydrogen production from wind energy. Int. J. Hydrogen Energy **45**(15), 8429–8439 (2020)

22. D. Zhang et al., Recent developments in thermal energy storage systems for solar energy applications. Renew. Sustain. Energy Rev. **94**, 109–119 (2018)
23. G. Panayiotou et al., Renewable energy integration: system architectures, case studies and future trends. Renew. Sustain. Energy Rev. **74**, 504–513 (2017)
24. F. Kretschmer et al., Energy storage and sector coupling in a 100% renewable energy system. Energy **199**, 117–125 (2020)
25. T. Buckley, Electricity storage in utility grids, Energy, (2023)
26. H.W. Lund et al., Smart energy systems: a new generation of energy systems. Energy **202**, 117–132 (2020)
27. O.K. Ajiboye, C.V. Ochiegbu, E.A. Ofosu, S. Gyamfi, A review of hybrid renewable energies optimisation: design, methodologies, and criteria, (in English). Int. J. Sustain. Energy Rev. **42**(1), 648–684 (2023)
28. M.D.A. Al-falahi, S.D.G. Jayasinghe, H. Enshaei, A review on recent size optimization methodologies for standalone solar and wind hybrid renewable energy system. Energy Convers. Manag. **143**, 252–274 (2017)
29. X. Yin, Z. Zhao, W. Yang, Optimizing cleaner productions of sustainable energies: a co-design framework for complementary operations of offshore wind and pumped hydro-storages, (in English). J. Clean. Prod. **396**, 135832 (2023)
30. T.F. Agajie et al., Techno-economic analysis and optimization of hybrid renewable energy system with energy storage under two operational modes, (in English). Sustain. Switz. **15**, 11735 (2023)
31. O. Pedram, E. Asadi, B. Chenari, P. Moura, M. Gameiro da Silva, A review of methodologies for managing energy flexibility resources in buildings, (in English). Energ. Rev. **16**(17), 6111 (2023)
32. M.M. Rahman, A.O. Oni, E. Gemechu, A. Kumar, Assessment of energy storage technologies: a review, (in English). Energy Convers. Manag. Rev. **223**, 113295 (2020)
33. R. Siddaiah, R.P. Saini, A review on planning, configurations, modeling and optimization techniques of hybrid renewable energy systems for off grid applications. Renew. Sustain. Energy Rev. **58**, 376–396 (2016)
34. H. Rezk, G.M. Dousoky, Technical and economic analysis of different configurations of stand-alone hybrid renewable power systems—a case study. Renew. Sustain. Energy Rev. **62**, 941–953 (2016)
35. Y. Sawle, S.C. Gupta, A.K. Bohre, Review of hybrid renewable energy systems with comparative analysis of off-grid hybrid system. Renew. Sustain. Energy Rev. **81**, 2217–2235 (2018)
36. D. Guangqian, K. Bekhrad, P. Azarikhah, A. Maleki, A hybrid algorithm based optimization on modeling of grid independent biodiesel-based hybrid solar/wind systems. Renew. Energy **122**, 551–560 (2018)
37. T.M. Adeyemi-Kayode, S. Misra, R. Damaševičius, Impact analysis of renewable energy based generation in West Africa—a case study of Nigeria, (in English). Probl. Ekorozwoju **16**(1), 67–78 (2021)
38. O.M. Babatunde, J.L. Munda, Y. Hamam, A comprehensive state-of-the-art survey on hybrid renewable energy system operations and planning, (in English). IEEE Access **8**, 75313–75346 (2020)
39. J.P. Hoffstaedt et al., Low-head pumped hydro storage: a review of applicable technologies for design, grid integration, control and modelling", (in English). Renew. Sustain. Energy Rev. Rev. **158**, 112119 (2022)
40. S.M. Alawad, R.B. Mansour, F.A. Al-Sulaiman, S. Rehman, Renewable energy systems for water desalination applications: a comprehensive review, (in English). Energy Convers. Manag. Rev. **286**, 117035 (2023)
41. B.F. Silinto, C. van der Laag Yamu, C. Zuidema, A.P.C. Faaij, Hybrid renewable energy systems for rural electrification in developing countries: a review on energy system models and spatial explicit modelling tools. Renew. Sustain. Energy Rev. **207**, 114916 (2025)

42. B. Alqahtani, J. Yang, M.C. Paul, Design and performance assessment of a pumped hydro power energy storage connected to a hybrid system of photovoltaics and wind turbines, (in English). Energy Convers. Manag. **293**, 117444 (2023)
43. L. Stoyanov, I. Bachev, Z. Zarkov, V. Lazarov, G. Notton, Multivariate analysis of a wind–pv-based water pumping hybrid system for irrigation purposes, (in English). Energies **14**(11), 3231 (2021)
44. M.R. Elkadeem, S. Wang, A.M. Azmy, E.G. Atiya, Z. Ullah, S.W. Sharshir, A systematic decision-making approach for planning and assessment of hybrid renewable energy-based microgrid with techno-economic optimization: a case study on an urban community in Egypt, (in English). Sustain. Cities Soc. **54**, 102013 (2020)
45. S.L.Y. Lo, B.S. How, W.D. Leong, S.Y. Teng, M.A. Rhamdhani, J. Sunarso, Techno-economic analysis for biomass supply chain: a state-of-the-art review, (in English). Renew. Sustain. Energy Rev. Rev. **135**, 110164 (2021)
46. L. Canale, A.R. Di Fazio, M. Russo, A. Frattolillo, M. Dell'Isola, An overview on functional integration of hybrid renewable energy systems in multi-energy buildings, (in English). Energ. Rev. **14**(4), 1078 (2021)
47. M. Sharafi, T.Y. ElMekkawy, Stochastic optimization of hybrid renewable energy systems using sampling average method. Renew. Sustain. Energy Rev. **52**, 1668–1679 (2015)
48. V.R. Rajendran Pillai, R. Rajasekharan Nair Valsala, V. Raj, M.I. Petra, S.K. Krishnan Nair, S. Mathew, Exploring the potential of microgrids in the effective utilisation of renewable energy: a comprehensive analysis of evolving themes and future priorities using main path analysis, (in English). Des. Rev. **7**(3), 58 (2023)
49. M.R. Chakraborty, S. Dawn, P.K. Saha, J.B. Basu, T.S. Ustun, System profit improvement of a thermal–wind–CAES hybrid system considering imbalance cost in the electricity market, (in English). Energies **15**(24), 9457 (2022)
50. N.K. Singh, C. Koley, S. Gope, S. Dawn, T.S. Ustun, An economic risk analysis in wind and pumped hydro energy storage integrated power system using meta-heuristic algorithm, (in English). Sustain. Switz. **13**(24), 13542 (2021)
51. P. Jian, W. Hu, E. Valipour, S. Nojavan, Risk analysis of integrated biomass and concentrated solar system using downside risk constraint procedure, (in English). Sol. Energy **238**, 44–59 (2022)
52. S.L. Lo, Y. et al. Techno-economic analysis for biomass supply chain: a state-of-the-art review. Renew. Sustain. Energy Rev. 135. (2021)
53. I. Petkov, P. Gabrielli, Power-to-hydrogen as seasonal energy storage: an uncertainty analysis for optimal design of low-carbon multi-energy systems. Appl. Energy, 274. (2020)
54. M. Rossi, et al. Energy hub and micro-energy hub architecture in integrated local energy communities: enabling technologies and energy planning tools. Energies, 17(19). (2024)
55. N.K. Singh et al., Optimal bidding strategy for social welfare maximization in wind farm integrated deregulated power system using artificial gorilla troops optimizer algorithm. IEEE Access **10**, 71450–71461 (2022)
56. L. Zhang, et al. How to achieve comprehensive carbon emission reduction in ports? A systematic review. J. Marine Sci. Eng. 12(5). (2024)

Chapter 9
Deep Energy Retrofit of Office Buildings with the Aim of Reaching an NZEB Performance Level in a Tropical Context

René I. Ariza, Luis A. Martínez, Melissa Ruíz, Saúl Vela, and Javier Castillo

Abstract This paper explores deep energy retrofits for office buildings in El Salvador, aiming for net-zero energy performance while keeping acceptable indoor thermal comfort conditions for occupants. While much research focuses on new construction, retrofitting older buildings is crucial for sustainability, especially in tropical climates. Using energy modeling and optimization, the study defines a baseline office building and applies both passive and active energy-saving measures. Key strategies include improving ventilation, lighting, and HVAC systems and enhancing the building envelope through glazing and shading upgrades. These interventions reduced the building's energy consumption from 285.41 kWh/m2 to 58.87 kWh/m2 annually. A photovoltaic system was added to achieve a net-zero energy balance. The study provides actionable insights for sustainable retrofitting, demonstrating the potential for significant energy savings while maintaining indoor comfort in tropical climates.

Keywords Deep energy retrofit · Net zero energy · Tropical climate buildings

R. I. Ariza · L. A. Martínez (✉)
Department of Energy and Fluid Sciences, Central American University "José Simeón Cañas" Boulevard de Los Próceres, Antiguo Cuscatlán, La Libertad, El Salvador
e-mail: lamartinez@uca.edu.sv

M. Ruíz
Department of Spatial Organization, Central American University "José Simeón Cañas" Boulevard de Los Próceres, Antiguo Cuscatlán, La Libertad, El Salvador

S. Vela · J. Castillo
Faculty of Engineering, Don Bosco University, La Libertad, El Salvador

D. Rossit et al. (eds.), *Proceedings of the VII Ibero-American Congress of Smart Cities, ICSC-Cities 2024, 12-14 November, San Carlos, Costa Rica*, Springer Proceedings in Energy, https://doi.org/10.1007/978-981-96-4301-1_9

9.1 Introduction

Significant effort is being made to improve the sustainability of buildings. The environmental impact of buildings can be substantial, and the urgency to ensure that new buildings meet higher sustainability standards is understandable. The emphasis is often put on a building's energy balance, with the goal of achieving a net zero energy balance, understood as the possibility of buildings generating their energy needs on-site, which should be substantially reduced. In many climatic contexts, many academic publications have been produced in the last decades, exploring the design features and performance of high-performing buildings at the net zero energy level.

However, disproportionate literature is focused on achieving net zero energy performance for new construction, with limited publications for existing structures. Deep energy retrofits can transform an existing, inefficient building into a high-performing, net zero energy building. Arguably, turning most existing buildings into high-performing ones is a more pressing issue than ensuring that new constructions are built with that aim in mind. The absence of research on deep energy retrofits in specific contexts like El Salvador can hinder the much-needed transformation of standard buildings to more sustainable ones.

Research reviews present the state of the art of building energy retrofits. Illustratively, Al Mindeel et al. [1] reviewed eighty-two research documents on the topic, published between 2013 and 2022, and found the following: (1) Most studies use one or more of these objective functions: energy consumption, thermal comfort, and air quality; (2) Most studies focus on residential and office buildings, (3) Coupling simulation tools with optimization algorithms is the approach most applied, with the genetic algorithm being used more often.

Harkouss et al. [2] also conducted a literature review on building energy retrofits, focusing on optimization methodologies for various climate zones; their results provided guidelines on how to achieve the net zero energy balance in each climate zone, and flowcharts to summarize a three-stage procedure to reach net zero energy performance. The previous authors indicated that building envelope features are least explored in the literature; in reality, changing the materials and envelope systems of an existing building is complicated. However, Kamel et al. [3] produced a review paper focusing on residential building retrofit options based on building envelopes using conventional and advanced material solutions. They found that building envelope-focused retrofits are more effective in cold climate zones. Several studies have explored specific cases where a methodology that often involves energy modeling and optimization leads to the best retrofit options.

We provide some notable examples as follows. Hamza et al. [4] modeled twelve office buildings in Cairo, Egypt, studying the effects of retrofit measures, which decreased energy consumption between 46 and 65%. Albatayneh [5] developed modeling and sensitivity analysis for a building in sub humid Mediterranean climate. In a representative study in the tropics, Costa et al. [6] studied 240 office buildings in Brasilia, Brazil, performing a morphological characterization and energy modeling aimed at retrofit. Results yielded a 46% reduction in total energy consumption, concluding that it is possible to achieve net zero energy performance in office buildings of up to four stories in their context. Similar studies have been conducted in Panama City [7, 8] for office and residential buildings.

As mentioned, many studies focus on determining the best strategies for retrofitting buildings from the perspective of energy consumption metrics, applying a multi-variable optimization. However, a growing body of literature focuses more specifically on single elements of the building. For example, Abdou et al. [9] conducted a retrofit particular to the west fenestration facade of a university building in UAE. Three fundamental parameters were identified (window-to-wall ratio, glazing type, and external shading). A similar approach was followed by Al-Tamini [10], who conducted calibrated simulations of office buildings in Saudi Arabia, coupled with parametric analysis, to retrofit the building envelope system. Additionally, Arauz et al. [11] and Mirrahimi et al. [12] also conducted investigations on building envelope retrofits in tropical climates.

Other retrofit studies focus on different areas of interest. For instance, Sofías et al. [13] focused their efforts on retrofitting the lighting performance, considering shading effects using CAD tools. Many other publications explore retrofit from specific contexts, with some variations in their approach [14, 15]. Recently, a new perspective has been researched: the life cycle effect of a retrofit. Fahlstedt et al. [16] reviewed fifty-four research articles published between 2005 and 2022, finding that retrofit options are limited due to restrictions set by decisions made decades ago, i.e. some changes cannot be made afterward.

In this paper, we explore the viability of deep energy retrofits using energy modeling and optimization techniques in the context of office buildings in El Salvador. First, we define the deep energy retrofit methodology and base case model building for the Metropolitan Area of San Salvador (AMSS) and then successively and iteratively apply energy efficiency measures to improve the building energy performance until the net zero energy balance is accomplished while keeping the indoor thermal comfort acceptable for occupants. By doing this, we believe it is possible to identify the most successful retrofit strategies in this scenario, which can inform building operators about the interventions that can improve existing office buildings sustainability.

9.2 Methodology

9.2.1 Characterization of the Office Building for the AMSS

In El Salvador, there is currently no updated census characterizing national buildings, with the most recent data coming from the 2007 census, which only details construction materials of residential buildings. The capital, San Salvador, and its Metropolitan Area, also known as AMSS, is the most densely populated area in the

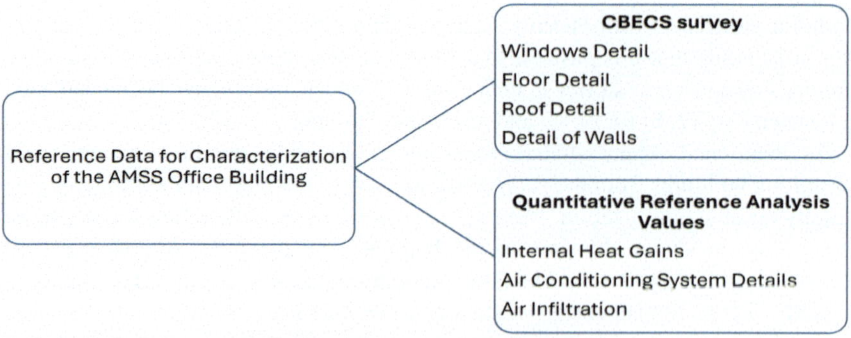

Fig. 9.1 Origin of the values used in the energy analysis of the base model

country, having a large number of old buildings, which are of interest for future energy retrofits, especially if incentives are available in the future. In addition, literature on user habits or energy profiles of buildings, which is essential for creating a baseline energy simulation, is lacking [17].

Due to the lack of data on local buildings and energy profiles, this research defined a base case office building characterized in Fig. 9.1. The "Commercial Buildings Energy Consumption Survey" (CBESC) developed by the U.S. Energy Information Administration in [18] was adapted to qualitatively characterize buildings in El Salvador. This survey, modified to exclude locally irrelevant information such as heating or propane gas use, provided qualitative insights into energy types and construction characteristics, though it only offers data on annual total fuel consumption.

The survey was administered to 38 local construction professionals in the AMSS, selected for their experience in designing, constructing, or maintaining office buildings. Selection criteria included specialized knowledge of materials, practical experience in local projects, and the variety of projects they had worked on. This survey provides an updated view of trends and preferences in building materials, facilitating informed decision-making tailored to El Salvador's specific geographic, climatic, and economic conditions.

For accurate energy modeling with quantitative characteristics in DesignBuilder v7, CBESC data was supplemented with additional aspects based on research by Al Saadi and Al Jabri [19] and an energy analysis of a building in Chihuahua, Mexico [20] adapting these values with international energy efficiency standards.

9.2.2 Retrofit Methodology

The energy retrofit methodology applied to the case study, which was an energy model of the baseline office building simulated in DesignBuilder v7 and is summarized in Fig. 9.2, following the process proposed by Chen et al. [8]. The methodology employs parametric simulations, sensitivity analysis, and optimizations to enhance the building and reduce the number of iterations required, thereby decreasing simulation times.

The retrofit methodology applied in this work began with the existing office building. In step 1, the passive strategy was improved by using natural ventilation instead of continuous air conditioning, switching to mixed ventilation. In step 2, the efficiency of lighting systems and miscellaneous loads was optimized based on standards such as ASHRAE 90.1 (2019), the "Advanced Energy Design Guide for Small to Medium Office Buildings" [21] and the IESNA Lighting Handbook [22]. In step 3, a partial parametric analysis of glazing options and shading devices was performed. In step 4, the efficiency of the HVAC system was improved to ensure it was not oversized after significantly reducing the thermal load with the previous steps. Finally, in step 5, a sensitivity analysis was conducted to identify the most suitable simulation variables for optimizing the building envelope and the air conditioning setpoint.

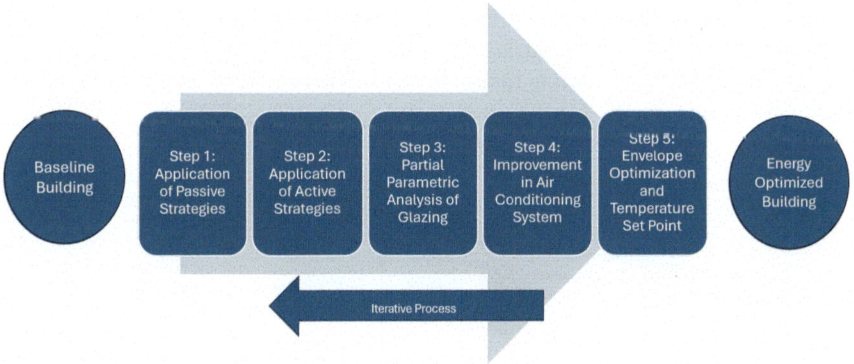

Fig. 9.2 Workflow for energy retrofit

9.3 Results

9.3.1 *Baseline Typical AMSS Office Building*

The survey conducted enabled an initial characterization of the building, shown in Fig. 9.3 as the first result of this work, identifying it as an administrative office and providing geometric and occupancy information, with the main values summarized in Table 9.1.

The values for electricity consumption and Energy Use Intensity (EUI) were taken from an administrative office model developed by SolArq in Chihuahua, Mexico [20], due to its similarity to the building studied in the AMSS. Load values such as Lighting Power Density (LPD) and miscellaneous loads (Power Density, PD) are used to define the building's baseline consumption, but they were adjusted during the retrofit to meet energy efficiency standards. A COP of 1 was used, which, from the authors' experience and the professionals consulted, aims to represent a low-efficiency system with many years of operation, as is the case for many office buildings in the AMSS. This low-efficiency baseline could reveal potential savings when adopting energy efficiency standards.

The building envelope was modeled using materials identified in the survey, focusing on its thermal rather than structural performance, with a roof inclined 8% toward the south. The envelope's thermal transmittance values (U-values), listed in Table 9.2, include additional layers in the roof and floor to simulate internal and external finishes. According to the "Advanced Energy Retrofit Guide for Office Buildings" by the Pacific Northwest National Laboratory [23], a deep retrofit first

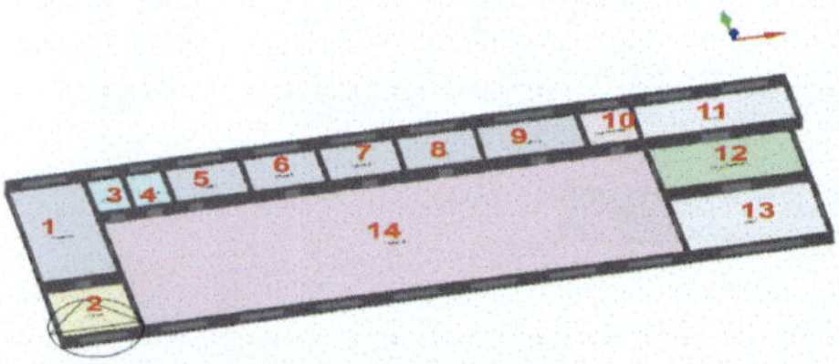

Number	Area
1	General Office
2	Cafeteria
3	Bathroom 1
4	Bathroom 2
5	Office 5
6	Office 4
7	Office 3
8	Office 2
9	Office 1
10	Photocopier Room
11	Stairs
12	Management Office
13	Meeting Room
14	Open Offices/ Reception (1st Floor)

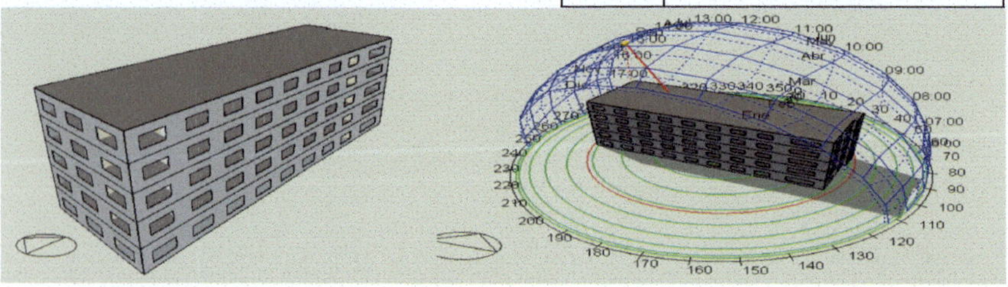

Fig. 9.3 Characterized baseline office building

Table 9.1 Baseline building general characteristics

Building feature	Detail
Location	San salvador
Built area	4 247.01 m^2
Roof slope	8%
Building footprint shape	Wide rectangle
Facade orientation	Facing north
% of Window/door glass area	25%
Building levels	5
Typical floor-to-ceiling height	4.00 m
Window glass type	Single-pane glass
Most used lighting type in the building	LED
Occupancy control installed for lighting	Yes
Activity type	Administrative office
LPD	7.5 W/m^2
Minimum lighting required	400 Lux
PD	11.77 W/m^2
Cooling temperature setpoint	22 °C
Cooling temperature setback	28 °C
COP	1

Table 9.2 Thermal characteristics of the baseline building envelope

Building element	Thermal transmittance value (W/m2 K)	Solar heat gain coefficient
External walls	4.142	–
Roof	1.238	–
Internal floor	1.302	–
External floor	0.259	–
Windows glass	3.835	0.768
Windows frame	5.881	–

requires reducing infiltration before adjusting the U-value with exterior insulation, which will optimize the building's performance.

9.3.2 Baseline Energy Consumption

The Meteonorm 7 weather station model was used to perform the energy calculations for San Salvador's climate, but the data was modified based on percentiles (0.4%, 1%, 2%) to improve accuracy. The model identified that the area belongs to

ASHRAE climate zone 1A and Koppen classification Aw, allowing for the use of values suggested by ASHRAE 90.1 standards for that zone.

In addition to the climatic data used to define the building's climate boundary, the following considerations were taken into account in the software model:

- Natural ventilation was set to the "calculated" option, which estimates airflows at each building opening using the airflow network method.
- Solar gains were modeled as complete on the exterior.
- The "Conduction Transfer Function (CTF)" was used for heat transfer, without considering the movement of moisture and its impact on the thermal properties of the material.
- The building was assumed to have good airtightness, with a constant infiltration rate of 0.25 ac/h.
- Thermal bridges were not considered.
- The thermal behavior of the soil was modeled, but its influence was excluded from subsequent analyses because, in tropical climates, the soil acts as a heat sink, showing an inverse relationship between its conductivity and the building's energy consumption.

The 5-story building's annual energy consumption is 1,212,139.12 kWh, equivalent to 285.41 kWh/m2 per year, exceeding by 286.64% the ASHRAE [21] recommended value for net-zero energy buildings in climate zone 1A (73.8174 kWh/m2 per year). Figure 9.4 shows the building's annual energy consumption and the behavior of miscellaneous, lighting, and cooling systems.

The base building has its highest energy load in the cooling system, representing 72% of the annual load, while lighting accounts for 10% and miscellaneous loads for 17%. This behavior is similar to that described by Yan et al. [24] for an office

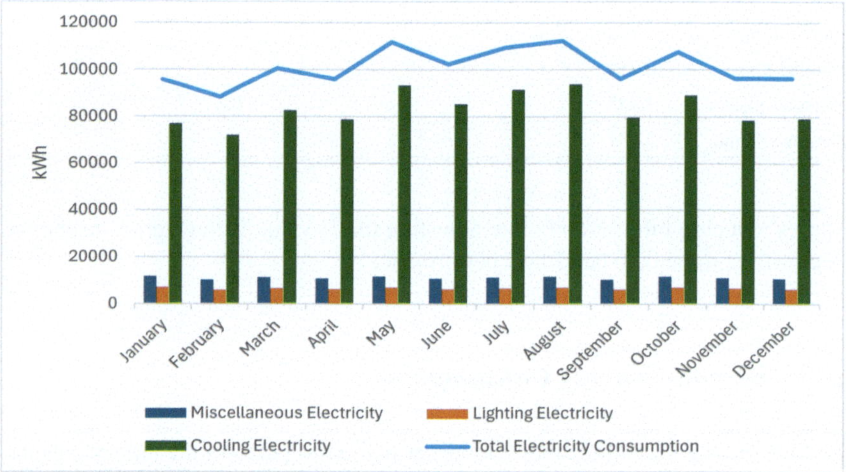

Fig. 9.4 Monthly energy consumption of the baseline office building

building in Singapore, also located in ASHRAE climate zone 1A, where cooling and ventilation account for 70% of the energy load. The thermal discomfort inside the building totals 1,019.62 h per year (11.64%), slightly above the 10% allowed by ASHRAE 55 standards. From May to August, energy demand increases due to higher cooling loads and high outdoor air temperatures, which reduce discomfort in these months due to the greater use of the air conditioning system.

9.3.3 Application of Retrofit Measures

Figure 9.5 shows the impact on energy consumption at each step of the retrofit, highlighting a slight increase between steps 4 and 5 to improve comfort without compromising the energy balance. The retrofit steps were applied, and the building's energy use decreased from 285.41 kWh/m2 per year to 58.87 kWh/m2 per year at the end of the simulated retrofit process.

1. **Application of Passive Strategies:** The building was configured in Design-Builder to operate in mixed mode, combining natural ventilation and air conditioning when conditions allow. The results shown in Fig. 9.5 indicate that this configuration reduces energy consumption by 9.46% but increases discomfort hours to 1,992.89 (22.75% annually), exceeding the ASHRAE 55 [25] standard which is 20% for mixed ventilation buildings. In step 5 of the retrofit, the temperature set point was adjusted to reduce discomfort hours to acceptable levels.
2. **Application of Active Strategies:** The improvement of active strategies in the simulations involved replacing lighting fixtures and electrical equipment with more efficient versions, reducing cooling, lighting, and miscellaneous loads, and decreasing the building's overall energy consumption. Figure 9.5 summarizes that after applying step 2 of the retrofit methodology, energy consumption was reduced by 24.3%, from 258.4 kWh/m2 to 195.62 kWh/m2. The new model configurations are shown in Table 9.3 and are based on the IESNA Lighting Handbook [22], using the "Space by Space" method for light levels and lighting power density, and the ASHRAE Handbook 2021 [26] for power density.

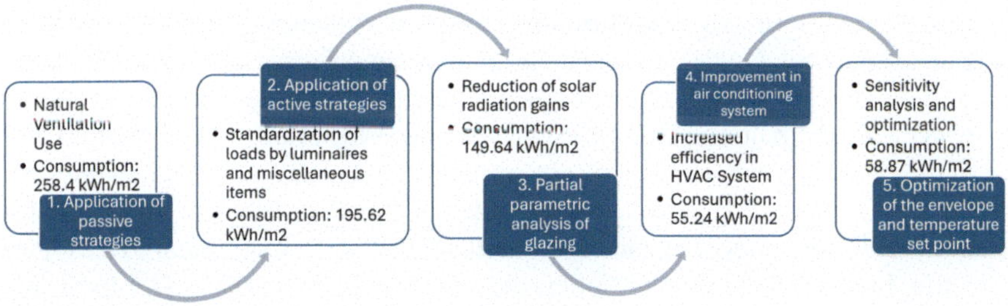

Fig. 9.5 Application of retrofit measures

Table 9.3 Standardized values of energy consumption for lighting and miscellaneous loads

Zone	Base case illuminance (Lux)	Illuminance according to IESNA (Lux)	Lighting power density (W/m^2)	Lighting power density according to IESNA (W/m^2)	Base case power density (W/m^2)	Power density according to ASHRAE (W/m^2)
General office	400	300	7.5	4.52	11.77	5.83
Reception	400	200	7.5	8.29	11.77	3.67
Meeting room	400	300	7.5	8.29	11.77	3.67
Cafeteria	400	200	7.5	5.27	11.77	NA
Lobby	400	200	7.5	3.66	11.77	NA
Machine room	400	200	7.5	4.52	11.77	NA
Stairs	400	50	7.5	5.27	11.77	NA
Office 5	400	300	7.5	4.52	11.77	5.83
Office 4	400	300	7.5	4.52	11.77	5.83
Office 3	400	300	7.5	4.52	11.77	5.83
Office 2	400	300	7.5	4.52	11.77	5.83
Office 1	400	300	7.5	4.52	11.77	5.83
Restroom 1	400	100	7.5	5.49	11.77	NA
Restroom 2	400	100	7.5	5.49	11.77	NA
Management	400	300	7.5	4.52	11.77	5.83
Open Offices	400	300	7.5	3.33	11.77	3.67

1. **Partial Parametric Analysis of Glazing:** The parametric variables were tested for optimizing glazing; including six types of glazing, seven exterior curtains, six shades, and eight lateral overhangs, selecting only the most efficient option from each group without combining variables. The best outcome was achieved with double-layer absorbent reflective glazing, As30 adjustable-angle shades, and 100 cm lateral overhangs, while exterior curtains were discarded due to lower efficiency. After implementing these improvements, the building's energy consumption was reduced by 23.5%, decreasing from 195.62 kWh/m2 to 149.64 kWh/m2 annually, significantly enhancing its performance.
2. **Improvement in Air Conditioning System:** According to the "Advanced Energy Retrofit Guide" [23], HVAC system improvements should be made after reducing lighting, miscellaneous loads, and applying natural ventilation to decrease the required HVAC system size. After lowering these loads, the COP of the HVAC system was improved to a value of 4, based on the "Achieving Zero Energy: Advanced Energy Design Guide for Small to Medium Office Buildings," [21] which recommends a minimum SEER of 15 (equivalent to a COP of 3.61). Figure 9.5 shows a 63% improvement compared to the model from step 3, achieving an energy consumption of 55.24 kWh/m2 per year, exceeding the

performance of a net-zero energy office building in climate zone 1A (73.82 kWh/m2 per year).

Optimization of the Envelope and Temperature Setpoint: The sensitivity analysis examined the influence of discrete variables shown in Table 9.4 on thermal discomfort and total energy consumption. The DesignBuilder sensitivity analysis module was used, and 200 samples (simulations) were taken for the analysis. It is recommended to use at least 10 times the number of independent variables in the model; however, more data were included to improve the model's power. The Latin hypercube sampling (LHS) method was selected for sampling. Based on the generated simulations, a multiple regression was performed to evaluate the influence of the variables on the results. The quality of the model was evaluated using the coefficient of determination R2 and the study of the error (residuals). Two models were studied: the first focused on assessing the total energy consumption of the building, and the second focused on the discomfort level. The model results were analyzed by using the statistical software SPSS®. To assess how well the model explains the dependent variable, an ANOVA analysis was performed, besides the determination coefficient analysis (R2).

The variable found to have a statistically significant influence on both models is the setpoint temperature, at a 95% confidence level. The results of the sensitivity analysis for each model are summarized in Table 9.5. For model 1, the full model can

Table 9.4 Variables used in sensitivity and optimization analysis

Variable category	Variable	Units	Distribution category	Distribution curve	Distribution parameters	Variable type
Independent	Roof insulation	W/ m^2-K	Discrete	Uniform (discrete)	• Energy code (U = 0.34) • Typical Ref. (U = 0.346) • Uninsulated (U = 1.54)	Nominal
	Walls insulation	W/ m^2-K	Discrete	Uniform (discrete)	•Energy code (U = 0.251) • Typical Ref (U = 0.351) • Uninsolated (U = 2.071)	Nominal
	Thermal mass	–	Discrete	Uniform (discrete)	• Light • Medium	Nominal
	Temperature setpoint	°C	Continuous	Normal curve	Min: 16 Max: 28	Scalar
Dependent	Discomfort ASHRAE 55 (all clo)	h	Continuous	Normal curve	Estimated	Scalar
	Total site energy	kWh/ m^2	Continuous	Normal curve	Estimated	Scalar

account for 96.43% of the total variability in the data at a 95% confidence level. It was found that, of all the independent variables, only the setpoint temperature significantly influenced the total consumption (p-value < 0.05); the rest of the variables, through the t-test, were found to have no significant influence on the dependent variable. An additional ANOVA analysis was conducted on the model to identify if any of the levels within the categorical variables had an influence on energy consumption, but no statistically significant relationships were found either. Model 2 can account for 72.82% of the total variability in the data, implying that discomfort can only be partially explained, either because more data is needed or because other variables influence discomfort but were not considered in the model. According to the regression coefficients, only the setpoint temperature and the intercept were estimated to be significant for the model. An ANOVA analysis was conducted to disseminate the influence of the different levels of the categorical variables on discomfort, but again, no significant relationship was found for a 95% confidence level.

Even if the cooling set point temperature had the most statistically significant influence in the sensitivity analysis, showing that is the most significant variable for both dependent variables, an optimization analysis was performed in DesignBuilder using the genetic algorithm method, finding the optimal values shown in Table 9.6, these values made the simulated building achieve the total energy site target and thermal discomfort target of annual discomfort hours.

The impact of each retrofit stage on the building's energy consumption and the cumulative savings from their application are shown in Fig. 9.6. A slight increase

Table 9.5 Results of the sensitivity analysis of each model

Variable	Model 1: total energy site		Model 2: thermal disconfort	
	Regression coeficient	p-value	Regression coeficient	p-value
Intercept	9085.5259	0.0000	− 7.2396	0.0004
Cooling set-point temperature (°C)	− 139.9816	0.0000	1.8517	0.0000
Thermal mass (No Units)	1.3983	0.7621	0.2339	0.2289
Insulation-walls (W/m^2-K)	− 1.9688	0.5505	− 0.0918	0.5090
Insulation-roof (W/m^2-K)	− 4.5446	0.1679	0.0817	0.5543

Table 9.6 Optimized building envelope variables

Building element	Optimized value
External walls insulation (W/m^2 K)	0.351
Roof insulation (W/m^2 K)	0.341
Thermal mass	Medium
Cooling temperature setpoint (°C)	24.2

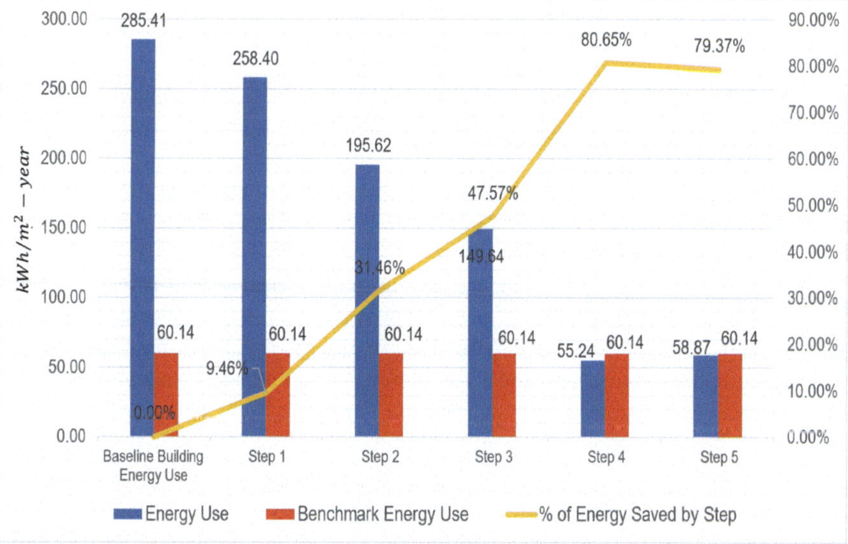

Fig. 9.6 Changes in the energy use of the building during the retrofit process

in consumption is observed between steps 4 and 5 due to improvements in indoor comfort while maintaining the annual energy balance within the net-zero energy limit. The steps with the most significant impact on reducing energy consumption were step 2, related to the efficiency of lighting and miscellaneous loads, and step 4, which involved optimizing the air conditioning system. The saved energy percentage is shown; it represents the energy reductions step by step, starting from the baseline energy consumption, showing the total energy saved after the application of the whole retrofit, totaling an energy use reduction of 79.37% in step 5.

9.3.4 Energy Balance of the Optimized Building

After the energy retrofit, the building is projected to consume 58.87 kWh/m2 annually, while its rooftop photovoltaic system will generate 60.14 kWh/m2 per year, resulting in a net surplus of 1.28 kWh/m2 annually. Achieving net-zero energy requires optimizing the building's energy consumption and adjusting the photovoltaic system to match the energy demand. The calculated power for the photovoltaic system was 162.15 kWp, commonly referred to as peak power in the photovoltaic field. Although control and automation were not included in the analysis because the building already had them, they are recommended in other cases.

9.4 Conclusions

The optimization process must consider various elements, as they can affect thermal comfort when reducing the energy consumption of climate control systems. As shown in step 4 of the retrofit, reducing the air conditioning size decreased energy consumption but increased discomfort inside the building. Optimization allows temperature set points to be identified for acceptable comfort levels and control energy consumption to preserve the net zero energy definition.

The first measure that should be carried out in an energy retrofit is the reduction of equipment consumption. This conclusion was reached in the analyzed case after applying step 2 of the retrofit, which involved reducing miscellaneous load densities and lighting consumption. This measure alone achieved an energy savings of 22%.

In the analyzed case, energy consumption is more influenced by the air conditioning temperature set point than wall or roof insulation, according to the sensitivity analysis conducted in step 5 of the retrofit. The combination of measures such as glass replacement and the use of exterior blinds and overhangs can reduce energy consumption by 16.11%. In comparison, natural ventilation can save up to 9.46% annually, although with a possible increase in discomfort hours.

A deeper analysis of occupant behavior and operational strategies could be integrated with the proposed retrofit methodology to evaluate the impact of retrofit strategies across multiple scenarios.

It is recommended that retrofit strategies include cost–benefit evaluations and analyses of their impact on the building's long-term performance, maintenance requirements, and life cycle costs.

Acknowledgements This publication takes elements from a Master's thesis developed at Universidad Don Bosco. Software provided by Universidad Centroamericana "José Simeón Cañas" was used for the modeling. We are grateful to the professionals who provided feedback through the survey. Also, we appreciate the support from the RIBIERSE-CYTED network in funding this submission. We also thank the TRAPECIO-CYTED network for their support.

The authors have no competing interests.

References

1. T. Al Mindeel, E. Spentzou, M. Eftekhari, Energy, thermal comfort, and indoor air quality: multi-objective optimization review, (2024) https://doi.org/10.1016/j.rser.2024.114682
2. F. Harkouss, F. Fardoun, P.H. Biwole, Optimization approaches and climates investigations in NZEB—A review, (2018). https://doi.org/10.1007/s12273-018-0448-6
3. E. Kamel, A.M. Memari, Residential building envelope energy retrofit methods, simulation tools, and example projects: a review of the literature, (2022). https://doi.org/10.3390/buildings12070954
4. M. Hamza, S. Alsaadani, M. Fahmy, Exploring the potential of nearly zero energy retrofitting for generic office buildings in Cairo Egypt. Energy Rep. **8**, 116–122 (2022). https://doi.org/10.1016/j.egyr.2022.01.089

5. A. Albatayneh, Sensitivity analysis optimisation of building envelope parameters in a sub-humid Mediterranean climate zone. Energy Explor. Exploit. **39**, 2080–2102 (2021). https://doi.org/10.1177/01445987211020432
6. J.F. Walter Costa, C.N. David Amorim, J.C. Ribeiro Silva, Retrofit guidelines towards the achievement of net zero energy buildings for office buildings in Brasilia. J. Build. Eng. 32, (2020). https://doi.org/10.1016/j.jobe.2020.101680
7. M.C. Austin, C. Carpino, D. Mora, N. Arcuri, A Methodology to identify appropriate refurbishment strategies towards zero energy buildings in a hot and humid climate. In: Journal of Physics: Conference Series. Institute of Physics (2022). https://doi.org/10.1088/1742-6596/2385/1/012020
8. C. Carpino, M. Chen Austin, D. Mora, N. Arcuri, Retrofit measures for achieving NZE single-family houses in a tropical climate via multi-objective optimization. Buildings. 14, (2024). https://doi.org/10.3390/buildings14030566
9. Y. Abdou, Y.K. Kim, A. Abdou, R. Anabtawi, Energy optimization for fenestration design: evidence-based retrofitting solution for office buildings in the UAE. Buildings. 12, (2022). https://doi.org/10.3390/buildings12101541
10. N. Al-Tamimi, Building envelope retrofitting strategies for energy-efficient office buildings in Saudi Arabia. Buildings. 12, (2022). https://doi.org/10.3390/buildings12111900
11. J. Araúz, D. Mora, M. Chen Austin, Assessment of different envelope configurations via optimization analysis and thermal performance indicators: a case study in a tropical climate. Sustainability. **14**, 2013 (2022)
12. S. Mirrahimi, M.F. Mohamed, L.C. Haw, N.L.N. Ibrahim, W.F.M. Yusoff, A. Aflaki, The effect of building envelope on the thermal comfort and energy saving for high-rise buildings in hot–humid climate. Renew. Sustain. Energy Rev. **53**, 1508–1519 (2016)
13. K. Sofias, Z. Kanetaki, C. Stergiou, S. Jacques, Combining CAD modeling and simulation of energy performance data for the retrofit of public buildings. Sustain. Switz. 15, (2023). https://doi.org/10.3390/su15032211
14. F. Ascione, N. Bianco, T. Iovane, G.M. Mauro, D.F. Napolitano, A. Ruggiano, L. Viscido, A real industrial building: modeling, calibration and pareto optimization of energy retrofit. J. Build. Eng. 29, (2020). https://doi.org/10.1016/j.jobe.2020.101186
15. E. Liapopoulou, T. Theodosiou, Energy performance analysis and low carbon retrofit solutions for residential buildings. In: IOP Conference Series: Earth and Environmental Science. Institute of Physics Publishing (2020). https://doi.org/10.1088/1755-1315/410/1/012026
16. O. Fahlstedt, F.N. Rasmussen, A. Temeljotov-Salaj, L. Huang, R.A. Bohne, Building renovations and life cycle assessment—a scoping literature review, (2024). https://doi.org/10.1016/j.rser.2024.114774
17. J. González, D. Mora, M. Chen Austin, Energy consumption difference found between typical and standard occupancy in residential buildings in a tropical developing country. Buildings. 13, (2023). https://doi.org/10.3390/buildings13092235
18. U.S. Energy Information Administration: Commercial Buildings Energy Consumption Survey (CBECS), (2018)
19. S.N. Al-Saadi, K.S. Al-Jabri, Optimization of envelope design for housing in hot climates using a genetic algorithm (GA) computational approach. J. Build. Eng. 32, (2020). https://doi.org/10.1016/j.jobe.2020.101712
20. Arturo Ordoñez, Design Builder Básico: Ejercicio 03-Cálculos y Simulaciones, (2023)
21. ASHRAE, Advanced energy design guide for small to medium office buildings: achieving zero energy. ASHRAE, ATLANTA. (2019)
22. M.S. Rea, *IESNA Lighting Handbook: Reference and Application* (Illuminating Engineering Society, Canada, 2000)
23. Pacific Northwest National Laboratory, Advanced Energy Retrofit Guide Office Buildings. U.S. Department of Energy. (2011)
24. K. Yan, A. Chong, Y. Mo, Generative adversarial network for fault detection diagnosis of chillers. Build Environ. 172, (2020). https://doi.org/10.1016/j.buildenv.2020.106698
25. R. American Society of Heating, A.C.I.: ASHRAE STANDARD 55–2010. (2010)
26. R. American Society of Heating, A.C. E.Inc. (ASHRAE), 2021 ASHRAE® Handbook—Fundamentals (SI Edition). (2021)

Chapter 10
Evaluating the Social Impact of Electrical Vehicle Towards a Sustainability Transition: A Critical Analysis of Tesla, BYD, and EU Environment Reports

Diogo Santos Cardoso da Silva, Gabriel do Nascimento Barboza, Pedro Henrique González, and Vanessa de Almeida Guimarães

Abstract A great part of the literature advocates that the electric vehicle is one of the solutions towards more sustainable development in a low-carbon era. Nevertheless, apart from the environmental issues caused by the final disposal of the battery at the end of the EVs useful life, it is important to discuss the social impacts caused by the mining to produce these batteries. Therefore, the main goal of this paper is to evaluate if and how the main players in the electrical vehicle industry (BYD and TESLA) as well as the European Union are addressing the social impacts caused by cobalt mining, with emphasis on the exploitation that take place on Democratic Republic of Congo (DCR). Content research, as in Bardin, was performed in the 2023 sustainable reports published by BYD and Tesla (private sector perspective), and the same methodological approach was adopted to analyze the UE reports (governmental point of view). Strategies to relieve the identified social burden were also gathered from the selected reports. As the main results, we identified that the efforts are still preliminary and do not completely address the negative impacts in DRC.

Keywords Electric vehicle · Social impact · Cobalt mining · Sustainability · Democratic Republic of Congo

D. S. C. da Silva · G. do Nascimento Barboza · V. de Almeida Guimarães (✉)
COPPEAD School of Business, Federal University of Rio de Janeiro, Rio de Janeiro, Brazil
e-mail: vanessa.guimaraes@coppead.ufrj.br

P. H. González
System Engineering and Computer Science Program, Alberto Luiz Coimbra Institute for Graduate Studies and Research Engineering, Federal University of Rio de Janeiro, Rio de Janeiro, Brazil

D. Rossit et al. (eds.), *Proceedings of the VII Ibero-American Congress of Smart Cities, ICSC-Cities 2024, 12-14 November, San Carlos, Costa Rica*, Springer Proceedings in Energy, https://doi.org/10.1007/978-981-96-4301-1_10

10.1 Introduction

Following the COP21 agreement in 2015, numerous nations have committed to exploring innovative methods to curb carbon dioxide (CO_2) emissions. Given the significant contribution of the transportation sector to CO_2 emissions, transitioning from fuel engines to electric engines has become a primary objective to mitigate air pollution [1]. Several automotive firms have shifted their research and development focus toward enhancing battery efficiency and durability. This strategic move aims to address key customer concerns regarding the range and charging speed of electric vehicles, which have been significant barriers to widespread adoption [2].

The electric vehicle (EV) boom requires a shift for the automobile manufacturers. Unlike combustion engines, EV demands various minerals, such as lithium, cobalt, nickel, and manganese, for producing lithium-ion batteries (LIB). This means rethinking their supply chains and finding new partners to source these essential minerals. However, ethical sourcing and geographic concentration of supply present challenges. EV production relies upon a "new" supply chain that demands "new" critical raw materials, some of which are extracted from areas suffering political instability, labor rights insecurity, and other problems [3, 4]. One such critical raw materials is cobalt, with more than 50% of the world's supply coming from Democratic Republic of Congo (DRC) [5], while most refinement is conducted in China [6].

It brings the following research question: do EV key players acknowledge and act effectively upon the impact of this new supply chain? Therefore, the main goal of this paper is to identify if and how the main players in the electrical vehicle industry (BYD and TESLA) as well as the European Union (EU) are addressing the social impacts caused by the cobalt mining, with emphasis on the exploitation that take place on DRC.

Although China has the most significant EV market share, most information is not available in English. The EU is the second largest market, with the US following in third [7]. In addition, EU is known for its Green Deal, is committed with the sustainable development and is recognized as a standard-setter. It helps to justify the analysis of the EU reports as a governmental part. From the private sector perspective, Tesla holds the first position among manufacturers of battery electric vehicles (BEV), followed closely by BYD [7]. Since 45% of the global cobalt is used to produce batteries for EVs and 26% for portable batteries [8], the focus of the paper relies on the EV sector, even though all the sectors should be held accountable.

From this Introduction, this paper is divided into five sections: a literature overview of the cobalt exploitation in Congo (Sect. 10.2), followed by the methodological procedures (Sect. 10.3). Section 10.4 presents the main findings while recommendations are proposed in Sect. 10.5. Finally, there are the final considerations.

10.2 The Cobalt Exploitation in Congo (DRC)

China's dominance in controlling LIB-related minerals has raised concerns for Western countries, pressuring them to prioritize the acquisition of these minerals, as they will be strategic resources to meet the future growing demand for batteries. As a result, a frenetic run to acquire mines caused much damage to the countries that owned those resources. Instead of mutually developing their economies, many countries have experienced the opposite. Nations have been subjected to extreme exploitation, resulting in significant distress for both their inhabitants and the environment.

Regarding cobalt mining, the Democratic Republic of Congo (DCR) stands out. Situated in the African continent, this country provides over 50% of the world's cobalt production [9]. However, this mining boom did not result in a positive development for the Congolese people [10]. More than 63% of Congolese citizens live below the national poverty line of less than $1 per day. Essential services are also scarce, with only 26% of people having access to clean water and just 6% having access to electricity. Additionally, the country also suffers from health issues, facing contagious diseases, such as HIV, and infectious diseases, such as Malaria [4].

Regarding the mining practices, Kara [3] explains that numerous individuals are subjected to substandard working conditions, that often leads to health issues due to inadequate safety equipment. They are also exposed to hazardous situations primarily because many miners operate outside the law. In addition, most locations are plagued by other violations, including child labor, forced labor, physical harm, and environmental damage [3]. Still according to [3], sometimes due diligence is not conducted correctly, and many so-called supervisory entities are corrupted, not performing their functions properly.

In addition, the Artisanal and Small-scale Mining (ASM) is a legally recognized practice in DRC. IGF [11] defines ASM as a complex and diversified sector that "ranges from informal individual miners seeking a subsistence livelihood, to small-scale formal commercial mining entities producing minerals in a responsible way". But, according to [3] it is mainly a process where individuals use rudimentary tools and methods to extract and refine mineral substances. The miners are predominantly adult men who create underground tunnels using basic tools like shovels, chisels, and mallets. These tunnels can extend up to thirty meters deep, where they mine copper and cobalt beneath the surface.

In 2015, the production of cobalt from ASM surpassed that of any other individual large-scale mine. Despite this, the financial condition of ASM miners is such that they often cannot afford essential equipment like a ladder, resorting to carving additional steps into the pit for access. Safety measures are frequently neglected, with miners rarely wearing helmets or protective gear. Sometimes, they do not even have essential tools like shovels or hammers, leading them to dig with their bare hands. A typical day for an ASM worker involves earning between $1 and $3 and producing about 30–50 kg of raw ore [4]. As for comparison, 1 kg of cobalt is sold by $24,3 in 2024 [12]. Figure 10.1 represents a typical cobalt supply chain in RDC.

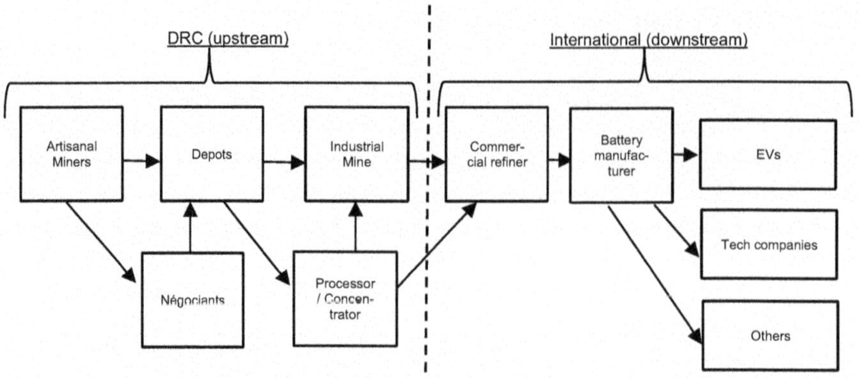

Fig. 10.1 Cobalt supply chain. Adapted from [3]

According to [3], the artisanal cobalt enters the "official supply chain" through an informal network of traders, known as *négociants*, and *comptoirs*, known as depots. Négociants are autonomous entities that operate in and around small-scale mining sites to buy cobalt from these miners. After acquiring their loads, the *négociants* transport to the depots for selling. In some more significant small-scale mining regions, depots are on-site, allowing miners to sell directly to them. Once the depots buy from the négociants or the miners, they sell their stock to industrial mining firms and processing plants. Beyond this point, it becomes impossible to distinguish between small-scale and industrial production [3].

The geographical remoteness and political instability of some mineral-rich countries make it difficult to hold EV companies accountable for the social issue caused by exploitation [13–15]. These challenges result from the multi-layered nature of global supply chains, limited oversight by international organizations of multinational supply chains, and low supervision of global entities.

10.3 Methodological Procedures

This is a descriptive research, based on multiple cases. The data was evaluated qualitatively, through the content analysis proposed by [16] as described in Sect. 3.1. Section 3.2 details the data selection process.

10.3.1 Content Analysis and Research

According to Bardin [16], content analysis is a set of methodologies applied to diversified content to extract and understand models based on deduction and inference. Its main objective is to balance objectivity and subjectivity in qualitative research but in a structured, continuous, and progressive way. It is composed of three main stages: pre-analysis, material exploration, and treatment, inference, and interpretation of results.

Pre-analysis is the organization phase, usually characterized by three main goals: document selection, formulation of hypothesis and objectives, and drafting key indicators to help support findings later. There is no definitive or clear order in pursuing these primary goals, but they are intricately connected and could often change or redefine the others. For example, objectives could shape document selection criteria, which could change objectives and research questions, but also the other way around. The pre-analysis must be exhaustive, representative, homogeny, and relevant. During the pre-analysis phase, we searched for keywords such as cobalt, Congo, DRC, Africa, LIB (lithium-ion battery), NMC (nickel-manganese-cobalt), NCA (nickel–cobalt-aluminum), LFP (lithium-iron-phosphate), EV (electric vehicle), greenhouse gas emissions (GHG), and carbon. This process had many cycles, as we had to move back and forth many times as new insights required us to reassess previous assumptions, insights, beliefs, and even keywords, as expected for the pre-analysis phase.

Material exploration is about organizing, coding, decomposing, and enumerating the pre-analysis content according to established rules. Treatment, inference, and interpretation of results, by its turn, involves applying adequate techniques to establish results, diagrams, figures, and models that better represent all findings.

10.3.2 Data Collection

Document selection was done by visiting the official websites of the EU, BYD, and Tesla. For BYD and Tesla, the 2023 Sustainability Reports were selected to be analyzed. As for the EU, we searched for sustainability policies regarding carbon-free emissions, greenhouse gas emissions, and so on that were published, updated, or effective in 2023 or later. Although mostly centered around the European Green Deal, there were found many proposed regulations, plans, and strategies for different sectors and areas that will be presented in Sect. 10.4.

10.4 Main Findings

In this Section, we present the main results related to the content analysis of the EU documents (governmental point of view)—Sect. 4.1—and the BYD and Tesla sustainable reports (private standpoint)—Sects. 4.2 and 4.3, respectively. Later, some recommendations are made based on the findings (Sect. 4.4).

10.4.1 EU Perspective

"The European Commission today presented The European Green Deal—a roadmap for making the EU's economy sustainable by turning climate and environmental challenges into opportunities across all policy areas and making the transition just and inclusive for all" [17].

This statement was presented in 2019 by The European Commission, setting in motion many strategies, plans, legislations, research, and discussions about the future of the EU [18]. The first significant commitment was signed in 2021 under the EU Climate Law, binding all economy sectors to reduce emissions by, at least, 55% until 2030 and neutral up to 2050, in comparison to 1990 levels [19–21].

The European Green Deal is about transforming the EU sustainably, addressing people, planet, and profit, and aligning with the Paris Agreement goals. This transformation is critical to foster innovation and economic growth, as well as it could be understood as a "protection measure" from unreliable partners and politics, such as Russia—Ukraine war and Europe's critical dependence on Russia as an energy supplier [19], and as a protection against global instabilities (as Covid-19) [21–24].

Those strategies aim to strengthen the EU's position as a responsible global leader, as stated in the European Commission Strategy and Policy's priorities: "The European Commission champions multilateralism and a rules-based global order through a more active role and a stronger voice for the EU in the world. A strong, fair, and open trade agenda makes Europe an attractive place for business. This is key to strengthening the EU's role as a global leader while ensuring the highest standards of climate, environmental, and labor protections. European leadership also means working closely with neighbouring countries, introducing a comprehensive strategy on Africa, and reaffirming the European perspective of the countries of the Western Balkans. The Commission seeks a coordinated approach to external action—from development aid to the Common Foreign and Security Policy—that secures a stronger and more united voice for Europe in the world" [18].

Although many actions are in motion for the EU's sustainability plan, this analysis focused on the second-highest GHG emitter sector: transportation 23.8% of the GHG emissions in 2022 [47]. To deal with this, the EU has set a specific goal in its Green Deal plan, "ensuring all new cars and vans registered in Europe will be zero-emission by 2035" [25, 26]. The main objective of this new directive is to have an intermediary goal toward neutral emissions by 2050, and EVs, among other strategies, are essential to achieve this goal [27–29]. In this sense, Verneque et al. [30] present some policies dedicated to incentivizing EVs around the world.

However, considering that: (i) EV production relies upon new critical raw materials, some extracted from areas suffering with social, issues as the case of DRC presented in Sect. 10.3; and, (ii) new policies incentivizing the adoption of EVs around the world might pressure the exploitation of cobalt while there are no other substitutes sources fully available, it becomes even more important to address the social impacts caused by its extraction if we are aiming at a sustainable approach.

As presented in the literature review, the DRC is one of the regions suffering from bad practices surrounding cobalt extraction, something already well-known to the EU. To deal with this problem, EU have established a set of regulations and research agenda, such as:

- EU Critical Raw Materials Act—EU CRMA (2023 regulation) [31].
- EU Global Gateway—EU GG (2023 regulation) [32].
- EU Corporate Sustainability Due Diligence—EU CSDD (2024 regulation) [33].
- Research on refinery in DRC [34].

EU-CRMA's primary goal is to reduce strategic dependencies on critical raw materials, reduce risk, and improve sustainability and circularity. The key goals for EU domestic suppliers to be met by 2030 are:

- At least, 10% of the EU's annual consumption for extraction.
- At least, 40% of the EU's annual consumption for processing.
- At least, 25% of the EU's annual consumption for recycling.
- No more than 65% of the EU's annual consumption from a single third country.

EU-CRMA also recognizes international trade as key to achieving diversification in the supply of critical materials. To ensure this, they established the EU-GG, with its first forum in October 2023. The focus was "on green energy, critical raw materials, connectivity, health, and education" [32]. This first forum has established a Strategic Partnership on Critical Raw Materials with DRC, signed on the same day [35], with roadmaps and joint actions to be implemented by 2030 [36]. There is a dedicated page for this partnership [37], from where we quote: "To diversify raw materials supply chains, as well as to develop local value addition and transforming trafficking of mineral resources into legitimate trade, the EU and DRC are setting up a Strategic Partnership on sustainable critical raw materials value chains (…) Sustainable and responsible production, sourcing, and valorization of critical and strategic raw materials, including through strengthening governance, due diligence and traceability, cooperation in the fight against illegal international trafficking in raw materials and aligning with international ESG standards."

EU-CSDD aims to "foster sustainable and responsible corporate behavior in companies' operations and across their global value chains" [33, 38]. Although previous regulations have already dealt with this sort of accountability, the EU's novelty proposition is that organizations will be responsible for due diligence across the whole supply chain, not just in its immediate suppliers. This new regulation was approved in May 2024. Still, its first effects will take some time until local EU governments transpose the directive into national law, and only then companies will have a gradual phase-in over the next five years [33].

In addition, there was also discussion about building a refinery in DRC, which could become part of the Strategic Partnership and a course of action in EU-GG. However, "local complications arising from political instability, conflict, corruption, irregular artisanal mining and Chinese ownership of mines" in DRC [34] may threaten this course of action. It is important to highlight that dealing with this situation is challenging: (i) stopping exploiting the cobalt could cause a great negative social impact related to the reallocation of the artisanal miners to other economic activities—but, at the same time, better conditions must be ensured if we are aiming at a more sustainable future; (ii) promoting the development of the refinery sector could bring knowledge of the process and more qualified positions, but would also face the mentioned local challenges; (iii) at the same time, the search for a substitute for cobalt also poses as potential negative social impact related to reallocating these artisanal miners in the economic sector.

10.4.2 BYD Sustainable Report

This section draws exclusively on the findings and data from BYD [39]. Founded in 1995, BYD Company Limited is a Chinese firm that operates primarily in four major industries: automobiles, rail transit, renewable energy, and electronics. The company employs over 703,504 people worldwide. In 2023, BYD sold more than three million EVs and generated over $6 billion in revenue.

In 2023, BYD released its annual corporate social responsibility report, which discusses major ESG topics such as corporate governance, social responsibility, and environmental equilibrium. The company's goal is clear: to reduce carbon emissions and waste of material and energy. Accordingly, the BYD report shows different types of strategies on that front.

For example, BYD has invested in finding alternative ways to recycle batteries by improving the circular economy. Furthermore, they have consistently diminished carbon emissions by offering alternative solutions for nonrenewable energy sources. AURO-N, their solar panel product, is their state-of-the-art product. It has good low-light power generation performance and a module efficiency of up to 22.45%.

The social section of the report primarily addresses employees. BYD has implemented numerous initiatives to enhance safety, quality of life, gender equality, and opportunities for training and development. For instance, BYD has achieved a record of zero work-related fatalities, and the average lost days due to occupational injuries is less than one, indicating that any injuries sustained are minor.

Regarding the supply chain, BYD is very vague about what measures are being taken to ensure that the raw material used in their production comes from responsible suppliers. Although they affirm compliance with international and applicable national and regional laws and social responsibility organizations' standards, such as the United Nations Global Compact (UNGC), they do not give more information about the selection and name of the suppliers.

The report shows BYD has over 10,000 cooperative suppliers and issued 583 supplier recruitments in 2023. Furthermore, they have institutionalized a risk committee responsible for collecting the risk management information of the supply chain and conducting risk assessments. Just in 2023, BYD introduced more than 1,700 new suppliers. For each supplier, BYD will investigate to check if they have violated any laws or regulations.

Regarding the mineral supply chain, BYD both states that the LFP battery is cobalt-free but also that it: "carries out a mineral supply chain survey on suppliers of tantalum, tin, tungsten, gold, and cobalt materials that are included in products or used in manufacturing processes. The survey results showed that metals used in the supply chain came from conflict-free mineral smelters and refineries recognized by CFSI" [39].

Here, BYD seems to leave room for interpretation regarding whether the company trusts the responses provided through forms or if they conduct a more thorough organizational assessment. No information about other types of due diligence were found.

10.4.3 TESLA Sustainable Report

This section draws exclusively on the findings and data from Tesla [40]. Tesla, Inc., founded in 2003, is an American EV and clean energy company headquartered in Palo Alto, California. Renowned for its innovation, Tesla designs and manufactures electric cars, battery energy storage systems, solar products, and related products and services. The company aims to accelerate the world's transition to sustainable energy, with popular models such as the Model S, Model 3, Model X, and Model Y leading the EV market.

Tesla's impact report addresses the company's efforts and achievements in promoting sustainability and reducing environmental impact. Key topics typically covered in the report include Environmental Sustainability, Energy Consumption and Production, Social Responsibility, and Corporate Governance.

In the EV supply chain chapter, Tesla states the essential minerals used to produce its car fleet. These include lithium, nickel, and cobalt, which are necessary for battery production. Aluminum is used for the car's body structure and various battery components. Tin, tantalum, and tungsten are critical for the electronic components within the vehicle. Finally, steel and iron are employed in the frame and overall structure of the car.

In 2022, Tesla published the first "Battery Passport" pilot [41, 42], a blockchain technology that enables digital traceability of shipments of cobalt, GHG emissions, and human rights data from mine to the final consumer. Tesla owners will be able to access the carbon footprint generated by manufacturing their vehicles.

Tesla has heightened its efforts to trace its supply chain in response to concerns about forced labor and child labor. Given the DRC's significant role as a supplier of crucial ores, Tesla has established specific objectives and implemented an action plan to enhance conditions within the country. As part of this initiative, Tesla states that exclusively procures materials from large-scale mining operations that undergo rigorous third-party audits. These audits aim at ensuring that no child or forced labor occurs within these mines and that no materials from unauthorized sources are utilized.

Tesla has launched a publicly available satellite monitoring system for its Kamoto Copper Company in the DRC to bring more transparency to this commitment. High-resolution images are updated monthly and allow anyone to see the state of the mine and its workers. Furthermore, Tesla has required its suppliers to implement risk mitigation measures to enhance the region's overall economy and quality of life. For instance, suppliers are tasked with improving air quality and ensuring everyone in the area can access clean drinking water. They must also increase transparency within their operations and conduct third-party audits.

Regarding the DRC issues, Tesla has acknowledged its responsibility and seems to be working to demonstrate to stakeholders the measures being taken to address some of the supply chain problems. However, Tesla admits that, as of 2023, only 55% of their cobalt usage is sourced directly from mines and refiners. Although they claim that cobalt from non-direct sources undergoes the same mapping and due diligence processes, they have not provided details about the origins of the remaining 45%. Further information on efforts to audit these additional suppliers has not yet been disclosed. Also of note is that Tesla is currently using alternative technologies that are either cobalt-free (LFP) or use less cobalt for its standard-range EV products.

10.4.4 Discussion and Recommendations

Considering what was presented in Sects. 4.1, 4.2 and 4.3, the current DRC situation requires interventions on many fronts (e.g., social, economic, political, governance, infrastructure, etc.). However, the results suggest that the key players are either not engaged enough with DRC or their actions are not impactful enough to help develop the country and a more sustainable cobalt supply chain.

Tesla and BYD's sustainability reports lack real impact on their critical supplier: the DRC. BYD has already transitioned to another technology (LFP) that is cobalt-free. However, its report indicates that the company still uses cobalt, although to a lesser extent and not in its new battery. Regarding the report, BYD has more than 10,000 suppliers, and its due diligence is a survey answered by the suppliers.

On the other hand, Tesla still uses cobalt in its EVs but already uses LFP for its standard-range EVs. Just like BYD, its report seems to be lacking. Although they have launched an audit program, it currently targets just 55% of Tesla's suppliers. About the achievements, Tesla says it has accomplished key goals: eighteen children working in ASM were enrolled in a remediation program; twenty-one local communities' savings and loans were deployed; and 5,206 ASM were trained on safety standards. But, there is no baseline index to compare those numbers to understand how relevant they are to the local community.

Considering how relevant cobalt mining is to DRC and to EV manufacturers, one might think that it should be one of the material themes to be prioritized on their sustainable report. In the BYD materiality matrix, the theme "supply chain management" has a degree of importance of, approximately, 6.5 to the stakeholders and

8.5 to the company (in a 1–10 scale, being 10 the highest importance). Nevertheless, the ASM seems to be superficially addressed along the report. The dimension "occupation health and safety" scores below 5 to stakeholders, but almost 9 to BYD. However, it seems to address only the companies' employees.

Regarding the Tesla materiality matrix, there is no mention to supply chain management among the 20 addressed topics. Nonetheless, the topics "child labor & forced labor" as well as "responsible sourcing" are the only ones highlighted to be assessed throughout the supply chain. In addition, the materiality "chart" is not clear about whether the topics are ranked according to their importance to the stakeholders x company or simply listed. The item "occupational health and safety" also appears among the most relevant themes, but focused on the internal employees only as in the BYD case. So, the question is why are not their stakeholders considering it a material theme?

Regarding the EU, they have plans to enforce more responsibility over the supply chain, with the EU-CSDD regulation gradual phase-in. But history also shows that this sort of regulation could create an exodus of companies from the DRC, just like in the past as "Congolese and international researchers have shown that Dodd-Frank has had unintended and negative consequences. Some industry players avoided sourcing minerals from the Democratic Republic of Congo through a quasi-boycott. As a result, artisanal miners, their families, and people in mining-related industries have been affected both economically through loss of income and livelihoods, and indirectly through, for example, increased infant mortality" [43].

There is already a research trend on innovative technologies, and both Tesla and BYD already have cobalt-free batteries in EVs. But this is not a problem exclusively of the EV industry. Although BYD affirms not to use cobalt in their EVs anymore, many other high-tech companies still do it directly or indirectly. Given that DRC provides more than 50% of the world's cobalt, one possible solution would be to build a refinery in DRC. By doing so, the country could benefit from the value creation in refining it and its population would benefit from job creation, improved infrastructure around the refinery, and better essential sanitation services.

Nevertheless, some challenges might arise due to the fact that DRC ranks 164 out of 174 countries on the 2020 Human Capital Index. An estimated 97% of 10-year-olds in DRC cannot read or understand simple text. Therefore, developing its local workforce can be a challenge. The country's issues regarding education might force them to import highly skilled laborers to work in the refinery, which could reduce the positive impacts on the DRC's population.

Besides, only 9% of the population has access to energy [44]. Having a reliable source of energy is essential to build the refineries. It would require investment in infrastructure, that should also be renewable. DRC has many potential hydroelectric sites and other renewable energy sources.

Although the EU has acknowledged its responsibility toward the DRC and even stated that its "responsibility is to make sure that this transition is a just transition, and that nobody is left behind as we deliver the European Green Deal" [17], not much seems to have changed from 2019 declaration to now. The Strategic Partnership on Critical Raw Materials was signed by the end of 2023, and the EV industry is moving away from the cobalt conflict and tightening regulations efforts, but not discussing how it would impact the DRC. For example, the EU research on building a refinery [34] was not included in EU-GG.

Nevertheless, DRC seems to be pursuing this challenge [45]: "Buenassa company announced the receipt of a US$3.5 million from the Government of the Democratic Republic of Congo (DRC). This grant from DRC's Industry Promotion Funds represents the first installment for the financing of the development of Buenassa's copper and cobalt refinery project (hydro-metallurgical plant). (...) The company's plan indicates that the project will be developed in the province of Lualaba and should be operational by the end of 2027. During the first phase, it envisions the annual production of 30,000 tonnes of copper cathodes and 5000 tonnes of cobalt sulfate."

If this project becomes successful, DRC would add value to its cobalt, helping develop its economy; the EU would have access to refined cobalt directly from DRC, reducing both its heft dependency upon China cobalt's refinery and possibly reducing transportation's CO_2 emission; and BYD and Tesla will have access to a better-developed supplier and maybe even lower prices [34]. In this sense, The World Bank [46] makes a provocation: why not plan on building a refinery and a battery plant?

10.5 Final Considerations

This paper highlights a relevant discussion regarding the sustainable role of the EV in the low-carbon economy. Although it is pointed out as one of the solutions to reduce GHG emissions, it has been criticized for the environmental impacts caused by battery disposal at the end of its life cycle. But, beyond that, there are social impacts that seem not to be taken into account. Therefore, this paper focused on evaluating if and how the main players in the EV industry (BYD and TESLA) as well as the EU are addressing the social impacts caused by cobalt exploitation, with emphasis on DRC.

We could identify that the EU has been addressing the topic recently, although more efforts (and actions) are necessary from the governmental point of view. From the private sector perspective, it seems not to be considered a "relevant material topic" in their sustainable reports, even though cobalt is a fundamental raw material in their supply chain. The research also identified efforts from the DRC government to develop a refinery in the country, in order to aggregate value to the product and improve work conditions. Nonetheless, a policy that involves building refineries might be constrained by political instability, funding restrictions, and the social issues that already exist (as corruption, education, and energy shortages).

As limitations, the article was based on secondary data and did not evaluate policies from other governments placed out of the EU. In addition, it is important to emphasize that all sectors that rely on cobalt to produce the batteries of their products should also be concerned about the sustainability of their value chains. So, as future research, interviews might be conducted with experts from academia, government and private sector that produce EVs and the investigations could be extended to other sectors. Besides, policies from other countries, as China and US, must be addressed.

Acknowledgements The authors thank the Programa de Apoio a Docente Recém Doutor—Edital ALV (23079.229096/2023-82 and 23079.227622/2023-70).

Disclosure of Interests The authors have no competing interests.

References

1. R. Wang, N. Mirza, D.G. Vasbieva, Q. Abbas, D. Xiong, The nexus of carbon emissions, financial development, renewable energy consumption, and technological innovation: What should be the priorities in light of COP 21 Agreements? J Environ Manage. 271, (2020). https://doi.org/10.1016/j.jenvman.2020.111027
2. R.R. Kumar, K. Alok, Adoption of electric vehicle: a literature review and prospects for sustainability. J Clean Prod. 253, (2020). https://doi.org/10.1016/j.jclepro.2019.119911
3. S. Kara, Cobalt red: how the blood of the congo powers our lives. St. Martin's Publishing Group. (2023)
4. B.K. Sovacool, The precarious political economy of cobalt: balancing prosperity, poverty, and brutality in artisanal and industrial mining in the democratic republic of the Congo. Extr. Ind. Soc. **6**, 915–939 (2019). https://doi.org/10.1016/j.exis.2019.05.018
5. R. Deberdt, The Democratic Republic of the Congo (DRC)'s response to artisanal cobalt mining: the Entreprise Générale du Cobalt (EGC). Extr. Ind. Soc. 8, (2021). https://doi.org/10.1016/j.exis.2021.101013
6. M. Grohol, Study on the critical raw materials for the EU 2023-Final report. (2023). https://doi.org/10.2873/725585
7. I.E.A., Global EV Outlook 2024, https://www.iea.org/reports/global-ev-outlook-2024. Last accessed 2024/07/12
8. A. Fleck, More than two-thirds of cobalt is used in battery production, https://www.statista.com/chart/33122/share-of-worldwide-cobalt-demand/. Last accessed 2024/11/21
9. H. Ren, D. Mu, C. Wang, X. Yue, Z. Li, J. Du, L. Zhao, M.K. Lim, Vulnerability to geopolitical disruptions of the global electric vehicle lithium-ion battery supply chain network. Comput Ind Eng. 188, (2024). https://doi.org/10.1016/j.cie.2024.109919
10. J. Elbel, S. Bose O'Reilly, R. Hrzic, A European union corporate due diligence act for whom? Considerations about the impact of a European Union due diligence act on artisanal and small-scale cobalt miners in the Democratic Republic of Congo. Resour. Policy **81**, 103241 (2023). https://doi.org/10.1016/j.resourpol.2022.103241
11. D. Paget, N. Garrett, A. Crawford, IGF Guidance for Governments: Managing artisanal and small-scale mining, Winnipeg: IISD? (2017)
12. Trading Economics, Cobalt—price—historical data, https://tradingeconomics.com/commodity/cobalt. Last accessed 2024/11/21
13. R.T. Nguyen, R.G. Eggert, M.H. Severson, C.G. Anderson, Global electrification of vehicles and intertwined material supply chains of cobalt Copper and Nickel. Resour. Conserv. Recycl **167**, 105198 (2021). https://doi.org/10.1016/J.RESCONREC.2020.105198
14. B. Ballinger, M. Stringer, D.R. Schmeda-Lopez, B. Kefford, B. Parkinson, C. Greig, S. Smart, The vulnerability of electric vehicle deployment to critical mineral supply. Appl. Energy **255**, 113844 (2019). https://doi.org/10.1016/J.APENERGY.2019.113844
15. R. Deberdt, P.L. Billon, Conflict minerals and battery materials supply chains: a mapping review of responsible sourcing initiatives. Extr. Ind. Soc. **8**, 100935 (2021). https://doi.org/10.1016/J.EXIS.2021.100935

16. L. Bardin, Análise de Conteúdo. Edições 70 (2011)
17. European Commission, The European green deal sets out how to make Europe the first climate-neutral continent by 2050, boosting the economy, improving people's health and quality of life, caring for nature, and leaving no one behind, https://ec.europa.eu/commission/presscorner/detail/en/ip_19_6691. Last accessed 2024/07/12
18. European Commission, A Stronger Europe in the World, https://commission.europa.eu/strategy-and-policy/priorities-2019-2024/stronger-europe-world_en. Last accessed 2024/07/12
19. European Commission, The European Green Deal, https://commission.europa.eu/strategy-and-policy/priorities-2019-2024/story-von-der-leyen-commission/european-green-deal_en, last accessed 2024/07/12
20. European Commission, The European Commission's priorities, https://commission.europa.eu/strategy-and-policy/priorities-2019-2024_en. Last accessed 2024/07/12
21. European Commission, The European Green Deal: a growth strategy that protects the climate, https://ec.europa.eu/stories/european-green-deal/. Last accessed 2024/07/12
22. European Commission, The European Green Deal Priorities, https://commission.europa.eu/strategy-and-policy/priorities-2019-2024/european-green-deal_en. Last accessed 2024/07/12
23. European Commission, Industry 5.0, https://research-and-innovation.ec.europa.eu/research-area/industrial-research-and-innovation/industry-50_en. Last accessed 2024/07/12
24. European Commission, Recovery plan for Europe, https://commission.europa.eu/strategy-and-policy/recovery-plan-europe_en. Last accessed 2024/07/12
25. European Commission, Zero emission vehicles: first 'Fit for 55" deal will end the sale of new CO2 emitting cars in Europe by 2035,' https://ec.europa.eu/commission/presscorner/detail/en/ip_22_6462. Last accessed 2024/07/12
26. European Commission, European Green Deal: Commission proposes transformation of EU economy and society to meet climate ambitions, https://ec.europa.eu/commission/presscorner/detail/en/ip_21_3541. Last accessed 2024/07/12
27. European Environment Agency, Electric vehicles, https://www.eea.europa.eu/en/topics/in-depth/electric-vehicles?activeAccordion=. Last accessed 2024/07/12
28. European Environment Agency, Europe's road transport needs to shift gears towards sustainability, https://www.eea.europa.eu/media/europes-road-transport-needs-to. Last accessed 2024/07/12
29. European Commission, Transport and the Green Deal, https://commission.europa.eu/strategy-and-policy/priorities-2019-2024/european-green-deal/transport-and-green-deal_en. Last accessed 2024/07/12
30. F.H.A. Verneque, P.H.G. Silva, M.A. Martinez, V.A. Guimarães, The role of the electrical vehicle in sustainable supply chains: a review. Presented at the (2020)
31. European Commission, Critical Raw Materials Act, https://single-market-economy.ec.europa.eu/sectors/raw-materials/areas-specific-interest/critical-raw-materials/critical-raw-materials-act_en#actions-under-the-critical-raw-materials-act. Last accessed 2024/07/12
32. European Commission, Global Gateway, https://commission.europa.eu/strategy-and-policy/priorities-2019-2024/stronger-europe-world/global-gateway_en. Last accessed 2024/07/12
33. European Commission, Corporate sustainability due diligence, https://commission.europa.eu/business-economy-euro/ng-business-eu/sustainability-due-diligence-responsible-business/corporate-sustainability-due-diligence_en. Last accessed 2024/07/12
34. V.W. Kjeld, European parliament think tank: will the EU help build a cobalt refinery in the Democratic Republic of the Congo?, https://www.europarl.europa.eu/thinktank/en/document/EPRS_ATA(2024)757828. Last accessed 2024/07/12
35. European Commission, Memorandum of Understanding EU-Democratic Republic of Congo on sustainable raw materials, https://single-market-economy.ec.europa.eu/publications/memorandum-understanding-eu-democratic-republic-congo-sustainable-raw-materials_en. Last accessed 2024/07/12
36. European Commission, Global Gateway: EU signs strategic partnerships on critical raw materials value chains with DRC and Zambia and advances cooperation with US and other key partners to develop the 'Lobito Corridor, https://ec.europa.eu/commission/presscorner/detail/en/ip_23_5303

37. European Commission, Democratic Republic of Congo, https://international-partnerships.ec.europa.eu/countries/democratic-republic-congo_en. Last accessed 2024/07/12
38. European Commission, Commission welcomes political agreement on rules enforcing human rights and environmental sustainability in global supply chains, https://ec.europa.eu/commission/presscorner/detail/en/ip_23_6599. Last accessed 2024/07/12
39. BYD, BYD Corporate Social Responsibility Report. (2023)
40. Tesla, Impact Report. (2023)
41. Battery Passport, https://www.globalbattery.org/battery-passport/. Last accessed 2024/09/21
42. Tesla, Impact report. (2022)
43. J. Hönke, L. Skender, Cobalt as 'conflict mineral'? On the opportunities and limits of new supply chain laws, https://infraglob.eu/2022/12/20/cobalt-as-conflict-mineral-on-the-opportunities-and-limits-of-new-supply-chain-laws/. Last accessed 2024/07/12
44. Democratic Republic of Congo Overview, Development news, research, data | World Bank, https://www.worldbank.org/en/country/drc/overview. Last accessed 2024/09/21
45. BEV Initiative, The DRC government and Buenassa company to set up a cobalt and copper refinery in Lualaba, https://www.uneca.org/stories/bev-initiative-the-drc-government-and-buenassa-company-to-set-up-a-cobalt-and-copper. Last accessed 2024/07/13
46. The World Bank, DRC—Country Economic Memorandum—Case Study 1: Mining (EV Battery) Value Chains. (2023)
47. European Environment Agency, EEA greenhouse gases—data viewer, https://www.eea.europa.eu/data-and-maps/data/data-viewers/greenhouse-gases-viewer. Last accessed 2024/07/12

Chapter 11
Parabolic Cylinder Collector for Roasting Cocoa Beans

K. M. Aguilar-Castro, P. R. Torres-Hernández, E. V. Macias-Melo, P. García-Alamilla, C. E. Torres-Aguilar, I. Hernández-Pérez, and J. Serrano Arellano

Abstract This paper presents the development and evaluation of a cocoa solar roaster using a Parabolic Cylinder Collector (PCC) with two degrees of freedom for solar tracking. The design, construction, and instrumentation of the PCC were divided into three main sections: solar collection area, absorber tube, and solar tracking system. The PCC can roast 0.5 kg of cocoa and monitor ambient temperature, solar irradiance, grain temperature, indoor air temperature, and absorber tube surface temperature. To analyze the roasting process, functionality tests without cocoa and experimental tests with cocoa were conducted. Efficiency was determined, and grain quality was analyzed by measuring moisture content. The experimental tests with cocoa were conducted for roasting times of 40, 55, and 70 min, during solar irradiance between 10:00 and 14:00. Maximum toasting temperatures of 125–137 °C were reached, with average solar irradiance of 685.7–930.5 Wm−2 and ambient temperatures of 31–38.4 °C. The minimum moisture content obtained was 2.19%, close to the optimal value (1–2%). Temperature and roasting time significantly affected moisture reduction. These results suggest the PCC prototype can roast cocoa beans in less than 40 min.

Keywords Parabolic cylinder collector · Roasting cocoa · Solar roaster

K. M. Aguilar-Castro · E. V. Macias-Melo (✉) · C. E. Torres-Aguilar · I. Hernández-Pérez
División Académica de Ingeniería y Arquitectura (DAIA) de la Universidad Juárez Autónoma de Tabasco (UJAT), Universidad Juárez Autónoma de Tabasco, Cunduacán, Tabasco, México
e-mail: edgar.macias@ujat.mx

P. R. Torres-Hernández
Maestría en Ciencias en Ingeniería, DAIA-UJAT, Cunduacán, Tabasco, México

P. García-Alamilla
Laboratorio de Ingeniería de Procesos, División Académica de Ciencias Agropecuarias, UJAT, Centro, Tabasco, México

J. S. Arellano
División de Postgrado e Investigación, Instituto Tecnológico de Pachuca-TecNM, Pachuca de Soto, Hidalgo, México

D. Rossit et al. (eds.), *Proceedings of the VII Ibero-American Congress of Smart Cities, ICSC-Cities 2024, 12-14 November, San Carlos, Costa Rica*, Springer Proceedings in Energy, https://doi.org/10.1007/978-981-96-4301-1_11

11.1 Introduction

Currently, climate change, decreasing hydrocarbon reserves, and increasingly complex extraction processes have caused instability in the energy market. This has led countries to seek efficient, competitive, and sustainable energy alternatives [1]. Renewable energies, such as solar energy, present an attractive solution due to their low environmental impact and economic potential. Solar energy can be used either as photovoltaic or solar thermal energy, with solar thermal energy converting radiation into thermal energy for various processes [2].

Solar thermal energy is categorized by temperature: low (< 100 °C), medium (100–350 °C), and high (> 350 °C) [1]. Medium-temperature processes, such as those in the chemical, paper, textile, and food industries, can reduce fossil fuel use and environmental impact [3]. Parabolic cylindrical collectors (PCCs) are used for medium-temperature processes, reaching up to 350 °C [4–6]. These devices concentrate solar radiation onto a small area, enhancing thermal energy absorption [7]. PCCs offer lower operating costs compared to fossil fuels, particularly in the food industry, where processes like roasting and storage require significant energy, typically between 100 and 500 °C [3, 8]. PCC applications in the food industry include shrimp processing, poultry feed production, and guava dehydration [9-11]. Cocoa roasting, crucial for chocolate production, fits within this temperature range (90–170 °C) [12, 13].

The roasting process involves moisture reduction, which is critical for determining flavor and aroma [14-16]. The optimal moisture content for cocoa is 1–2% [12, 17]. Though solar roasting using heliostats has been studied, no research has used PCCs for cocoa roasting, despite their potential for greater heat distribution [3, 18, 19]. PCCs, which can be built cost-effectively for small-scale use, may provide a feasible solution for reducing energy consumption in cocoa processing while minimizing environmental impact. This paper presents the development and thermal evaluation of a cocoa solar roaster using a PCC. The design, construction, and testing of the prototype are discussed, with a focus on optimizing solar energy gains and analyzing the quality of roasted beans.

11.2 Design and Construction of the Prototype

Three main sections present the design and development of the PCC: solar collection area, the absorber tube or grain container, and the solar tracking system.

11.2.1 Solar Collection Area

The sizing of the solar collector area of the PCC was determined by calculating the parabola's focus using the commercial dimensions of the reflective aluminum sheet

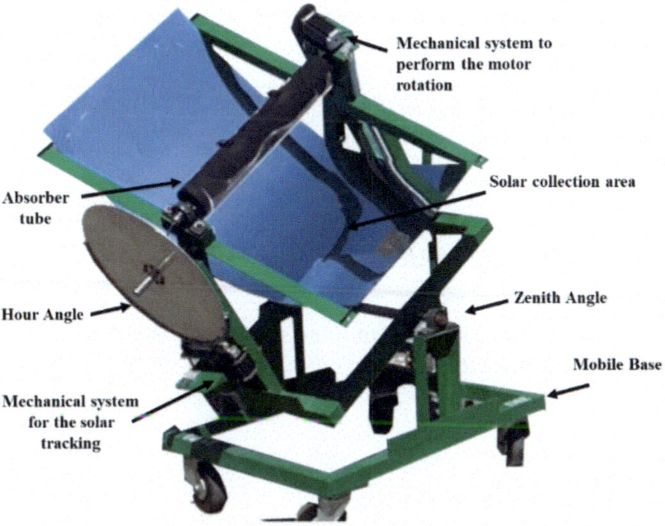

Fig. 11.1 Photograph of the PCC with an absorber tube

(122 cm × 61 cm) and the parabola equation. The optimal focal length and opening angle for maximum solar uptake were 30 cm and 61 cm, respectively. Reflective Miro-Sun aluminum plates (0.05 cm thick, 90% reflective) were used, supported by an aluminum structure with steel angular supports.

Material selection and the rotation system were based on structural design methodology, comparing functional alternatives using a decision matrix. Criteria included thermal properties, resistance, durability, availability, and cost. The solar collection area support has 2.54 cm rowlocks for zenith angle mobility, while the parabola tracks the sun's path (hour angle). Figure 11.1 shows the PCC and solar collection area.

Absorber tube

The design was based on an energy balance, considering solar radiation, energy losses, and heat gain inside the absorber tube. The tube, designed to hold 0.5 kg of cocoa beans, allows for even distribution and free movement for uniform roasting. It was cut transversely and secured with hinges and a snap, with two internal dividers to separate the beans. The absorber tube, made from 304-grade stainless steel (10 cm diameter, 60 cm length), features three 20 cm chambers to prevent bean accumulation when inclined. Stainless steel supports were added at each end for rotation during roasting. Figure 11.1 shows the tube open and closed.

Solar tracking system

The tracking system includes a support structure and a solar tracking system for zenith and azimuth angles. The design allows two degrees of freedom to follow the sun's apparent movement. Zenith tracking is manual since it only changes daily, based on the test site's latitude and earth's declination. Hour angle tracking is automatic, with adjustable supports for rotating the absorber tube to maximize solar capture.

The support and tracking structures were designed for strength and mobility, using steel Rectangular Tubular Profiles (RTP) of 5.08 cm (2 in) and 2.54 cm (1 in) rowlocks for movement. Figure 11.1 shows the tracking system for zenith and hour angles.

11.3 Instrumentation of the PCC

The instrumentation was carried out in three sections of the PCC: (1) the motion automation system for solar tracking, (2) the measurement of the temperature in the absorber tube and the cocoa beans, and (3) the meteorological variables (ambient temperature and global solar irradiance).

11.3.1 Temperature Measurement in the PCC

T-type thermocouples with an uncertainty of ± 0.5 °C were used to measure the temperature inside the duct. Nine thermocouples were calibrated and installed: three in each chamber (C1, C2, C3) of the absorber tube. In each chamber, one thermocouple measured the tube's interior wall, another measured air temperature, and the third was placed inside a cocoa bean to measure its temperature, see Fig. 11.2.

To avoid cable entanglement during rotation, the thermocouples were inserted through a hole in the shaft. The motors alternated between two clockwise and two counterclockwise turns, repeating this cycle until the roasting process was completed.

Measurement of meteorological variables

Ambient temperature and global solar irradiance were measured during the experiment. Temperature was measured with an RTD sensor (PT100 class B) with ± 0.3 °C uncertainty, calibrated per ASTM E-230–2017 [20]. The sensor was housed to protect it from direct radiation and air streams. Global solar irradiance was measured with a

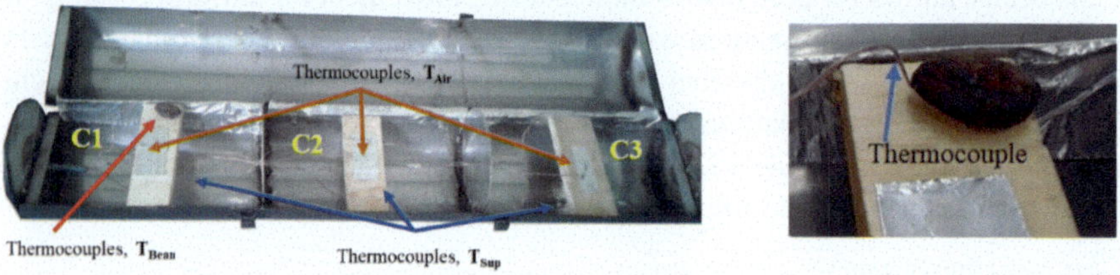

a) b)

Fig. 11.2 **a** Instrumentation of the absorber tube, and **b** instrumented cocoa bean

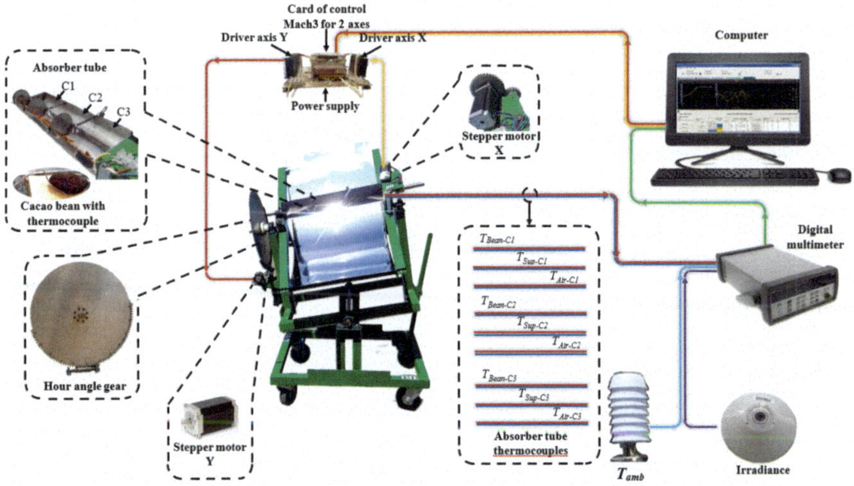

Fig. 11.3 Experimental set up of the PCC

first-class pyranometer with $\pm$ 20Wm^{-2} uncertainty. The temperature of the absorber tube, cocoa beans, and meteorological variables were monitored using a Keysight 34972A multimeter at 1-min intervals.

Automation system for solar tracking

The system's movement was designed using a set of helical, worm, conical, and straight gears for tracking zenith and azimuth angles. Helical and worm gears transmitted movement between two shafts at a 90° angle. The screws were made on a lathe, and the gears were 3D-printed. One helical gear was coupled to the tracking system with a Nylamid support and worm screw for hour-angle tracking, while the other gears rotated the absorber tube (360°) for uniform roasting.

Two Nema 23 bipolar motors were used for automatic movement, each powered by a 24 V source and controlled by Arduino. The first motor tracked the sun at 15° per hour, while the second rotated the absorber tube at 0–8 rpm for uniform roasting. Both motors were programmed with Mach3 software. Figure 11.3 shows the experimental setup and the automatic movement system.

11.4 Experimental Procedure

The experimental procedure is divided into two parts: roasting tests with solar irradiance and physicochemical tests to assess quality. During roasting, the cocoa bean temperature is monitored, and their quality is analyzed afterward.

The cocoa beans used were of the Trinitario Mexicano variety, fermented and sun-dried, from the Integral Consultancy and Advisory Center of Villahermosa, Tabasco.

Beans were selected by weight and size from a 15 kg batch for consistency, with individual weights between 1.1 and 1.5 g and lengths of 21–25 mm. For each test, 0.5 kg of beans were roasted, and a temperature sensor was placed in one bean in each chamber of the absorber tube (Fig. 2b). Two functional tests were conducted to evaluate the heating potential of the prototype. Test 1F lasted 30 min, and the temperature was monitored with a FLUKE Ti400 infrared camera. Test 2F lasted 39 min, with temperatures monitored using the described instrumentation.

Three roasting tests were conducted at 10:00, 12:00, and 14:00 h, with intervals of 40, 55, and 70 min. The schedules were based on maximum solar irradiance, and roasting times were selected considering literature values (5–65 min) and the results of functional tests.

During the experiments, the PCC was positioned for optimal solar irradiance, and the tracking system was adjusted for zenith and hour angles. A 5-min preheating was done before the cocoa beans were introduced and the rotation system activated. The tests were stopped after the designated times (40, 55, and 70 min). Temperatures in the PCC, meteorological data, and the mass of cocoa beans before and after roasting were recorded to determine moisture content (MC). Additionally, pH and total acidity were measured. All tests were performed in triplicate.

11.4.1 Moisture Content

The moisture content (MC) of the roasted cocoa beans was measured by taking a sample of 10 grains, weighing them, and drying them in a Felisa hot air oven at 103 °C for 24 h. After drying, the sample was cooled to room temperature and reweighed using a Denver Instrument analytical balance with ± 0.1 mg accuracy. The MC (%) was calculated using Eq. (11.1), and its uncertainty was determined based on the general law of error propagation [21, 22] with a coverage factor of $k = 3$.

$$MC = \frac{m_i - m_f}{m_i} \times 100 \tag{11.1}$$

Thermal efficiency

The thermal efficiency of the roasting process was determined using Eq. (11.2), which represents the relation between the energy used for roasting the 0.5 kg of cocoa beans and the energy received by the PCC.

$$\eta = \frac{Q_u}{A_\alpha \int_0^{t_f} Idt} = \frac{mCp\Delta T + m_{ev}h_{ev}}{A_\alpha \int_0^{t_f} Idt} \tag{11.2}$$

11.5 Results and Discussion

11.5.1 Functionality Tests: Without Cocoa

Table 11.1 summarizes the general data and maximum temperatures reached in the absorber tube for Test 1F. Figure 11.4 shows thermographic images at 5, 10, 15, and 20 min, selected based on the minimum reported cocoa roasting time (5–65 min). A homogeneous temperature distribution is observed, with lower temperatures at the tube ends due to convection and lack of direct solar irradiance. Preheating exceeded 90 °C, even under clear and partially cloudy conditions, consistent with reported roasting temperatures (90–170 °C). In the first 5 min, a maximum temperature of 97.5 °C was reached.

Test 2F was conducted with an average solar irradiance of 900 Wm^{-2}, starting at 10:30 am in Cunduacán, Tabasco. Figure 11.5 shows the ambient temperature, solar irradiance, and temperatures inside and on the surface of the absorber tube. Ambient temperature ranged between 34 and 36 °C, with irradiance averaging 900 Wm^{-2}. Surface temperatures of the absorber tube reached up to 150 °C, particularly in Chamber 3, due to its optimal azimuth angle for solar capture. Interior temperatures exceeded 130 °C in Chamber 3, while Chamber 1 reached temperatures above the minimum required for cocoa roasting (90 °C). After 15 min, a brief drop in temperature was observed, likely due to air currents, but recovered within 10 min. Results

Table 11.1 Summary of initial parameters and behavior of the temperature, Test 1F

Conditions of the test		
Average solar irradiance		900 Wm^{-2}
Surface of the PCC		0.744 m^2
Total reflectance		0.9
Average T_{amb}		37 °C
Emissivity of the absorber tube		0.94
Location		Cunduacán, Tabasco
Starting hour		10:24 a.m
Angular speed (steps)		200 r.p.m
t (min)	Maximum T_{Sup} (°C)	Conditions of the sky
01	97.2	Clear
02	97.5	Clear
03	90.9	Partially cloudy
04	91.0	Partially cloudy
05	97.4	Clear
10	97.5	Clear
15	91.0	Partially cloudy
20	88.1	Cloudy
25	85.9	Cloudy
30	88.3	Cloudy

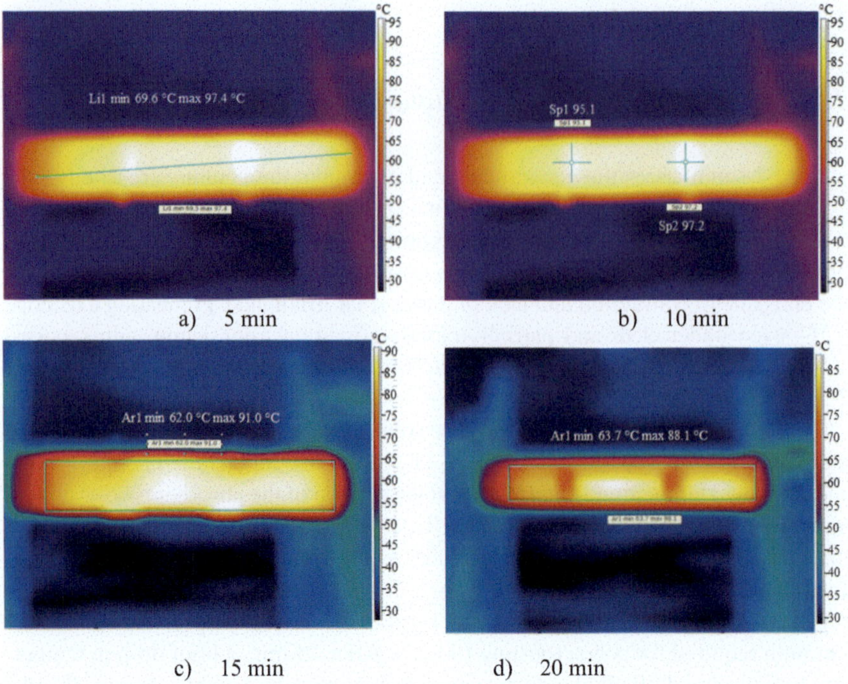

Fig. 11.4 Thermographic image of the temperature distribution of the absorber tube for the period with highest temperature

from Tests 1F and 2F confirm that the system can reach the necessary temperatures for cocoa roasting.

Experimental tests of roasting with Cocoa

Table 11.2 summarizes the experimental tests, considering roasting times and the hours of highest solar incidence at the study site. Three tests were conducted with increasing roasting times. Figure 11.3 shows the PCC's location and orientation during the test, aligned with the zenith angle (following the sun's path) and azimuthal angle (following the sun's height). Figure 11.6, present the temperature behavior of the cocoa beans, indoor air, and absorber tube surface in each of the three chambers, along with solar irradiance and room temperature for Tests 1R.

Figure 11.6 shows the results obtained from Test 1R, this figure indicates that the maximum ambient temperature was 34 °C, while the maximum solar irradiance was 830 Wm^{-2}. These environmental conditions allowed to reach a maximum T_{Sup}, T_{Air}, and T_{Bean} of 130, 120, 125 °C, respectively. As expected, the temperature of the absorber tube surface was the highest; however, it is interesting to note that the cocoa beans reached temperatures slightly higher than the interior air temperature. Figure 6b and c show that the behavior of T_{Air} had a similar behavior in the three chambers, while the T_{Bean} was higher in C2 and C3. This behavior of the T_{Bean}

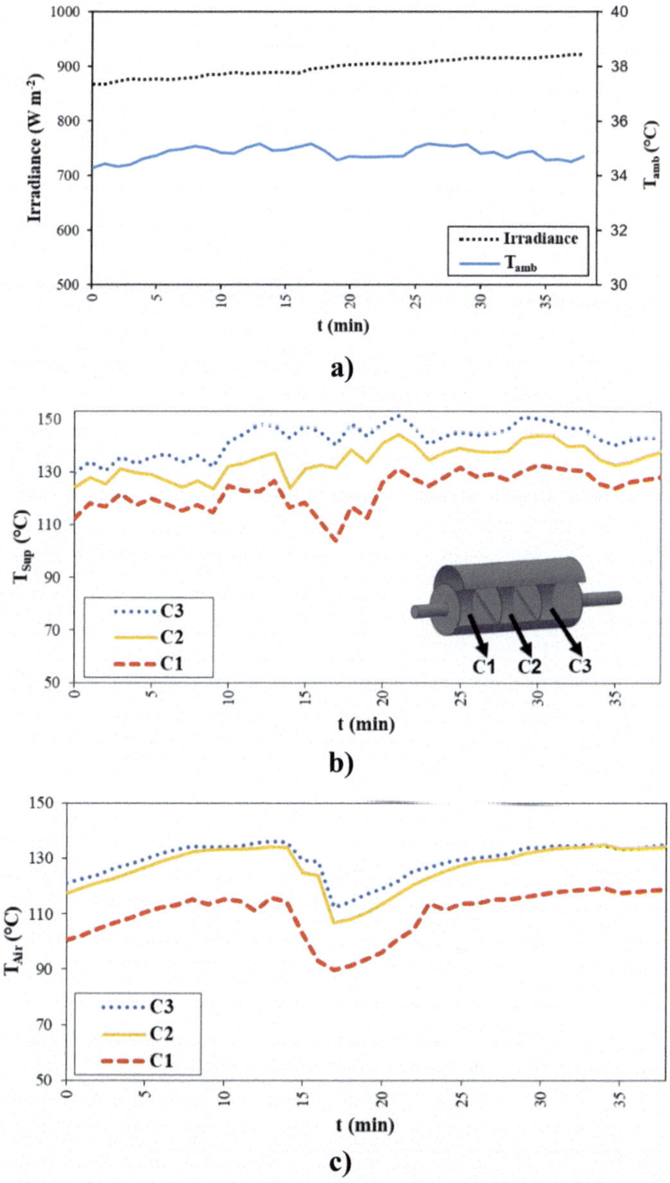

Fig. 11.5 Behavior of the variables during the test 2F, **a** meteorological variables, **b** surface temperature of the absorber tube, **c** temperature of the air inside the absorber tube

Table 11.2 Summary of initial parameters and behavior of the temperature, Test 1F

Test	Time	Schedule
1R	40	10:00
2R	55	12:00
3R	70	14:00

is because the energy captured on the surface of theabsorber tube of the C1 had maximum temperature of 90 °C, as shown in Fig. 6d. In general, it can be observed that for an average irradiance smaller than 800 Wm^{-2} it was possible to toast the cocoa beans.

The 2R test and 3R test showed similar behavior to the IR test, so their results are presented in a concentrated manner in Table 11.3.

Table 11.3 presents a summary of the average values of the meteorological variables, T_{Sup}, T_{Air}, T_{Bean}, and MC_f of the cocoa beans for each chamber of the absorber tube. The $M\ C_f$ is the final moisture content of the roasted cocoa bean, and the MC_i is the initial moisture content of the cocoa bean before the roasting process, both were determined with Eq. (11.2). The MC_i is used as a reference for the analysis of the results and it was estimated on average at 6.0%, considering all tests. The average maximum temperature in the chamber surfaces was obtained in C2 and C3 of Test 2R, with values of 131.0 and 134.8 °C, respectively. These results occurred because Test 2R was carried out during the peak solar incidence time, with an average solar irradiance of 930.5 Wm^{-2}. While the average maximum T_{Air} and T_{Bean} the average maximum T_{Bean} were obtained in C2 and C3 during the Test 3R. Although the average maximum T_{Sup} was obtained in the Test 2R, it was not in this test with the highest temperatures of the T_{Air} and the T_{Bean}, this is because in the Test 3R the roasting process was performed with a heating of 70 min, which allowed an accumulation of heat of the indoor air in the absorber tube, and in the cocoa bean due to its thermal mass. This effect allowed the cocoa bean to reach average temperatures of up to 126.7 °C in the roasting process, although the Test 3R was the one with the lowest average solar irradiance 685.7 Wm^{-2}.

On the other hand, regarding the values of the MC_f it was observed that the Test 3R had the lowest values of humidity content, being 2.19% the minimum value obtained in the C2. This value is close to 1–2% reported by [18, 23] as the optimum moisture content for roasted cocoa for chocolate making. It is important to note that this moisture value was obtained even when the roasting time was 70 min and the T_{amb} was the highest of the test set, 38.4 °C. The solar irradiance contributed to having a slow roasting process, where the mass of water evaporated from the roasted grain was only 0.013 kg, as shown in Table 11.4.

Efficiency of the PCC for the toasting process

The efficiency of the roasting process was estimated based on the energy used to remove moisture and heat the cocoa beans, as shown in Eq. (3). Table 11.4 presents the data and calculated efficiency for the three tests. Test 1R achieved the highest efficiency (7.45%) due to the initial temperature and test duration. Although this is

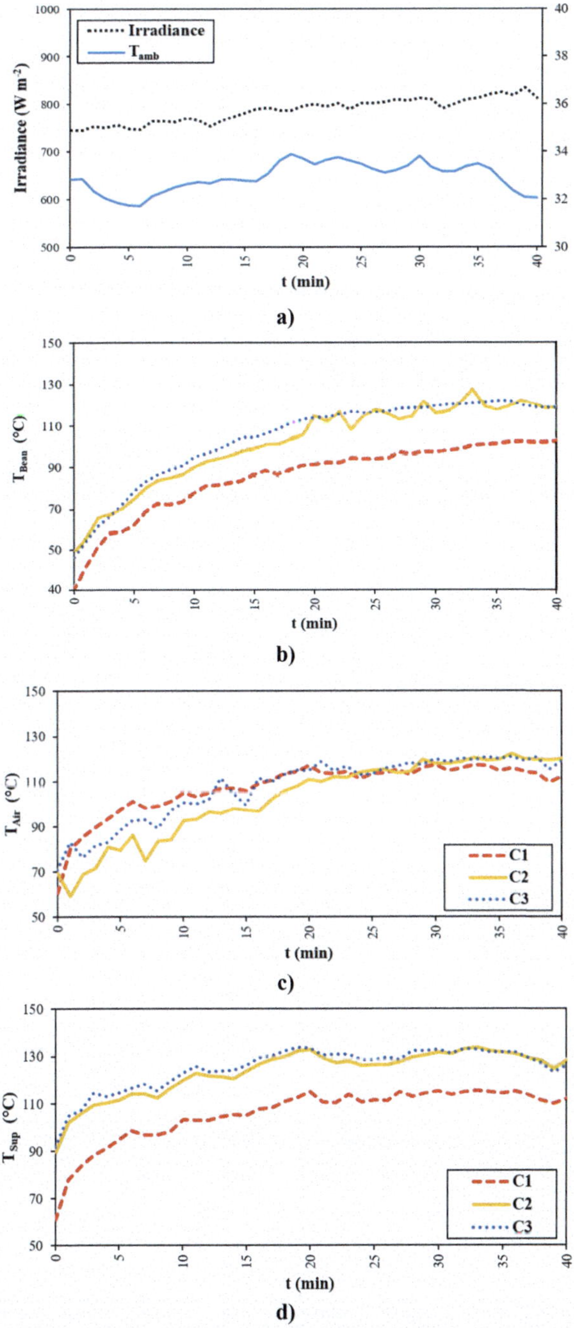

Fig. 11.6 Behavior of the variables in test 1R, **a** meteorological variables, **b** temperature of the cocoa bean, **c** temperature of the air in the absorber tube, and **d** surface temperature of the absorber tube

Table 11.3 Summary of initial parameters and behavior of the temperature, Test 1F

Test		1	2	3
I (Wm^{-2}) ± 20 Wm^{-2}		785.9	930.5	685.7
T_{amb} (°C) ± 0.3 °C		33.0	35.5	36.8
T_{Sup} (°C) ± 0.5 °C	C1	112.3	122.7	117.9
	C2	123.2	131	127.2
	C3	113.7	134.8	130
T_{Air} (°C) ± 0.5 °C	C1	107.2	109.1	117.9
	C2	102.5	109.1	123.2
	C3	107.2	118.3	121.6
T_{Bean} (°C) ± 0.5 °C	C1	88.8	102.5	116.2
	C2	103.4	117.1	123.4
	C3	104.9	115	126.7
MC_f (%) ± 0.092	C1	4.34	3.73	2.88
	C2	3.81	3.17	2.19
	C3	3.5	3.06	2.71

Table 11.4 Data used to calculate the efficiency

Data	Test 1R	Test 2R	Test 3R
η (%)	7.45	3.84	3.83
m_c (kg)	0.5		
c_p (J kg^{-1} °C)	1,653.5		
h_{ev} (J kg^{-1})	2,333,300		
MC_i (g)	6.0		
A_a (m^2)	0.744		
T_f (°C)	118.3	125.0	130.7
T_i (°C)	46.7	65.7	68.0
m_{ev} (kg)	0.019	0.017	0.013
I_{avg} (W m^{-2})	785.9	930.5	685.7
$t(s)$	2400	3300	4200
t(min)	40	50	70
MC_f (%) avg	3.88	3.32	2.60
η (%) ideal	25*		
m_{ev} (kg) ideal	0.005		
MC_f (%) ideal	1.0		
t(s) ideal	484.8	350.7	497.9
t(min) ideal	8.1	5.8	8.3

*[10]

lower than the 25–77% efficiency reported for other PCC systems [5, 11, 24], the comparison is limited as this study focused only on energy used for roasting, not other intermediate elements. Additionally, the small sample size (0.5 kg) means much of the captured energy maintained the temperature level.

Despite Test 3R achieving an ideal moisture content, its efficiency was low. An analysis was done to estimate the ideal roasting time and efficiency (25%) to achieve a 1% moisture content under the test conditions. The results indicated that roasting would require only 5.8–8.3 min to reach optimum conditions for chocolate production.

The best temperature behavior was observed in tests between 12:00 and 14:00. However, results suggest that roasting was complete after 10 min, as only 0.005 kg of water needed to be removed for a 1% moisture content. In the tests, 0.0013 to 0.0019 kg of water was removed, nearly four times the ideal value.

11.6 Conclusions

A solar roaster with two degrees of freedom for solar tracking was developed, and roasting tests for cocoa beans were performed. The process efficiency and cocoa bean quality were evaluated. In tests without cocoa beans, the absorber tube reached 150 °C in chamber 3 with an average irradiance of 900 Wm^{-2}. Under cloudy conditions, it still reached 91 °C, within the range for cocoa roasting (90–170 °C). In tests with cocoa beans, chamber 3 reached 137 °C, reducing moisture to 2.19% in 70 min with an irradiance of 685.7 Wm^{-2}. Test 3R achieved moisture levels close to the ideal 1–2%. To achieve 25% efficiency, roasting should be stopped after 10 min for quality cocoa beans. The results indicate that the PCC is suitable for cocoa roasting, but roasting time must be reduced to maintain moisture under 2%. Redesigning the absorber tube could improve heat distribution and prevent cocoa bean clumping. Scaling up the PCC could reduce costs by minimizing fossil fuel use.

Acknowledgements P.R. Torres-Hernández acknowledges to the National Council of Science and Technology (CONACYT) the grant for his Master Degree studies. The authors also acknowledge to the Program for Teaching Development (PRODEP) for the funds given to the project UJAT-PTC-251 (Development and Evaluation of a Cocoa Roaster in the Tabasco Region) and to Oscar May García for his collaboration in this project.

Abbreviations

Nomenclature

A_a	Opening area (m^{-2})
C1	Chamber one
C2	Chamber two
C3	Chamber three
c_p	Specific heat (J Kg^{-1} $°C^{-1}$)
f	Focal point of the parabola (m)
h	Enthalpy (J kg^{-1})
I	Solar Irradiance (Wm^{-2})
k	Covering factor (J)
m	Mass (kg)
MC	Moisture content (%)
Q_u	Useful heat (J)
R	Electric resistance (Ω)
T	Temperature (°C)
t	Time (s)

Greek

η	Efficiency

Subscripts

Air	Air
amb	Ambient
Bean	Cocoa bean
c	Cocoa
ev	Evaporation
f	Final
i	Initial
Sup	Absorber tube surface

References

1. S. Romero, O. Romero, D. Wood, Energías renovables: impulso político y tecnológico para un México sustentable USAID/ITAM, Tech. Rep. (2011)
2. W. Jara, Introducción a las energías renovables no convencionales. Empresa Nacional de Electricidad S.A., Tech. Rep. (2008)
3. E. Venegas-Reyes, O.A. Jaramillo, J.O. Aguilar, F. Sosa-Montemayor, Design, construction, and testing of a parabolic trough solar concentrator for hot water and low enthalpy steam generation. J. Renew. Sustain. Energy **4**, 053103 (2012). https://doi.org/10.1063/1.4751461
4. V.K. Jebasingh, G.M.J. Herbert, J.O. Aguilar, F. Sosa-Montemayor, A review of solar parabolic trough collector. Renew. Sustain. Energy Rev. **54**, 1085–1109 (2016). https://doi.org/10.1016/j.rser.2015.10.043
5. E. Bellos, C. Tzivanidis, V. Belessiotis, Daily performance of parabolic trough solar collectors. Sol. Energy **158**, 663–678 (2017). https://doi.org/10.1016/j.solener.2017.10.038
6. C. Cadena, L. Saravia, J. Franco, Multiple use communal solar cookers. Sol. Energy **77**, 217–223 (2004). https://doi.org/10.1016/j.solener.2004.03.026
7. J.A. Duffie, W.A. Beckman, Solar engineering of thermal processes, J. Wiley and Sons 4th (ed). (2013)
8. H.T. Retta, M.H. Hailu, A.T. Baheta, M.G. Haile, Design of an Injera baking system using parabolic trough solar collectors at Mekelle University Cafeteria. Heliyon **10**(17), 36864 (2024). https://doi.org/10.1016/j.heliyon.2024.e36864
9. P.D. Tagle, A. Agraz, C.I. Rivera, Parabolic trough solar collector technology in Mexican industry. Energy Proced. **91**, 661–667 (2016)
10. Y. Yang, A.G. Darwish, I. El-Sharkawy, Q. Zhu, S. Sun, J. Tan, Rapid determination of the roasting degree of cocoa beans by extreme learning machine (ELM)-based imaging analysis. J. Agric. Food Res. **10**, 100437 (2022). https://doi.org/10.1016/j.jafr.2022.100437
11. J.T. Liu, M. Li, Q.F. Yu, D.L. Ling, A novel parabolic trough concentrating solar heating for cut tobacco drying system. Int. J. Photoenergy **10**, 1–10 (2014). https://doi.org/10.1155/2014/209028
12. P. García-Alamilla, L.M. Lagunes-Gálvez, J. Barajas-Fernández, R. García-Alamilla, Physicochemical changes of cocoa beans during roasting process, J. Food Q. pp. 1–11. (2017). https://doi.org/10.1155/2017/2969324
13. E.O. Afoakwa, Chocolate science and technology, John Wiley & Sons (2016)
14. W. Zzaman, T.A. Yang, Moisture, color and texture changes in cocoa beans during superheated steam roasting. J. Food Process. Preserv. **38**, 1364–1370 (2014). https://doi.org/10.1111/jfpp.12098
15. J.A. Hernández, B. Heyd, B. Valdovinos, G. Trystram, Analysis of the heat and mass transfer during coffee batch roasting. J. Food Eng. **78**, 1141–1148 (2007). https://doi.org/10.1016/j.jfoodeng.2005.12.041
16. M. Hinneh, E.E. Abotsi, D. Van de Walle, D.A. Tzompa-Sosa, A. De Winne, J. Simonis, K. Messens, J. Van Durme, E.O. Afoakwa, L. De Cooman, K. Dewettinck, Pod storage with roasting: a tool to diversifying the flavor profiles of dark chocolates produced from "bulk" cocoa beans? (Part ii: quality and sensory profiling of chocolates). Food Res. Int. **132**, 109116 (2020). https://doi.org/10.1016/j.foodres.2020.109116
17. A.C. Aprotosoaie, S.V. Luca, A. Miron, Flavor chemistry of cocoa and cocoa products-an overview. Compr. Rev. Food Sci. Food Safety **15**, 7391 (2016). https://doi.org/10.1111/1541-4337.12180
18. A.R. Tarek, I.S. Fahim, K.M. Mahran, Investigative review of design techniques of parabolic trough solar collectors. Front. Heat Mass Trans. **22**(1), 317–339 (2024)
19. L. Patoda, V. Parashar, Solar energy for domestic and small industries with help of parabolic trough solar concentrator. Int. J. Eng. Res. Technol. **2**, 2828–2832 (2013)
20. ASTM International, Standard specification and temperature electromotive force (emf) tables for standardized thermocouples, ASTME 230–17, ASTM International, West Conshohocken, PA

21. W.A. Schmid, R.J. Lazos Martínez, Guía para estimar la incertidumbre de la medición, Centro Nacional de Metrología (CENAM), El Marques, Querétaro, México. http://depa.fquim.unam.mx/amyd/archivero/EstimaciondelaIncertidumbre_32954.pdf
22. E. Martínez, L. Lira, Evaluación de la Incertidumbre de Algunos Factores de Influencia en la Determinación del Contenido de Humedad en Granos, Centro Nacional de Metrología, Querétaro, México 1096_1–1096_5 (2008). https://www.cenam.mx/simposio2008/sm_2008/memorias/M2/SM2008{M219{1096}.pdf
23. SAGARPA, Planeación agrícola nacional 2017–2030, SAGARPA https://www.gob.mx/agricultura/documentos/planeacion-agricola-nacional-2017-2030?state=published (Visited on September 22, 2024, 13:00 h)
24. M.E. Soudani, K.E. Aiadi, D. Bechki, S. Chihi, Experimental and theoretical study of parabolic trough collector (ptc) with a at glass cover in the region of algerian sahara (ouargla). J. Mech. Sci. Technol. **31**(8), 4003–4009 (2017). https://doi.org/10.1007/s12206-017-0747-3

Chapter 12
Fairness Index Analysis in Local Energy Communities Considering Electric Vehicles and Energy Storage

José Almeida, João Soares, Ehsan Aliyan, Steffen Limmer, Ricardo Faia, and Sérgio Ramos

Abstract This paper is centered on community-based energy sharing with a fairness analysis, focusing on integrating electric vehicles (EV) and energy storage systems. This work utilizes mixed integer linear programming optimization methods to decrease the overall energy expenses of the entire community by optimizing the energy distribution among its members. Multiple benefit distribution strategies were compared and analyzed in terms of fairness using the Jain index, F, and participation index to determine the best-performing strategies. This research enhances comprehension of the potential variations in community-based energy sharing based on the distribution method and fairness score. Three cases are proposed: PV without storage and EV forecasted demand, PV with storage still with no EV control, and PV, storage, and EV control. The influence of transactions in the energy community allows for a cost-benefit of around 7–14%. The benefit distribution approaches are also analyzed, and the fairness of each case is examined.

Keywords Benefit allocation · Electric vehicles · Energy storage systems · Fairness · Game theory · Local energy communities

J. Almeida (✉) · J. Soares · E. Aliyan · R. Faia · S. Ramos
GECAD—Research Group on Intelligent Engineering and Computing for Advanced Innovation and Development, LASI—Intelligent Systems Associate Laboratory, ISEP, Polytechnic of Porto, Porto, Portugal
e-mail: jorga@isep.ipp.pt

J. Soares
e-mail: jan@isep.ipp.pt

E. Aliyan
e-mail: ehsal@isep.ipp.pt

R. Faia
e-mail: rff@isep.ipp.pt

S. Ramos
e-mail: scr@isep.ipp.pt

S. Limmer
Honda Research Institute Europe, Offenbach, Germany
e-mail: steffen.limmer@honda-ri.de

D. Rossit et al. (eds.), *Proceedings of the VII Ibero-American Congress of Smart Cities, ICSC-Cities 2024, 12-14 November, San Carlos, Costa Rica*, Springer Proceedings in Energy, https://doi.org/10.1007/978-981-96-4301-1_12

12.1 Introduction

The decentralization of the electric power system, with the integration of multiple distributed energy resources, mainly photovoltaic (PV) generation, energy storage systems (ESS), and electric vehicles (EVs), allows a bidirectional flow of electricity through the injection of overproduction in the grid in a certain period. As such, in this new paradigm, traditional consumers can become prosumers with more active participation in the system's flexibility [7]. Local energy communities (LEC), often referred to as local electricity markets (LEMs), are gaining a lot of attention in the energy domain, allowing local transactions between community members at more profitable rates than transacting with the retailer [3, 6].

Multiple buildings constitute a LEC inside a specific radius depending on the region and the voltage level, enabling energy sharing, where the benefits obtained from forming a community are distributed among players through cost/benefit-sharing models [2]. Still, pricing strategies can be implemented, such as mid-market ratio (MMR), bill sharing (BS), and others, to determine the price of the transactions between users inside the community in a LEM trading mechanism [14]. Other approaches based on game theory are applied for cost/benefit distribution allowing fairer and more stable allocation of these benefits since they consider each combination (coalition) of players when forming a community [12]. Even though these processes can allocate these benefits more properly, fairness can still not be guaranteed [13]. The work in [8] proposes a novel energy-sharing model by the name of virtual net-billing and assesses the fairness of a LEC using multiple energy-sharing methods, some of them based on game theories like Shapley value (SV). It uses the F and Jain's fairness indexes to evaluate the validity of these methods. In [15], an overall performance index based on the economic and technical performance of multiple sharing mechanisms like supply and demand ratio (SDR), BS, and MMR is formulated in peer-to-peer (P2P) trading methodology.

This paper compares and analyzes multiple fairness indexes taken from [13] for multiple benefit distribution strategies using game theory. Different LEC pricing approaches such as BS, MMR, and SDR are implemented and compared to specific game theory allocation strategies like the SV and the Nucleolus, the first considered fair and the second considered stable [10]. Three different cases regarding the considered technologies are implemented, encompassing the integration of energy storage systems (ESS) and electric vehicles (EVs) for a total of 10 users, where two models are implemented regarding only grid transactions and transactions between users in a community manner.

Following an introduction to the topic and contributions of this work in Sect. 12.1, the mathematical formulation concerning LECs and a game theoretic approach is presented in Sect. 12.2. This formulation includes ESS and EV control optimization and benefit allocation strategies. The formulation covering the selected fairness indexes used to evaluate the fairness of the allocation techniques is described in Sect. 12.3. The case study of the LEC regarding the generated instances is given in Sect. 12.4.

Section 12.5 provides the findings of this study on the distribution of LEC benefits in various scenarios, together with the related fairness analysis. Section 12.6 contains the key conclusions and potential future topics for further investigation.

12.2 Proposed Formulation

This section presents the mathematical formulation used regarding the game theory approach for the LEC so that benefit allocation strategies such as the SV and Nucleolus and a simple equal benefit split (EQS) can be formulated. In addition, multiple pricing strategies, namely BS, MMR, and SDR, have been implemented, and their benefit distribution has been compared. EV and ESS control optimization is also shown in this section for each generated coalition in the game theory formulation.

12.2.1 Game Theory in Local Energy Communities

Multiple users, as denoted by the indices i, comprise a LEC. These users can be consumers or prosumers if they have production capabilities. In some instances, EV and ESS can be integrated for each user (or some), so more resources can be available to satisfy user demand and sell the overproduction. As such, the net generation and net consumption of each user are computed by $b_{i,t}^{netCon} = max(b_{i,t}, 0)$ and $b_{i,t}^{netGen} = min(b_{i,t}, 0)$. The independent cost of each user by only trading with the electricity grid at retail tariffs is given by Eq. (12.1) [10].

$$C_i = \sum_{t \in T} (pr_t^{\mathrm{buy}} \cdot b_{i,t}^{netCon} + pr_t^{\mathrm{sell}} \cdot b_{i,t}^{netGen}) \; \forall i \in \mathcal{N} \tag{12.1}$$

where pr_t^{buy} and pr_t^{sell} are the retail buy and sell prices when transacting with the electricity grid. The power balance for each user $b_{i,t}$ at each time step t is formulated in Eq. (12.2).

$$b_{i,t} = l_{i,t} - g_{i,t} + ess_{i,t}^{\mathrm{ch}} - ess_{i,t}^{\mathrm{dch}} + \sum_{k \in K_i} ev_{i,k,t}^{\mathrm{Hch}} - \sum_{k \in K_i} ev_{i,k,t}^{\mathrm{Hdch}} \; \forall i \in \mathcal{N} \; \forall t \in T \tag{12.2}$$

where the parameter $l_{i,t}$ represents the load demand of user i for the period t, the parameter $g_{i,t}$ describes the forecasted PV generation of the user i for the period t and T demonstrates the integer set of the total number of time steps (96 total). The decision variables associated with the ESS of the user i for the period t for charge and discharge are given by $ess_{i,t}^{\mathrm{ch}}$ and $ess_{i,t}^{\mathrm{dch}}$, respectively. Similarly, variables $ev_{i,k,t}^{\mathrm{Hch}}$ and $ev_{i,k,t}^{\mathrm{Hdch}}$ give the home charging and discharging of EV k owned by user i for each period t.

Note that some users are only consumers and do not have PV and ESS capabilities, so the parameters associated with these technologies become zero.

When a LEC $\mathcal{N}$ is formed with a total of N users and energy sharing is allowed, the community cost is determined by Eq. (12.3), where the community balance is calculated using cooperative ESS and EV technologies.

$$C_{\mathcal{N}} = \sum_{t \in T} (pr_t^{\text{buy}} \cdot \sum_{i \in \mathcal{N}} b_{i,t}^{netCon} + pr_t^{\text{sell}} \cdot \sum_{i \in \mathcal{N}} b_{i,t}^{netGen}) \tag{12.3}$$

where the benefit obtained by forming a LEC compared to independently transacting with the grid is described by Eq. (12.4).

$$v(\mathcal{N}) = \sum_{i \in \mathcal{N}} C_i - C_{\mathcal{N}} \tag{12.4}$$

where $v(\mathcal{N})$ represents the total community benefit, C_i are the singleton costs of each user i and $C_{\mathcal{N}}$ are the costs of the community $\mathcal{N}$.

Using a game theoretic approach, the representation of the community can be interpreted as a cooperative game, where coalitions (multiple sub-communities) between users are made that cooperate to maximize the coalition benefits. A transferrable utility game is a pair $(\mathcal{N}, v)$, where $\mathcal{N}$ is a finite set of users in the community and v in Eq. (12.5) is formulated to represent the benefits of each coalition.

$$v(S) = \sum_{i \in S} C_i - C_S \tag{12.5}$$

where $v(S)$ represents the benefit of coalition S, C_i the singleton costs of each user i inside each coalition and C_S the total cost of each coalition S. The total number of possible sub-coalitions is $2^N - 1$.

A payoff allocation vector, whose entry x_i for each coalition represents the benefit for each user $i \in S$. With these concepts introduced, it is possible to calculate the satisfaction or, in our case, the dissatisfaction of each user inside the coalition, represented by the excess value in Eq. (12.6), which needs to be minimized.

$$e(x, S) = v(S) - \sum_{i \in S} x_i^* \tag{12.6}$$

where x_i^* is the benefit allocation distributed to user i for the benefit distribution strategy $*$. If $e(x, S) \geq 0$, the coalition S accrues more advantages independently than inside the community. Otherwise, the coalition is satisfied inside the community.

12.2.2 Optimization Considering EVs

Implementing a schedule optimization model is necessary when considering technologies such as EVs and ESS. The total energy cost is minimized by operating ESS and EVs in each coalition [9]. Equation (12.7) formulates the objective function in each coalition.

$$min\ C_S = \sum_{t \in T} (pr_t^{\text{buy}} \cdot \sum_{i \in S} b_{i,t}^{netCon} + pr_t^{\text{sell}} \cdot \sum_{i \in S} b_{i,t}^{netGen}) \tag{12.7}$$

This objective function is subject to many constraints that can be found in [1]. The constraints are the maximum import and export limits imposed for each user, the charging and discharging limits associated with each ESS, the operation state of each, their state-of-charge (SoC), and associated limits. Concerning EV constraints, we also consider the maximum charge and discharge bounds, the EV battery SoC, the corresponding limits, and the home charging and discharging equations.

12.2.3 Benefit Allocation

When energy sharing is possible, and grid transactions are reduced, various benefit distribution systems inside a LEC for each user must be evaluated to distribute the obtained benefits fairly. In this context, we apply the equal benefit split (EQS), the BS, MMR, and SDR mechanisms; these last three are based on pricing schemes and two game theory approaches, namely the SV and nucleolus.

Equal benefit split The equal benefit split (EQS) given in Eq. (12.8), equally divides the payoffs between each user inside the LEC so that each user will receive the same amount of benefits in reducing the electricity bill.

$$x_i^{EQS} = v(\mathcal{N})/N \ \forall i \in \mathcal{N} \tag{12.8}$$

Bill sharing Using a pro-rata mechanism, the BS method allocates the overall coalitional deficit power consumption payment or surplus electricity generation revenue to individual prosumers based on their fraction of net energy demand or energy generation in the entire community. The buy and sell LEC prices are determined by Eqs. (12.9)–(12.10). In this context, users without overproduction have to pay according to the buy price, and users with overproduction are paid according to the sell price. The motivation for this strategy is regarding the lower buy and higher sell prices that these users have in the LEC context in contrast to the retail prices.

$$pr_t^{\text{L,buy}} = \frac{\sum_{t \in T} pr_t^{\text{buy}} \cdot max(\sum_{i \in \mathcal{N}} b_{i,t}, 0)}{\sum_{t \in T} \sum_{i \in \mathcal{N}} b_{i,t}^{netCon}} \tag{12.9}$$

$$pr_t^{\text{L, sell}} = \frac{\sum_{t \in T} pr_t^{\text{sell}} \cdot |min(\sum_{i \in \mathcal{N}} b_{i,t}, 0)|}{\sum_{t \in T} \sum_{i \in \mathcal{N}} |b_{i,t}^{netGen}|} \tag{12.10}$$

Mid-market rate Regarding the MMR pricing strategy, an average pricing between the buy and sell retail prices. If the total community balance is zero $\sum_{i \in \mathcal{N}} b_{i,t} = 0$ (demand matches generation), the LEC buy and sell prices are given by Eq. (12.11).

$$pr_t^{\text{MMR}} = \frac{pr_t^{\text{buy}} + pr_t^{\text{sell}}}{2} \quad \forall t \in T \tag{12.11}$$

When the balance is positive ($\sum_{i \in \mathcal{N}} b_{i,t} > 0$), the selling price for users with positive net generation is the pr_t^{MMR} and the buying price for users with a positive net consumption is formulated in Eq. (12.12).

$$pr_t^{\text{L,buy}} = \frac{pr_t^{\text{MMR}} \cdot |\sum_{i \in \mathcal{N}} b_{i,t}^{netGen}| + pr_t^{\text{buy}} \cdot \sum_{i \in \mathcal{N}} b_{i,t}}{\sum_{i \in \mathcal{N}} b_{i,t}^{netCon}} \quad \forall t \in T \tag{12.12}$$

In contrast, when the LEC balance is negative ($\sum_{i \in \mathcal{N}} b_{i,t} < 0$), the buying price for users with a positive net consumption is given by pr_t^{MMR} and the selling price for users with a positive net generation is now presented in Eq. (12.13).

$$pr_t^{\text{L,sell}} = \frac{pr_t^{\text{MMR}} \cdot \sum_{i \in \mathcal{N}} b_{i,t}^{netCon} + pr_t^{\text{sell}} \cdot |\sum_{i \in \mathcal{N}} b_{i,t}|}{|\sum_{i \in \mathcal{N}} b_{i,t}^{netGen}|} \quad \forall t \in T \tag{12.13}$$

Supply-demand rate The SDR pricing [4] strategy sets the internal community price according to a ratio between the community net generation and net demand, formulated by Eq. (12.14).

$$r_t^{\text{sdr}} = \frac{|\sum_{i \in \mathcal{N}} b_{i,t}^{ng}|}{\sum_{i \in \mathcal{N}} b_{i,t}^{nc}} \quad \forall t \in T \tag{12.14}$$

The pr_t^{SDR} can be obtained by following the formulation in [4] for the calculations of the electricity bill of each user can be computed. Using the BS, MMR, and SDR pricing strategies, the benefit of each user in the LEC is calculated in Eq. (12.15).

$$x_i^* = C_i - \sum_{t \in T} (pr_t^{\text{L,buy}} \cdot \max(b_{i,t}, 0) + pr_t^{\text{L,sell}} \cdot \min(b_{i,t}, 0)) \; \forall i \in \mathcal{N} \tag{12.15}$$

Shapley value The SV in game theory allocates the benefits among users according to their marginal contribution. The method evaluates the significance of each player i in a game by computing a weighted average of their incremental contributions towards all possible coalitions they are allowed to join. In the context of LECs, users who contribute more to the benefit of the community get a higher share of it. The benefit given by SV for each player is formulated in Eq. (12.15).

$$x_i^{SV} = \sum_{S \subseteq \mathcal{N} \setminus \{i\}} \frac{|S|!(|\mathcal{N}| - |S| - 1)!}{|\mathcal{N}|!} \cdot (v(S \cup \{i\}) - v(S)) \quad (12.16)$$

In game theory, SV is considered a fair approach since it meets four different criteria: additivity, efficiency, null player, and symmetry, as [8] demonstrates.

Nucleolus The Nucleolus in a cooperative game lexicographically minimizes the excess values; that is, it optimizes player dissatisfaction, as shown in Eq. (12.6). By defining the excess vector $\epsilon(x) = e(x, S_j) \in K \backslash N, \forall j = 1, ..., 2^N - 2$, the Nucleolus is the optimization of j linear programming (LP) models. The formulation in [11] was used to find the payoff allocation vector using Nucleolus using a single LP optimization.

12.3 Fairness Indexes

Different fairness indexes are employed as measures or instruments to quantify and assess the fairness of a process or system. The first index we employ is the Jain's Fairness Index (JFI), often used to estimate the fairness of distributions or allocations. Equation (12.17) displayed the formulation of this index, assuring equality regarding the benefit allocation distribution among community users and preventing the monopolization of gains by one user.

$$JFI = \frac{(\sum_{i=1}^{N} x_i^*)^2}{N \cdot \sum_{i=1}^{N} x_i^{*2}} \quad (12.17)$$

where N is the total number of users in the community, x_i benefit allocation vector of the distribution strategy. Another index we use is the F fairness index shown in Eq. (12.19), which compares each benefit allocation strategy with the SV strategy, which in game theory is considered as fair. So, with a smaller value, the strategy is similar to the SV distribution being more fair.

$$F = \sum_{i-1}^{N} \left| \frac{x_i^*}{\sum_{i=1}^{N} x_i^*} - \frac{x_i^{SV}}{\sum_{i=1}^{N} x_i^{SV}} \right| \quad (12.18)$$

where x_i^{SV} represents the allocation of the SV method. The third fairness metric encompasses the willingness of users to participate in the energy community, and it is depicted in Eq. (12.19), varying between zero and one with a higher value representing a higher willingness to participate in the LEC.

Table 12.1 Number of considered cases considered in the experiments made

Cases	Technologies	Model
Case 1	PV + EV^f	BAU
		LEC
Case 2	PV + EV^f + ESS	BAU
		LEC
Case 3	PV + EV + ESS	BAU
		LEC

f—EV forecasted demand with predicted load values

$$PI = \frac{N_{CostReduction}}{N} \tag{12.19}$$

where $N_{CostReduction}$ represents the number of users that reduced their costs in the LEC trading paradigm compared to trading directly to the grid for a given benefit allocation strategy.

12.4 Case Study

The dataset in [5] was utilized for the simulations implemented, where a total of 10 users were taken from the dataset together with the technology data, such as consumption, PV generation data, ESS, and EV data. Three different cases are proposed based on the technologies considered, which are shown in Table 12.1.

The first case considers only PV generation for auto consumption and EV forecasted demand with predetermined values, where the charging is made in the first periods of the time horizon T until the charging meets the needed demand. In the second case, ESS are introduced with control optimization to determine the optimal charging and discharging, and the third case considers EV control optimization together with V2H capabilities. Two models are considered for each case, the business-as-usual (BAU) and the LEC, wherein the first, the users can only trade with the electricity grid, that is, only trade with the retailer. In the second model, users form a LEC, where energy can be traded between users inside the community.

Figure 12.1 shows the total aggregated PV generation and load consumption for 10 users during a single day with 96 periods. Between periods 43 and 57, there is an overproduction in the community, meaning that this electricity can be traded with the grid in the BAU model and shared between users in the LEC model.

The retail tariffs are presented in Fig. 12.2, showing the feed-in tariff for exporting the exceeding electricity back into the grid and the time-of-use (ToU) buying tariff used when importing electricity from the grid. A constant value of 0.05 €/kWh was

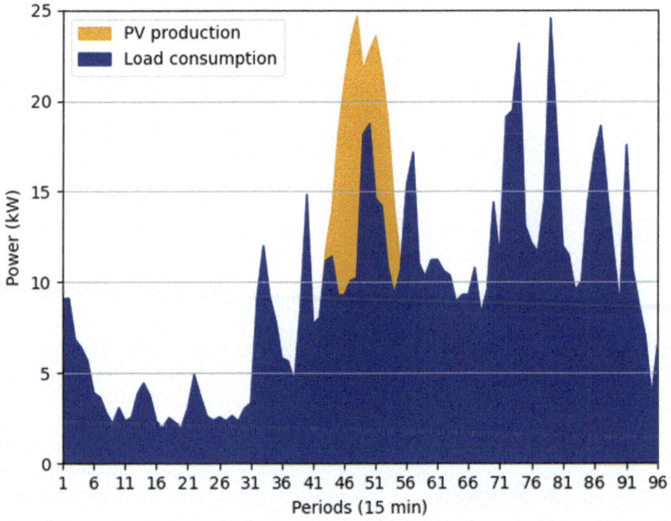

Fig. 12.1 Total aggregated community load demand and PV production

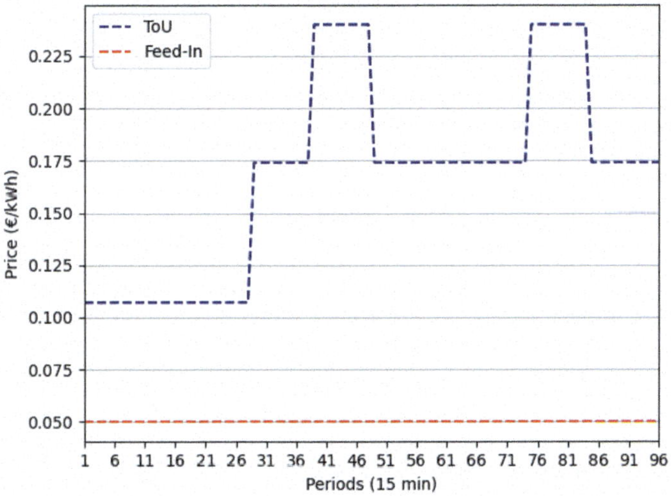

Fig. 12.2 Retail tariffs for electricity buy and sell to the grid

considered for the feed-in, and a variation of values between 0.1072 €/kWh and 0.24 €/kWh were considered for the ToU, taken from a Portuguese retailer.[1]

[1] https://sueletricidade.pt/.

Simulations were done on a Windows 11 Pro workstation with a 13th Gen Intel® Core™i9-13900 2.00 GHz CPU and 32 GB of RAM using the Python programming language, with the Gurobi optimization solver for the BES and EV optimization control.

12.5 Results

This section shows the results of the proposed methodologies, comparing the benefit distribution strategies and a fairness analysis via three different indexes described previously.

12.5.1 Benefit Allocation

Concerning the results, Table 12.2 presents the total costs for each proposed case for ten users in both BAU and LEC models. The table also shows the total benefits when the LEC is formed, local energy-sharing between users is possible, and the cost reductions between cases for each model. When analyzing case 1 with just PV generation and forecasted EV demand without any control and ESS, the costs for the BAU model are higher due to the absence of interactions between users. Yet, when users establish a LEC with energy-sharing capabilities, the expenses are decreased by 3.13 €. The introduction of ESS results in an 11.64% reduction in costs in the BAU model and a 16.21% reduction in the LEC model, compared to the case when PV is the only form of production. A benefit of 4.55 € is observed in the costs within the community between the BAU and LEC in the second case. The incorporation of EV control optimization with V2H capacity allows a reduction in costs of 13.41% and 23.49% in the BAU model compared to cases 1 and 2, respectively. The same analysis is made for the LEC model, where reductions of 15.37 and 29.09% are verified when having V2H. In this case, a benefit of 4.58 € was obtained when a LEC is in place.

The 10-user community balance when incorporating ESS and EV is shown in Fig. 12.3, where positive values imply the net demand and negative values the net

Table 12.2 Overall cost results for each generated case

	Case 1 (€)	Case 2 (€)	Case 3 (€)	Reduction 1–2 (%)	Reduction 2–3 (%)	Reduction 1–3 (%)
BAU	42.11	37.21	32.22	11.64	13.41	23.49
LEC	38.98	32.66	27.64	16.21	15.37	29.09
Benefits	3.13	4.55	4.58	–	–	–

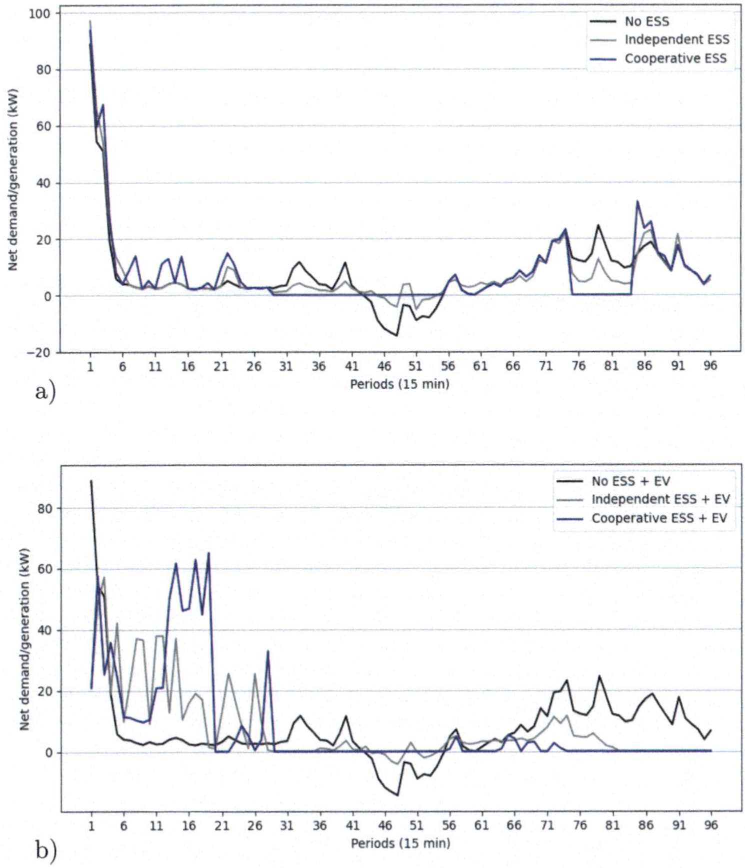

Fig. 12.3 Electricity balance of a 10-user game **a** using no ESS, independent ESS, and cooperative ESS; **b** using no ESS and EV, independent ESS and EV and cooperative ESS and EV

generation. Figure 12.3a) shows the case where users do not operate ESS units (black line), the case where users operate these devices independently (grey line), only transacting with the grid, and the case where users use cooperation in the operation of their ESS to minimize coalition costs (blue line). Large initial values in net demand can be seen due to the non-controlled charging of EVs and also in the cooperative ESS operation. The exceeding PV generation is stored in the afternoon on the ESS units and discharged during nightly periods when the buy price is high. Figure 12.3b) is similar to the addition of EVs with V2H capabilities. In this case, the introduction of EV control reduced the initial high net demand values and the V2H capabilities in the cooperative control, allowing EVs to be discharged when the electricity buy price from the grid is high, making the net demand null in those periods.

The benefit distributions for each case and each implemented strategy are illustrated in Fig. 12.4. The recommended solutions in case 1 (upper subplot) appear to

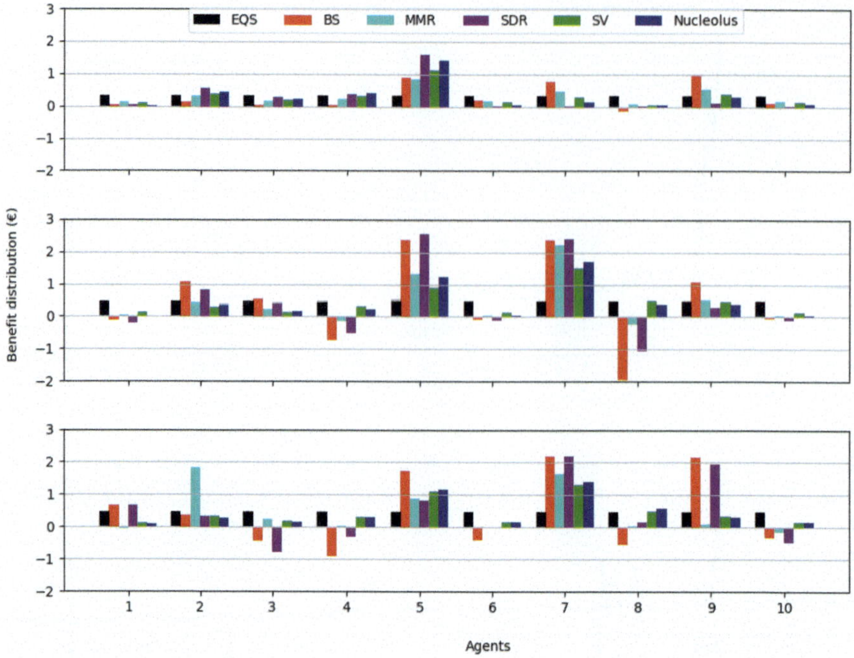

Fig. 12.4 Benefit distributions according to the considered mechanisms for case 1 (upper), case 2 (center), and case 3 (lower)

benefit user 5 most, particularly when the SDR and Nucleolus methods are employed. In this case, the BS method presents negative benefits for the eighth user, meaning that this user will have a higher electricity bill when inside the LEC. In the second case (center subplot), only EQS, SV, and Nucleolus guarantee that users do not have negative benefits, with BS and SDR having negative benefits for users 1, 4, 6, 8, and 10 and MMR for users 4 and 8. The fifth and seventh users obtain the highest benefits for all strategies since they have high load consumption values and benefit the most inside the LEC, trading electricity at lower prices. A similar analysis can be made for the last scenario, where in this situation, BS presented negative benefits for users 3, 4, 6, 8, and 10, MMR presented for users 1, 6, and 10, and SDR presented negative benefits for users 3, 4, 6 and 10. These situations, where users get negative benefits, mean they would be better off outside the LEC since, with the technologies available, the users can meet their demand, and the remaining excess energy is not transacted at attractive prices in these strategies.

Table 12.3 Fairness index analysis for each allocation method and proposed case

Strategies	Case 1			Case 2			Case 3		
	JFI	F	PI	JFI	F	PI	JFI	F	PI
EQS	**1.00**	0.60	**1.00**	**1.00**	0.69	**1.00**	**1.00**	0.67	**1.00**
BS	0.40	0.73	0.90	0.11	1.82	0.50	0.14	1.71	0.50
MMR	0.65	**0.30**	**1.00**	0.28	0.64	0.80	0.30	0.82	0.70
SDR	0.31	0.54	**1.00**	0.14	1.51	0.50	0.19	1.33	0.60
SV	0.54	**0.00**	**1.00**	0.54	**0.00**	**1.00**	0.58	**0.00**	**1.00**
Nucleolus	0.39	0.31	**1.00**	0.42	**0.28**	**1.00**	0.53	**0.09**	**1.00**

12.5.2 Fairness Analysis

When it comes to the fairness analysis through the implementation of three different indexes (JFI, F, and PI), Table 12.3 presents the values obtained for each index for each case and benefit allocation strategy to evaluate their fairness according to these indexes. In the case of Jain's index, the EQS method achieves the perfect scores for each case since this index represents equality in the payoffs for each user, with MMR for case 1 and SV for the remaining also presenting somewhat good equality payoffs.

Concerning the F index, SV is considered the ideal value since, in game theory, it is considered a fair approach. As such, the values for this method are null. In case 1, the method closer to SV is MMR, with Nucleolus coming in a close second. In the second and third cases, the method that archives similar fairness to SV is Nucleolus, which, in the final case, is almost equal in fairness to SV with a value close to zero.

The final fairness index, PI, shows that a higher value between [0, 1] represents that users are more willing to participate in the LEC since they obtain lower energy bills. In the first case, only BS obtains a lower value, meaning that not all users obtain lower costs in the LEC, as confirmed by Fig. 12.4, where user eight pays more for being in an LEC. In the other two cases, only EQS, SV, and Nucleolus obtained perfect scores, with all users willing to participate in the LEC. The other methods showed that some users pay more for being inside the LEC and would be better off trading only with the electricity grid.

12.6 Conclusions

The present paper addressed many benefit distribution techniques for LECs using game-theoretic methodologies, assessing the fairness of these distributions via different fairness indexes proposed in the literature. Three cases were created: one focusing on users having PV equipment to trade, another incorporating ESS with discharging capacities, and in the final case, EVs were integrated with V2H capabilities, allowing battery discharge. The forecasted demand for EVs was considered in the first two cases, with no EV charging control.

The findings indicate that implementing ESS resulted in a decrease in user costs ranging from 11 to 16%. The introduction of EV control optimization reduced the overall costs between 13 and 29%, depending on the case and implemented model (BAU or LEC). When analyzing the benefit distribution methods implemented, it was possible to observe that EQS, SV, and Nucleolus were the ones that guaranteed that, in all cases, the users obtained a positive benefit with a decrease in the electricity bill. Regarding fairness analysis, depending on the case and fairness index, different methods seem to be more fair than others, with SV and Nucleolus finding a good balance in all the indexes used.

Further extension of the presented work is possible. The authors want to use the shown techniques on a broader scope, such as with energy communities consisting of a large number of participants with other instances, using approximation approaches to address both stability and fairness concerns. The authors also intend to incorporate more realistic strategies for energy communities, such as regulatory considerations and network limitations, to effectively organize a group of interested nearby members to enhance benefits and address social and environmental concerns.

Acknowledgements This work was supported by the Interreg Atlantic Area Programme through the European Regional Development Fund (EAPA_0019/2022—SAtComm). The work has also been supported by Honda Research Institute Europe. In addition, the Portuguese team acknowledges the research facilities and conditions provided by GECAD, partially funded by Fundacao para a Ciencia e Tecnologia (FCT), namely grant UIDB/00760/2020 (https://doi.org/10.54499/UIDB/00760/2020). Joao Soares has an FCT grant, namely CEECIND/00420/2022 to support his research work. José Almeida is supported by FCT with Ph.D. Grant 2022.09590.BD.

Disclosure of Interests No competing interests to declare.

References

1. J. Almeida, J. Soares, S. Limmer, E. Aliyan, R. Faia, F. Lezama, S. Ramos, Z. Vale, Community-based energy sharing using game theory approaches for benefit distribution, in *2024 22nd International Conference on Intelligent Systems Applications to Power Systems (ISAP)*. pp. 1–6 (2024). https://doi.org/10.1109/ISAP63260.2024.10744299
2. E. Barabino, D. Fioriti, E. Guerrazzi, I. Mariuzzo, D. Poli, M. Raugi, E. Razaei, E. Schito, D. Thomopulos, Energy communities: a review on trends, energy system modelling, business models, and optimisation objectives. Sustain. Energy, Grids Netw. **36**, 101187 (2023). https://doi.org/10.1016/j.segan.2023.101187, https://linkinghub.elsevier.com/retrieve/pii/S2352467723001959
3. A. Bartolini, F. Carducci, C.B. Muñoz, G. Comodi, Energy storage and multi energy systems in local energy communities with high renewable energy penetration. Renew. Energy **159**, 595–609 (2020). https://doi.org/10.1016/j.renene.2020.05.131, https://linkinghub.elsevier.com/retrieve/pii/S0960148120308351

4. M.F. Dynge, K. Berg, S. Bjarghov, U. Cali, Local electricity market pricing mechanisms' impact on welfare distribution, privacy and transparency. Appl. Energy **341**, 121112 (2023). https://doi.org/10.1016/j.apenergy.2023.121112, https://linkinghub.elsevier.com/retrieve/pii/S0306261923004762
5. R. Faia, C. Goncalves, L. Gomes, Z. Vale, Dataset of an energy community with prosumer consumption, photovoltaic generation, battery storage, and electric vehicles. Data Brief **48**, 109218 (2023). https://doi.org/10.1016/j.dib.2023.109218, https://linkinghub.elsevier.com/retrieve/pii/S2352340923003372
6. R. Faia, F. Lezama, J. Soares, T. Pinto, Z. Vale, Local electricity markets: a review on benefits, barriers, current trends and future perspectives. Renew. Sustain. Energy Rev. **190**, 114006 (2024). https://doi.org/10.1016/j.rser.2023.114006, https://linkinghub.elsevier.com/retrieve/pii/S136403212300864X
7. Z. Foroozandeh, S. Limmer, F. Lezama, R. Faia, S. Ramos, J. Soares, A MBNLP method for centralized energy pricing and scheduling in local energy community, in *2022 IEEE PES Generation, Transmission and Distribution Conference and Exposition—Latin America (IEEE PES GTD Latin America)*. IEEE, La Paz, Bolivia, pp. 1–6 (2022). https://doi.org/10.1109/IEEEPESGTDLatinAmeri53482.2022.10038287, https://ieeexplore.ieee.org/document/10038287/
8. V.Z. Gjorgievski, S. Cundeva, N. Markovska, G.E. Georghiou, Virtual net-billing: a fair energy sharing method for collective self-consumption. Energy **254**, 124246 (2022). https://doi.org/10.1016/j.energy.2022.124246, https://linkinghub.elsevier.com/retrieve/pii/S0360544222011495
9. J. Li, Y. Ye, D. Papadaskalopoulos, G. Strbac, Computationally efficient pricing and benefit distribution mechanisms for incentivizing stable peer-to-peer energy trading. IEEE Internet Things J. **8**(2), 734–749 (2021). https://doi.org/10.1109/JIOT.2020.3007196, https://ieeexplore.ieee.org/document/9134398/
10. S. Limmer, Empirical study of stability and fairness of schemes for benefit distribution in local energy communities. Energies **16**(4), 1756 (2023). https://doi.org/10.3390/en16041756, https://www.mdpi.com/1996-1073/16/4/1756
11. J. Puerto, F. Perea, Finding the nucleolus of any n-person cooperative game by a single linear program. Comput. Oper. Res. **40**(10), 2308–2313 (2013). https://doi.org/10.1016/j.cor.2013.03.011, https://linkinghub.elsevier.com/retrieve/pii/S0305054813000877
12. A. Safdarian, P. Astero, M. Baranauskas, A. Keski-Koukkari, A. Kulmala, Coalitional game theory based value sharing in energy communities. IEEE Access **9**, 78266–78275 (2021). https://doi.org/10.1109/ACCESS.2021.3081871, https://ieeexplore.ieee.org/document/9435374/
13. J. Soares, F. Lezama, R. Faia, S. Limmer, M. Dietrich, T. Rodemann, S. Ramos, Z. Vale, Review on fairness in local energy systems. Appl. Energy **374**, 123933 (2024). https://doi.org/10.1016/j.apenergy.2024.123933, https://linkinghub.elsevier.com/retrieve/pii/S0306261924013163
14. C. Zhang, J. Wu, Y. Zhou, M. Cheng, C. Long, Peer-to-peer energy trading in a microgrid. Appl. Energy **220**, 1–12 (2018). https://doi.org/10.1016/j.apenergy.2018.03.010, https://linkinghub.elsevier.com/retrieve/pii/S0306261918303398
15. Y. Zhou, J. Wu, C. Long, Evaluation of peer-to-peer energy sharing mechanisms based on a multiagent simulation framework. Appl. Energy **222**, 993–1022 (2018). https://doi.org/10.1016/j.apenergy.2018.02.089, https://linkinghub.elsevier.com/retrieve/pii/S0306261918302149

Part III
Electric Systems for Smart Cities

Chapter 13
Clustering of Real Load Curves at a Public University: Steps Towards a Sustainable Campus

Walquiria N. Silva, Bruno H. Dias, Luís H. B. Bandória, Madson C. de Almeida, and Jesús. C. Hernández

Abstract With the growing importance of efficient energy management in educational institutions, this study employs the K-means clustering technique with the Soft-DTW metric to classify active and reactive power measurements from service transformers at the Federal University of Juiz de Fora (UFJF) campus. The primary goal is to identify typical load curve profiles according to the characteristics of the loads, thereby characterizing the nature of usage of the buildings served by these transformers. The resulting typical curves enable more realistic network studies, such as energy procurement, transformer sizing, photovoltaic plant allocation, and electric vehicle charger placement. This approach addresses the gap in existing research by providing a robust method for clustering load profiles that can be applied to other universities where load curves are not measured. Clustering the load curves makes it feasible to design low-cost measurement systems by sampling only a small subset of transformers with similar load characteristics. The results demonstrated the superior performance of Soft-DTW over Euclidean distance in aligning and grouping time series with varied profiles. This enhanced clustering accuracy is fundamental for effective load profiling, supporting informed decision-making

W. N. Silva (✉)
Institute of Energy and Environment (IEE), University of São Paulo (USP), São Paulo, Brazil
e-mail: walquiria.nasc.20@gmail.com

B. H. Dias
Department of Electrical Energy, Federal University of Juiz de Fora—UFJF, Juiz de Fora, MG, Brazil
e-mail: bruno.dias@ufjf.edu.br

L. H. B. Bandória · M. C. de Almeida
Department of Systems and Energy, Universidade Estadual de Campinas-Unicamp, Campinas, SP, Brazil
e-mail: l156449@dac.unicamp.br

M. C. de Almeida
e-mail: madson.almeida@unicamp.br

Jesús. C. Hernández
Department of Electrical Engineering, University of Jaen—UJA, Jaén, Spain
e-mail: jcasa@ujaen.es

D. Rossit et al. (eds.), *Proceedings of the VII Ibero-American Congress of Smart Cities, ICSC-Cities 2024, 12-14 November, San Carlos, Costa Rica*, Springer Proceedings in Energy, https://doi.org/10.1007/978-981-96-4301-1_13

in power system management and planning. In conclusion, the advanced clustering techniques applied in this research provide a deeper understanding of electrical load management, suggesting potential extensions to other facilities and broader factors influencing consumption patterns. Ultimately, this research contributes to optimizing power systems to meet better the demands of a dynamic and evolving energy landscape.

Keywords Clustering · Load profile · Power system · Smart Campus · Smart Grids

13.1 Introduction

Faced with growing concerns about climate change, the scarcity of supplies from fossil fuels [18], and the increasing environmental impacts of traditional energy generation [8], there has been a significant push to replace conventional sources of energy generation in the electricity matrix with generation systems that use renewable energy sources, such as wind and solar [1]. This transition of the energy matrix to more sustainable bases directly reflects environmental challenges. It is, therefore, a strategy in line with energy transition policies that seek a global reformulation of energy systems [9].

This context of energy transition intensely involves the development and improvement of research in the areas of Distributed Energy Resources (DERs). These resources include renewable energies, distributed generation systems, electric mobility, and other fundamental aspects for a more sustainable energy matrix [10]. In addition, the need for energy storage, the implementation of energy efficiency policies, demand response, and energy conservation are complementary elements that enrich this scenario [13]. Together, these mechanisms provide a favorable environment for the emergence of sustainable initiatives on university campuses [4]. These initiatives aim to reduce operating costs and mitigate environmental impacts by developing and implementing solutions that maximize energy use efficiency, contributing significantly to sustainability [5].

From the perspective of reducing and optimizing energy consumption, the essential starting point for any effective strategy involves the detailed characterization and quantification of electricity consumption profiles [14, 15]. This critical information is obtained by analyzing load curves, which map energy consumption throughout the day or at specific time intervals [16, 17]. Through a thorough analysis of these curves, it is possible to identify particular usage patterns, offering essential insights that improve consumers' energy performance [12].

This approach aims to identify areas of inefficiency, highlighting critical points where consumption can be effectively reduced. Based on these considerations, it is possible to target the implementation of energy-saving measures that can be adapted to users' real needs, resulting in more efficient and personalized energy consumption management, resource optimization, and energy sustainability.

With advances in smart grid technologies and the widespread collection of data by smart meters and meters installed on distribution transformers, the opportunity has arisen to cluster consumers based on their characteristic energy demands [11]. This clustering process aims to improve understanding of the various types of energy demand and is a catalytic tool for improving the planning and operation of the electricity system. Clustering thus becomes a fundamental pillar in distribution system planning, as it allows loads to be predicted and classified in such a way as to ensure both stability and reliability in energy delivery. Furthermore, detailing the patterns of magnitude and temporal variability of demand enables effective demand management, facilitating optimization in energy contracting and the sizing of infrastructure such as transformers. This process contributes significantly to the efficient allocation of distributed energy resources, improving the overall efficiency of the electricity system [2].

In this context, this work aims to explore load curves collected on a university campus to identify distinct energy consumption profiles between the various buildings. This analysis will allow for a more detailed understanding of the specific energy demands in diverse academic environments. To carry out this study, the knowledge Discovery in Databases (KDD) methodology will be simplified, focusing on the data mining stage, consisting of applying the K-means clustering algorithm. This algorithm is recognized for its robustness and widespread use in identifying natural groupings within multidimensional data sets [3]. The curves obtained are essential as they provide a solid basis for developing efficient and cost-effective energy monitoring and analysis methodologies.

The article is structured as follows: this section presents the context and objective of this study. Section 13.2 presents the methodology for data clustering analysis, and Sect. 13.3 discusses the results obtained from the clustering analysis on a university campus as a case study. Finally, Sect. 13.4 presents the conclusions drawn from this study.

13.2 Methodology

This section describes the methodology applied in this study, which follows the knowledge Discovery in Databases (KDD) process proposed by [6, 7]. This process involves interconnected steps that transform raw data into proper knowledge. The proposed methodology is illustrated in Fig. 13.1, summarizing the methodological steps adopted.

As seen in Fig. 13.1, the methodology is divided into five main stages, with particular attention to data mining through the clustering technique. The division of these stages aims to specifically address the characteristics and complexities of the data collected from energy meters. The stages are as follows:

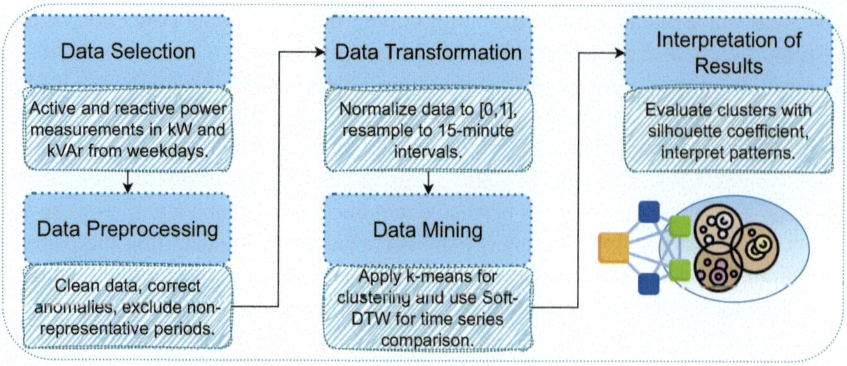

Fig. 13.1 Proposed methodology simplified flowchart

1. **Data Selection**: The selected data includes measurements of active (P) and reactive (Q) power in kW and kVAr from low-voltage transformer electrical measurements through the connection of meters. Each dataset contains several measurements recorded over time. For this study, only weekdays were selected as they represent the analyzed case study more.
2. **Data Preprocessing**: This stage involves cleaning and preparing the data, including correcting anomalous or missing values and excluding non-representative periods such as holidays and maintenance. Abnormal values can be corrected using statistical methods, such as replacing outliers with the median or mean of the data. Excluding non-representative periods ensures that atypical events do not influence subsequent analysis.
3. **Data Transformation**: Data normalization and resampling are performed in this stage to ensure consistency and facilitate comparative analyses. Normalization is a process that adjusts data values to a standard range, usually [0,1], and can be performed using min-max normalization, as Eq. 13.1:

$$x' = \frac{x - x_{min}}{x_{max} - x_{min}}, \tag{13.1}$$

where x' is the normalized value, x is the original value, x_{min} is the minimum value of the dataset, and x_{max} is the maximum value of the dataset.
The transformation includes resampling the data to 15-min intervals, resulting in 96 data points for each daily curve. This resampling ensures that the data is consistently formatted and allows for detailed temporal analysis of the power consumption patterns throughout the day. The active and reactive power measurements are organized into separate arrays, and each resulting matrix is normalized to the [0,1] range.
4. **Data Mining**: This research applies the K-means algorithm to cluster load curves to identify distinct energy consumption patterns. K-means was selected for its effectiveness in grouping high-dimensional data and applicability in time series

studies. The K-means algorithm partitions the data into k clusters, minimizing the sum of squared distances between data points and the cluster centroid, as per Eq. 13.2:

$$J = \sum_{i=1}^{k} \sum_{j=1}^{n} ||x_j^{(i)} - c_i||^2, \tag{13.2}$$

where J is the objective function, $x_j^{(i)}$ represents the data points in the i-th cluster, and c_i is the centroid of the i-th cluster.
To enhance the analysis and assess clustering effectiveness, two distance metrics are compared: the traditional Euclidean distance and the Soft-DTW (Dynamic Time Warping) metric. The Soft-DTW is particularly suitable for time series analysis because it accounts for shifts and distortions in time, making it highly recommended for identifying subtle patterns and nuances in load curves. Soft-DTW is defined by Eq. 13.3:

$$\text{soft-DTW}(X, Y) = \min_{\pi} \sum_{(i,j)\in\pi} d(x_i, y_j) \tag{13.3}$$

where X and Y are the time series to be compared, $d(x_i, y_j)$ is the distance between points x_i and y_j, and π is an alignment path that minimizes the sum of distances.
The analysis will test different values of k clusters, specifically with k varying from 3 to 5, to evaluate clustering quality across different numbers of clusters.

5. **Interpretation of Results**: The silhouette coefficient, which measures clustering quality, evaluates the clusters formed. The silhouette coefficient is calculated by Eq. 13.4:

$$s(i) = \frac{b(i) - a(i)}{\max\{a(i), b(i)\}} \tag{13.4}$$

where $a(i)$ is the average distance between i and all other points in the same cluster, and $b(i)$ is the minimum average distance from i to all points in any other cluster. The silhouette coefficient ranges from -1 to 1, with values closer to 1 indicating better clustering quality. Higher values imply that data points are well-matched to their own cluster and poorly matched to neighboring clusters. Interpreting the found patterns helps understand energy consumption dynamics and identify representative load curves.

Implementing this methodology aims to obtain a detailed and applicable understanding of energy consumption profiles, facilitating management and planning in energy distribution systems. This approach comprehensively analyzes energy behavior and supports operational and strategic decisions based on concrete data and robust evaluations.

13.3 Results and Discussions

This section first addresses the main characteristics of the case study, followed by the presentation of the primary results concerning the clustering of data related to active power, reactive power, and the clustering of the concatenated curve (a sequential form of active and reactive power) for the evaluated case study. These results were obtained by applying the proposed methodology in Sect. 13.2. All proposed analyses were conducted using the *Python* programming language.

13.3.1 Case Study

The proposed case study analyzes the load curve profiles of the Juiz de Fora University campus, part of the Federal University of Juiz de Fora (UFJF). The analyzed electrical network does not have advanced metering infrastructure; however, data from a measurement campaign conducted in 2018 from April to August were used. Power quality analyzers collected these data, specifically Qualimeter *Schneider Model Power Logic ION8600*, connected to the network's busbars. Figure 13.2 shows a simplified diagram of the UFJF network.

Given the loads' similarities, different university campus buildings may exhibit similar energy consumption profiles. A typification of the consumption profiles, it is emphasized that this study aims to cluster so that the algorithm can classify the types of loads that comprise the university campus. The main characteristics of the analyzed network are:

- The distribution utility services the substation: Cemig Distribuidora S.A. with a supply voltage of 23.1 kV;
- The network has three main circuits, totaling 24 busbars, one of which is dedicated to a photovoltaic system;

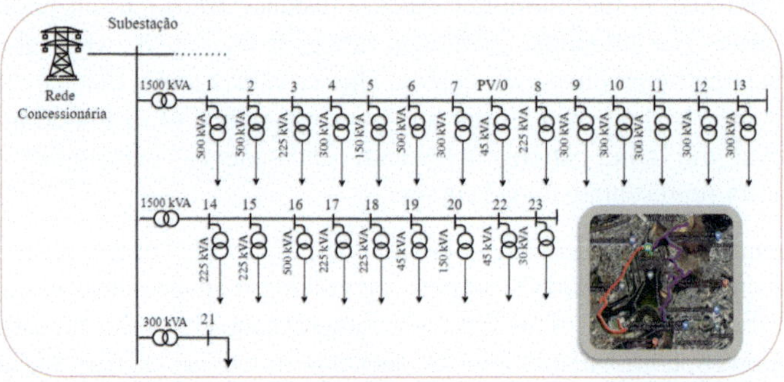

Fig. 13.2 Simplified diagram of the UFJF network

– Each busbar corresponds to a transformer that supplies different buildings, which mainly consist of classrooms, laboratories, and administrative offices, among others;
– The load curves for each busbar were collected over one week.

13.3.2 Clustering Evaluation

The K-means algorithm necessitates the prior specification of the number of clusters (k). This study aims to identify the optimal number of clusters for each scenario considered, with k ranging from 3 to 5. A representative curve is generated for each identified cluster, derived from the average behavior of all the curves within that cluster. The silhouette index was employed to evaluate the clustering quality for each number of clusters. Additionally, a visual assessment of the representative curve's behavior relative to the curves constituting each cluster was utilized to determine the optimal number of clusters.

13.3.2.1 Clustering Based on Active Power

In this analysis, the K-means algorithm, utilizing the Soft-DTW metric, was applied to the set of all normalized active power load curves corresponding to weekdays. This classification aimed to categorize the loads according to their active power consumption profiles. The representative curves for each cluster and the grouped curves for k = 3, k = 4, and k = 5 are presented in Fig. 13.3.

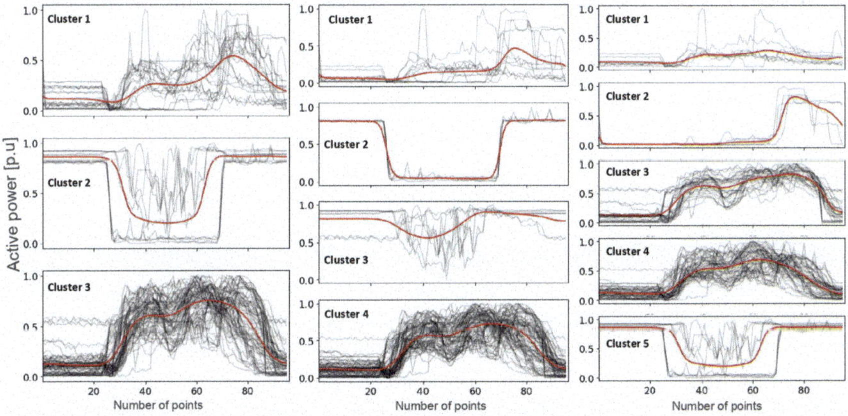

Fig. 13.3 Clustering of active power considering k = 3; k = 4; and k = 5

When analyzing Fig. 13.3, there is a representation of the grouping of active power curves in p.u, considering k = 3, k = 4, and k = 5. The centroid of the clusters in each case is represented in red, summarizing the typical load behavior within each cluster. The gray curves represent the samples collected through measurements and allocated into clusters based on the methodology adopted. These curves represent daily load profiles, with the number of points corresponding to the 24-hour daily cycle (96 points).

For k = 3, the method successfully identified the profiles of singular-use buildings, public lighting, and administrative/classroom buildings. Furthermore, it is noted that the profile of the photovoltaic panel was grouped with the public lighting profile, and the university cafeteria profile was grouped with the administrative/classroom buildings. Indeed, as presented in Sect. 13.2, the university campus comprises five distinct consumption profiles. Thus, it is expected that these profiles cannot be adequately identified with only three clusters.

For k = 4, the method identified the profiles of singular-use buildings, public lighting, photovoltaic panels, and administrative/classroom buildings, respectively. It is observed that among the five previously identified profiles, the consumption profile of the university refectory could have been distinctly identified with k=4, as it was grouped with the administrative/classroom buildings. This classification likely occurs due to the refectory's operating hours from 08:00 to 20:00, with peak usage during meal times, thus resembling the continuous usage profile of administrative/classroom buildings.

For k = 5, the method distinguished the singular-use building profiles into two distinct groups (Clusters 1 and 2) and the administrative/classroom building profiles into two distinct groups (Clusters 3 and 4) while grouping the photovoltaic panel profile with the public lighting profile (Cluster 5).

The silhouette index obtained using the Soft-DTW metric for (i) k = 3 was 0.49592, (ii) k = 4 was 0.64826, and (iii) k = 5 was 0.38240. Consequently, upon analyzing the silhouette index and visually inspecting the curves, it is determined that, despite the inability to differentiate the cafeteria profile from the administrative/classroom building profiles, the optimal number of clusters for classifying the consumption profiles considering only the active power demand is k=4.

13.3.2.2 Clustering Based on Reactive Power

Analogous to the clustering of active power, the K-means algorithm was utilized with the Soft-DTW metric on the set of normalized reactive power load curves for weekdays. The objective is to classify the loads according to their reactive power consumption profiles. The representative curves for each cluster and the grouped curves for k = 3, k = 4, and k = 5 are presented in Fig. 13.4.

Similarly to the analysis of active power curves, Fig. 13.4 shows the clustering of reactive power curves in p.u. For k = 3, Cluster 1 exhibits an increase in reactive power throughout the day. In contrast, Cluster 2 decreases over the day, while Cluster 3 displays a relatively constant behavior. When considering k = 4, a similar profile

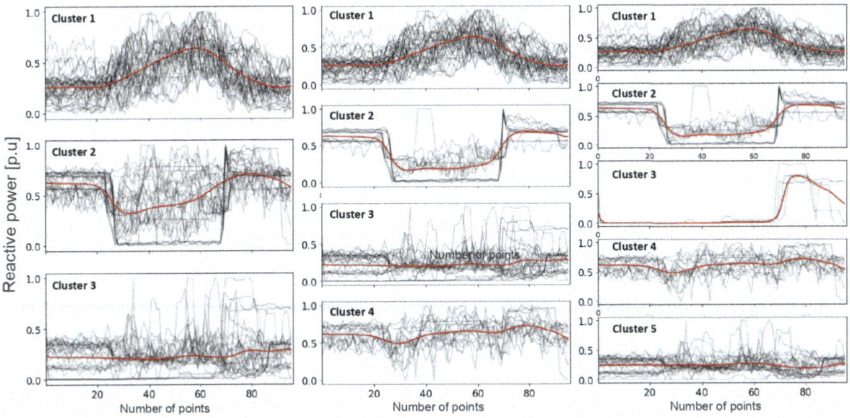

Fig. 13.4 Clustering of reactive power considering k = 3; k = 4; and k = 5

to the k = 3 scenarios is obtained, with an additional group exhibiting a constant load consumption profile identified in Cluster 4. For k = 5 clusters, a new group with reactive power consumption only at the end of the day is determined (Cluster 3).

The silhouette index obtained using the Soft-DTW metric for (i) k = 3 was 0.42465, (ii) k = 4 was 0.46675, and (iii) k = 5 was 0.48972. Despite achieving relatively good silhouette indices, it is challenging to determine typical reactive power consumption profiles, except for the public lighting consumption profile, as presented in the Cluster 2 curve, which exhibits a well-defined characteristic profile. The profiles can be associated with inductive and capacitive reactive power consumption in other cases.

The inductive reactive power consumption profile is characterized by representative load consumption curves with positive values, indicating that the power flow is in the first or second quadrants. This is typically due to transformers and motors, which consume inductive reactive power. In this study, data normalization was performed in the [0,1] range. Therefore, the obtained curves with an upward concavity can be associated with inductive reactive power consumption profiles.

On the other hand, the capacitive reactive power consumption profile is characterized by representative load consumption curves with negative values, indicating that the power flow is in the third or fourth quadrants. This is generally due to computers, capacitor banks, and fluorescent lamps, which consume capacitive reactive power. As previously mentioned, the data normalization was done in the [0,1] range. Consequently, the obtained curves with a downward concavity can be associated with capacitive reactive power consumption profiles.

In electrical engineering, reactive power plays a critical role in the operation and stability of power systems. Inductive loads, such as motors and transformers, consume reactive power to establish magnetic fields, while capacitive loads, such as capacitor banks and specific lighting systems, provide reactive power. Proper reactive power management is essential to maintain voltage levels within specified limits, improve power factors, and reduce losses in the transmission and distribution networks. Analyzing reactive power consumption profiles provides valuable insights into the types of loads and their impact on the power system, aiding in effective system planning and operation.

13.3.2.3 Clustering Based on Combined Active and Reactive Power

In this analysis, the K-means algorithm utilizing the Soft-DTW metric was applied to all normalized concatenated P+Q load curves for weekdays to classify the load set according to their combined active and reactive power consumption profiles. The representative curves for each cluster and the grouped curves for k = 3, k = 4, and k = 5 are presented in Fig. 13.5.

Upon analyzing Fig. 13.5, the active and reactive power curves concatenated in p.u., considering k = 3; k = 4; and k = 5. The centroid of the clusters in each case is represented in red, summarizing the typical load behavior within each cluster. The gray curves represent the samples collected through measurements, which are allocated to different clusters based on the methodology adopted. In this figure, the curves represent concatenated daily load profiles, where the first 96 points correspond to active power and points 97 to 180 correspond to reactive power, totaling 180 points for each daily curve.

For k = 3, the method identified in Cluster 1 is similar to administrative/classroom buildings combined with a reactive power profile that exhibits an increase (upward concavity) during the day. The characteristic curve obtained in Cluster 2 has a profile similar to that of public lighting. The characteristic curve obtained in Cluster 3 represents the consumption curves of singular-use buildings. It is also noted that the active power curve of the photovoltaic panel was classified within this group. In this scenario, the reactive power profile has a load curve with downward concavity.

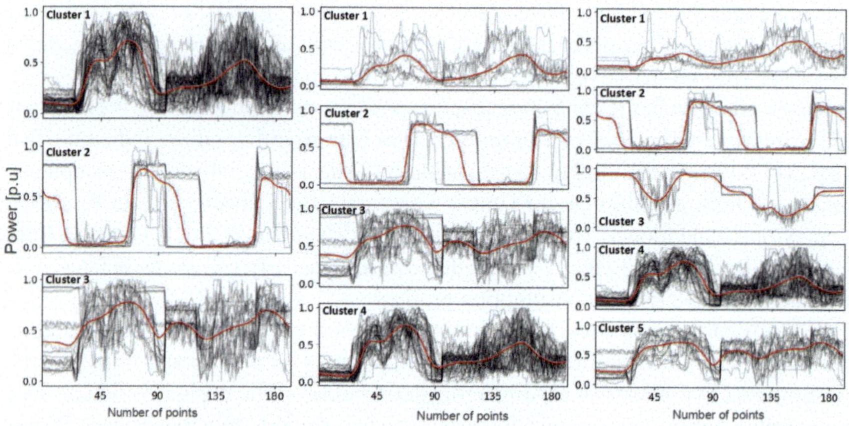

Fig. 13.5 Clustering of power (P+Q), considering k = 3; k = 4; and k = 5

For k = 4, the obtained curves exhibit profiles similar to those observed in the k = 3 analysis. However, the curve corresponding to Cluster 1 (k = 3), which presented a profile identical to administrative/classroom buildings, was divided into two new clusters (Cluster 1 and Cluster 4). For k = 5, the method successfully identified the consumption profile associated with the photovoltaic panel (Cluster 3) while maintaining the other previously identified clusters.

The silhouette index obtained using the Soft-DTW metric for (i) k = 3 was 0.45869, (ii) k = 4 was 0.34332, and (iii) k = 5 was 0.39225. Thus, based on the silhouette index and the visual inspection of the curves, and analogous to the results obtained in the analyses of active and reactive power in Sects. 13.3.2.1 and 13.3.2.2, respectively, it is concluded that the number of clusters that best represents the analysis of active power is k = 4. However, it is not possible to identify typical consumption profiles for reactive power; rather, it is possible to associate the consumption with inductive and capacitive reactive power.

13.3.3 Euclidean Versus Soft-DTW

One of the hyperparameters to be adjusted in the K-means algorithm is the distance metric, which determines the data's proximity and grouping. The Euclidean distance metric is the most widely used in the literature; however, the data transformation step in the applied methodology led this study to group time series with different rhythms or typologies. Therefore, this work summarizes in Table 13.1 the results of the silhouette validation coefficient, which demonstrates the superiority of the Soft-DTW metric over the Euclidean distance. This superiority is due to the optimization in the non-linear alignment of sequences of numerical values, which allows for finding patterns among time series with variations in their profiles.

The results presented in Table 13.1 indicate the superior performance of the Soft-DTW metric compared to the Euclidean distance across different clustering scenarios (k = 3, k = 4, and k = 5). The silhouette coefficients, which measure the cohesion

Table 13.1 Silhouette coefficient analysis

Variables	Euclidean			Soft-DTW		
	k = 3	k = 4	k = 5	k = 3	k = 4	k = 5
P	0.37559	0.41017	0.27767	0.49592	**0.64826**	0.38240
Q	0.23884	0.24557	0.20660	0.42465	0.46675	**0.48972**
P + Q	0.22595	0.21201	0.17974	0.45869	0.34332	**0.39225**

The bold values represent the highest silhouette coefficient for each row (i.e., for each variable: P, Q, and P + Q). These values indicate the best clustering performance among the tested configurations (Euclidean vs. Soft-DTW and different values of k) for each variable. Silhouette coefficient values range from −1 to 1, where a higher value indicates better-defined and more distinct clusters. Therefore, the bold values highlight the optimal combination of distance metric and number of clusters (k) for each variable based on silhouette score and separation of the clusters, are consistently higher for the Soft-DTW metric. The silhouette coefficient ranges from -1 to 1, with values closer to 1 indicating better-defined and well-separated clusters. Higher values suggest that the data points are well-matched to their own clusters and poorly matched to neighboring clusters. This improved clustering quality with the Soft-DTW metric can be attributed to its ability to non-linearly align sequences, thereby accommodating variations in the temporal dynamics of the data more effectively than the Euclidean distance.

In the context of electrical load profiling, where consumption patterns can exhibit significant temporal shifts and irregularities, Soft-DTW is advantageous. The Euclidean distance, which assumes a linear alignment, may fail to capture these subtleties, leading to less coherent clusters. On the other hand, Soft-DTW aligns the sequences optimally by considering possible distortions and shifts, thus providing a more accurate representation of the underlying patterns.

These findings are essential for power system management and planning applications, where understanding the nuanced consumption behaviors of different load types is essential. The enhanced clustering quality facilitated by Soft-DTW enables more precise identification of consumption profiles, which can inform targeted interventions for demand-side management, efficiency improvements, and load forecasting.

The consistent superiority of the Soft-DTW metric across various clustering configurations underscores its robustness and suitability for time series clustering in the context of electrical load analysis. Future studies could further explore the application of Soft-DTW in other domains where temporal alignment is essential for accurate data analysis.

13.4 Conclusions

This research applied the K-means clustering technique using the Soft-DTW metric to classify active and reactive power measurements in distribution transformers on the UFJF campus. The main objective was to identify typical load curve profiles based on load characteristics, allowing a more accurate characterization of the use of the buildings served by these transformers. These typical profiles enable more realistic studies for the electrical system, such as energy acquisition planning, transformer sizing, photovoltaic plant allocation, and installation of electric vehicle chargers.

The results show that applying K-means with the Soft-DTW metric offers a significant advantage in identifying complex temporal patterns, especially when aligning time series with subtle variations. However, it is essential to note that Soft-DTW, although practical, has a considerably higher computational cost than traditional K-means with Euclidean metric. This trade-off between accuracy and computational efficiency should be carefully evaluated in the context of the size of the dataset and the available computational resources. On the other hand, K-means is widely discussed in the literature and widely used due to its simplicity and efficiency, presenting results similar to Soft-DTW in several scenarios. Thus, the choice of the ideal technique depends on the nature of the data and the specific need for accuracy or computational efficiency, ensuring that researchers are well-informed about the practical considerations of their research.

Future lines of research may explore the application of other partitioning algorithms, such as k-medoids and DBSCAN, which may offer greater robustness in scenarios with outliers and non-uniformly distributed data. In addition, new cluster validity indices, such as the Dunn index, the Davies-Bouldin index, and the Scatter and Discriminant Index, may provide a more complete assessment of the quality of the clusters formed. Another relevant possibility is the inclusion of additional variables, such as environmental variables, to enrich the analysis and facilitate a more accurate and detailed characterization of consumption patterns.

Another analysis that can be incorporated is investigating the impact of different temporal granularities, such as 15-min, 1-h, and 1-day profiles. This can reveal how the temporal scale influences the quality of the clusters and the representativeness of the profiles formed. Applying the methodology in different contexts, such as industrial or commercial networks, can also validate the effectiveness of the models and identify necessary adaptations. Finally, a hybrid approach that combines supervised and unsupervised techniques can improve modeling capacity, especially for predicting and classifying profiles based on historical data and new scenarios.

Acknowledgements The authors thank the Coordenação de Aperfeiçoamento de Pessoal de Nível Superior (CAPES) under Grant 001, Conselho Nacional de Desenvolvimento Científico e Tecnológico (CNPq) under grants, Fundação de Amparo à Pesquisa do Estado de Minas Gerais (FAPEMIG) under grant APQ-03609-17 (Instituto Nacional de Ciência e Tecnologia em Energia Elétrica-INERGE) and the Smart4grids research group. Additionally, the authors acknowledge the Centro Paulista de Estudos da Transição Energética (CPTEn), São Paulo Research Foundation (FAPESP), and especially on the Red para la Integración a gran Escala de EERR en los sistemas eléctricos (RIBIERSE-Cyted).

Disclosure of Interests The authors have no competing interests to declare relevant to this article's content.

References

1. T.Z. Ang, M. Salem, M. Kamarol, H.S. Das, M.A. Nazari, N. Prabaharan, A comprehensive study of renewable energy sources: classifications, challenges and suggestions. Energy Strat. Rev. **43**, 100939 (2022). https://doi.org/10.1016/j.esr.2022.100939, https://www.sciencedirect.com/science/article/pii/S2211467X2200133X
2. S. Balasubramanian, P. Balachandra, Characterising electricity demand through load curve clustering: a case of karnataka electricity system in india. Comput. Chem. Eng. **150**, 107316 (2021). https://doi.org/10.1016/j.compchemeng.2021.107316
3. E.C. Bobric, G. Cartina, G. Grigoras, Clustering techniques in load profile analysis for distribution stations. Adv. Electr. Comput. Eng. **9**(1), 63–66 (2009)

4. J.G. Cypriano, L.F. Pinto, L.C. Machado, L.C. Da Silva, L.S. Ferreira, Energy management methodology for energy sustainable actions in university of campinas—Brazil, in *IOP Conference Series: Earth and Environmental Science*, vol. 257 (Institute of Physics Publishing, 2019). https://doi.org/10.1088/1755-1315/257/1/012034
5. L.C.P. Da Silva, J. Gama, I. Assumpção, E. Tomaz, *Campus sustentável: um modelo de inovação em gestão energética para América Latina e o Caribe* (Synergia, 2022)
6. U.M. Fayyad, G. Piatetsky-Shapiro, P. Smyth et al., Knowledge discovery and data mining: towards a unifying framework. KDD **96**, 82–88 (1996)
7. D.G. Ferrari, L.N.d.C. Silva, *Introdução a mineração de dados* (Saraiva Educação SA, 2017)
8. D. Gayen, R. Chatterjee, S. Roy, A review on environmental impacts of renewable energy for sustainable development. Int. J. Environ. Sci. Technol. **21**(5), 5285–5310 (2024)
9. D. Gielen, F. Boshell, D. Saygin, M.D. Bazilian, N. Wagner, R. Gorini, The role of renewable energy in the global energy transformation. Energy Strat. Rev. **24**, 38–50 (2019). https://doi.org/10.1016/j.esr.2019.01.006, https://www.sciencedirect.com/science/article/pii/S2211467X19300082
10. B. Igliński, U. Kiełkowska, M.B. Pietrzak, M. Skrzatek, G. Kumar, G. Piechota, The regional energy transformation in the context of renewable energy sources potential. Renew. Energy **218**, 119246 (2023). https://doi.org/10.1016/j.renene.2023.119246, https://www.sciencedirect.com/science/article/pii/S0960148123011618
11. A.G. Kazaki, T.A. Papadopoulos, Cluster analysis of university campus smart meter data, in *2018 53rd International Universities Power Engineering Conference (UPEC)* (IEEE, 2018), pp. 1–6
12. J. Le Dréau, R.A. Lopes, S. O'Connell, D. Finn, M. Hu, H. Queiroz, D. Alexander, A. Satchwell, D. Österreicher, B. Polly, A. Arteconi, F. de Andrade Pereira, M. Hall, T. Kırant-Mitić, H. Cai, H. Johra, H. Kazmi, R. Li, A. Liu, L. Nespoli, M.H. Saeed, Developing energy flexibility in clusters of buildings: a critical analysis of barriers from planning to operation. Energy Build. **300**, 113608 (2023). https://doi.org/10.1016/j.enbuild.2023.113608, https://www.sciencedirect.com/science/article/pii/S0378778823008381
13. R.N. Muniz, C.T. da Costa Júnior, W.G. Buratto, A. Nied, G.V. González, The sustainability concept: a review focusing on energy. Sustainability **15**(19) (2023). https://doi.org/10.3390/su151914049, https://www.mdpi.com/2071-1050/15/19/14049
14. T. Nacht, R. Pratter, J. Ganglbauer, A. Schibline, A. Aguayo, P. Fragkos, E. Zisarou, Modeling approaches for residential energy consumption: a literature review. Climate **11**(9) (2023), https://www.mdpi.com/2225-1154/11/9/184
15. S. Panda, S. Mohanty, P.K. Rout, B.K. Sahu, M. Bajaj, H.M. Zawbaa, S. Kamel, Residential demand side management model, optimization and future perspective: a review. Energy Rep. **8**, 3727–3766 (2022). https://doi.org/10.1016/j.egyr.2022.02.300, https://www.sciencedirect.com/science/article/pii/S2352484722005479
16. J.F.M. Pessanha, V.L. Xavier, M.R.d.S. do Amaral, L. da Costa Laurencel, Construindo tipologias de curvas de carga com o programa r. Pesquisa Operacional para o Desenvolvimento **7**(1), 29–54 (2015)
17. A. Satre-Meloy, M. Diakonova, P. Grünewald, Cluster analysis and prediction of residential peak demand profiles using occupant activity data. Appl. Energy **260**, 114246 (2020). https://doi.org/10.1016/j.apenergy.2019.114246, https://www.sciencedirect.com/science/article/pii/S0306261919319336
18. J. Wang, W. Azam, Natural resource scarcity, fossil fuel energy consumption, and total greenhouse gas emissions in top emitting countries. Geosci. Front. **15**(2), 101757 (2024). https://doi.org/10.1016/j.gsf.2023.101757, https://www.sciencedirect.com/science/article/pii/S1674987123002244

Chapter 14
Coordinated Tuning of STATCOM-POD Controllers Using PSO and NBA: A Case Study in DIgSILENT PowerFactory

Marta Bernal-Sancho, Marta Muñoz-Lázaro, María Paz Comech, and Noemí Galán-Hernández

Abstract The increasing integration of renewable energy sources into power grids has led to an increasing prevalence of weakly interconnected grids, which are more susceptible to stability issues such as inter-area oscillations. This study addresses these challenges by investigating the coordinated tuning of STATCOM-POD controllers using Particle Swarm Optimization and the Novel Bat Algorithm within DIgSILENT PowerFactory. Two network models are employed: a simple two-area system and the PST16 bus system, both characterized by weak interconnections that exacerbate stability issues. The optimization process aims to improve the damping of inter-area oscillations by fine-tuning multiple POD controllers, evaluated by temporal and eigenvalue analysis. The findings underscore the critical role of advanced optimization techniques in enhancing stability and damping performance in power systems increasingly impacted by weak interconnections and variable generation sources.

Keywords STATCOM-POD · Inter-area oscillation · Coordinated tuning

14.1 Introduction

The global shift towards renewable energy sources (RES), such as wind and solar power plants, has significantly transformed the power generation landscape. This transition poses severe challenges for grid stability, particularly in weakly interconnected networks that are more vulnerable to disturbances. The inherent variability

M. Bernal-Sancho (✉) · M. Muñoz-Lázaro · N. Galán-Hernández
Electrical Systems Department, CIRCE Technology Centre, Zaragoza, Spain
e-mail: mbernal@fcirce.com

M. P. Comech
ENERGAIA Mixed Research Institute, Zaragoza, Spain

D. Rossit et al. (eds.), *Proceedings of the VII Ibero-American Congress of Smart Cities, ICSC-Cities 2024, 12-14 November, San Carlos, Costa Rica*, Springer Proceedings in Energy, https://doi.org/10.1007/978-981-96-4301-1_14

and lack of inertia associated with RES exacerbate these challenges, increasing susceptibility to inter-area oscillations and threatening overall system stability [1].

Weakly interconnected networks are particularly prone to stability issues due to their limited ability to transfer power between areas effectively. Disturbances in these networks can quickly propagate, amplifying oscillatory behavior and compromising system stability. Additionally, the absence of robust damping mechanisms in such grids underscores the urgent need for practical solutions, such as power oscillating damping (POD) controllers integrated into flexible AC transmission systems (FACTS). However, successfully integrating these controllers requires precise tuning to mitigate power oscillations without adverse interactions effectively. Poor coordination among multiple controllers, even in conventional grids, can reduce damping or exacerbate oscillations, making precise tuning crucial, especially in weakly interconnected networks. This highlights the growing need for optimization techniques to improve the performance and coordination of POD controllers in these vulnerable grid environments [2].

Extensive research has been conducted on various methods of tuning PSS and POD controllers. For instance, studies have employed Particle Swarm Optimization (PSO), developed in 1995, to parametrize PSS [3, 4] or POD accurately [5, 6]. Many articles also use PSO to coordinate PSS and FACTS-POD controllers, aiming to avoid control interactions, such as coordinating PSS with Thyristor-Controlled Series Capacitors (TCSC) [7, 8] or Static Synchronous Series Compensators (SSSC) [9].

Other algorithms, such as the Genetic Algorithm (GA), have also been used for similar tuning and coordination tasks. GA has been applied to tune PSS parameters [10, 11] and coordinate PSS with TCSC-POD controller parameters [12–14]. Further developments include Genetic Local Search (GLS) and Real-Code Genetic Algorithm (RCGA) techniques for tuning PSS and coordinating STATCOM-POD with PSS [15]. Hybrid algorithms like GA-PSO have been proposed to overcome the limitations of each approach [16]. Alternative algorithms have also been explored, including automated tuning processes using linear matrix inequality solvers [17] or methods combining system participation factor analysis with pole placement for tuning PSS [18].

More recently, in 2010, Yang developed the Bat Algorithm (BA) [19]. It has shown notable effectiveness in tuning PSS parameters [20, 21] and has been compared to other methods like Ant Colony Optimization (ACO) and GA in mitigating low-frequency oscillatory modes [22]. A Novel Bat Algorithm (NBA) developed around 2015 [23] has been evaluated against the original BA [24] and PSO [25], showing promising potential for small-signal stability analysis. These algorithms collectively provide robust tools for enhancing damping performance in power systems.

The main objective of this article is not to propose a new method for tuning POD controllers, given that numerous algorithms already exist. Instead, this study aims to demonstrate the effectiveness and comparative performance of two fundamental methods, one widely used, PSO, and one more recent, NBA, within the advanced DIgSILENT PowerFactory (DSPF) simulation environment. An already widely used method (PSO) and a new optimized method (NBA) provide two different results, highlighting PSO's proven reliability and NBA's enhanced flexibility and development. Unlike previous studies focusing on coordinating a single PSS with a FACTS-POD, this research proposes an innovative approach by investigating the coordinated tuning of multiple FACTS devices, providing a deeper understanding of their collective behavior and interactions in complex network environments.

The study utilizes two different network models. First, a simple two-area system with a single STATCOM device is tested to individually assess the algorithms' performance on damping inter-area oscillations. Second, the PST16 bus system, characterized by weak interconnections and complex dynamics, is simulated to parametrize and coordinate three STATCOMs. The final objective is to compare which optimization method provides better tuning effectiveness in mitigating inter-area oscillations, a critical factor for maintaining stability in weak and complex power systems. This dual-network approach allows for a broader assessment of the effectiveness of the optimization methods. By leveraging the powerful simulation capabilities of DSPF, including time (RMS) and frequency domain (eigenvalues) analysis, the findings are further validated and enhanced in credibility, offering a robust approach to optimizing damping controllers in complex power systems.

The article is organized as follows. Section 14.2 describes the structure of the POD controller, detailing its integration into the STATCOM. Section 14.3 outlines the solution methodology, including an overview of the optimization problem and a detailed explanation of the PSO and NBA used for tuning the POD controllers. Section 14.4 provides a comprehensive description of the test networks used for the case studies, including a simple two-area system and the PST16 bus system. Section 14.5 presents the results of applying the PSO and NBA to both test networks, comparing the algorithms' performance in damping ratios. Finally, Sect. 14.6 summarizes the main findings and discusses the study's practical implications.

14.2 Structure of the POD Controller

Figure 14.1 presents the block diagram of the POD model integrated into the STATCOM. The POD control is introduced as a supplementary control signal to the reactive control loop, enhancing the reference sent to the STATCOM, *POD_out*. The POD controller is structured with the same three main blocks as the PSS, with the stabilizer gain *Kpod*, a washout filter with time constant *Tw*, and two-phase compensation blocks with time constants *Tb1*, *Ta1*, *Ta2*, and *Tb2* being the parameters to be set. The POD receives input containing oscillation information from the power system, such as the active power, voltage, or frequency [26]. This article focuses on the terminal frequency where the STATCOM is connected as input (*frequency*). The output of the POD control (*POD_out*) serves as the command for damping the oscillation, acting as an additional reference added to the STATCOM reactive power reference.

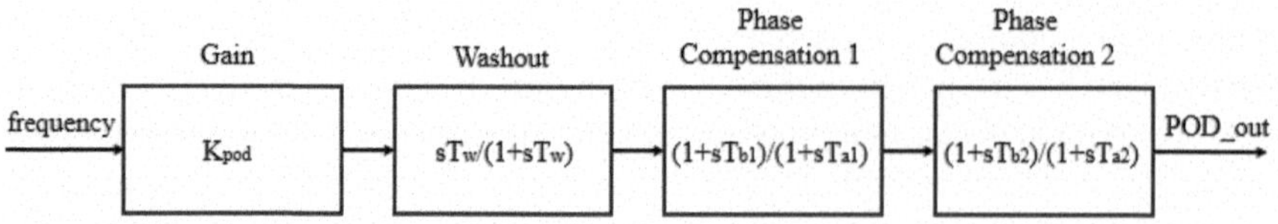

Fig. 14.1 POD model block diagram

14.3 Solution Methodology Description

14.3.1 Overview of the Optimization Problem

As the main function of the POD is to suppress low-frequency oscillations, in this study, the damping ratio of the system is chosen as the objective function for the optimization of the POD parameters, as shown in Eq. (14.1) [6, 27]. According to the Spanish operating procedure [28] a threshold of 5% damping ratio determines system's stability. Hence, the higher the damping ratio, the more stable the system.

$$J = \sum (\xi_0 - \xi_{min})^2 \tag{14.1}$$

where ξ_0 is the lower bound of the expected damping ratio for the system, and ξ_{min} is the minimum value of all damping ratios.

For the STATCOM-POD, the parameters must be set in a range defined in Eq. (14.2) [16].

$$C\min \leq C \leq C\max \tag{14.2}$$

where C is each of the parameters of the STATCOM-POD controller depicted on Fig. 14.1, which is defined as $C = [KPOD, Tw, Tb1, Ta1, Tb2andTa2]$. $Cmin$ and $Cmax$ constrain the minimum and maximum limits of the POD controller. For the article, the range of the parameters are set according to [26].

14.3.2 Particle Swarm Optimization (PSO)

Particle Swarm Optimization (PSO) is a nature-inspired optimization algorithm developed by Kennedy and Eberhart in 1995 [29]. PSO emulates the social behaviors observed in birds and fish. Over the years, it has become a widely used tool for solving optimization problems in several fields due to its simple implementation and computation. In PSO, each particle in the swarm represents a solution and moves through the solution space dynamically, adjusting its position and velocity to minimize the value of the objective function [4].

At each iteration, particles navigate an n-dimensional search space, influenced by their historical best positions, $p_{best,i}$, and the globally recognized best position within the swarm, g_{best}. This strategy causes particles to converge towards promising regions of the search space. An inertia coefficient, ω, is introduced to regulate the balance between exploration and exploitation, thereby modulating the impact of the particle's previous velocity and improving the search dynamics.

The position and velocity of each particle are represented by vectors $x_i(t)$ and $v_i(t)$, respectively. Moreover, the velocity and position update equations are shown in Eqs. (14.3) and (14.4), respectively [16].

$$v_i(t+1) = \omega \cdot v_i(t) + c_1 \cdot r_1 \cdot \left(p_{best,i} - x_i(t)\right) + c_2 \cdot r_2 \cdot \left(g_{best} - x_i(t)\right) \quad (14.3)$$

$$x_i(t+1) = x_i(t) + v_i(t+1) \quad (14.4)$$

where c_1 and c_2 are acceleration coefficients that indicate the influence of cognitive and social factors. r_1 and r_2 are random variables in the range 0 and 1, designed to prevent premature convergence and maintain diversity within the swarm [25].

This study configures the PSO algorithm with 30 particles and 50 iterations. The inertia coefficient is set to 0.7 and both acceleration coefficients c_1 and c_2 are assigned a value of 2.

Figure 14.2a represents the PSO flowchart and shows the procedure for solving the optimization problem by minimizing the objective function [4, 6, 9, 16].

14.3.3 Novel Bat Algorithm (NBA)

The NBA is an improved version of the original BA developed by Yang in 2010 [19]. The BA is inspired by the echolocation behavior of bats, which use sound waves to navigate and hunt prey. The NBA was developed around 2015 by Meng [30]. It incorporates modifications to improve exploration and exploitation capabilities, making it more robust and efficient for complex optimization problems, such as controller tuning in power systems. The improvements introduced are that all bats can be in different habitats, depending on stochastic selection, and can compensate for the Doppler effect of echoes. They can adaptively adjust the compensation rate depending on their targets' proximity.

The bats (particles) are initialized with random positions, x_{ij}^t, velocities $v_{i,j}^t$, frequencies f_{ij}, pulse rates r_i, and loudness levels A_i.

Bats' habitat selection depends on several random phenomena. For simplicity, habitat selection is modeled as a stochastic decision, P, in the range [0, 1]. If a uniform random number in the interval [0, 1] is less than P, bats will choose quantum behavior; otherwise, they will select mechanical foraging behavior in limited habitats.

In the NBA algorithm, bats with quantum behavior can forage in various habitats. When one bat finds a feeding site, others quickly follow suit. This behavior allows

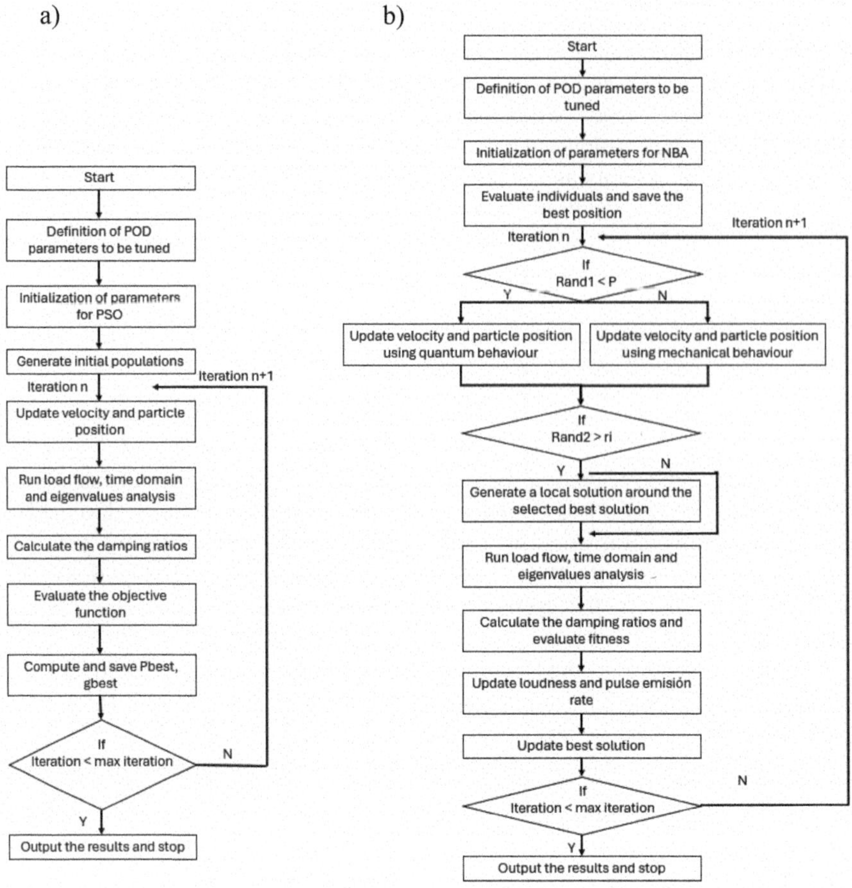

Fig. 14.2 **a** Flow chart PSO **b** flow chart NBA

the best global position in the bat swarm to serve as an attractor. The position of the virtual bats with quantum behavior is formulated in Eq. (14.5).

$$x_{ij}^{t+1}\begin{cases} g_j^t + \theta \cdot \left|mean_{ij}^t - x_{ij}^t\right| \cdot \ln\left(\frac{1}{u_{ij}}\right), \, if\, rand_j(0,1) < 0.5 \\ g_j^t + \theta \cdot \left|mean_{ij}^t - x_{ij}^t\right| \cdot \ln\left(\frac{1}{u_{ij}}\right), \, otherwise \end{cases} \tag{14.5}$$

where u_{ij} is a random number between 0 and 1, $mean_{ij}^t$ is the mean position of the bats and θ is the contraction–expansion coefficient.

The mechanical behavior considers the Doppler effect. Each bat's frequency update is influenced by three factors: the interval between the minimum and maximum frequencies (f_{min} and f_{max}), the Doppler effect and the Doppler compensation rates of the bats. When bats chase prey, they adjust their frequency as a function of position relative to the global best solution, g_j^t. Compensation rates, C_i, vary between subjects, allowing adaptive behavior to effectively manage the Doppler effect in echoes.

In addition, NBA introduces an inertia weight, w, into the velocity update equation, which controls the influence of the previous velocity on the current movement, improving control over the bat's speed and direction. These adjustments improve the bats' ability to balance exploration and exploitation during optimization.

$$f_{ij} = f_{min} + (f_{min} - f_{max}) \cdot \beta \tag{14.6}$$

$$f_{ij}' = \frac{c + v_{i,j}^t}{c + v_{g,j}^t} \cdot f_{ij} \cdot \left(1 + C_i + \frac{g_j^t + x_{i,j}^t}{\left|g_j^t + x_{i,j}^t\right| + \varepsilon}\right) \tag{14.7}$$

$$v_{ij}^{t+1} = w \cdot v_{ij}^t + \left(g_i^t - x_{ij}^t\right) \cdot f_{ij} \tag{14.8}$$

$$x_{i,j}^{t+1} = x_{i,j}^t + v_{i,j}^t \tag{14.9}$$

where β is a random number uniformly distributed between 0 and 1, c is the speed of sound in air, and ε is a small constant to avoid division by zero.

Bats increase their pulse frequency and decrease their loudness to approach prey more quietly, which increases their chances of capture. For simplicity, individual bats' relative loudness compared to all bats' average loudness influences their local search for prey. This ratio guides how each bat adjusts its position when searching near the prey, generating a local solution around the best-selected solution.

$$x_{i,j}^{t+1} = g_j^t \cdot \left(randn\left(0, \sigma^2\right)\right) \tag{14.10}$$

$$\sigma^2 = \left|A_i^t - A_{mean}^t\right| + \varepsilon \tag{14.11}$$

where rand $n\left(0,\ \sigma^2\right)$ is a Gaussian distribution with mean 0 and standard deviation σ^2 and A_{mean}^t is the mean loudness of all bats at time step t.

Finally, the loudness A_i^t and pulse rate r_i are updated following Eqs. (14.12) and (14.13):

$$A_i^{t+1} = \alpha \cdot A_i^t \tag{14.12}$$

$$r_i^{t+1} = r_i^0 \cdot \left(1 - e^{-\gamma t}\right) \tag{14.13}$$

where α and γ are constants that control the reduction of loudness and the increase of the pulse rate, respectively [23].

In this study, the population size of bats is set to 30, with the optimization running for 50 iterations. The loudness and pulse rate are initialized at 0.5, while the values of alpha and gamma are fixed at 0.9. The probability of selecting the quantum habitat

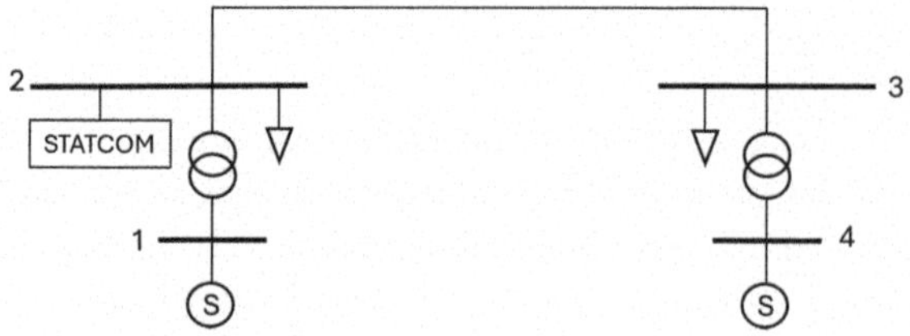

Fig. 14.3 Simple two-area system

is set to 0.5. The inertia weight and the compensation rates for the Doppler effect in echoes are randomly assigned within the range [0, 1]. The contraction–expansion coefficient is defined as 0.2.

Figure 14.2b represents the NBA flowchart and shows the procedure for solving the optimization problem by minimizing the objective function [22, 23, 30].

14.4 Network Description

14.4.1 Simple Two-Area System

The simple two-area system is taken from the Spanish grid code [31] and represents a small equivalent of the Spanish and French networks. It consists of two synchronous generators and a power line with a high reactance value. This network model has been specifically designed to replicate inter-area oscillations, which makes it ideal for evaluating the effectiveness of the tuning methods in managing inter-area oscillations.

As shown in Fig. 14.3, the STATCOM of 150 MVA rated power is connected at node 1. The nominal power of the synchronous generator at node 1 is 1500 MVA, generating 1350 MW, while the load at node 2 is 1250 MW. The synchronous generator at node.

4, representing the external system to which it is connected, has a rated power of 5000 MW, generating 3900 MW. The power consumed by the load connected to terminal 3 is 4000 MW. Therefore, the line transports 100 MW from node 1 to node 4. The synchronous machines controls are based on DSPF library models [32], and the parameters characterization is defined in [31]. Furthermore, the PSSs of the synchronous machines are disabled, as only the performance of the STATCOM-POD control to damp the oscillation is of interest.

14.4.2 PST16 Bus System

Figure 14.4 shows the PST16 bus system. This system was developed based on characteristic parameters of the European interconnected power system and is a robust setup for studying various stability problems, particularly low-frequency oscillations.

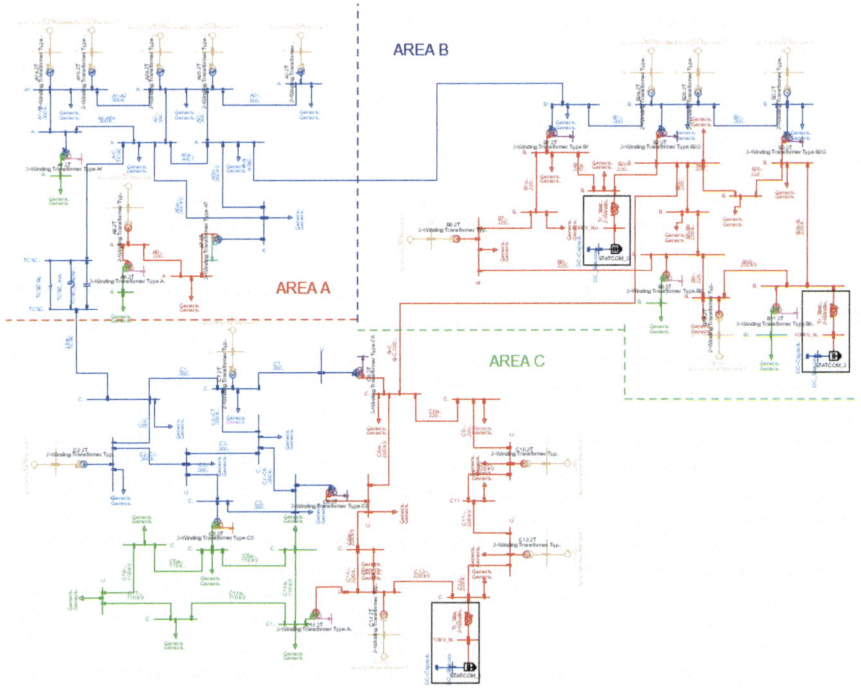

Fig. 14.4 PST16 bus system

It comprises three strongly meshed areas, 66 buses, 16 generators, 28 transformers, and 51 transmission lines, with the interconnecting lines between areas being weak due to their length of more than 100 km and up to 300 km [33].

The system generators include hydro, nuclear, and thermal power plants. The total system-wide installed capacity is about 16.13 GW, and the total demand is 15.75 MW. The transmission network comprises the voltage levels 380, 220, and 110 kV. Hydro, thermal, and nuclear generators are rated at 220, 247, and 259 MVA, respectively. Most loads are concentrated in area C, and power is transferred from areas A and B to area C through two long tie-lines [34].

Most PST16 bus system components are modeled using the built-in DSPF library models [32]. The STATCOMS of 150 MVA rated power are connected randomly, since the objective of the study is to optimize the parameterization. Two STATCOMS are connected in area B, and the other is in area C, as shown in the black squares of Fig. 14.4. For this article, all PSS units of the synchronous generators are deactivated, isolating the effect of the STATCOM-POD controllers.

14.5 Simulation Results

The tuning results obtained by the two algorithms will first be evaluated in the simple two-area system, with a single STATCOM-POD, and then in the PST16 model, where three STATCOM-PODs will be coordinated. The calculations will be performed in DSPF in time (RMS) and frequency (eigenvalues) domain, giving the results obtained more robustness and reliability. The eigenvalue diagram is a valuable tool for oscillation analysis. It aids in the identification of the system's most critical modes, which are characterized by their minimal damping ratios (%), which is the objective function established in the two optimization algorithms and will be compared in both networks.

14.5.1 Simple Two-Area System

A power increase is applied to the load connected to terminal 3 (see Fig. 14.3) to induce the desired oscillatory event at $t = 10$ s. This event sets the conditions necessary to test the effectiveness of the two tuning methods in managing inter-area oscillation.

The event created is an inter-area oscillation. The simulation results when the POD is disabled are shown in Fig. 14.5. The frequency at terminals 2 and 3 oscillate in opposition, with a frequency of 0.33 Hz, as shown in the RMS plot in Fig. 14.5a. The green line represents the frequency at terminal 2 while the red line is that of terminal 3. Frequency analysis results are shown in the eigenvalue plot of Fig. 14.5b. Each green X represents one system mode, and the lowest damping ratio is 2.11%.

The PSO and NBA algorithms are employed to parameterize the POD controller integrated within the STATCOM, with each algorithm configured to run for 50 iterations. The minimum damping ratio of the system with the activated and tuned POD controller, optimized by both methods, is detailed in Table 14.1. Multiple simulations are conducted to ensure the robustness and reliability of the results, and average values

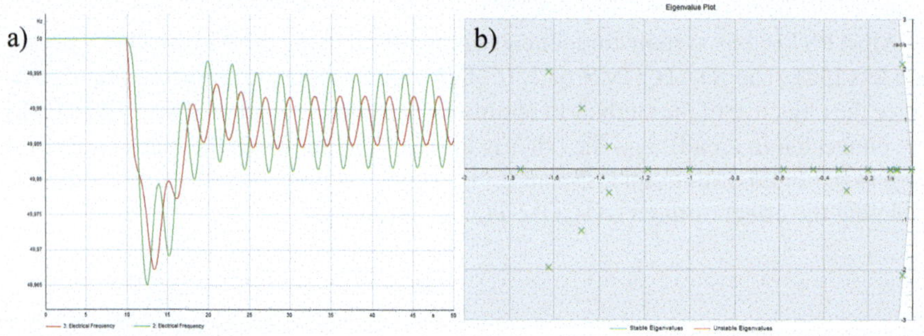

Fig. 14.5 Results after load event **a** RMS **b** Eigenvalues

Table 14.1 Simple two-area system results

Simulation	Damping ratio (%)	Execution time (min: seg)
Initial	2.11	–
PSO (50 iter.)	40.31	2:50
NBA (50 iter.)	41.85	2:51
NBA (40 iter.)	41.76	2:20

are computed, providing a more realistic assessment of the algorithms' performance than relying on single-run results. Both algorithms have similar execution times, due to their inherently rapid nature, making time differences minimal. Both methods significantly improved the initial values regarding damping ratios, with NBA showing a superior enhancement. Additional validation is performed by running NBA with 40 iterations, which still resulted in higher damping ratios compared to PSO at 50 iterations. This result demonstrates that NBA offers greater computational efficiency and produces parameters that significantly improve system stability when a single control is tuned.

Figures 14.6 (PSO) and 14.7 (NBA) illustrate the graphical outcomes of the activated POD controller, which has been tuned using both optimization algorithms with 50 iteractions. In Figs. 14.6a and 14.7a, the RMS response (solid line) of the system with the active POD control is shown alongside the baseline scenario without active control (refer to Fig. 14.5a, dotted line), highlighting the comparative impact of the tuned controllers. The green line indicates the frequency response at terminal 2, while the red line corresponds to terminal 3 (see Fig. 14.3). Figures 14.6b and 14.7b present the frequency domain results via the eigenvalue plots, demonstrating the system's modal behavior under the influence of the optimized controllers. The activation of the controls results in a notable enhancement of the damping ratios, achieving values slightly above 40%. This improvement is consistent with the observed damping effects in the RMS plots, confirming the effectiveness of the controllers' parameterization in mitigating oscillations.

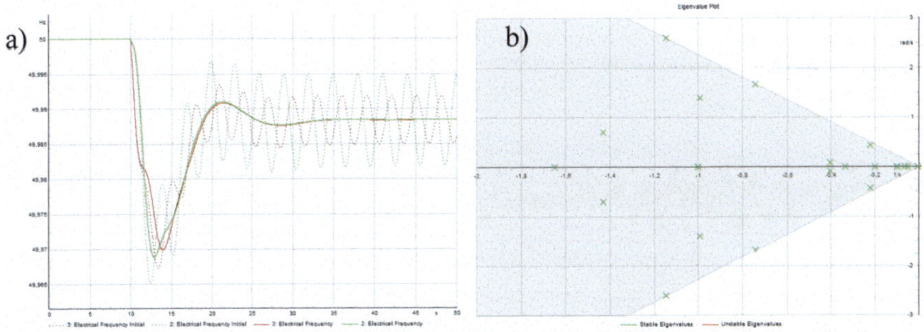

Fig. 14.6 Results after POD controller, adjusted by PSO **a** RMS **b** EV

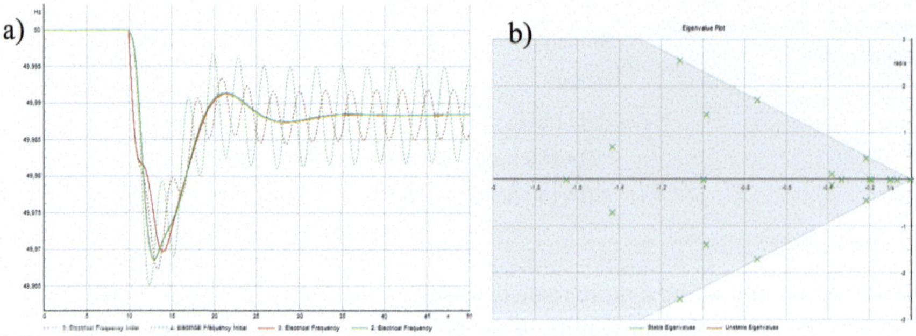

Fig. 14.7 Results after POD controller, adjusted by NBA **a** RMS **b** EV

14.5.2 PST16 Bus System

In the PST16 bus system, a short circuit of 25 ms on a transmission line in Area C (see Fig. 14.4) is applied to induce the desired oscillatory event at $t = 1$ s. The results when the PODs controllers are disabled are depicted in Fig. 14.8. Given the weak interconnections within the system, this fault induces an inter-area oscillation, as illustrated in the RMS graph of Fig. 14.8a. Area B (red) oscillates in opposition to Area A (blue) and C (green) with an oscillation frequency of 0.41 Hz. The graphical results indicate that the frequencies of all terminals within the same area exhibit similar dynamics; therefore, for improved visualization of the activated POD control effects, only one terminal per zone is presented in future results graphs. Frequency domain analysis is also conducted, displaying the eigenvalue plot in Fig. 14.8b, where the lowest damping ratio is 0.17%, reflecting the system's vulnerability due to its weak interconnections.

The optimization of three POD controllers using PSO and NBA improved the damping of inter-area oscillations in the PST16 bus system. This scenario presents a more complex challenge due to the simultaneous coordination of three POD controllers. As for the simple two-area system, multiple simulations are conducted

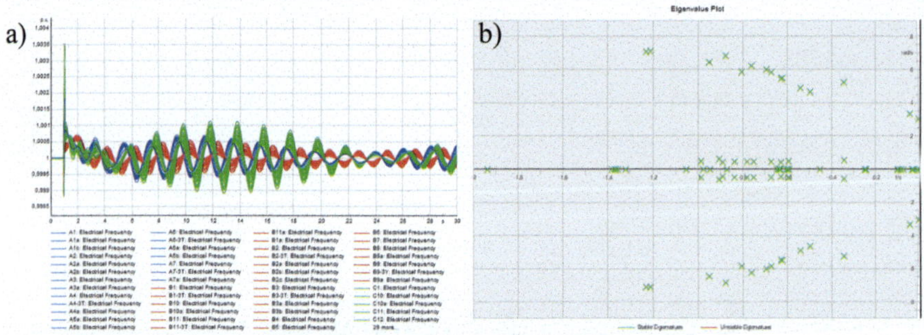

Fig. 14.8 Results after short circuit **a** RMS **b** Eigenvalues

Table 14.2 PST16 Bus system results

Simulation	Damping ratio (%)	Execution time (min:seg)
Initial	0.17	–
PSO (50 iter.)	6.06	12:15
NBA (50 iter.)	6.64	12:45
NBA (40 iter.)	6.51	10:25

to ensure result robustness and reliability. The simulation results, summarized in Table 14.2, demonstrate that PSO and NBA effectively improved the damping ratio of the system despite the network's inherent complexity and weak interconnections. Although NBA required slightly more execution time than PSO, it delivered a marginally better improvement in damping performance. Additional validation is performed by running NBA with 40 iterations, which still resulted in higher damping ratios compared to PSO at 50 iterations, as in the simple two-area system. This result confirms that, as was the case when fitting a single POD (Sect. 5.1), the NBA offers computational efficiency and produces parameters that significantly improve system stability.

Figures 14.9 (PSO) and 14.10 (NBA) illustrate the dynamic performance of the POD controllers optimized by both algorithms. Figures 14.9a and 14.10a show the RMS response (solid line) of the system with active POD controllers compared to the baseline scenario without active control (dashed line, as seen in Fig. 14.8a), effectively highlighting the impact of the optimized tuning. The blue, red, and green lines represent the frequencies of terminals in Areas A, B, and C, respectively. Figures 14.9b and 14.10b provide the eigenvalue plots. The activation of the POD controllers significantly enhances the damping coefficients, pushing the values above the 5% threshold mandated by Spanish stability regulations. This improvement is critical as it ensures all system modes are relocated to the stable region. The consistency between the time and frequency domain results indicates effective damping across the system, reducing oscillations and enhancing overall stability. While the damping response in the simple two-area system is more immediate, the PST16 bus system exhibits gradual yet effective damping that stabilizes the system, confirming the robustness and reliability of the optimized coordinated controllers in managing inter-area oscillations in complex, weakly interconnected grids.

14.6 Conclusions

The paper presents a comprehensive analysis of STATCOM-POD controller tuning and coordination using PSO and NBA algorithms, focusing on weakly interconnected networks with significant stability problems. The study uses two network models, a simple two-area system and PST16 bus system, to evaluate the effectiveness of these optimization methods in improving damping and mitigating inter-area oscillations.

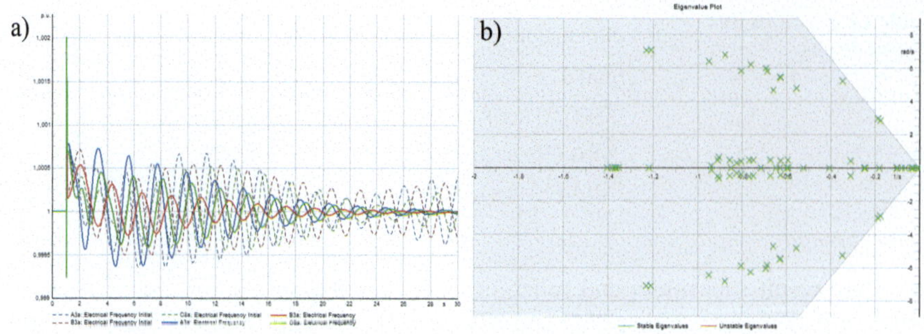

Fig. 14.9 Results after POD controller, adjusted by PSO **a** RMS **b** EV

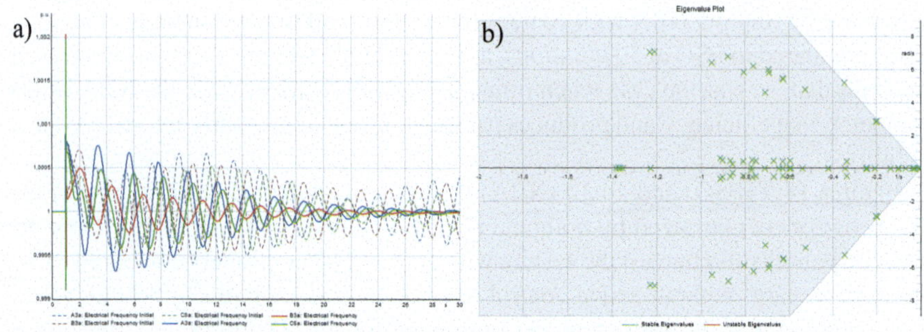

Fig. 14.10 Results after POD controller, adjusted by NBA **a** RMS **b** EV

The POD control, the PSO and NBA tuning methods, and the two networks used are defined in detail throughout the article.

PSO and NBA significantly improve the damping coefficient in the simple two-area system, reaching values above 40%, demonstrating their ability to stabilize networks with weak interconnections and oscillatory events. In the case of the more complex PST16 bus system, characterized by intricate dynamics and weak interconnections, both algorithms increase the damping coefficients above 6%, effectively stabilizing the system. However, NBA performs better than PSO, achieving superior damping performance with only slightly more computational time and delivering even better results with fewer iterations. The superior performance of NBA can be attributed to the faster convergence due to its improved exploration of the solution space, allowing it to find better parameter sets for complex systems.

Overall, the study confirms that advanced optimization techniques like PSO and NBA are highly effective in tuning and coordinating multiple STATCOM-POD controllers in weakly interconnected grids influenced by renewable energy. These methods enhance system stability and prevent inter-area oscillatory behavior, crucial in modern power systems facing increasing renewable integration. Future research will explore optimization approaches and real-time data analysis to improve the performance of POD controllers further, addressing the changing stability needs of power systems with high renewable energy penetration. Furthermore, the scalability of these methods suggests their applicability to larger networks, though computational complexity must be managed.

Acknowledgements This work has been funded by the European Union's Horizon Europe Energy Research and Innovation program under the project eFORT (Grant Agreement no. 101075665). The content of the article reflects the views only of the authors. The European Commission is not re-sponsible for any use that may be made of the information contained therein.

Disclosure of Interests The authors have no competing interests to declare that are relevant to the content of this article.

References

1. M.I. Saleem, S. Saha, T.K. Roy, Y.S.K. Ghosh, Assessment and management of frequency stability in low inertia renewable energy rich power grids. IET Gener. Trans. Dist. (2024). https://doi.org/10.1049/gtd2.13129
2. D. Wu, A.M. Aldaoudeyeh, M. Javadi, F. Ma, J. Tan, Y.J.N. Jiang, A method to identify weak points of interconnection of renewable energy resources, Int. J. Elect. Power Energy Syst., Vol. 110, (2019). https://doi.org/10.1016/j.ijepes.2019.03.003
3. M. Eslami, H. Shareef, Y.A. Mohamed, Optimal tuning of power system stabilizers using modified particle swarm optimization
4. A. Patel, Y.P.R. Gandhi, Damping oscillations in detail model of synchronous generator using PSO based PSS, IEEE 1st International Conference on Power Electronics, Intelligent Control and Energy Systems, IEEE, (2016). https://doi.org/10.1109/ICPEICES.2016.7853470
5. M. Edrah et al., Effects of POD control on a DFIG wind turbine structural system, IEEE Trans. Energy Convers., pp. 765–774, (2020). https://doi.org/10.1109/TEC.2020.2966181
6. Y. Liu, J. He, P. Wang, Y.H. Zhao, Particle swarm optimization for PSS parameters and application to DIgSILENT PowerFactory, 2023 International Conference on Energy, Electrical and Power Engineering, IEEE, (2023) https://doi.org/10.1109/CEEPE58418.2023.10165758
7. S. Mahapatra, A.N. Jha, PSS & TCSC coordinated design using particle swarm optimization for power system stability analysis», 2012 2nd International Conference on Power, Control and Embedded Systems, India, IEEE, (2012) https://doi.org/10.1109/ICPCES.2012.6508094
8. H. Hasanvand, M.R. Arvan, B. Mozafari, Y.T. Amraee, Coordinated design of PSS and TCSC to mitigate interarea oscillations. Int. J. Elect. Power Energy Syst. **78**, 194–206 (2016). https://doi.org/10.1016/j.ijepes.2015.11.097
9. J. Bhukya, Y.V. Mahajan, Optimization of damping controller for PSS and SSSC to improve stability of interconnected system with DFIG based wind farm, Int. J. Elect. Power Energy Syst. 108, (2019). https://doi.org/10.1016/j.ijepes.2019.01.017
10. Y.L. Abdel-Magid, Y.M.M. Dawoud, Tuning of power system stabilizers using genetic algorithms. Elect. Power Syst. Res. **39**(2), 137–143 (1996). https://doi.org/10.1016/S0378-7796(96)01105-4
11. H.T. Canales, F.C. Torres, Y.J.C. Silva Chavez, Tuning of power system stabilizer PSS using genetic algorithms, en 2014 IEEE International Autumn Meeting on Power, Electronics and Computing, IEEE, pp. 1-6. (2014) https://doi.org/10.1109/ROPEC.2014.7036350
12. S. Panda, Y.N.P. Padhy, Power system with PSS and FACTS controller: modelling, simulation and simultaneous tuning employing genetic algorithm, (2007)
13. B.R. Gamino, P.B. De Araujo, E. De V. Fortes, L.F.B. Martins, Y.E.L. Miotto, Coordinated tuning of PSS and TCSC-POD controller using an elite genetic algorithm, en 2016 12th IEEE

International Conference on Industry Applications (INDUSCON). Curitiba: IEEE, pp. 1–8. (2016). https://doi.org/10.1109/INDUSCON.2016.7874511
14. A. Chirra, Y.D.R. Patil, Automatic tuning of PSS through FACTS by using genetic algorithm, (1996)
15. A. Aghazade, Y.A. Kazemi, Simultaneous coordination of power system stabilizers and STATCOM in a multi-machine power system for enhancing dynamic performance», en 4th International Power Engineering and Optimization Conference, Malaysia: IEEE, pp. 13–18. (2010). https://doi.org/10.1109/PEOCO.2010.5559221
16. P. He, Q. Fang, H. Jin, Y. Ji, Z. Gong, Y.J. Dong, Coordinated design of PSS and STATCOM-POD based on the GA-PSO algorithm to improve the stability of wind-PV-thermal-bundled power system, IJEPES, Vol. 141, (2022). https://doi.org/10.1016/j.ijepes.2022.108208
17. R.V. De Oliveira, R.A. Ramos, Y.N.G. Bretas, An algorithm for computerized automatic tuning of power system stabilizers. Control. Eng. Pract. **18**(1), 45–54 (2010). https://doi.org/10.1016/j.conengprac.2009.08.004
18. P. Marić, R. Kljajić, H.R. Chamorro, Y.H. Glavaš, Power system stabilizer tuning algorithm in a multimachine system based on s-domain and time domain system performance measures. Energies **14**, 18 (2021). https://doi.org/10.3390/en14185644
19. X.S. Yang, A new metaheuristic bat-inspired algorithm. In Nature inspired cooperative strategies for optimization, (NICSO 2010). Stud. Comput. Intell. **284**, 65–74 (2010)
20. L. Chaib, A. Choucha, Y.S. Arif, Optimal design and tuning of novel fractional order PID power system stabilizer using a new metaheuristic Bat algorithm. Ain Shams Eng. J. **8**(2), 113–125 (2017). https://doi.org/10.1016/j.asej.2015.08.003
21. D.K. Sambariya, Y.R. Prasad, Robust tuning of power system stabilizer for small signal stability enhancement using metaheuristic bat algorithm, IJEPES, Vol. 61, (2014). https://doi.org/10.1016/j.ijepes.2014.03.050
22. W. Peres, E.J. De Oliveira, J.A. Passos Filho, Y.I.C. Da Silva Junior, Coordinated tuning of power system stabilizers using bio-inspired algorithms, Int. J. Elect. Power Energy Syst. Vol. 64, (2015). https://doi.org/10.1016/j.ijepes.2014.07.040
23. X.B. Meng, X.Z. Gao, Y. Liu, Y.H. Zhang, A novel bat algorithm with habitat selection and Doppler effect in echoes for optimization. Expert Syst. Appl. **42**(17–18), 6350–6364 (2015). https://doi.org/10.1016/j.eswa.2015.04.026
24. E.L. Miotto, P.B.D. Araujo, L.M. Andre Takahashi, Y.E.D.V. Fortes, Tuning of the parameters of damping controllers for power electrical systems using Bioinspired Algorithms, en 2018 13th IEEE International Conference on Industry Applications, São Paulo, Brazil: IEEE, (2018). https://doi.org/10.1109/INDUSCON.2018.8627160
25. E.L. Miotto, P.B. De Araujo, E. De Vargas Fortes, B.R. Gamino, Y.L.F.B. Martins, Coordinated tuning of the parameters of PSS and POD controllers using bioinspired algorithms. IEEE Trans. Ind. Appl. **54**(4), 3845–3857 (2018). https://doi.org/10.1109/TIA.2018.2824249
26. Red Eléctrica, Redeia Guía para la implementación de controles POD, v1, (2024)
27. Artificial bee colony algorithm based approach for power system oscillation damping with PSS and STATCOM, Int. J. Res. Stud. Elect. Electron. Eng. Vol. 5, 2, (2019). https://doi.org/10.20431/2454-9436.0502004
28. Ministerio De Industria, Turismoy Comercio, Procedimiento de Operación 13.1. Criterios de Desarrollo de la Red de Transporte, (2005)
29. J. Kennedy, Y.R. Eberhart, Particle swarm optimization, en Proceedings of ICNN'95—International Conference on Neural Networks, Perth, WA, Australia: IEEE, pp. 1942–1948. (1995). https://doi.org/10.1109/ICNN.1995.488968
30. Y. Wang et al., A novel bat algorithm with multiple strategies coupling for numerical optimization. Mathematics **7**(2), 135 (2019). https://doi.org/10.3390/math7020135
31. REE, Aelec, Naturgy, Cide, Aseme, Norma técnica de supervisión de la conformidad de los módulos de generación de electricidad según el Reglamento UE 2016/631, (2021)
32. PowerFactory DIgSILENT. [En línea]. Disponible en: https://www.digsilent.de/en/powerfactory.html
33. Oscillatory stability assessment of power systems using computational.pdf

34. J. Mola-Jimenez, J.L. Rueda, A. Perilla, W. Da, P. Palensky, Y.M. Van Der Meijden, PowerFactory-Python based assessment of frequency and transient stability in power systems dominated by power electronic interfaced generation, Modeling and Simulation Cyber-Physical Energy Systems, Porto: IEEE, (2018). https://doi.org/10.1109/MSCPES.2018.8405403

Chapter 15
A Backstepping Control Design for Output Voltage Regulation in an Unknown Linear Load Interfaced through a Buck DC-DC Converter

Oscar Danilo Montoya, Federico Martin Serra, Francisco Daniel Esteban, Walter Gil-González, and Jesús C. Hernández

Abstract This research presents a nonlinear control design for a Buck DC-DC converter that powers an unknown linear load. The proposed controller utilizes the backstepping approach, which ensures asymptotic stability in the sense of Lyapunov during closed-loop operation. The backstepping control law developed for the Buck converter is general and adaptable to a wide range of operating conditions. To address the uncertainty in the linear load, an immersion and invariance approach is employed, enhancing the robustness of the control strategy in the face of load variations. Numerical simulations, along with comparisons against the extended feedback linearization control method, demonstrate the superior performance of the proposed controller design, as evidenced by an evaluation of the integral time square error, integral absolute error, and integral time absolute error indices. All numerical simulations were conducted in the PLECs simulation tool of the MATLAB/Simulink environment.

O. D. Montoya (✉)
Grupo de Compatibilidad e Interferencia Electromagnética (GCEM), Facultad de Ingeniería, Universidad Distrital Fracisco José de Caldas, Bogotá, D.C., Colombia
e-mail: odmontoyag@udistrital.edu.co

F. M. Serra · F. D. Esteban
Laboratorio de Control Automático (LCA), Facultad de Ingeniería y Ciencias Agropecuarias, Universidad Nacional de San Luis, CONICET, Villa Mercedes, San Luis, Argentina
e-mail: fmserra@unsl.edu.ar

F. D. Esteban
e-mail: festeban@unsl.edu.ar

W. Gil-González
Departamento de Ingneiería Eléctrica, Facultad de Ingenierías, Universidad Tecnológica de Pereira, Pereira, Colombia
e-mail: wjgil@utp.edu.co

J. C. Hernández
Department of Electrical Engineering, University of Jaén, Jaén, Spain
e-mail: jcasa@ujaen.es

D. Rossit et al. (eds.), *Proceedings of the VII Ibero-American Congress of Smart Cities, ICSC-Cities 2024, 12-14 November, San Carlos, Costa Rica*, Springer Proceedings in Energy, https://doi.org/10.1007/978-981-96-4301-1_15

Keywords Backstepping control design · Buck DC-DC converter · Unknown linear load · Iimmersion and invariant load estimator · Output voltage regulation

15.1 Introduction

15.1.1 General Context

In the development of smart cities, the integration of efficient and reliable power systems is critical to ensuring the sustainability and functionality of urban infrastructures [1]. Among the key components in these systems are DC-DC converters, which play a vital role in managing and optimizing power distribution across various applications, from renewable energy sources to electric vehicles and smart grids [2]. These converters are essential in regulating voltage levels, enhancing energy efficiency, and ensuring the stable operation of devices and systems within a smart city [3].

Given the importance of DC-DC converters in these contexts, designing effective controllers for them is crucial [4]. Controllers are responsible for maintaining the desired performance of DC-DC converters under varying load conditions and disturbances, directly impacting the overall efficiency and reliability of power systems in smart cities [5]. The design of these controllers must account for dynamic and sometimes unpredictable changes in the environment, making their role indispensable in the ongoing efforts to create resilient and adaptive urban power networks.

15.1.2 Motivation

The design of controllers for DC-DC converters poses several intricate challenges that must be carefully addressed in order to achieve reliable and efficient performance [6]. One of the primary challenges is dealing with the inherent nonlinearities in DC-DC converters. These nonlinear characteristics can complicate the control process, making it difficult to maintain a consistent performance under different operating conditions. Moreover, ensuring the stability of the system across a wide range of operating conditions is crucial [7]. This includes managing sudden load changes, input voltage variations, and other disturbances that can affect the converter's operation. The design must strike a delicate balance between robustness and flexibility, ensuring that the converter can adapt to varying conditions without compromising stability [8].

In addition to these technical challenges, optimizing controllers for both performance and cost is essential, particularly in the context of large-scale deployments such as smart cities [5]. The controller must be effective, cost-efficient, and capable of real-time responsiveness. The need for integration with other smart city systems adds further complexity to the design process, requiring controllers to communicate and coordinate seamlessly with other urban infrastructure components [9]. This integration demands innovative approaches that go beyond traditional control methods, incorporating advances in both the theoretical and practical aspects of power electronics and control systems [8]. Addressing these multifaceted challenges requires a deep understanding of the underlying principles, as well as a commitment towards developing cutting-edge solutions that meet the evolving demands of modern power systems.

15.1.3 Literature Review

Numerous control strategies have been developed to regulate the output voltage of Buck DC-DC converters under varying resistive loads, including finite-time control [10], sliding mode voltage control (SMVC) [11, 12], extended state observer control [13], model-predictive control (MPC) [14], disturbance observer-based control [15], adaptive control [16], and adaptive PI passivity-based control [6]. While these strategies have shown promising results, many face challenges in practice, such as complex computational requirements and high implementation costs. Techniques like MPC demand significant processing power due to their optimization-based approach, while SMVC, despite its robustness, often requires intricate tuning and can exhibit chattering, which introduces undesirable oscillations.

Moreover, not all proposed strategies ensure robust stability during closed-loop operation, particularly under varying load conditions or model uncertainties [6]. This failure to ensure stability is a pressing concern, especially in safety-critical applications where instability could lead to system failure or damage. Consequently, there is a dire need for control strategies combining high performance with ease of implementation, cost-effectiveness, and robust stability across various operating conditions [17]. Addressing these challenges continues to be a vital area of research in the field of power electronics.

15.1.4 Contributions and Scope

The design of a backstepping controller for a Buck converter with an unknown linear load represents a significant and innovative area of research within the field of power electronics and control systems [18]. Buck converters, widely used in various applications such as power supplies, battery chargers, and renewable energy systems, are essential for efficiently stepping down voltage levels to meet specific load requirements [19]. However, one of the key challenges in real-world applications is the uncertainty and variability of the load, which can significantly affect the performance and stability of the converter [6].

Traditionally, controllers for Buck converters are designed under the assumption that the load characteristics are known and remain constant or vary within a predictable range. However, in many practical scenarios, the load may be unknown or subject to sudden changes, as when powering devices with varying power demands or when external factors influence the load [8]. This uncertainty complicates the control problem, making it difficult to achieve robust performance through conventional control strategies.

The backstepping control approach offers a promising solution to this challenge. Backstepping is a recursive, systematic design method that allows constructing controllers for nonlinear systems by considering system dynamics in a step-by-step manner [20]. This approach is particularly well-suited for systems like Buck converters, where nonlinearity and load uncertainties are prevalent [18]. Through backstepping, it is possible to design controllers that can adapt to unknown load conditions, ensuring stable operation and optimal performance across a wide range of scenarios [21].

The novelty of this research lies in the application of backstepping control to a Buck converter with an unknown linear load. While backstepping has been successfully applied to other control problems, its application in power converters with uncertain loads is relatively unexplored. This research not only addresses a critical gap in the literature but also has the potential to significantly enhance the reliability and efficiency of power conversion systems in applications where load conditions are unpredictable. By developing and validating a backstepping controller for such scenarios, this work contributes to the advancement of robust control strategies in power electronics, with potential implications for a wide range of modern technologies.

15.1.5 Document Structure

The remainder of this document is organized as follows. Section 15.2 presents the general design of a backstepping control approach for two-dimensional dynamical systems by describing each step of its formulation, stability tests, and control implementation. Section 15.3 defines the general dynamical model of a Buck DC-DC converter feeding an unknown linear load. Section 15.4 describes the application of the backstepping controller to the Buck DC-DC converter. Section 15.5 illustrates the general design of a linear load estimator based on the inverse and invariance approach. Section 15.6 provides information on the test system used and presents all the numerical validations and comparative analyses carried out. Finally, Sect. 15.7 lists the main concluding remarks derived from this work, as well as some possible future works.

15.2 Backstepping Control Approach

Backstepping control theory is a methodical and recursive approach in control engineering. It is particularly effective for stabilizing nonlinear systems in strict-feedback form. This technique involves deconstructing complex control challenges into manageable sub-problems and progressively developing control laws and Lyapunov functions at each stage. The Lyapunov function is carefully selected to ensure system stability, with its derivative rendered negative-definite via the strategic design of control inputs throughout the process. Backstepping is particularly advantageous for nonlinear systems and can be adapted for scenarios where system parameters are unknown or change over time by incorporating parameter estimation techniques into the framework. Overall, backstepping offers a systematic method for the design of stabilizing controllers in a wide range of nonlinear systems.

15.2.1 Generalized Backstepping Control Design

The design of a controller using backstepping control theory allows analyzing the dynamics of the nonlinear system represented by Eqs. (15.1) and (15.2) without loss of generality.

$$\dot{\eta} = f(\eta) + g(\eta)\,\zeta, \tag{15.1}$$

$$\dot{\zeta} = h(\eta, \zeta) + j(\eta, \zeta)\,u. \tag{15.2}$$

Here, η and ζ are the state variables, with $f(\eta)$ and $h(\eta, \zeta)$ representing smooth nonlinear functions of the state variables, and $g(\eta)$ and $j(\eta, \zeta)$ denoting the control input matrices that relate the state variables to the control inputs.

It is worth mentioning that, in backstepping control theory, the state variable ζ serves as an auxiliary control input for the first component of the dynamical system described by Eq. (15.1).

To design a backstepping controller for the dynamical system described in (15.1) and (15.2), a systematic procedure must be followed, which is outlined below.

First step: selecting the ζ controller The primary goal of this stage is to design the structure of the auxiliary control input ζ, in order to stabilize the dynamical system described by (15.1), ensuring compliance with Lyapunov's stability theorem.

Theorem 1 (Lyapunov's stability theorem) A Dynamical System $\dot{x} = f(x)$ is considered to be asymptotically stable at the origin ($x = 0$) if there is a Lyapunov Candidate Function $\mathcal{V}(x)$ that satisfies the following conditions:

1. $\mathcal{V}(x)$ is positive-definite, meaning $\mathcal{V}(x) > 0$ for all $x \neq 0$, and $\mathcal{V}(0) = 0$
2. the Derivative $\dot{\mathcal{V}}(x)$ is negative-definite, i.e., $\dot{\mathcal{V}}(x) < 0$

The auxiliary control input ζ is defined as $\varphi(\eta)$, and a candidate Lyapunov Function $\mathcal{V}(\eta)$ is selected to satisfy Theorem 1. Under these conditions, the following must be ensured:

$$\begin{aligned}\dot{\mathcal{V}}(\eta) &= \frac{\partial \mathcal{V}}{\partial \eta}\dot{\eta} \leq -\mathcal{W}(\eta), \\ &= \frac{\partial \mathcal{V}}{\partial \eta}(f(\eta) + g(\eta)\varphi(\eta)) \leq -\mathcal{W}(\eta),\end{aligned} \tag{15.3}$$

where $\mathcal{W}(\eta)$ is a positive-definite function.

Remark 1 Note that, if Condition (15.3) is satisfied, the dynamical system defined by Eq. (15.1) is stable. If the inequality condition is replaced with a strict inequality, the Dynamical System (15.1) will be asymptotically stable at $\eta = 0$.

Second step: defining the equivalent system The next phase in designing the backstepping control involves reformulating the original Dynamical System (15.1)–(15.2) into an alternative form that uses the state variable η and a newly introduced variable z, which is defined as $z = \zeta - \varphi(\eta)$. By substituting this expression for z into Eqs. (15.1) and (15.2), the following system is obtained:

$$\dot{\eta} = f(\eta) + g(\eta)z + g(\eta)\varphi(\eta), \tag{15.4}$$

$$\dot{z} = h(\eta, \zeta) + j(\eta, \zeta)u - \dot{\varphi}(\eta) = \omega, \tag{15.5}$$

where ω denotes a new auxiliary control input.

Remark 2 The general control input u can be derived from (15.5), provided that the auxiliary control input ω is known. This is done as follows:

$$u = j^{-1}(\eta, \zeta)(\omega + \dot{\varphi}(\eta) - h(\eta, \zeta)), \tag{15.6}$$

assuming that $j(\eta, \zeta)$ is a non-singular function over the domain of η.

Third step: analyzing the stability of the new dynamical system To achieve stability in the equivalent Dynamical System (15.4)–(15.5), the auxiliary control input ω must be appropriately selected. This selection is carried out by applying Lyapunov's Stability Theorem 1 to the equivalent system. Consider the following candidate Lyapunov function:

$$\mathcal{H}(\eta, z) = \mathcal{V}(\eta) + \frac{1}{2}z^2, \tag{15.7}$$

which is positive-definite, with $\mathcal{H}(0, 0) = 0$. The time derivative of this function is given by

$$\dot{\mathcal{H}}(\eta, z) = \dot{\mathcal{V}}(\eta) + z\dot{z}. \tag{15.8}$$

Substituting (15.4) and rearranging the terms yields the following expression:

$$\dot{\mathcal{H}}(\eta, z) = \frac{\partial \mathcal{V}}{\partial \eta}\left(f(\eta) + g(\eta)\varphi(\eta)\right) + z\left(\frac{\partial \mathcal{V}}{\partial \eta} g(\eta) + \omega\right). \tag{15.9}$$

It is important to note that (15.10) assumes the existence of the Positive-Definite Function $\mathcal{W}(\eta)$, as defined by (15.3). Consequently, (15.10) can be expressed as

$$\dot{\mathcal{H}}(\eta, z) \leq \mathcal{W}(\eta) + z\left(\frac{\partial \mathcal{V}}{\partial \eta} g(\eta) + \omega\right). \tag{15.10}$$

With an appropriate choice of the auxiliary control input ω, the system will achieve asymptotic stability in the sense of Lyapunov.

Remark 3 The auxiliary control input ω can be selected as follows:

$$\omega = -k_p z - \frac{\partial \mathcal{V}}{\partial \eta} g(\eta). \tag{15.11}$$

This selection ensures that the time derivative of the Candidate Lyapunov Function $\mathcal{H}(\eta, z)$ satisfies the second condition of Theorem 1, where k_p is a positive constant.

Fourth step: calculating the generalized control input In backstepping control design, it is essential to determine the general structure of the original control input u as a function of the original variables η and ζ. To establish the structure of u, the definition provided in (15.6) is employed, substituting the structure of the auxiliary control input ω. This leads to the following result:

$$u = j^{-1}(\eta, \zeta)\left(-k_p z - \frac{\partial \mathcal{V}}{\partial \eta} g(\eta) + \dot{\varphi}(\eta) - h(\eta, \zeta)\right). \tag{15.12}$$

Recalling the definition of the Auxiliary State Variable $z = \zeta - \varphi(\eta)$, the Control Input (15.12) can be rewritten as:

$$u = j^{-1}(\eta, \zeta)\left(-k_p(\zeta - \varphi(\eta)) - \frac{\partial \mathcal{V}}{\partial \eta} g(\eta) + \dot{\varphi}(\eta) - h(\eta, \zeta)\right). \tag{15.13}$$

Remark 4 The general control input given in (15.13) includes the time derivative of the Function $\varphi(\eta)$, which can be expressed as $\dot{\varphi}(\eta) = \frac{\partial \varphi}{\partial \eta}\dot{\eta}$. By applying the original dynamical system definition from Eq. (15.1), the general control input u, derived through the backstepping design in (15.13), can be reformulated to obtain the final expression (15.14).

$$u = j^{-1}(\eta, \zeta)\left(-k_p\zeta + k_p\varphi(\eta) - \frac{\partial \mathcal{V}}{\partial \eta} g(\eta) + \frac{\partial \varphi}{\partial \eta}\left(f(\eta) + g(\eta)\zeta\right) - h(\eta, \zeta)\right). \tag{15.14}$$

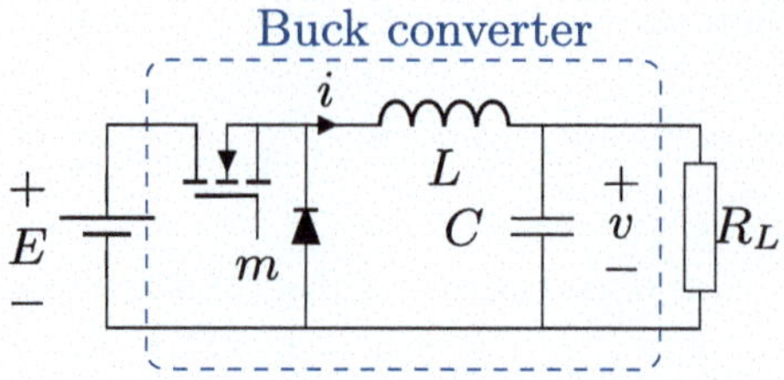

Fig. 15.1 Schematic representation of a buck converter feeding a resistive load

15.3 Converter Model

Figure 15.1 shows the schematic diagram of a Buck converter supplying a resistive load. In this diagram, E denotes the constant input voltage powering the Buck converter, while L and C represent the positive-definite parameters associated with the inductor and the capacitor, respectively. R_L signifies the unknown resistive load connected to the converter. Additionally, i and v correspond to the inductor current and the voltage at the the capacitor terminals, and m represents the discrete control input, with values of 0 (switch open) or 1 (switch closed).

To obtain the average dynamical model of the Buck converter supplying an unknown linear load (see Fig. 15.1), Kirchhoff's first and second laws are applied, where the control input m is relaxed as a continuous input u in the interval [0, 1] [6]. The equivalent dynamical model is defined by (15.15).

$$\begin{aligned} \frac{d}{dt}v &= \frac{1}{C}i - \frac{1}{R_L C}v, \\ \frac{d}{dt}i &= -\frac{1}{L}v + \frac{E}{L}u, \end{aligned} \tag{15.15}$$

where the equilibrium point is defined as follows:

$$\begin{aligned} i_\star &= \frac{1}{R_L}v_\star, \\ u_\star &= \frac{1}{E}v_\star, \end{aligned}$$

with $v_\star$ being the desired operating voltage output at the terminals of the resistive load R_L.

Remark 5 Note that, prior to applying the backstepping controller to the Buck converter feeding a resistive load R_L, the error dynamics must be determined. These are obtained by considering the Error Variables $\eta = v - v_\star$, $\zeta = i - i_\star$, and $\tilde{u} = u - u_\star$. With these definitions, the equivalent dynamical system under analysis takes the following form:

$$\dot{\eta} = \frac{1}{C}\zeta - \frac{1}{CR_L}\eta,$$
$$\dot{\zeta} = -\frac{1}{L}\eta + \frac{E}{L}\tilde{u}. \tag{15.16}$$

15.4 Backstepping Control Applied to the Buck Converter

To stabilize the studied Buck converter via backstepping control, the dynamical structure defined in (15.16) is compared to the generic structure defined in Eqs. (15.1) and (15.2), which allows identifying the following components:

$$f(\eta) = -\frac{1}{R_L C}\eta,$$
$$g(\eta) = \frac{1}{C},$$
$$h(\eta, \zeta) = -\frac{1}{L}\eta,$$
$$j(\eta, \zeta) = \frac{E}{L}. \tag{15.17}$$

To demonstrate the application of the backstepping controller discussed in Sect. 15.2, the next section outlines the controller selection process and presents the stability analysis of the Buck converter system.

15.4.1 Selection of the ζ Controller

To obtain a generalized controller to stabilize the converter, the Auxiliary Control Input $\zeta = \varphi(\eta)$ can be defined as follows:

$$\varphi(\eta) = -\alpha\eta^3, \tag{15.18}$$

where α is a positive-definite constant within the domain of real variables. To prove that the reduced dynamical system is stable, the Candidate Lyapunov Function $\mathcal{V}(\eta)$ is selected as a quadratic function:

$$\mathcal{V}(\eta) = \frac{1}{2}\beta\eta^2, \tag{15.19}$$

which implies that, as defined in Condition (15.3), and considering the definitions in (15.17),

$$\dot{\mathcal{V}}(\eta) = \frac{\partial \mathcal{V}}{\partial \eta}\left(f(\eta) + g(\eta)\varphi(\eta)\right) = -\frac{\beta}{C}\eta^2\left(\alpha\eta^2 + \frac{1}{R_L}\right) \leq 0. \tag{15.20}$$

This confirms that the first dynamical system is stable in the sense of Lyapunov for the selected $\varphi(\eta)$ function.

15.4.2 General Control Design

The design of the backstepping controller shown in Sect. 15.2 depends exclusively on the selected Functions $\varphi(\eta)$ and $\mathcal{V}(\eta)$, as defined in the generalized control law presented by (15.14). To obtain the general control law for the Buck converter feeding a resistive load, the partial derivatives of the Functions $\varphi(\eta)$ and $\mathcal{V}(\eta)$ are obtained:

$$\frac{\partial \mathcal{V}}{\partial \eta} = \beta\eta, \tag{15.21}$$

$$\frac{\partial \varphi}{\partial \eta} = -3\alpha\eta^2. \tag{15.22}$$

Remark 6 The general control law for the Buck converter system using the backstepping control design defined in (15.14), using the definitions in (15.21) and (15.22) and based on the structure of the converter defined by (15.17), takes the following form:

$$\tilde{u} = -\frac{L}{E}\left(k_p\left(\zeta + \alpha\eta^3\right) + \frac{\beta}{C}\eta + \frac{3}{C}\alpha\eta^2\left(\zeta - \frac{1}{R_L}\eta\right) - \frac{1}{L}\eta\right), \tag{15.23}$$

where α, β, and k_p are control parameters defined by the designer.

15.4.3 General Converter Control Input

Once the control input $\tilde{u}$ has been defined in order to regulate the behavior of the state variables through the incremental model, the definition of the converter variables is recovered, with the purpose of determining the General Control Input $u = u_\star + \tilde{u}$. Considering the definition of $\eta = v - v_\star$ and $\zeta = i - i_\star$, the general control input for the Buck converter feeding an unknown linear load takes the following structure:

$$u = \frac{1}{E}v_\star - \frac{L}{E}\begin{pmatrix} k_p\left((i - i_\star) + \alpha(v - v_\star)^3\right) + \frac{\beta}{C}(v - v_\star) + \\ \frac{3}{C}\alpha(v - v_\star)^2\left((i - i_\star) - \frac{1}{R_L}(v - v_\star)\right) - \frac{1}{L}(v - v_\star) \end{pmatrix}, \tag{15.24}$$

where $i_\star$ is defined as $i_\star = \frac{1}{R_L}v_\star$.

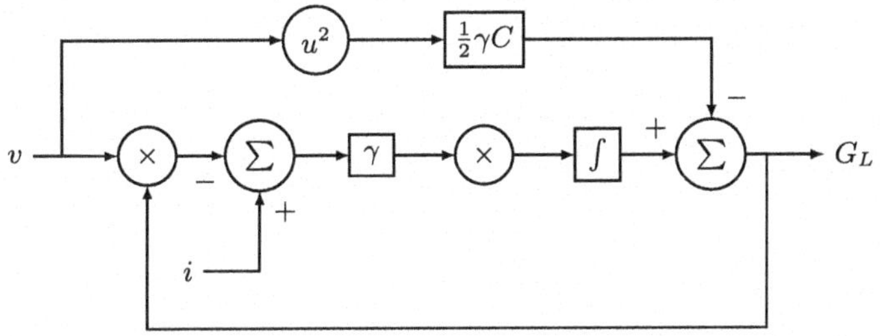

Fig. 15.2 Load estimation scheme

15.5 Load Estimation Approach

To successfully implement the proposed controller, it is crucial to accurately determine the value of the resistive load R_L. To this effect, a classical inverse and invariant (I&I) load estimator is employed [8]. This estimator is particularly advantageous since it can ensure global asymptotic convergence to the actual value of the resistive load. Remarkably, this convergence is achieved merely by measuring the voltage output at the capacitor terminals, as demonstrated in [17]. The I&I estimator is effectively represented by its Conductance Equivalent $G_L = \frac{1}{R_L}$, providing a straightforward approach to estimating R_L. The general implementation of this estimator for a resistive load is illustrated in Fig. 15.2, showcasing its integration into the overall control system.

Remark 7 Note that the load estimator in Fig. 15.2 ensures exponential convergence to the actual value of the conductance G_L from an initial value G_{L0}, simply by choosing γ, a positive constant [17].

15.6 Numerical Validations

This section evaluates the numerical performance of the proposed backstepping controller designed for output voltage regulation in Buck converters feeding unknown linear loads. The converter parameters are presented in 15.1.

Remark 8 For the sake of comparison, the proposed backstepping controller is compared against the extended feedback linearlization approach proposed by the authors of [8], where the control input was formulated as follows:

$$u = \frac{v_\star}{E} - \bar{K}_1 (i - i_\star) - \bar{K}_2 (v - v_\star). \tag{15.25}$$

Here, $\bar{K}_1$ and $\bar{K}_1$ are two positive control gains defined as 0.1567 and 0.2173, respectively.

Table 15.1 Buck converter and controller parameters

Parameter	Value	Unit
Input voltage	30	V
Output voltage reference	20	V
Capacitance	100	μF
Inductance	47	μH
Load resistance variation	8 to 16	Ω
Switching frequency	100	kHz

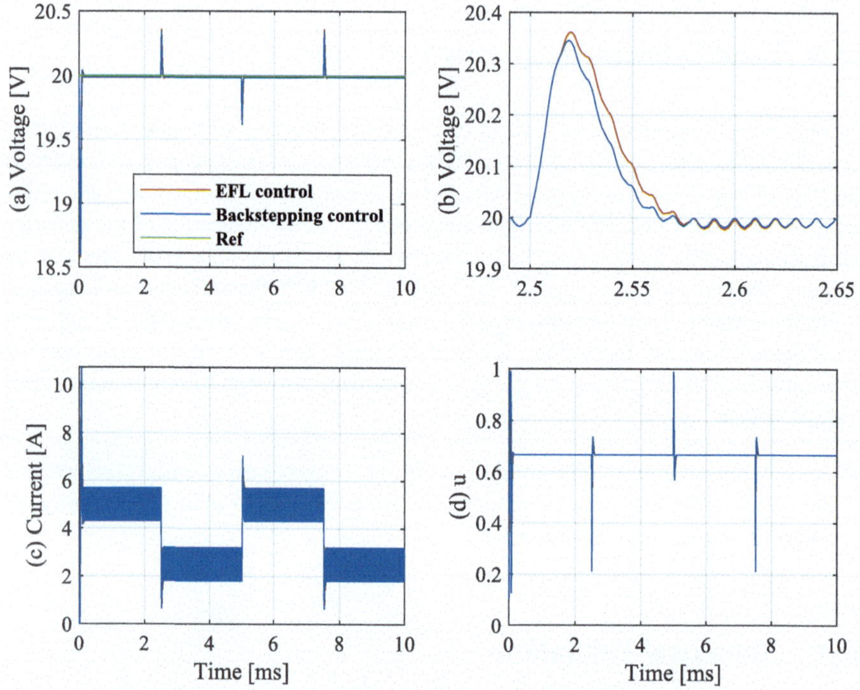

Fig. 15.3 Performance comparison of the backstepping and the EFL controllers

Note that, in the case of the backstepping controller, the control parameters α, β, and k_p were set as 1×10^{-5}, 80, and 1×10^{5}, respectively.

Figure 15.3 presents a comparative analysis of the proposed backstepping controller and the extended feedback linearization (EFL) approach.

In this figure, it is observed that:

i. Both controllers effectively regulate the voltage output at the terminals of the unknown linear load. The proposed backstepping controller is a little better than the EFL when the load changes; the maximum over-peak is lower in comparison, as shown in Fig. 15.3a and b.

Table 15.2 Summary of the analyzed indices

Control	ITSE	IAE	ITAE	T_s (ms)
EFL	6.025×10^{-8}	2.034×10^{-4}	6.658×10^{-7}	0.12
Backstepping	5.145×10^{-8}	1.963×10^{-4}	6.402×10^{-7}	0.10

ii. The behavior of the current consumed by the load is consequent with the nature of the load (see Fig. 15.3)c, i.e., a linear load, which implies that, when the load value increases, the current consumption increases, and *vice versa*. Thus, both controllers work very similarly, confirming the effective performance of the I&I estimator.
iii. The behavior of the control input (see Fig. 15.3d) shows that, regardless of load changes, it does not become saturated, which implies that both controllers ensure an asymptotically stable behavior concerning output voltage regulation at the terminals of the load. This is a significant result for practical applications.

Finally, in order to corroborate the effectiveness of the proposed controller when compared to the EFL approach, four performance indicators are listed in Table 15.2: the integral time square error (ITSE), the integral absolute error (IAE), the integral time absolute error (ITAE), and the stabilizing time.

The numerical results presented in Table 15.2 clearly demonstrate the effective performance of the proposed backstepping controller when compared to the EFL approach. Specifically, the ITSE index was reduced by approximately 14.61%, the IAE showed an improvement of about 3.49%, and the ITAE improved by 3.84%. Additionally, the settling time was reduced by approximately 200μs, further highlighting the enhanced performance of the proposed backstepping controller design in contrast to the EFL approach reported by the authors of [8].

15.7 Conclusions and Future Works

This research introduced a novel approach that employs backstepping control theory to achieve effective voltage regulation in a Buck converter supplying an unknown linear load. The proposed controller ensures asymptotic stability in the sense of Lyapunov during closed-loop operation. A key advantage of the backstepping controller is its use of three adjustable control gains, i.e., α, β, and k_p, which allow fine-tuning the controller to meet desired dynamic performance criteria. Additionally, the nonlinear structure of the control input enables a faster regulation of the incremental model, thereby reducing the settling time in comparison with the EFL approach.

Various performance indices demonstrated the effective effectiveness of the proposed backstepping approach when compared to the EFL approach. Notably, significant reductions were observed in all tested metrics, specifically ITSE, IAE, and ITAE, with decreases of approximately 14.61%, 3.49%, and 3.84%, respectively. Furthermore, the backstepping controller exhibited improved settling times, outperforming the EFL approach by 200 μs.

In future works, the following research directions can be pursued: (i) extending the design of the backstepping controller to Buck converters supplying unknown constant power loads, (ii) developing a generalized control input that considers a more comprehensive structure of the $\varphi(\eta)$ and $\mathcal{V}(\eta)$ functions, and (iii) exploring the application of the backstepping control approach to current source and voltage source converters for single-phase microgrid applications.

Acknowledgements The authors acknowledge the support provided by the Thematic Network 723RT0150, *Red para la integración a gran escala de energías renovables en sistemas eléctricos* (RIBIERSE-CYTED), funded by the 2022 call for Thematic Networks of the CYTED (Ibero-American Program of Science and Technology for Development). This research was also supported by project no. 6-24-9, titled *Desarrollo de una metodología de control secundario para microrredes de corriente continua aisladas empleando control predictivo basado en el modelo* and attached to Universidad Tecnológica de Pereira's Vicerrectoría de Investigación, Innovación y Extensión.
Disclosure of Interests The authors declare no competing interests that are relevant to the content of this article.

References

1. K.A. Mahafzah, M.A. Obeidat, A. Mansour, E.R. Sanseverino, G. Zizzo, A new smart grid hybrid dc-dc converter with improved voltage gain and synchronized multiple outputs. Appl. Sci. **14**(6), 2274 (2024)
2. J.S. Gómez-Chitiva, A.F. Escalante-Sarrias, O.D. Montoya, Voltage regulation in second-order dc-dc converters via the inverse optimal control design with proportional-integral action. TecnoLógicas **25**(55), e2369 (2022)
3. C. Wei, X. Li, in *2023 7th International Conference on Smart Grid and Smart Cities (ICSGSC)*, vol 22. Review of Bidirectional DC-DC Converters for Electric Vehicle Energy Management Systems (IEEE, 2023), pp. 432–436
4. N. Bajoria, P. Sahu, R. Nema, S. Nema, in *2016 International Conference on Electrical Power and Energy Systems (ICEPES)*, vol 5. Overview of Different Control Schemes Used for Controlling of DC-DC Converters (IEEE, 2016), pp. 75–82
5. F. Asadi, K. Eguchi, *Importance of Control in DC-DC Converters* (Springer International Publishing, 2018), pp. 11–88
6. Gil-González, Walter, Montoya, O. Danilo, G. Espinosa-Perez, *Adaptive Control for Second-Order DC-DC Converters: PBC Approach* (Elsevier, 2021), pp. 289–310
7. E. Campos-Mercado, E.F. Mendoza-Santos, J.A. Torres-Muñoz, E. Román-Hernández, V.I. Moreno-Oliva, Q. Hernández-Escobedo, A.J. Perea-Moreno, Nonlinear controller for the set-point regulation of a buck converter system. Energies **14**(18), 5760 (2021)
8. O.D. Montoya, W. Gil-González, F. Serra, C, De Angelo, J.C. Hernandez, in *2023 IEEE 14th Latin America Symposium on Circuits and Systems (LASCAS)*. Extended Feedback Linearization Control for Voltage Regulation in a Buck Converter with an Unknown Resistive Load (IEEE, 2023), pp. 1–4
9. P. Pandiyan, S. Saravanan, K. Usha, R. Kannadasan, M.H. Alsharif, M.K. Kim, Technological advancements toward smart energy management in smart cities. Energy Rep. **10**, 648–677 (2023)

10. J. Wang, C. Zhang, S. Li, J. Yang, Q. Li, Finite-time output feedback control for pwm-based dc-dc buck power converters of current sensorless mode. IEEE Trans. Control Syst. Technol. **25**(4), 1359–1371 (2017)
11. L. Ma, Y. Zhang, X. Yang, S. Ding, L. Dong, Quasi-continuous second-order sliding mode control of buck converter. IEEE Access **6**, 17859–17867 (2018)
12. R. Ling, D. Maksimovic, R. Leyva, Second-order sliding-mode controlled synchronous buck dc-dc converter. IEEE Trans. Power Electron. **31**(3), 2539–2549 (2016)
13. B. Wu, J. Yang, J. Wang, S. Li, in *Proceedings of the 33rd Chinese Control Conference*, vol 60. Extended State Observer Based Control for DC-DC Buck Converters Subject to Mismatched Disturbances (IEEE, 2014), pp. 8080–8085
14. L. Cavanini, G. Cimini, G. Ippoliti, A. Bemporad, Model predictive control for pre-compensated voltage mode controlled dc-dc converters. IET Control Theory Appl. **11**(15), 2514–2520 (2017)
15. J. Yang, B. Wu, S. Li, X. Yu, Design and qualitative robustness analysis of an dobc approach for dc-dc buck converters with unmatched circuit parameter perturbations. IEEE Trans. Circ. Syst. I Regular Papers **63**(4), 551–560 (2016)
16. C.H. Tsai, B.M. Chen, H.L. Li, Switching frequency stabilization techniques for adaptive on-time controlled buck converter with adaptive voltage positioning mechanism. IEEE Trans. Power Electron. **31**(1), 443–451 (2016)
17. W. Gil-González, O.D. Montoya, C. Restrepo, J.C. Hernández, Sensorless adaptive voltage control for classical dc-dc converters feeding unknown loads: a generalized pi passivity-based approach. Sensors **21**(19), 6367 (2021)
18. S.D. Gangula, T.K. Nizami, R.R. Udumula, Experimental investigation on backstepping control of dc-dc buck converter fed constant power load. IFAC-Papers OnLine **57**, 137–142 (2024)
19. A. Garcés-Ruiz, W. Gil-González, O.D. Montoya, Stability analysis for an ad-hoc model predictive control in dc/dc converters with a constant power load. Results Eng. **22**, 102262 (2024)
20. O.D. Montoya-Giraldo, C.A. Ramírez-Vanegas, F. Mesa, A phase-portrait stability analysis for a reaction wheel pendulum using a generalized backstepping control approach. Stat. Optim. Inf. Comput. (2024). http://www.iapress.org/index.php/soic/article/view/2164
21. T.K. Nizami, C. Mahanta, in *2014 Annual IEEE India Conference (INDICON)*, vol 5. Adaptive Backstepping Control for DC-DC Buck Converters Using Chebyshev Neural Network (IEEE, 2014), pp. 1–5

Chapter 16
Analyzing Economic Impacts of Variable Discount Rates on Medium Voltage Distribution Network Planning: A Sensitivity Analysis Considering Global Perspectives

Fábio Castro, Bruno Canizes, João Soares, Jesús C. Hernández, and Zita Vale

Abstract This paper dives into the complex world of economic analysis in the context of medium voltage optimal planning, examining the extensive consequences of various discount rates from a variety of countries and scenarios. With a particular emphasis on multi-period investment strategies, renewable resource remuneration, and the persistent presence of uncertainty, this study seeks to illuminate the interplay of these factors and their cumulative impact on decision-making processes. The research uses a thorough analytical approach that includes various discount rate scenarios to highlight the intricate trade-offs and ramifications for parties involved in medium voltage infrastructure development by studying significant indicators such as Net Present Value, Internal Rate of Return and Payback. This research provides practical insights critical for guiding informed policy formulations and strategic planning in the domain of energy infrastructure development. The model is applied to a 180-bus realistic network that is primarily reliant on renewable energy sources.

F. Castro (✉) · B. Canizes · J. Soares · Z. Vale
GECAD—Research Group on Intelligent Engineering and Computing for Advanced Innovation and Development, LASI—Intelligent Systems Associate Laboratory, ISEP – Porto School of Engineering, IPP – Polytechnic of Porto, Porto, Portugal
e-mail: fadsc@isep.ipp.pt

B. Canizes
e-mail: bmc@isep.ipp.pt

J. Soares
e-mail: jan@isep.ipp.pt

Z. Vale
e-mail: zav@isep.ipp.pt

J. C. Hernández
Department of Electrical Engineering, University of Jaén, Jaén, Spain
e-mail: jcasa@ujaen.es

D. Rossit et al. (eds.), *Proceedings of the VII Ibero-American Congress of Smart Cities, ICSC-Cities 2024, 12-14 November, San Carlos, Costa Rica*, Springer Proceedings in Energy, https://doi.org/10.1007/978-981-96-4301-1_16

The results indicate the discrepancy between highly and lesser developed countries, in the way they value their future investments. The high discount rates make it less appealing for investors to profit from long-term projects in lesser developed countries.

Keywords Discount rate · Distribution networks · Economical evaluation · Expansion planning · Localized financial output · Seasonal uncertainty

16.1 Introduction

In the rapidly evolving topic of electrical power systems, the planning and development of medium voltage (MV) distribution networks (DN) plays a critical role in ensuring the reliable and efficient delivery of electricity to consumers. As these networks are increasingly integrated with renewable energy sources (RES), the complexity of planning processes has surged. The convergence of technological advancements, environmental considerations, and economic constraints has brought about a significant shift in the strategic approach to MV DN planning [1].

One of the key factors influencing the planning and optimization of these networks is the discount rate (DR) [2]. The DR, which represents the time value of money, is pivotal in the economic assessment of investment decisions. It directly affects the valuation of future cash flows, thereby influencing decisions related to infrastructure investments, maintenance, and expansion of the DN. However, the challenge lies in the fact that DR are not static [3]; they vary across different regions, economic conditions, and scenarios. This variability introduces a layer of complexity into the planning process, making it essential to understand the implications of different DR scenarios on the overall economic viability of medium voltage distribution networks.

Moreover, the shift towards RES has added new dimensions to the planning process. The intermittency and unpredictability of RES generation necessitate a more robust and flexible approach to network design and operation. Investment strategies must account for the long-term uncertainties associated with these energy sources, further complicating the decision-making process. This has led to the adoption of multi-period investment strategies, where the timing and sequencing of investments are carefully optimized to achieve the best possible outcomes over the network's lifespan [4].

The consideration of reliability metrics is also crucial in this context. Ensuring high levels of reliability while optimizing costs is a delicate balancing act, especially in networks heavily reliant on RES [5]. The interplay between reliability, economic performance, and investment strategies underscores the importance of an analysis that incorporates the effects of varying DR.

This paper aims to address these challenges by exploring the impacts of variable DR on MV DN planning. By analyzing different scenarios and their effects on key financial indicators such as Net Present Value (NPV), Internal Rate of Return (IRR), and Payback Period, this study provides valuable insights into the trade-offs involved in planning MV infrastructure.

Many authors have worked around the topic, such as [6], which discusses the importance of the DR in energy investments, highlighting its role in determining project feasibility, calculating the levelized cost of electricity, and managing risk, with a focus on how different factors such as inflation, risk-free real return, and project risk influence the DR, and how it varies between government and private sector investments, particularly in the context of regulated utilities, nuclear projects, and renewable energy sources.

In [7], Ayodele et al. discuss the economic and environmental benefits of generating electricity from biogas. It compares two biogas recovery technologies: anaerobic digestion and landfill gas recovery. The study uses life cycle assessment to evaluate environmental impacts and various economic metrics to assess viability.

Manzhos, in [8] discusses the levelized cost of energy, particularly in the context of RES like photovoltaics. It argues that different rates should be used for borrowing and discounting, suggesting that the risk-free rate is more appropriate for discounting when comparing costs across different producers and technologies. The author critiques recent analyses for using the same rate for both borrowing and discounting, which can lead to underestimations of costs for risky borrowers and distortions in sensitivity analyses. The importance of solar-to-electricity conversion efficiency is re-emphasized under the proposed treatment of the DR. The paper also highlights the need for realistic cost estimates to guide resource allocation and establish research priorities, especially given the higher achievable costs of RES compared to traditional sources like hydro, nuclear, coal, and gas.

Miranda et al. [9] discuss the application of genetic algorithms to the optimal multistage planning of distribution networks. The paper highlights the limitations of previous static models and the advantages of using genetic algorithms, which can handle dynamic planning and multiple criteria. The authors present a model that incorporates realistic features such as multiple feeders, reuse of lines, and open loops for reliability. The genetic algorithm approach involves selection, crossover, and mutation to generate and evaluate potential solutions.

García-Gusano et al. [10] discuss the impact of social and technology-specific DRs on energy systems optimization models. They highlight that lower social DRs increase renewable energy contributions, while higher rates favor fossil fuels. The study recommends using a social DR of 4–5% for European countries and incorporating technology-specific hurdle rates to better reflect investment risks.

Shen et al. [11] present a multistage active DN planning model that integrates energy storage systems (ESS). It optimizes long-term investment and short-term operation strategies, enhancing power reliability and economic efficiency. The model's effectiveness is demonstrated through numerical cases, considering both centralized and distributed ESS options.

Xiao et al. [12] proposes a bi-level optimization model for determining the optimal site and capacity of battery ESS in DN. It aims to minimize the total NPV by optimizing BESS scheduling and capacity, considering battery lifetime. The method is validated on a modified IEEE 33-bus system.

In [5], Canizes et al. propose a stochastic methodology for planning the expansion of DN with high-RES penetration. It considers seasonal demand and RES variability, optimizing new power lines and energy storage systems. The approach aims to minimize costs, improve reliability, and reduce power losses.

The findings of this paper are intended to inform policymakers, engineers, and stakeholders involved in the energy sector, helping them make more informed decisions in the context of sustainable and resilient energy infrastructure development. The analysis is grounded in a case study of a 180-bus realistic network, predominantly powered by RES, offering practical relevance to the ongoing discourse on energy transition and infrastructure modernization.

As such, the main objectives of this work can be defined as follows:

- **Analyze the Impact of Variable DRs**: Investigate how different DRs reflective of various economic conditions and regions, influence the economic feasibility and planning of medium voltage distribution networks.
- **Evaluate Multi-Period Investment Strategies**: Examine the effects of adopting multi-period investment approaches on the long-term economic performance of distribution networks, particularly in scenarios heavily reliant on renewable energy sources.
- **Assess Key Financial Indicators**: Explore the implications of variable DRs on critical financial metrics such as NPV, IRR, and Payback Period, providing a comprehensive understanding of their role in medium voltage network planning.
- **Incorporate Reliability Analysis**: Integrate reliability metrics into the planning process to balance economic efficiency with network reliability.
- **Provide Practical Insights for Policy and Strategic Planning**: Offer actionable recommendations and insights to guide policymakers and stakeholders in formulating informed policies and strategies for the development and optimization of medium voltage distribution networks, particularly in the context of renewable energy integration.
- **Apply and Validate the Model on a Realistic Network**: Implement the proposed analytical framework on a 180-bus realistic network predominantly powered by renewable energy, to validate the findings and demonstrate the practical applicability of the study.

This paper is divided into the following sections: 1. Introduction, where the topic is introduced, relevant literature is analyzed, and the objectives and paper structure are outlined. Following is 2. Proposed Methods, where the DN model is shown, as well as the details of the economic analysis and the DR considerations. Then, in 3. Results and Discussions, there is an exposition of results and respective discussion, followed by 4. Conclusions.

16.2 Proposed Methods

16.2.1 Distribution Network Model Implications

The details of the application of DN expansion planning follow that of [4] strictly, apart from the consideration for risk aversion, which was not applied in this study, the reason being that there are different countries/locations being analyzed and risk analysis could vary according to the specific location.

Apart from all the details that are in [4], some details, such as wind curvature, solar irradiance and scenario division are extracted from [5].

Uncertainty, Resource Remuneration, Carbon Emissions, and Multi-Period

Since, as previously mentioned, most details are extracted from [4], this subchapter serves as a brief note on these topics.

Regarding the uncertainty, there are 16 total main scenarios, being the combinations of the four seasons: spring, summer, winter, fall; and the four periods of day: morning, peak, afternoon, night. For each of these scenarios a study of the PV/wind/load pattern from 2017 to 2022 was made, and the resulting factors were applied to the 16 main scenarios. For each of these 16 scenarios, 50,000 subscenarios are created and then reduced via a backward scenario reduction tool [13].

As for the resource remuneration values, they are an average of several sources: [14–20]. The resulting values can be seen in [21].

The carbon emission factors follow the same trend, with these sources: [22, 23]. Here, the resulting values can be seen in [20].

The multi-period implementation follows that of [4].

Mathematical Model

Regarding the detailed mathematical model, the objective function is as follows in Eqs. (16.1–16.4). (1) represents the summary of the objective, minimizing all costs. (2) and (3) represent the immediate and non-immediate decision variables, respectively. (4) represents the carbon emission consideration.

$$Minimize PC = PC_1 + PC_2 + PC_{CO_2} \tag{16.1}$$

$$PC_1 = \sum_{is \in \Omega_{IS}} \begin{bmatrix} \sum_{i \in \Omega_B} \sum_{j \in \Omega_B} \sum_{c \in \Omega_O} y_{(i,j,c,s)} \cdot LineCost_{(i,j,c)} + \\ \sum_{i \in \Omega_B} \sum_{j \in \Omega_B} \sum_{c \in \Omega_O} y_{(i,j,c,s)} \cdot LineMCost_{(i,j,c)} + \\ \sum_{u \in \Omega_{STO}} a_{(e,u,s)} \cdot StCost_u + \\ \sum_{u \in \Omega_{STO}} a_{(e,u)} \cdot StMCost_u + \end{bmatrix} \tag{16.2}$$

$$PC_2 = \sum_{s\in\Omega_S}\left[\left(\begin{array}{l}\left(\sum_{i\in\Omega_B}\sum_{c\in\Omega_O} Flow_{(i,j,c,s)}\cdot FOR_{(i,j,c)}\right)\cdot ENSCost+\\ \left(\sum_{i\in\Omega_B}\sum_{c\in\Omega_O} Flow_{(i,j,c,s)}\cdot LF_{(i,j,c)}\right)\cdot LossC+\\ \sum_{g\in\Omega_{DG}^{nd}} P_{PGC(g,s)}\cdot PCost_{PGC(g)}+\\ \sum_{lo\in\Omega_L^b} P_{LoadCut(lo,s)}\cdot CutCost_{lo}+\\ P_{Sub(bs)}\cdot Sub_{Price}+\sum_{bio\in Bio} P_{Bio(bio,s)}\cdot Bio_{Price}+\\ \sum_{pv\in PV} P_{PV(pv,s)}\cdot PV_{Price}+\sum_{wi\in WI} P_{WI(wi,s)}\cdot WI_{Price}+\\ \sum_{e\in E} P_{E(e,s)}\cdot ESS_{Price}\end{array}\right)\cdot\omega_s\right] \tag{16.3}$$

$$PC_{CO_2} = \sum_{s\in\Omega_S}\left[\left(\begin{array}{l}P_{Sub(bs)}\cdot\iota_{bs}+\sum_{bio\in Bio} P_{Bio(bio,s)}\cdot\iota_{Bio}+\\ \sum_{pv\in PV} P_{PV(pv,s)}\cdot\iota_{PV}+\sum_{wi\in WI} P_{WI(wi,s)}\cdot\iota_{WI}+\\ +\sum_{e\in E} P_{E(e,s)}\cdot\iota_E\end{array}\right)\cdot\omega_s\cdot EC_{CO_2}\right] \tag{16.4}$$

Equation (16.2) is dependent on the investment stages, and (3) and (4) on the scenarios.

16.2.2 Economic Analysis

The model analyzes the results and inputs them into a mathematical economic analysis, which outputs the following economic indexes: NPV, IRR, and Payback.

They follow Eqs. (16.5), (16.6), and (16.7), respectively. Additionally, the capital recovery factor that is then used in [4] follows Eq. (16.8).

$$\begin{gathered} NPV = \left(\sum_{t=1}^{n}\frac{Net\,Cash\,Inflow_t}{(1+r)^t}\right) - Initial_{Investment}\\ NPV = Savings - Investment\\ NPV > 0 \end{gathered} \tag{16.5}$$

$$Payback = \frac{Total\,Project\,\mathrm{Cost}}{AnnualCashInflows} \tag{16.6}$$

$$IRR = \sum_{t-1}^{t} \frac{Net\,Cash\,Inflow_t}{(1+r)^t} - Initial_{Investment} \tag{16.7}$$

$$CRF = \frac{t \cdot (1+t) \cdot n}{(1+t) \cdot n - 1} \tag{16.8}$$

All these calculations are dependent on the project lifetime (which will be 30 years in this case), the current project time, and the DR, which is the focus of this study. As previously mentioned in the Introduction, the DR varies substantially according to the economic state of a country or region and may heavily affect the project's outcome.

16.2.3 Discount Rate Considerations and Specific and Generalized Case Study

For the proposed study, several countries were considered. The main objective of this study is to use a real network, which is located in Leiria, Portugal, and understand the effects a different economic scenario would have on the results, if the DN would hypothetically be located there. In Table 16.1 are all the considered studies. All the DRs are extracted from CEICData [24]. Apart from Portugal, which also uses maximum and minimum, all DRs are dated to the 1st of January 2024. In parenthesis are the percentage changes between all DRs and the actual realistic one (Portugal current).

It is common for less developed countries to have a higher DR. When comparing, as an example, Ghana to Germany, Ghana has a 29.85% DR, suggesting higher opportunity cost of capital, indicating that there is a preference over present consumption, as opposed to future consumption. Germany, on the other hand, with a 2.10% DR, implies a much greater valuation of future consumption, essentially focusing more

Table 16.1 All considered discount rates (%)

Portugal current—2.86	Portugal min—0.03 (− 98.95%)	Portugal max–13.85(+ 384.27%)
Iceland—1 (− 65.03%)	USA—4.02 (+ 40.56%)	Germany—2.10 (− 26.57%)
Ghana—29.85 (+ 943.71%)	Brazil—6.53 (+ 128.32%)	China—2.57 (− 10.14%)
Russia—11.86 (+ 314.69%)	Switzerland—0.66 (− 76.92%)	South Korea—3.35 (+ 17.13%)
France—2.65 (− 7.34%)	European Union—2.90 (+ 1.4%)	Kenya—14.37 (+ 402.45%)

on a long-term vision. Given that this project being studied has a lifetime of 30 years, which is considerably long, having a broad spectrum of possible situations is crucial.

The DN being studied can be profoundly seen in [4], and its initial configuration and partial data can be found in [25]. It is a 180-bus network located in Leiria, Portugal. The accurate representation of its study would be the one with a DR of 2.86 (Portugal's current). It is clear from the DR values that there is a distinct correlation between how developed a country is and how much they value their future.

16.3 Results and Discussions

After running the full model of [4] with the different DRs, the final costs for each run are as seen in Table 16.2. Here, the focus is on final costs, after 30 years, and not on incremental multi-period investments (as in the original model), for simplicity's sake. Again, in parenthesis is the comparison to the actual situation (Portugal's current DR).

The monetary trend is as expected. On a situation where a country values their future more prominently, long-term projects are cheaper, whereas in countries that are less developed, and thus focus more on present issues as opposed to future targets, the project would be more expensive.

Figure 16.1 shows the trajectory of the cost according to the DR in this study.

While the trajectory is not completely linear, it is relatively straight-forward, once again supporting how difficult it is for lesser-developed countries to project future sights.

The analyses for NPV, IRR and Payback versus DR show the same trends as the ones in Fig. 16.1. Underdeveloped countries are always at a disadvantage.

In lesser-developed countries, long-term project planning often faces significant challenges due to high DRs. These elevated rates reflect an economic environment where present resources are heavily prioritized over future gains. For projects with long lifetimes, such as infrastructure investments with a 30-year horizon (which is

Table 16.2 Results for all runs (m.u.)

Portugal current—24,211,373	Portugal minimum—16,312,112 (− 32.63%)	Portugal maximum—40,913,366 (+ 68.98%)
Iceland—19,483,362 (− 19.53%)	USA—26,547,921 (+ 9.65%)	Germany—22,449,411 (− 7.28%)
Ghana—61,785,926 (+ 155.19%)	Brazil—30,697,250 (+ 26.79%)	China—23,560,082 (− 2.69%)
Russia—38,315,933 (+ 58.26%)	Switzerland—18,443,264 (− 23.82%)	South Korea—25,255,081 (+ 4.31%)
France—23,738,299 (− 1.95%)	European Union—24,284,966 (+ 0.30%)	Kenya—42,221,220 (+ 74.39%)

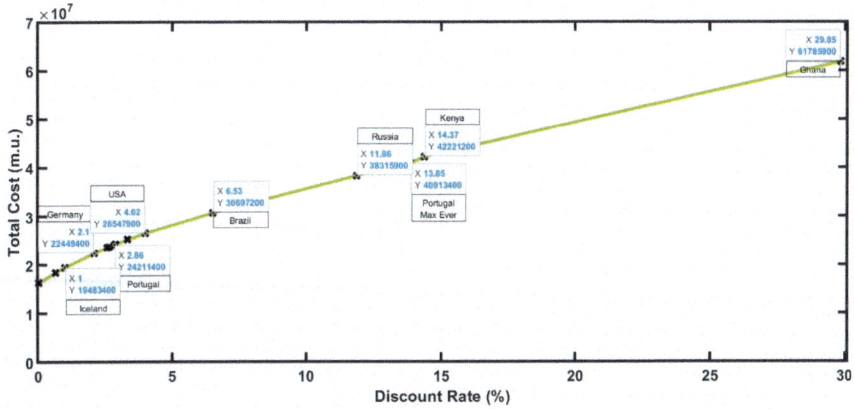

Fig. 16.1 Cost versus DR for the study

the case), this poses a fundamental issue. The high DR makes future benefits seem comparatively insignificant, leading to difficulties in justifying and securing funding for projects whose returns are primarily realized in the distant future.

In the case of infrastructure like MV DNs, the lifespan of the project spans several decades, and its value lies in the sustained benefits it provides over time—reliable power supply, reduced operational costs, and increased economic activity. However, in countries with higher DRs, these future gains are discounted heavily, making them appear less valuable in today's terms. This short-term economic focus often discourages long-term investments, despite their potential to drive sustainable development.

As a result, project developers in these environments must grapple with the challenge of convincing stakeholders to invest in initiatives whose true value only becomes evident many years down the road. Without a strong emphasis on future benefits, such essential projects risk being sidelined in favor of more immediate, short-term priorities.

Analyzing the specific case of Ghana versus Iceland, for example. In Iceland, the NPV would be 10,854,993 m.u., with an IRR of 25.39% and a Payback of 4.93 years, while in Ghana, the project is not economically viable. There is no Payback in the project's lifetime of 30 years, no IRR, and negative NPV, making these investments seem less appealing to potential investors.

16.4 Conclusions

The economic realities of lesser-developed countries, marked by high DRs, create significant challenges for long-term infrastructure planning and development. Projects like MV DNs, with lifespans extending over 30 years, are crucial for driving sustainable growth and ensuring access to essential services. However, the high DRs prevalent in these economies devalue future returns, making it difficult to justify investments that primarily yield benefits in the distant future. This short-term economic focus leads to underinvestment in critical infrastructure, perpetuating a cycle of limited growth and development.

To address this, a shift in approach is required. Policymakers and project developers must seek innovative solutions that account for both the immediate economic constraints and the long-term benefits of infrastructure projects. This could involve tapping into alternative funding sources, such as international development loans or public–private partnerships, to ease the financial burden and spread risk over time. Additionally, implementing policies that lower DRs or create incentives for long-term investments can help reshape the financial motives, making future benefits more tangible. Equally important is raising awareness among stakeholders—governments, investors, and the general public—about the broader economic and social gains that infrastructure projects bring. Education and advocacy efforts can emphasize how these projects not only improve immediate living standards but also create a foundation for economic resilience, innovation, and job creation over the long term.

In conclusion, overcoming the barriers posed by high DRs and a short-term economic focus is essential for enabling countries to build the infrastructure they need for sustainable development. Through targeted policy reforms, creative financing solutions, and a collective effort to value future returns, these nations can pave the way for long-term progress and growth.

16.5 Disclosure of Interests

The authors have no competing interests to declare that are relevant to the content of this article.

Acknowledgements This research has received funding from FEDER funds through the Operational Programme for Competitiveness and Internationalization (COMPETE 2020), National Funds through the FCT Portuguese Foundation for Science and Technology under project UIDB/000760/2020-(https://doi.org/10.54499/UIDB/00760/2020), CEECIND/00420/2022 and from the NGS Innovation Pact—New Generation Storage (C644936001-00000045), co-financed by NextGeneration EU, through the Incentive System Agendas for Business Innovation, within the scope of the Recovery and Resilience Plan (PRR). Fábio Castro, Bruno Canizes, João Soares, and Zita Vale are with GECAD—Research Group on Intelligent Engineering and Computing for Advanced Innovation and Development, LASI—Intelligent Systems Associate Laboratoy and Polytechnic of Porto, Portugal. Professor Jesús Hernández is with Escuela Politécnica Superior de Jaén. The authors also acknowledge the support provided by the Thematic Network 723RT0150 "Red para la integración a gran escala de energia renovables en sistemas electrics (RIBIERSE-CYTED)" financed by the call for Thematic Networks of the CYTED (Ibero-American Program of Science and Technology for Development) for 2022.

Nomenclatures

Indices and Sets

bio	Biomasses
bs	Substation
c	Line options
e	Energy storage system (ESS)
g	Distributed generators (pv + w)
i,j	Electrical buses
is	Current investment stage
lo	Loads
pv	Photovoltaic parks
s	Scenario
u	ESS options
wi	Wind farms
Ω_B	Set of buses
Ω_L^b	Set of load buses
Ω_{IS}	Set of investment stages
Ω_{DG}^{nd}	Set of non-dispatchable DG
Ω_S	Set of scenarios
Ω_{STO}	Set of ESS

Parameters

Bio_{Price}	Biomass generation price (m.u./MWh)
$CutCost$	Load curtailment cost (m.u./MWh)
$ENSCost$	Energy not supplied price (m.u./MWh)
$FOR_{(i,j,c)}$	Forced outage rate between bus *i* and *j*
$LF_{(i,j,c)}$	Loss factor between bus *i* and *j*
$LineCost$	Cost of line implementation
$LineMCost$	Line maintenance costs
$LossC$	Power loss costs (m.u./MWh)
PV_{Price}	Photovoltaic generation price (m.u./MWh)
Sub_{Price}	Substation generation price (m.u./MWh)
$StCost$	ESS cost (m.u.)
$StMCost$	ESS maintenance cost (m.u.)
WI_{Price}	Wind generation price (m.u./MWh)

Variables

$a(e,u)$	Binary variable for ESS, *e*, use
$Flow_{(i,j,s)}$	Power flow between bus *i*, *j*, in scenario, *s* (MW)
n	Project lifetime (years)
$p_{Bio(bio,s)}$	Power by biomass, bio, in scenario, *s* (MW)
$PCost_{PGC}$	Power generation curtailment cost (m.u./MWh)
$P_{LoadCut(lo,s)}$	Power load curtailment in load, lo, scenario, *s*
$P_{PGC(g,s)}$	Power curtailment, for DG, *g*, in scenario, *s* (MW)
$p_{PV(pv,s)}$	Power generated by PV, pv, in scenario, *s* (MW)
$p_{Sub(s)}$	Power delivered by substation in scenario, *s* (MW)
$p_{WI(wi,s)}$	Power generated by wind, *w*, in scenario, *s* (MW)
t	Current time, in '*n*' years
$y(i,j)$	Binary variable for line, *i*, *j*, use
ω_s	Scenario, s, probability

References

1. W. Wang, Y. Huang, M. Yang, C. Chen, Y. Zhang, X. Xu, Renewable energy sources planning considering approximate dynamic network reconfiguration and nonlinear correlations of uncertainties in distribution network. Int. J. Electr. Power Energy Syst. **139**, 107791 (2022). https://doi.org/10.1016/j.ijepes.2021.107791
2. J.D. Foppert, The discount rate differential required for harvesting-by-dice-rolling to outperform optimal rotation planning. For. Policy Econ. **160**, 103154 (2024). https://doi.org/10.1016/j.forpol.2024.103154
3. R.L. McDonald, S.M. Chilton, M.W. Jones-Lee, H.R.T. Metcalf, Evidence of variable discount rates and non-standard discounting in mortality risk valuation. J. Environ. Econ. Manag. **82**, 152–167 (2017). https://doi.org/10.1016/j.jeem.2016.11.005
4. F. Castro, B. Canizes, J. Soares, J. Almeida, Z. Vale, Comprehensive framework for distribution network multi-investment expansion planning: emissions, uncertainty, and resource remuneration integration. Energy Convers. Manag. **316**, 118734 (2024). https://doi.org/10.1016/j.enconman.2024.118734
5. B. Canizes, J. Soares, F. Lezama, C. Silva, Z. Vale, J.M. Corchado, Optimal expansion planning considering storage investment and seasonal effect of demand and renewable generation. Renew. Energy **138**, 937–954 (2019). https://doi.org/10.1016/j.renene.2019.02.006
6. H. Khatib, The discount rate-a tool for managing risk in energy investments
7. T.R. Ayodele, A.S.O. Ogunjuyigbe, M.A. Alao, Economic and environmental assessment of electricity generation using biogas from organic fraction of municipal solid waste for the city of Ibadan, Nigeria. J. Clean. Prod. **203**, 718–735 (2018). https://doi.org/10.1016/j.jclepro.2018.08.282
8. S. Manzhos, On the choice of the discount rate and the role of financial variables and physical parameters in estimating the levelized cost of energy. Int. J. Financ. Stud. **1**(3), 54–61 (2013). https://doi.org/10.3390/ijfs1030054
9. V. Miranda, J.V. Ranito, L.M. Proenca, Genetic algorithms in optimal multistage distribution network planning. IEEE Trans. Power Syst. **9**(4), 1927–1933 (1994). https://doi.org/10.1109/59.331452

10. D. García-Gusano, K. Espegren, A. Lind, M. Kirkengen, The role of the discount rates in energy systems optimisation models. Renew. Sustain. Energy Rev. **59**, 56–72 (2016). https://doi.org/10.1016/j.rser.2015.12.359
11. X. Shen, M. Shahidehpour, Y. Han, S. Zhu, J. Zheng, Expansion planning of active distribution networks with centralized and distributed energy storage systems. IEEE Trans. Sustain. Energy **8**(1), 126–134 (2017). https://doi.org/10.1109/TSTE.2016.2586027
12. J. Xiao, Z. Zhang, L. Bai, H. Liang, Determination of the optimal installation site and capacity of battery energy storage system in distribution network integrated with distributed generation. IET Gener. Transm. Distrib. **10**(3), 601–607 (2016). https://doi.org/10.1049/iet-gtd.2015.0130
13. F. Castro, B. Canizes, J. Soares, Z. Vale, Hourly uncertainty applied to optimal expansion planning considering energy storage and seasonal network impacts, in 2023 IEEE PES Innovative Smart Grid Technologies Europe (ISGT EUROPE), IEEE, pp. 1–5. (2023). https://doi.org/10.1109/ISGTEUROPE56780.2023.10407911
14. BloombergNEF, Cost of new renewables temporarily rises as inflation starts to bite. Accessed: (2024). [Online]. Available: https://about.bnef.com/blog/cost-of-new-renewables-temporarily-rises-as-inflation-starts-to-bite/
15. M. Armstrong, The falling cost of renewable energy. Accessed, Mar. 14, (2024). [Online]. Available: https://www.statista.com/chart/26085/price-per-megawatt-hour-of-electricity-by-source/
16. WindEurope, Wind energy is the cheapest source of electricity generation. Accessed, Mar. 24, (2024). [Online]. Available: https://windeurope.org/policy/topics/economics/
17. Knoema, A decade of change: how renewables became competitive with fossil fuels. Accessed, Mar. 14, (2024). [Online]. Available: https://knoema.com/infographics/fmzvyg/a-decade-of-change-how-renewables-became-competitive-with-fossil-fuels
18. Irena, Renewable power generation costs in 2021. Accessed, Mar. 14, (2024). [Online]. Available: https://www.irena.org/-/media/Files/IRENA/Agency/Publication/2022/Jul/IRENA_Power_Generation_Costs_2021.pdf?rev=34c22a4b244d434da0accde7de7c73d8
19. Lazard, Levelized cost of energy, Levelized cost of storage, and Levelized cost of hydrogen. Accessed, Mar. 14, (2024). [Online]. Available: https://www.lazard.com/perspective/levelized-cost-of-energy-levelized-cost-of-storage-and-levelized-cost-of-hydrogen-2020/
20. F. Castro, B. Canizes, J. Soares, J. Almeida, Z. Vale, Optimizing smart distribution network expansion incorporating the seasonal impacts, CO2 Emissions and Resources Remuneration: A Multi-Stage Stochastic Framework, in Energy Proceedings, (2024)
21. F. Castro, B. Canizes, J. Soares, M. J. Rider, Z. Vale, Distributed Resources Remuneration on a Medium Voltage Network with Uncertainty and Seasonal Impacts, pp. 158–174. (2024). https://doi.org/10.1007/978-3-031-48652-4_11
22. IEA, IEA energy emission rates. Accessed, Mar. 14, (2024). [Online]. Available: https://www.iea.org/data-and-statistics/data-tools/greenhouse-gas-emissions-from-energy-data-explorer
23. T.D. De Lima, A. Tabares, N.B. Arias, J.F. Franco, A stochastic programming model for the planning of distribution systems considering renewable distributed generation and CO_2 emissions, in 2019 IEEE PES Conference on Innovative Smart Grid Technologies, ISGT Latin America 2019, Institute of Electrical and Electronics Engineers Inc., (2019). https://doi.org/10.1109/ISGT-LA.2019.8895395
24. CEIC, CEICData. Accessed, Sep. 11, (2024). [Online]. Available: www.ceicdata.com/
25. F. Castro, B. Canizes, J. Soares, J. Almeida, B. Francois, Z. Vale, 180-Bus medium voltage distribution network data. Accessed, Jan. 15, (2024). [Online]. Available: https://zenodo.org/records/10371848

Chapter 17
Single-Phase Cascaded Multilevel Inverter Operating as a Grid Forming Converter Under dVOC Control Using SOGI Strategy

Esteban I. Marciel, Carlos R. Baier, José M. Rojas, Agustín Quinteros, and Francisco D. Esteban

Abstract Grid-forming converters (GFM) offer an alternative to traditional converters by providing additional features such as operation in island mode and grid-connected mode, emulating a synchronous generator or providing inertia support, which helps enhance grid reliability. Although most control schemes need to operate in the dq or $\alpha\beta$ reference frames, some strategies can be used to implement these type of control in single-phase converters. This paper discusses the use of a second-order generalized integrator (SOGI) strategy to implement a dispatchable virtual oscillator control (dVOC) as a GFM strategy in a single-phase cascaded H-bridge multilevel inverter (CHB-MLI). The control scheme enables the converter to operate both in island mode and while connected to the grid, managing voltage and frequency in island mode, and controlling the power delivered when connected to a larger grid. Results are presented for the converter's operation when supplying loads on island mode, injecting energy into a grid, and during transitions between island and grid-connected modes.

Keywords Grid-forming inverters · Single-phase · dVOC · SOGI

17.1 Introduction

Currently, power inverters are widely used to inject energy into electrical grids, where there is an increasing effort to incorporate renewable energy sources that rely on these converters. However, a high proportion of traditional inverters compared to synchronous generators in power grids can negatively impact system stability

E. I. Marciel · C. R. Baier (✉) · J. M. Rojas · A. Quinteros
Departamento de Ingeniería Eléctrica, Universidad de Talca, Talca, CL, Chile
e-mail: cbaier@ieee.org

F. D. Esteban
Laboratorio de Control Automático (LCA), Facultad de Ingeniería y Ciencias Agropecuarias, Universidad Nacional de San Luis, CONICET, San Luis, Argentina

D. Rossit et al. (eds.), *Proceedings of the VII Ibero-American Congress of Smart Cities, ICSC-Cities 2024, 12–14 November, San Carlos, Costa Rica*, Springer Proceedings in Energy, https://doi.org/10.1007/978-981-96-4301-1_17

due to the inherent requirements and limitations of inverters and their traditional control schemes. For instance, inverters typically require PLL algorithms for grid synchronization and lack of inertia contribution [1]. As a result, grid-forming converters (GFM) are emerging as a promising alternative to traditional or grid-following converters (GFL). Unlike GFLs, which depend on other element as primary source to supply the grid while focusing solely on energy injection, GFMs can operate autonomously, providing power to a grid independently [2].

GFM converters aim to emulate the behavior of synchronous generators, enabling them to function as the primary power source in a grid or as a supplementary energy supplier. To achieve this, GFMs incorporate control loops that do not rely on PLL strategies for voltage reference generation. Instead, they synchronize with the grid by other means, allowing them to operate with or without a voltage reference to follow in the grid, and even to perform black-start operations autonomously [3]. Additionally, certain control strategies enable GFMs to contribute additional services, such as inertia response, which is characteristic of synchronous generators, thereby enhancing grid robustness [4].

One of the most widely studied GFM control strategies is droop control [5], which provides a simple yet effective approach to regulating frequency and voltage amplitude. This method uses an active power loop to generate the converter's output voltage frequency and angle, while a reactive power loop controls the voltage amplitude. The droop control strategy requires measuring voltage and current in the dq reference frame to determine the power delivered by the converter [6].

Alternatively, virtual oscillator control (VOC) strategies enable converters to mimic weak nonlinear oscillators, generating near-sinusoidal signals in steady-state conditions [7]. A key feature of VOC is its ability to self-synchronize with the grid under various operating conditions, eliminating the need for additional control loops to regulate delivered voltage and frequency. Since VOC operates in the time domain, unlike other GFM strategies, it can respond more quickly to dynamic changes and contingencies [8].

The implementation of VOC strategies, such as the Andronov-Hopf Oscillator (AHO), allows for simultaneous power and voltage control, and strategies with this feature are known as dVOC [9, 10]. However, implementing this type of control in single-phase circuits poses challenges, as both the dVOC and power control strategies require signals in the $\alpha\beta$ reference frame, typically derived from three-phase signals using the Park transform [11]. This issue can be addressed by applying a Second-Order Generalized Integrator (SOGI), which generates orthogonal components from a single-phase signal, enabling the application of dVOC control with power regulation in single-phase grids [12].

Various studies have explored the performance of GFM converters using VOC strategies. In [13], a modified AHO strategy is proposed, leveraging the rapid synchronization capabilities of VOC to perform pre-synchronization, ensuring safe converter connection to the grid. In [14], local VOC control is applied to each inverter module of a three-phase cascaded converter, facilitating automatic adaptation for converters with numerous modules. In [9], power control of a converter is achieved using a Van-der-Pol oscillator, which allows control of both active and reactive power, though it requires current and voltage measurements from the three-phase system.

In [15], a control strategy is developed to operate a converter powered by a photovoltaic system under dVOC control with a current limiter and MPPT capability. This work adds a DC voltage control to the AHO strategy, influencing the active power reference. Internal current loops in the dq plane are also implemented, allowing the converter's current to be limited during contingencies, thereby enhancing reliability. However, DC and AC voltage measurements, as well as output current from the converter and the LCL filter, are necessary for proper operation. A dual-dVOC strategy for negative sequence current control is presented in [16], where two parallel controls manage the positive and negative sequences of the output current. The combined modulation improves converter performance under grid imbalances. In [17], virtual inertia is introduced into a dVOC based on AHO, enabling the converter to provide an inertial frequency response to power changes while maintaining typical GFM behavior.

In [18], single-phase inverter synchronization using VOC is evaluated with a dead-zone oscillator, which requires only the α current as input. This simplifies single-phase converter implementation by using the converter's output current as the α current. It is demonstrated that converters synchronize effectively and share loads according to their capacities. Similarly, [19] analyzes islanded inverter control with dVOC based on a Van-der-Pol oscillator, using only the α current for control. Experimental results show that the control achieves power set-points with the help of PI regulators, eliminating steady-state errors by adjusting oscillator gains.

From a topological perspective, CHB-MLIs offer two key advantages over traditional voltage-source inverters (VSI) for GFM operation. First, CHB-MLIs consist of multiple modules connected in series, dividing the output voltage across the modules and allowing the use of lower-rated components compared to traditional converters. Second, their modularity increases fault tolerance, as faulty cells can be isolated without shutting down the entire converter [20], a crucial feature for a GFM.

This work presents the development of a CHB-MLI operating in GFM mode with dVOC control based on the Andronov-Hopf Oscillator (AHO), where a three-phase system control is implemented in a single-phase converter using the SOGI strategy while only the output current is measured. This approach allows enables the benefits of dVOC control in small systems without three-phase grids, while ensuring stable performance under grid frequency variations. The work includes system description, detailing dVOC control, the SOGI strategy, and converter characteristics, followed by simulation results. Three case studies are presented, concluding with final observations.

17.2 System Description

The proposed system scheme is presented in Fig. 17.1, where the Andronov-Hopf dVOC is the GFM control, and the SOGI strategy is used to generate the current in

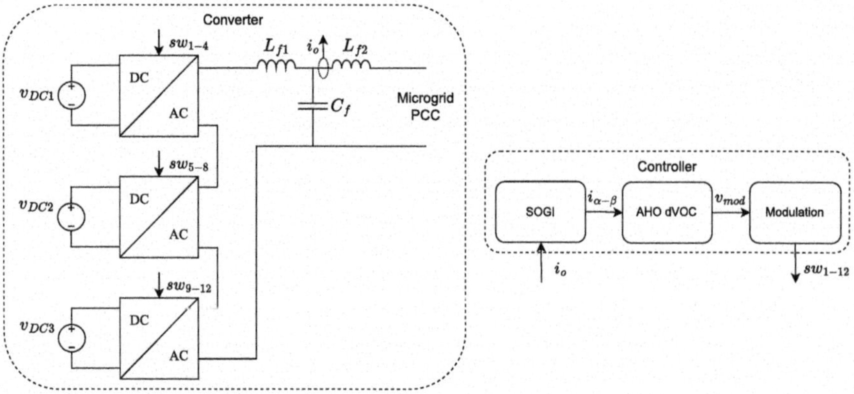

Fig. 17.1 Proposed system

the $\alpha\beta$ reference frame, necessary for the AHO. The converter used is a cascaded multilevel converter composed of three H-bridges, powered by balanced and isolated DC links, while an LCL filter at the converter output connects the converter to point of common coupling (PCC) that connects the converter to the loads and the grid.

17.2.1 Virtual Oscillator Control

AHO is a nonlinear control in the time domain that is implemented in a stationary two-phase $\alpha\beta$ reference frame. As shown in Fig. 17.2, the controller mimics an LC resonant circuit with nonlinear sources, where the inputs are the calculated current errors, and the outputs are the α - β voltages, which are sent to the modulation stage. The LC circuit has a natural frequency that is selected as the nominal frequency of the grid in which the converter will operate, and this frequency is represented by Eq. 17.1.

$$\omega_n = \frac{1}{\sqrt{LC}} \tag{17.1}$$

The circuit of an AHO can be represented as a system with two state variables, the voltage v_C and the current i_L, while the system input is the error between the converter's output current ($i_{\alpha-\beta}$) and its reference ($i^*_{\alpha-\beta}$), which is generated by the reference current estimator. This system can be described by Eqs. 17.2 and 17.3.

$$C\dot{v_C} = i_m - i_L - u_\alpha \tag{17.2}$$

$$L\dot{i_L} = v_m - v_C - \epsilon u_\beta \tag{17.3}$$

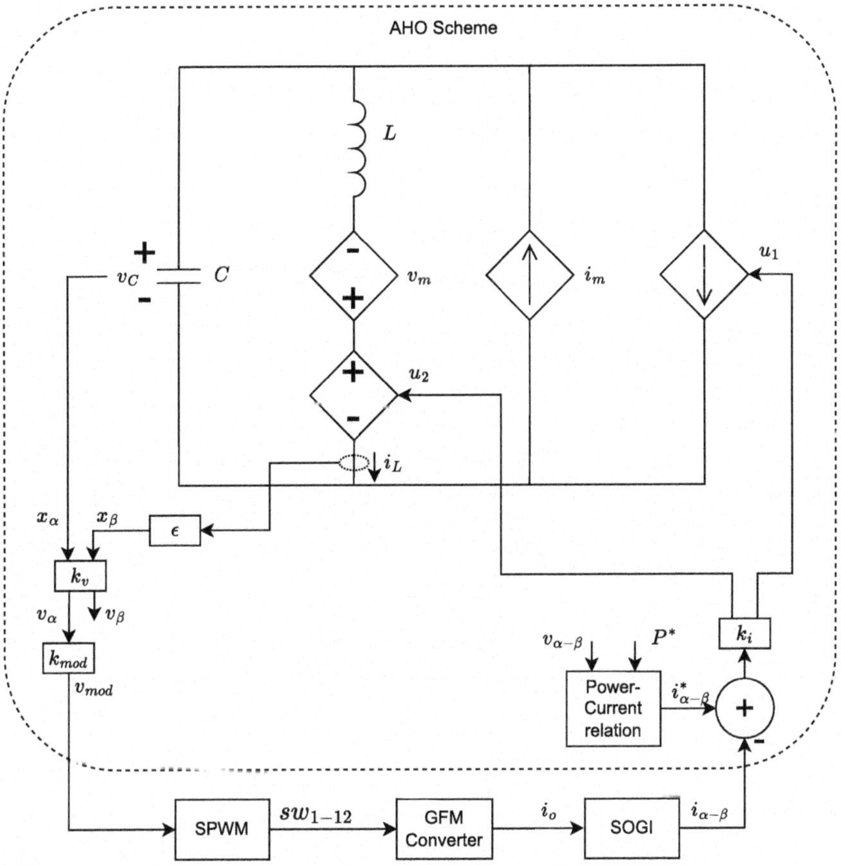

Fig. 17.2 Andronov-Hopf oscillator and control scheme

From the above equations, the terms v_C and i_L can be presented as the variables x_α and x_β/ϵ, respectively, which in turn are expressed as v_α and v_β according to the expression in 17.4. Additionally, the term ϵ is the characteristic impedance of the oscillator and is defined in 17.5.

$$\begin{bmatrix} v_\alpha \\ v_\beta \end{bmatrix} = k_v \begin{bmatrix} x_\alpha \\ x_\beta \end{bmatrix} \tag{17.4}$$

$$\epsilon = \sqrt{\frac{L}{C}} \tag{17.5}$$

In Eqs. 17.2 and 17.3, the values v_m and i_m represent the nonlinear sources present in the oscillator. These source values are defined by the expressions in Eqs. 17.6 and 17.7. In these equations, the term ξ is a constant that dictates the convergence speed of the states, V_n is the nominal RMS value of the converter, and V is the estimated RMS value, as defined in expression 17.8.

$$v_m = \frac{\xi}{\omega_n}(2V_n^2 - V^2)x_\beta \tag{17.6}$$

$$i_m = \frac{\xi}{\omega_n}(2V_n^2 - V^2)x_\alpha \tag{17.7}$$

$$V^2 = v_\alpha^2 + v_\beta^2 \tag{17.8}$$

Additionally, the terms u_α and u_β are derived from the product in Eq. 17.9. In this expression, the rotation matrix depends on the rotation angle (φ) to adjust the relationship between power control and nominal voltage and frequency control, while k_i is the scaling factor for the measured current. Moreover, in this expression, the terms i_α and i_β correspond to the current measured at the converter, while i_α^* and i_β^* are the reference currents generated by the set-point of active and reactive power, as defined in expression 17.10.

$$\begin{bmatrix} u_\alpha \\ u_\beta \end{bmatrix} = k_i \begin{bmatrix} \cos\varphi & -\sin\varphi \\ \sin\varphi & \cos\varphi \end{bmatrix} \begin{bmatrix} i_\alpha - i_\alpha^* \\ i_\beta - i_\beta^* \end{bmatrix} \tag{17.9}$$

$$\begin{bmatrix} i_\alpha^* \\ i_\beta^* \end{bmatrix} = \frac{2}{3(v_\alpha^2 + v_\beta^2)} \begin{bmatrix} v_\alpha & v_\beta \\ v_\beta & -v_\alpha \end{bmatrix} \begin{bmatrix} P^* \\ Q^* \end{bmatrix} \tag{17.10}$$

Finally, by substituting the expressions 17.6, 17.7, and 17.9 into 17.2 and 17.3, the Eq. 17.11 is obtained, which is also depicted in the diagram in Fig. 17.3. Once the $v_{\alpha-\beta}$ values are obtained, only the v_α value is sent to the modulation stage. However, the value of v_{mod} must be adjusted depending on the voltage of the converter's DC sources (v_{DC}) and the number of CHB-MLI modules (n_{mod}), as shown in Eq. 17.12.

$$\begin{bmatrix} \dot{v_\alpha} \\ \dot{v_\beta} \end{bmatrix} = \begin{bmatrix} \frac{\xi}{k_v^2}(2V_n^2 - V^2) & -\omega_n \\ \omega_n & \frac{\xi}{k_v^2}(2V_n^2 - V^2) \end{bmatrix} \begin{bmatrix} v_\alpha \\ v_\beta \end{bmatrix} - \frac{k_v k_i}{C} \begin{bmatrix} \cos\varphi & -\sin\varphi \\ \sin\varphi & \cos\varphi \end{bmatrix} \begin{bmatrix} i_\alpha - i_\alpha^* \\ i_\beta - i_\beta^* \end{bmatrix} \tag{17.11}$$

$$v_{mod} = v_\alpha \frac{1}{v_{DC} n_{mod}} \tag{17.12}$$

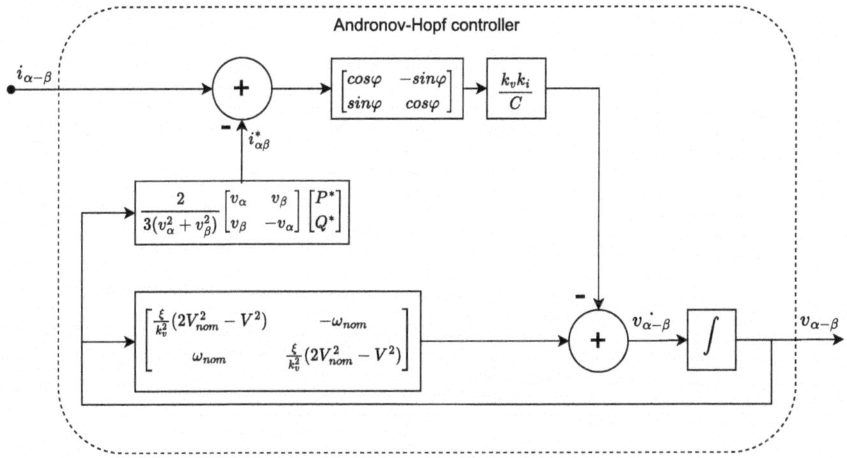

Fig. 17.3 dVOC control scheme

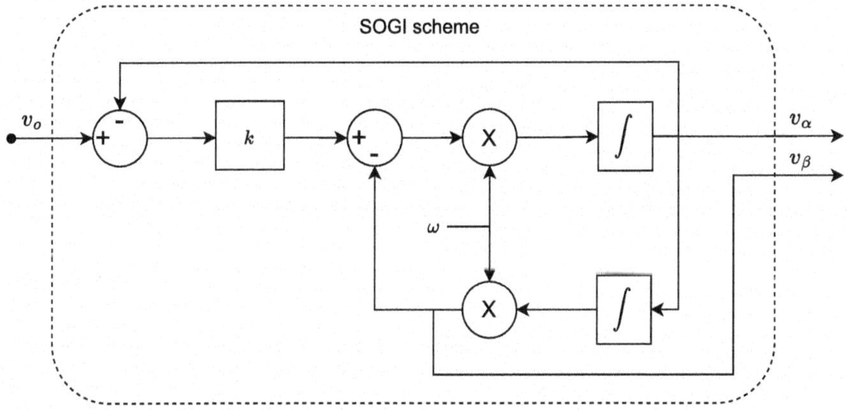

Fig. 17.4 SOGI strategy scheme

17.2.2 Single-Phase SOGI PLL

The SOGI strategy generates a signal in the $\alpha - \beta$ frame from a single-phase signal, and its operating diagram is shown in Fig. 17.4. The input signal i_{in} corresponds to the single-phase input signal—in this case, the converter's output current—while i_α and i_β are the orthogonal output signals sent to the VOC control. Additionally, the system requires the frequency (ω) at which the converter will operate. Equation 17.13 shows how the frequency is obtained from the signals used in the AHO. By using the frequency estimated by the control instead of the nominal frequency, the control can adapt to frequency variations in weak grids.

$$\omega = \frac{v_\alpha \dot{v_\beta} - \dot{v_\alpha} v_\beta}{2V^2} \tag{17.13}$$

17.2.3 Modulation

Just as control signals v_α and v_β are obtained in three-phase operations, modulation for a single-phase signal is generated from the v_{mod} signal, which depends only on v_α. This signal is fed into a Phase shift pulse width modulation (PS-PWM) modulation strategy, where each H-bridge receives the modulation signal compared to a triangular waveform shifted according to the formula in Eq. 17.14.

$$\theta_{ph} = \frac{\pi}{n_{mod}} \tag{17.14}$$

17.3 Results

Simulation tests of the proposed system have been conducted, with the converter and control characteristics detailed in Table 17.1. Additionally, the power system diagram can be seen in Fig. 17.5.

The tests conducted to evaluate the system are divided into three stages: one assessing the converter's ability to control power injection into the grid while changing references, another evaluating the converter's operation when disconnected from

Table 17.1 Converter and control parameters

Item	Value	Item	Value
V_{grid}	620 V	V_n	625 V
ω_{grid}	100π rad/s	ω_n	100π rad/s
L_g	2.7 mH	$V_{DC(1,2,3)}$	300 V
R_{Lg}	0.2 Ω	f_{sw}	500 Hz
L_{f1}	4.3 mH	C	0.108
R_{Lf1}	0.2 Ω	k_v	40
C_f	60 μ F	k_i	0.3
L_{f2}	0.37 mH	ξ	220
R_{Lf2}	0.1 Ω	φ	$\pi/4$
R_{Load}	20 Ω	P^*	10000 W

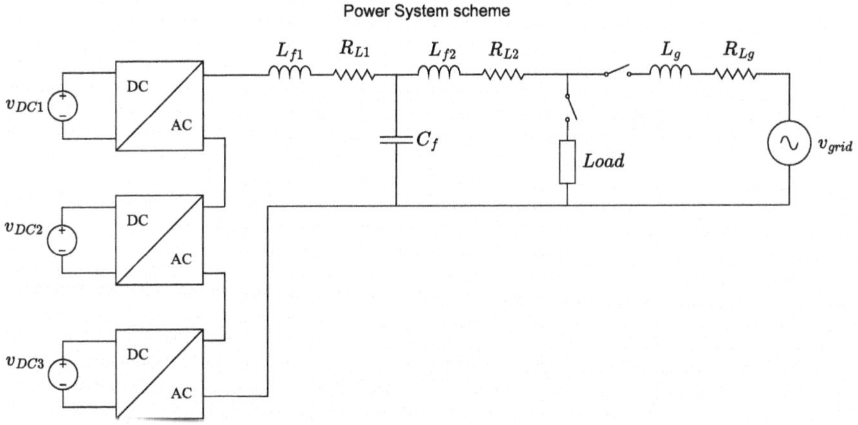

Fig. 17.5 Power system scheme

the grid, and a third assessing the converter's behavior during grid connection and disconnection stages, ensuring that the converter can operate during transitions.

17.3.1 Power Injection Operation

This section presents the results of operating the converter connected to an electrical grid, then a change in the active power reference is made to evaluate if the system maintains correct power injection control.

Figure 17.6 shows the results of a change in the injected power reference from 10 to 15 kW, in Fig. 17.6a the power injected by the converter effectively tracks the reference change, reaching 15 kW in 0.2 s after the change. In Fig. 17.6b the grid frequency shows a sudden increase of 0.7%, after which it returns to the nominal value. In Fig. 17.6c the current delivered by the converter exhibits an increase in amplitude from 32 to 48 A, corresponding to the increased power delivered, while in Fig. 17.6d the voltage across the output filter capacitor remains unaffected by the change in injected power.

17.3.2 Island Mode Operation

This section presents the results of operating the converter in island mode, solely powering passive loads. In this case, the goal is to evaluate the converter's ability to autonomously operate a microgrid.

Figure 17.7 shows the results of operating the converter in island mode, starting with the connection of a 10 kW load, followed by the addition of a second load with

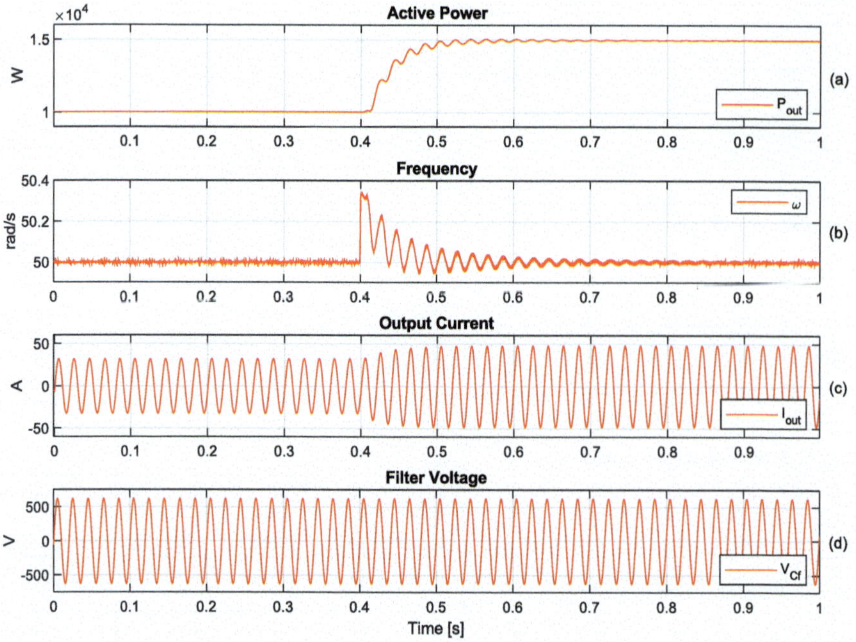

Fig. 17.6 GFM output power, frequency, current and voltage in response to a variation in the injected power reference

the same nominal consumption. From these results, in Fig. 17.7b can be observed that connecting the first load causes a slight frequency drop from 50.53 to 49.95 Hz. Similarly, Fig. 17.7d shows a voltage amplitude drop from 636 to 627 V, which is characteristic of grid-forming (GFM) behavior. When the second load is connected, both frequency and voltage decrease further to 49.46 Hz and 615 V, respectively, indicating that the power delivered by the converter in Fig. 17.7a does not effectively reach 20 kW, but approximately 90% of its nominal value. This is due to the observed drops in frequency and voltage amplitude, a behavior typical of synchronous generators as well. Finally, Fig. 17.7c shows how the current increases from 0 A to approximately 31 A with the connection of the first load, and further rises to 61 A when the second load is connected.

17.3.3 Transition Between Island and Grid-Connected Mode

This section presents results from transitioning the converter between connecting and disconnecting from a microgrid, aiming to assess the converter's ability to alternate these states autonomously and maintaining variables in safe operating ranges.

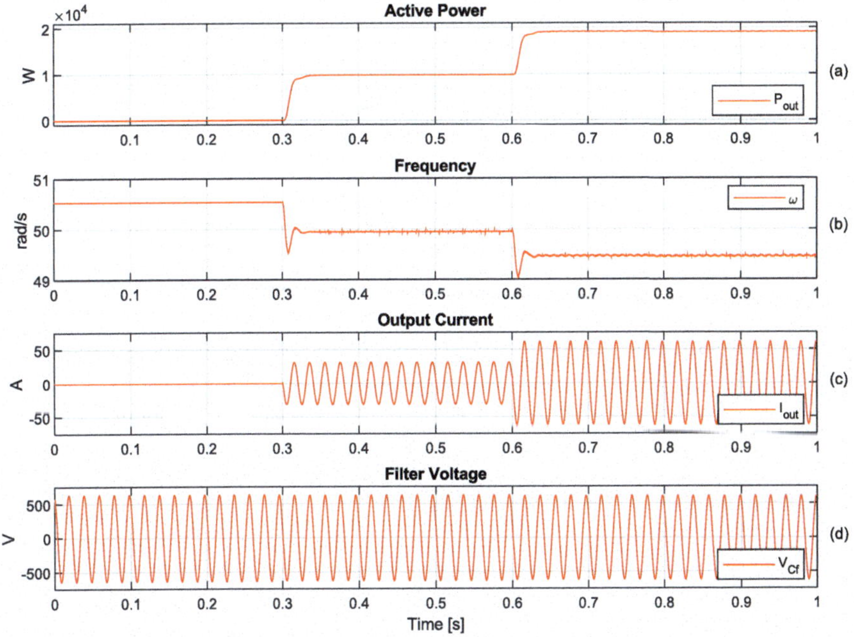

Fig. 17.7 GFM output power, frequency, current and voltage when a load is connected

Figure 17.8 shows the results of connecting and disconnecting the converter to a grid. In this case, Fig. 17.8a illustrates how the converter initially powers a 10 kW load before being connected to the grid, and then switches to a 10 kW power injection reference for the grid. During the connection phase, at 0.1 s, a sudden variation in power can be observed, caused by the control synchronization period with the main grid phase and because the power is calculated based on $v_{\alpha-\beta}$, which must synchronize with the grid. After 0.2 s, the power adjusts correctly to the reference. In Fig. 17.8b, it can be seen that the frequency remains around 50 Hz, although at the moment of connection to the grid, the frequency briefly increases to 51.5 Hz before returning to the grid's nominal value. The frequency generated by the AHO control loop shows instability due to variations in the $v_{\alpha-\beta}$ values, as the grid maintains a stable frequency at the point of common coupling. However, in Fig. 17.8c, the current remains within safe amplitude values during the connection, decreasing its amplitude momentarily, and after 0.15 s, it returns to the nominal value. Finally, in Fig. 17.8d, the voltage does not show any amplitude variations during the process.

At 0.8 s, the grid disconnection occurs, where the variation is minimal, with power and frequency maintaining their expected values with minor fluctuations, as they no longer follow the grid reference, similar to the current and voltage.

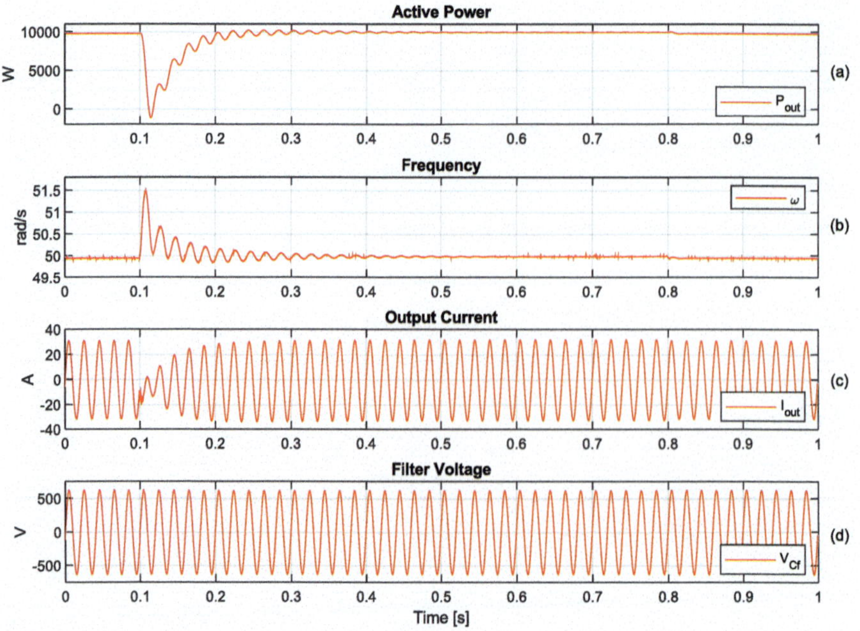

Fig. 17.8 GFM output power, frequency, current and voltage when the converter is connected and disconnected to a grid

17.4 Conclusions

The presented work introduces a single-phase CHB-MLI operating as a GFM using an AHO-type dVOC and a SOGI strategy to work in the $\alpha\beta$ framework. This system allows the converter to operate effectively in island and grid-connected modes, keeping key variables close to their nominal values. Notably, this is achieved with the control only requiring the converter's output current.

The results demonstrate that the converter behaves as a GFM in island mode, maintaining voltage and frequency within nominal ranges. However, when loads are connected, there is a slight drop in frequency and amplitude, which is characteristic of GFM operation.

The converter successfully injects power in grid-connected mode while maintaining its reference values. Additionally, it keeps current and voltage within safe limits during grid connection and disconnection, with the controller ensuring a smooth transition between operating modes.

Acknowledgements This research was funded by a grant from the Council of Andalucía (Junta de Andalucía, Consejería de Transformación Económica, Industria, Conocimiento y Universidades, Secretaría General de Universidades, Investigación y Tecnología) under Project ProyExcel_00381. Additional support was provided by the Thematic Network RIBIERSE-CYTED (723RT0150). Additional support was provided by ANID-Subdirección de Capital

Humano/Doctorado Nacional/2023-21231115. Finally the authors gratefully acknowledge the support from the Energy Conversion Technology Center at Universidad de Talca.

References

1. H. Zhang, W. Xiang, W. Lin, J. Wen, Grid forming converters in renewable energy sources dominated power grid: control strategy, stability, application, and challenges. J. Modern Power Syst. Clean Energy **9**(6), 1239–1256 (2021)
2. R. Rosso, X. Wang, M. Liserre, X. Lu, S. Engelken, Grid-forming converters: control approaches, grid-synchronization, and future trends-a review. IEEE Open J. Ind. Appl. **2**, 93–109 (2021)
3. A. Peña Asensio, S. Arnaltes Gómez, J.L. Rodriguez-Amenedo, Black-start capability of pv power plants through a grid-forming control based on reactive power synchronization. Int. J. Electr. Power Energy Syst. **146**, 108730 (2023). https://www.sciencedirect.com/science/article/pii/S0142061522007268
4. Q. Hu, R. Han, X. Quan, Z. Wu, C. Tang, W. Li, W. Wang, Grid-forming inverter enabled virtual power plants with inertia support capability. IEEE Trans. Smart Grid **13**(5), 4134–4143 (2022)
5. E. Ebinyu, O. Abdel-Rahim, D.E.A. Mansour, M. Shoyama, S.M. Abdelkader, Grid-forming control: advancements towards 100 review. Energies **16**(22) (2023). https://www.mdpi.com/1996-1073/16/22/7579
6. R. Rosso, X. Wang, M. Liserre, X. Lu, S. Engelken, Grid-forming converters: an overview of control approaches and future trends, in *2020 IEEE Energy Conversion Congress and Exposition (ECCE)*, pp 4292–4299 (2020)
7. M. Lu, Virtual oscillator grid-forming inverters: state of the art, modeling, and stability. IEEE Trans. Power Electron. **37**(10), 11579–11591 (2022)
8. V. Gurugubelli, A. Ghosh, A.K. Panda, Design and implementation of optimized Andronov-Hopf oscillator control method for parallel inverters in standalone microgrid. IEEE Trans. Ind. Appl. **59**(6), 7013–7026 (2023)
9. D. Raisz, T.T. Thai, A. Monti, Power control of virtual oscillator controlled inverters in grid-connected mode. IEEE Trans. Power Electron. **34**(6), 5916–5926 (2019)
10. G.S. Seo, M. Colombino, I. Subotic, B. Johnson, D. Groß, F. Dörfler, Dispatchable virtual oscillator control for decentralized inverter-dominated power systems: analysis and experiments, in *2019 IEEE Applied Power Electronics Conference and Exposition (APEC)*, pp. 561–566 (2019)
11. B.B. Johnson, M. Sinha, N.G. Ainsworth, F. Dörfler, S.V. Dhople, Synthesizing virtual oscillators to control islanded inverters. IEEE Trans. Power Electron. **31**(8), 6002–6015 (2016)
12. Q. Weijie, Z. Zhongfeng, L. Jiapeng, G. Yiyun, W. Jianhua, Software phase locked loop based on sogi transform, in *2022 IEEE 6th Conference on Energy Internet and Energy System Integration (EI2)*, pp 428–432 (2022)
13. M. Lu, An inrush current limiting strategy for virtual-oscillator-controlled grid-forming inverters. IEEE Trans. Energy Convers. **38**(3), 1501–1510 (2023)
14. M. Lu, S. Dutta, B. Johnson, Self-synchronizing cascaded inverters with virtual oscillator control. IEEE Trans. Power Electron. **37**(6), 6424–6436 (2022)
15. M. Lu, R. Mallik, B. Johnson, S. Dhople, Dispatchable virtual-oscillator-controlled inverters with current-limiting and mppt capabilities, in *2021 IEEE Energy Conversion Congress and Exposition (ECCE)*, pp. 3316–3323 (2021)
16. A. Acharya, R. Ayyanar, Dual-dvoc based controlled negative sequence current injection for grid-forming inverters under asymmetrical grid conditions, in *2023 IEEE Power & Energy Society General Meeting (PESGM)*, Orlando, FL, JUL 16–20 (2023)

17. S. Luo, W. Chen, X. Li, Z. Hao, A new virtual inertial strategy for Andronov-Hopf oscillator based grid-forming inverters. IEEE J. Emerg. Select. Top. Power Electron. **12**(2), 1995–2005 (2024)
18. B.B. Johnson, S.V. Dhople, A.O. Hamadeh, P.T. Krein, Synchronization of parallel single-phase inverters with virtual oscillator control. IEEE Trans. Power Electron. **29**(11), 6124–6138 (2014)
19. M. Ali, H.I. Nurdin, J.E. Fletcher, Dispatchable virtual oscillator control for single-phase islanded inverters: analysis and experiments. IEEE Trans. Ind. Electron. **68**(6), 4812–4826 (2021)
20. M. Malinowski, K. Gopakumar, J. Rodriguez, M.A. Pérez, A survey on cascaded multilevel inverters. IEEE Trans. Ind. Electron. **57**(7), 2197–2206 (2010)

Chapter 18
A PI-PBC Method for Voltage Support Using a PWM-CSC with Adaptive Load Estimation

Angélica Mercedes Nivia-Vargas, Oscar Danilo Montoya, Walter Gil-Gonzalez, Federico Martin Serra, and Jesús C. Hernández

Abstract This article presents the development and implementation of a proportional-integral passivity-based control (PI-PBC) method for voltage support in energy conversion systems, specifically utilizing a pulse-width modulation current source converter (PWM-CSC). The increasing demand for highly efficient and reliable energy conversion systems has driven the exploration of advanced control strategies, particularly in applications with variable loads and nonlinear dynamics. The proposed approach incorporates an adaptive load estimation mechanism, allowing the control system to dynamically adjust its parameters in response to real-time load variations. This enhances both system stability and converter performance, ensuring efficient operation across a wide range of load conditions. Simulations show that the PI-PBC method delivers precise voltage regulation, which is critical for the integration of renewable energy systems and microgrids. The results demonstrate PI-PBC's potential as a robust solution for nonlinear control in power electronics applications, offering stability and adaptability in demanding operational environments. This work advances the field of energy systems by proposing

A. M. Nivia-Vargas · O. D. Montoya (✉)
Grupo de Compatibilidad e Interferencia Electromagnética (GCEM), Facultad de Ingeniería, Universidad Distrital Fracisco José de Caldas, Bogotá, Colombia
e-mail: odmontoyag@udistrital.edu.co

A. M. Nivia-Vargas
e-mail: amniviav@udistrital.edu.co

F. M. Serra
Laboratorio de Control Automático (LCA), Facultad de Ingeniería y Ciencias Agropecuarias, Universidad Nacional de San Luis, CONICET, Villa Mercedes, San Luis, Argentina
e-mail: fmserra@unsl.edu.ar

W. Gil-Gonzalez
Department of Electrical Engineering, Universidad Tecnológica de Pereira, Pereira, Colombia
e-mail: wjgil@utp.edu.co

J. C. Hernández
Department of Electrical Engineering, University of Jaén, Campus Lagunillas s/n, Edificio A3, 23071 Jaén, Spain
e-mail: jcasa@ujaen.es

D. Rossit et al. (eds.), *Proceedings of the VII Ibero-American Congress of Smart Cities, ICSC-Cities 2024, 12-14 November, San Carlos, Costa Rica*, Springer Proceedings in Energy, https://doi.org/10.1007/978-981-96-4301-1_18

a theoretical and practical framework that can be leveraged in future research and technological developments.

Keywords Passivity-based control · Current source converter · Adaptive load estimation · Nonlinear control design · Passivity-based control design

18.1 Introduction

The growing demand for highly efficient and reliable power conversion systems has driven research on advanced control strategies for power electronic converters. Among the numerous converter topologies, single-phase pulse-width modulation current source converters (PWM-CSC) have garnered significant attention [1], primarily due to their intrinsic capability to regulate current with high precision, which makes them particularly suitable for applications characterized by varying load conditions. Nevertheless, the control of such converters poses considerable challenges, especially in the presence of nonlinear dynamics and unknown or time-varying loads [7]. These complexities have prompted the study of nonlinear control techniques such as passivity-based control (PBC), which provide robust performance in the face of such demanding operational conditions.

Proportional-integral passivity-based control (PI-PBC) has emerged as a powerful approach for addressing the control challenges of nonlinear systems. PI-PBC employs the principles of passivity, which ensure system stability by guaranteeing the dissipation of energy rather than its generation [13]. This control technique is particularly well-suited for power converters, where maintaining stability while optimizing performance is a necessity. By integrating proportional and integral actions, PI-PBC provides enhanced a dynamic response, ensuring both steady-state error correction and robustness against disturbances [14].

In the context of single-phase PWM-CSC, the application of PI-PBC has yielded promising results in improving system stability and performance, particularly in scenarios involving uncertain or adaptive loads [3]. Load variations, which are common in real-world applications, can result in instability or reduced performance if not adequately addressed. An adaptive load estimation mechanism, integrated into the PI-PBC framework allows the control system to dynamically adjust its parameters in response to real-time load alterations [15]. This adaptation enhances the overall system performance, ensuring that the converter operates efficiently across a wide range of load conditions.

One of the most significant challenges in the control of PWM-CSCs is their inherent nonlinearity, a formidable obstacle in the design of conventional linear controllers. It is often the case that traditional control methods are unable to perform adequately when applied to systems with highly nonlinear behaviors [8]. The nonlinear nature of single-phase PWM-CSCs, when considered in conjunction with load uncertainties, necessitates the use of advanced control strategies such as PI-PBC in order to ensure robust and stable operation [16]. By exploiting the energy-shaping properties of passivity-based control, PI-PBC is able to effectively address these nonlinearities and provide a structured approach to achieving the desired control objectives.

This paper presents an in-depth analysis of the application of PI-PBC to a single-phase PWM-CSC with adaptive load estimation. It presents a theoretical framework for PI-PBC as well as a control strategy and adaptive load estimation mechanism, and it discusses their design and implementation. The simulations carried out demonstrate the effectiveness of the proposed control approach in ensuring stable operation under varying load conditions. The results highlight the potential of PI-PBC as a robust solution for nonlinear control in power electronics applications, offering both stability and adaptability in challenging operating environments.

18.2 Dynamic Modeling of a PWM-CSC

18.2.1 PWM-CSC Model

A PWM-CSC is a type of power electronic converter that regulates the output current to a load or to a system using the principle of pulse-width modulation. Unlike voltage source converters (VSCs), where the output voltage is the main controlled variable, PWM-CSCs control and maintain a constant current by adjusting the switching of semiconductor devices such as IGBTs or GTOs [5]. The electrical topology of a PWM-CSC is shown in Fig. 18.1. This device is used to modulate the magnitude and direction of the current, with the modulation index (m) governing this variation.

The system includes an output filter that absorbs harmonic currents, represented by a capacitor C_o. State variables are defined for the inductor current (i_s) and the capacitor voltage (v_c). By applying Kirchhoff's voltage and current laws, the system dynamics can be described by Eqs. (18.1) and (18.2):

$$L_s \frac{di_s}{dt} = V_s - R_s i_s - m v_c, \tag{18.1}$$

$$C_o \frac{dv_c}{dt} = m i_s - G_L v_c. \tag{18.2}$$

where V_s represents the input DC voltage, L_s is the inductor with series resistance R_s, and G_L represents the conductance of a linear load, which is defined as $G_L = 1/R_L$.

Expressed in terms of the state variables x_1 and x_2, the system equations take the following form:

$$L_s \dot{x}_1 = V_s - R_s x_1 - u x_2, \tag{18.3}$$

$$C_o \dot{x}_2 = u x_1 - G_L x_2. \tag{18.4}$$

where the control input u is determined by the PWM signal, which activates the switches based on a comparison between a reference and a carrier signal.

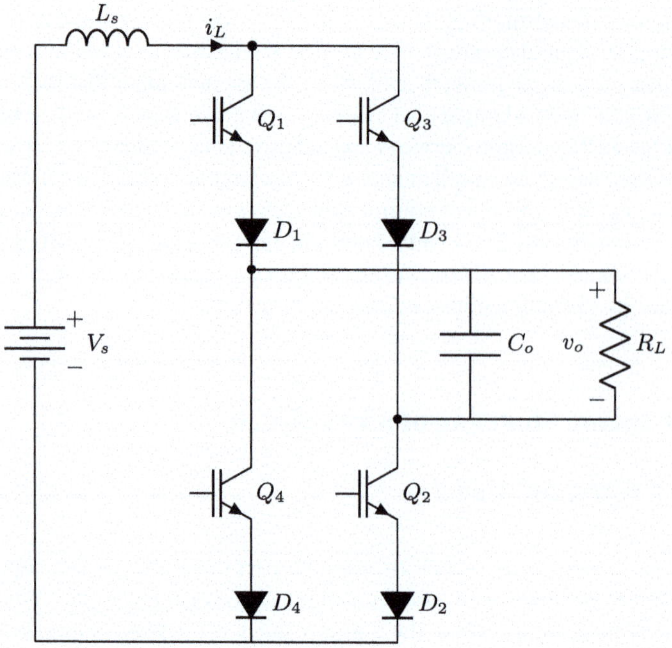

Fig. 18.1 PWM-CSC configuration

The operation of the system is constrained by the fact that only two switches can be active at any given time. The control of the transistors is governed by the PWM signal, which is generated through the comparison between a reference signal and a triangular carrier signal, with the latter adjusted to a frequency of $10kHz$. This modulation scheme ensures the desired system response, expressed as $x_2^* = V_p \sin(\omega t + \theta)$, where V_p denotes the peak voltage, ω the angular frequency, and θ the phase angle.

18.3 Proportional-Integral Passivity-Based Control

The PI-PBC approach is a well-established and robust control theory for addressing a subclass of nonlinear dynamical systems with a bilinear structure [2]. This approach leverages classical PI controllers to design an asymptotically stable controller in the sense of Lyapunov, using port-Hamiltonian theory applied to passive networks [12].

Remark 1 The authors of [4] have demonstrated that the PI-PBC approach is a highly effective control design tool for power electronic converter applications with linear loads, as it stabilizes these converters at their equilibrium point by applying a PI controller to the system's passive output.

Note that a PWM-CSC can be represented as a port-Hamiltonian system, as defined below:

$$\begin{bmatrix} L_s & 0 \\ 0 & C_o \end{bmatrix} \begin{bmatrix} \dot{x}_1(t) \\ \dot{x}_2(t) \end{bmatrix} = \left(\begin{bmatrix} 0 & -u(t) \\ u(t) & 0 \end{bmatrix} - \begin{bmatrix} R_s & 0 \\ 0 & G_L \end{bmatrix} \right) \begin{bmatrix} x_1(t) \\ x_2(t) \end{bmatrix} + \begin{bmatrix} V_s \\ 0 \end{bmatrix} \tag{18.5}$$

$$\mathcal{D}\dot{x} = (\mathcal{J}(u) - \mathcal{R})\, x + \vartheta, \tag{18.6}$$

where $\mathcal{D}$ corresponds to the energy storage matrix; $\mathcal{J}(u)$ is a skew-symmetric interconnection matrix, implying that $\mathcal{J}(u) + \mathcal{J}^{\top}(u) = 0$; $\mathcal{R}$ corresponds to the dissipation matrix; and ϑ represents the vector of external inputs. Note that the structure of each of the aforementioned matrices can be obtained by comparing Eqs. (18.5) and (18.6).

Remark 2 To design a PI-PBC controller, the incremental model must be defined. This implies that the system must have an equilibrium trajectory or operating point - in the case of the PWM-CSC, there is an equilibrium trajectory.

18.3.1 Defining an Incremental Model

Definition 1 *The Dynamical System* (18.6) *admits an equilibrium trajectory* $x^{\star}$, *such that it has a unique solution:*

$$\mathcal{D}\dot{x}^{\star} = \left(\mathcal{J}\left(u^{\star}\right) - \mathcal{R}\right) x^{\star} + \vartheta, \tag{18.7}$$

where $\dot{x}^{\star}$ *represents the time derivative of the admissible trajectory (which is different from zero for single-phase converter applications), and* $u^{\star}$ *is a soft control input.*

To obtain an incremental model, *i.e.*, the error model's dynamics, the error of the state variables is defined as $\tilde{x} = x - x^{\star}$ and $\tilde{u} = u - u^{\star}$. Considering this definition, the following result is obtained by subtracting the desired Dynamical System (18.7) from (18.6) and making some algebraic manipulations:

$$\mathcal{D}\dot{\tilde{x}} = \mathcal{J}(\tilde{u})\, x^{\star} + (\mathcal{J}(u) - \mathcal{R})\, \tilde{x}, \tag{18.8}$$

Note that the Incremental Model (18.8) can be asymptotically stabilized by defining a PI control law, as demonstrated by the authors of [11].

18.3.2 Controller Design

Theorem 1 (PI-PBC law) *The incremental model defined in* (18.8) *reaches a stable equilibrium point at* $\tilde{x} = 0$ *when the control input* $\tilde{u}$ *is defined as follows:*

$$\begin{aligned}\tilde{u} &= -k_p\tilde{y} - k_i\tilde{z},\\ \dot{\tilde{z}} &= \tilde{y},\end{aligned} \tag{18.9}$$

with k_p and k_i being two positive definite control gains and $\tilde{y}$ the passive output of the incremental model.

Proof To demonstrate that the control input scheme defined by (18.9) allows stabilizing the Incremental Model (18.8), Lyapunov's stability conditions must be applied.

Theorem 2 (Lyapunov's stability theorem) *A dynamical system $\dot{\tilde{x}} = f(\tilde{x})$ is considered to be asymptotically stable at the origin ($\tilde{x} = 0$) if there is a Lyapunov candidate function $\mathcal{V}(\tilde{x})$ that satisfies the following conditions:*

1. *$\mathcal{V}(\tilde{x})$ is positive-definite, meaning that $\mathcal{V}(\tilde{x}) > 0$ for all $\tilde{x} \neq 0$ and $\mathcal{V}(0) = 0$*
2. *the derivative $\dot{\mathcal{V}}(\tilde{x})$ is negative-definite, i.e., $\dot{\mathcal{V}}(\tilde{x}) < 0$.*

Considering Theorem 2, the following candidate Lyapunov function is defined:

$$\mathcal{V}(\tilde{x}, \tilde{z}) = \frac{1}{2}\tilde{x}^\top \mathcal{D}\tilde{x} + \frac{1}{2}k_i\tilde{z}^2, \tag{18.10}$$

whose time derivative is as follows:

$$\dot{\mathcal{V}}(\tilde{x}, \tilde{z}) = \tilde{x}^\top \mathcal{D}\dot{\tilde{x}} + k_i\tilde{z}\dot{\tilde{z}}. \tag{18.11}$$

Note that, by substituting the Dynamic Incremental Model (18.8) into Eq. (18.11), the following result is obtained:

$$\begin{aligned}\dot{\mathcal{V}}(\tilde{x}, \tilde{z}) &= \tilde{x}^\top \left(\mathcal{J}(\tilde{u})\, x^\star + (\mathcal{J}(u) - \mathcal{R})\,\tilde{x}\right) + k_i\tilde{z}\tilde{y}\\ &= -\tilde{x}^\top\mathcal{R}\tilde{x} + \tilde{x}^\top\mathcal{J}(\tilde{u})\, x^\star + k_i\tilde{z}\tilde{y}\end{aligned} \tag{18.12}$$

Note that Eq. (18.12) considers the properties of the skew-symmetric matrix, such that $\tilde{x}^\top\mathcal{J}(\tilde{u})\,\tilde{x} = 0$ [9]. In addition, taking the structure of this interconnection matrix into account, it is observed that:

$$\mathcal{J}(\tilde{u})\, x^\star = \mathcal{J}_1 x^\star\tilde{u} \implies \mathcal{J}_1 = \begin{bmatrix} 0 & -1 \\ 1 & 0 \end{bmatrix}$$

which can be substituted into Eq. (18.12) in conjunction with the general control input defined by (18.9), yielding the following results:

$$\begin{aligned}\dot{\mathcal{V}}(\tilde{x}, \tilde{z}) &= -\tilde{x}^\top\mathcal{R}\tilde{x} + \tilde{x}^\top\mathcal{J}_1 x^\star\tilde{u} + k_i\tilde{z}\tilde{y} &(18.13)\\ &= -\tilde{x}^\top\mathcal{R}\tilde{x} + \tilde{x}^\top\mathcal{J}_1 x^\star\left(-k_p\tilde{y} - k_i\tilde{z}\right) + k_i\tilde{z}\tilde{y}, &(18.14)\end{aligned}$$

Definition 2 *A port-Hamiltonian system with the structure defined by* (18.9) *has a passive output defined by* (18.15).

$$\tilde{y}^{\top} = \tilde{x}^{\top}\mathcal{J}_1 x^{\star} \implies \tilde{y} = x^{\star,\top}\mathcal{J}_1\tilde{x}. \tag{18.15}$$

Note that, considering the definition in (18.15), the time derivative of the Lyapunov Candidate Function (18.14) can be reduced as follows:

$$\begin{aligned}\dot{\mathcal{V}}(\tilde{x},\tilde{z}) &= -\tilde{x}^{\top}\mathcal{R}\tilde{x} - \tilde{y}^{\top}k_p\tilde{y} - k_i\tilde{y}^{\top}\tilde{z} + k_i\tilde{z}\tilde{y},\\ &= -\tilde{x}^{\top}\mathcal{R}\tilde{x} - \tilde{y}^{\top}k_p\tilde{y} \leq 0.\end{aligned} \tag{18.16}$$

Here, it is considered that $\tilde{y}^{\top} = \tilde{y}$, confirming that the PI-PBC Control Input (18.9) stabilizes the incremental model around the origin of coordinates, thus completing the proof. The authors of [2] presented an asymptotic stability test using LaSalle's invariance principle.

18.3.3 PWM-CSC Controller

Note that the main interest concerning the PWM-CSC application studied herein is to provide a sinusoidal voltage profile to a linear load (see Fig. 18.1). Here, the value of the desired variable $x_2^{\star}$ is fully known, *i.e.*, $x_2^{\star} = V_m \sin(\omega t)$, where V_m is the desired voltage amplitude and ω represents the desired angular frequency.

Now, considering the structure of the desired dynamical system in (18.7), as well as the structure of the PWM-CSC defined by (18.6), the desired control input $u^{\star}$ is defined as follows:

$$u^{\star} = \frac{1}{x_1^{\star}}\left(C_o\dot{x}_2^{\star} + G_L x_2^{\star}\right), \tag{18.17}$$

To complete the controller, *i.e.*, $u = \tilde{u} + u^{\star}$, the passive output $\tilde{y}$ is calculated, which produces:

$$\tilde{y} = x^{\star,\top}\mathcal{J}_1\tilde{x} = \begin{bmatrix} x_1^{\star} & x_2^{\star}\end{bmatrix}\begin{bmatrix} 0 & -1 \\ 1 & 0\end{bmatrix}^{\top}\begin{bmatrix}\tilde{x}_1 \\ \tilde{x}_2\end{bmatrix} \implies \tilde{y} = x_1^{\star}\tilde{x}_2 - x_2^{\star}\tilde{x}_1. \tag{18.18}$$

This yields the following PI-PBC controller:

$$u = \frac{1}{x_1^{\star}}\left(C_o\dot{x}_2^{\star} + G_L x_2^{\star}\right) - k_p\left(x_1^{\star}\tilde{x}_2 - x_2^{\star}\tilde{x}_1\right) - k_i\int_0^{t_f}\left(x_1^{\star}\tilde{x}_2 - x_2^{\star}\tilde{x}_1\right)d\tau. \tag{18.19}$$

Remark 1 To implement the PI-PBC control law defined by (18.19) in a PWM-CSC application, the trajectory followed by the inductor current (*i.e.*, $x_1^{\star}$) must be calculated. This trajectory is reached by solving the differential equation of the desired dynamical system in (18.8), which is presented below:

$$L_s \dot{x}_1^{\star} = V_s - R_s x_1^{\star} - u^{\star} x_2^{\star}. \tag{18.20}$$

18.3.4 Load Estimation

Considering the information provided by the state variables and the model described in Eqs. (18.3) and (18.4), an integral estimator is proposed, as shown in Eq. (18.25). This estimator has been specifically crafted to achieve asymptotic convergence to the load value by employing the corresponding conductance G_L [10].

$$\widehat{G}_L = G_{LI} + \alpha f(x_2), \tag{18.21}$$

The goal of the estimation process, represented by $\widehat{G}_L$, is to ensure its convergence towards G_L as the time t approaches infinity. To facilitate this convergence, an initial estimate is expressed as $\tilde{G}_L(t) = \tilde{G}_L(0)e^{-\alpha\beta t}$.

The estimation error is defined as:

$$\tilde{G}_L = \widehat{G}_L - G_L. \tag{18.22}$$

In designing the estimator, it is important that $\tilde{G}_L \to 0$ as $t \to \infty$. Consequently, the time derivative of the estimation error, as indicated in Eq. (18.22), can be expressed as follows:

$$\dot{\tilde{G}}_L = \dot{G}_{LI} - G_L + \alpha \dot{f}(x_2)\dot{x}_2, \tag{18.23}$$

From this, the formulation for $\dot{G}_{LI}$ is derived:

$$\dot{G}_{LI} = \alpha \dot{f}(x_2)\left(\frac{u(t)}{C}x_1 - \frac{\tilde{G}_L}{C}x_2\right) + \alpha(x_2^2 - \beta)\widehat{G}_L. \tag{18.24}$$

Thus, the estimation is articulated as

$$\widehat{G}_L = G_{LI} + \alpha f(x_2), \tag{18.25}$$

The parameters and functions G_{LI}, α, and $f(x_2)$ are specified in accordance with Eqs. (18.26) and (18.24). The estimator requires inputs including u, the control signal, and the state measurements x_1 and x_2, as depicted in Fig. 18.2.

$$f(x_2) = -\frac{1}{2}C_o x_2^2. \tag{18.26}$$

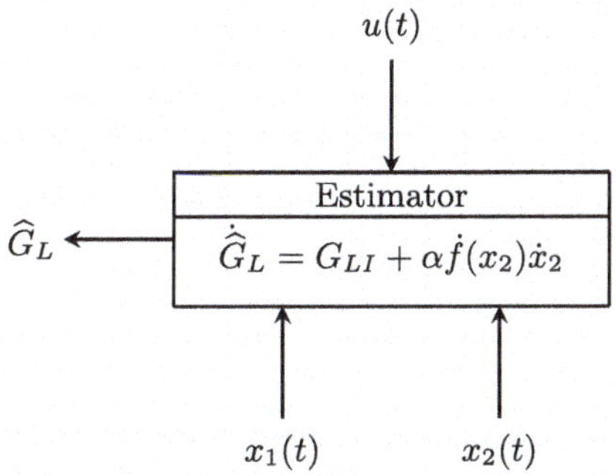

Fig. 18.2 Load estimator

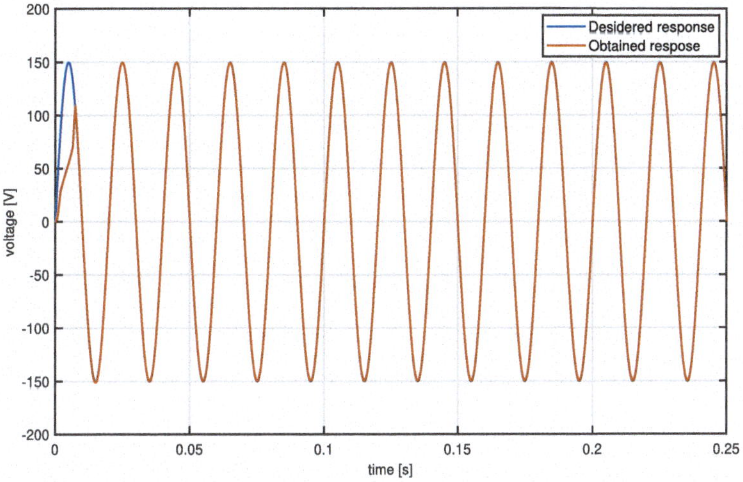

Fig. 18.3 Following the voltage reference signal

18.3.5 Simulation and Results

The controller described in Sect. 18.3 was implemented in Matlab, which allowed analyzing its performance. To this effect, the working values were adopted according to [6]: $V_p = 150$ V, $V_s = 48$ V, $L_s = 10$ mH, $C_o = 200\,\mu$F, and $R_s = 1\,\Omega$.

Initially, the control constants $K_p = 1.5$ and $K_i = 0.8$ were experimentally tuned to allow the PWM system to regulate the switching and achieve the desired response $x_2^*(t) = V_p \sin(\omega t + \theta)$V. Figure 18.3 compares the reference signal and that generated by implementing the control. Note that this signal reaches the reference quickly, adjusting to it in just the first half cycle and following it appropriately.

As for the current, the response is shown in Fig. 18.4. The steady state is reached in 56 ms, maintaining a state error of less than 1%.

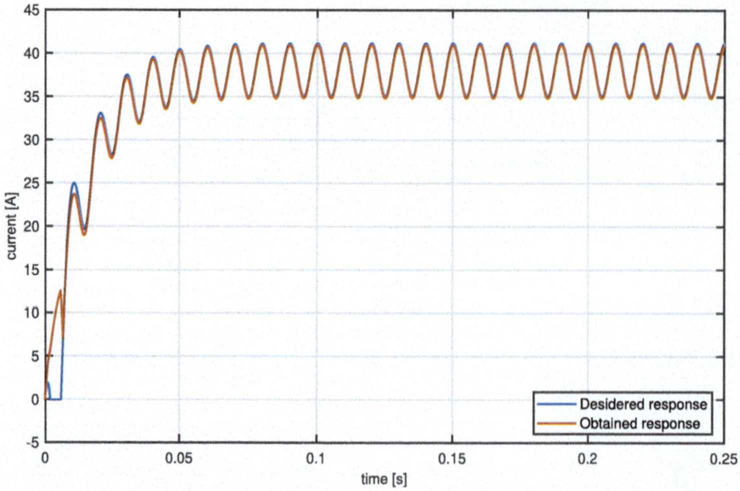

Fig. 18.4 Following the current reference signal

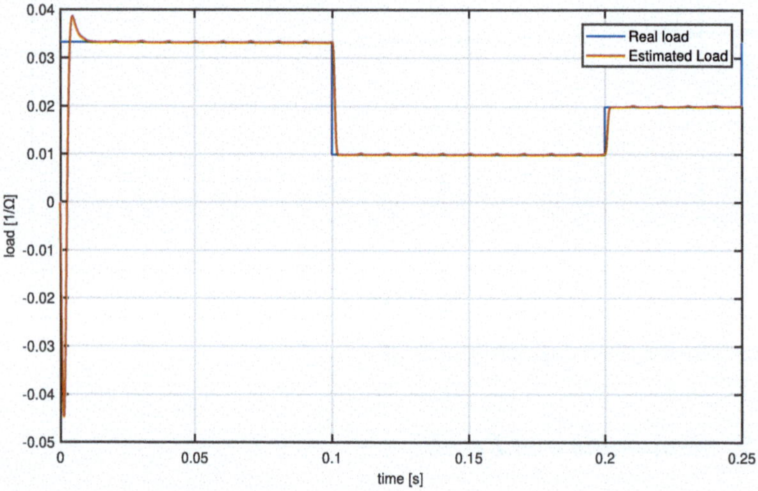

Fig. 18.5 Tracking load changes GL

The result is an effective value of 2.112 A, an alternating current of 2.987 A, and a total harmonic distortion (THD) value of 3.35%, which indicates outstanding performance characteristics that meet international standards regarding harmonic pollution.

Subsequently, the design and performance of the load estimator should be adjusted, which, in this work, involved conducting tests with a variable load, obtaining the results presented in Fig. 18.5 and demonstrating complete and rapid convergence in response to load variations.

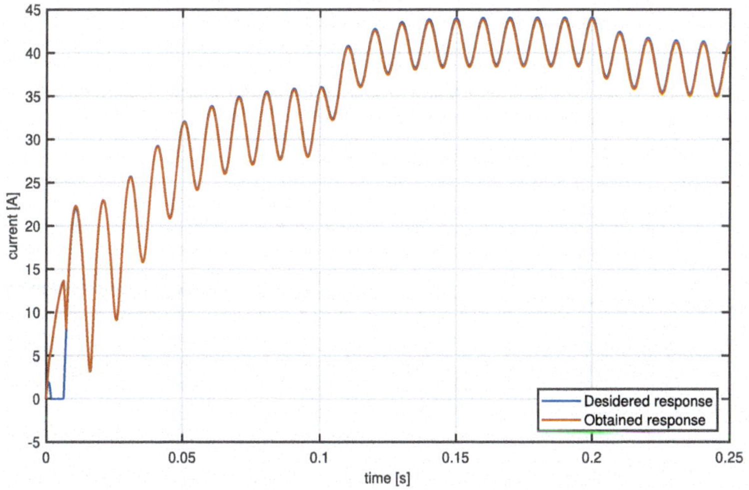

Fig. 18.6 Following the current reference signal in the face of load changes

The behavior of the estimator was analyzed under variable load conditions, showing that the adjusted response closely follows the load variation (see Fig. 18.6).

18.3.6 Conclusions

The study presented in this article demonstrates the effectiveness of the proportional-integral passivity-based control method for voltage support in pulse-width modulation current source converters. By integrating an adaptive load estimation mechanism, significant improvements in system stability and performance were achieved, even under variable load conditions and nonlinear dynamics.

Simulation results indicate that the PI-PBC approach enables precise voltage regulation, which is essential for the efficient operation of microgrids and the integration of renewable energy sources. This method stands out as a robust solution to the inherent challenges of nonlinearity in energy conversion systems, providing a theoretical and practical framework applicable in various industrial and research applications.

Moreover, our study highlights the importance of adaptability in the design of controllers for energy systems, suggesting that future research could explore the combination between PI-PBC and other advanced control techniques to further enhance resilience and efficiency. In conclusion, the proposed approach not only contributes to the advancement of knowledge in the field of power electronics but also opens new avenues for the development of more sustainable and efficient energy management technologies.

The modulation scheme employed in PWM-CSCs plays a crucial role in reducing harmonic distortion. By increasing the switching frequency, the dominant harmonics are shifted to higher frequencies, which are easier to filter. Additionally, inductive filters are often used to smooth the current waveform, further reducing the total harmonic distortion. This ensures that PWM-CSCs deliver high-quality current with minimal distortion, making them ideal for applications requiring precise current regulation and low electromagnetic interference.

As potential further work, key characteristics of PWM-CSCs could be explored, such as their suitability for interfacing with inductive loads. Given their ability to maintain a stable and continuous current, these converters are well-suited for applications involving high inductance, such as synchronous motors or transformers. This contrasts with voltage source converters, which may struggle with inductive loads due to the phase shift between voltage and current. PWM-CSCs' ability to directly control current also enhances their performance in such scenarios, providing a smooth operation and reducing voltage spikes.

Modern PWM-CSCs also support bidirectional power flows, enabling both rectification and inversion operations. This feature is particularly useful in regenerative systems, such as motor drives and energy storage applications, where power needs to flow in both directions depending on the operating mode. The ability to seamlessly switch between rectification and inversion enhances these converters' versatility, making them suitable for energy management systems such as electric vehicle chargers and grid-tied renewable energy installations.

References

1. I.A.M. Abdelsalam, Current source based wind energy conversion system (2015)
2. R. Cisneros, M. Pirro, G. Bergna, R. Ortega, G. Ippoliti, M. Molinas, Global tracking passivity-based pi control of bilinear systems: application to the interleaved boost and modular multilevel converters. Control Eng. Pract. **43**, 109–119 (2015). https://doi.org/10.1016/j.conengprac.2015.07.002
3. O. Danilo, M. Giraldo, Passivity-based analysis and control of ac microgrids: integration, operation and control of energy storage systems (2019). https://hdl.handle.net/11059/10132
4. W. Gil-González, O.D. Montoya, S. Riffo, C. Restrepo, J. Muñoz, A global tracking sensorless adaptive pi-pbc design for output voltage regulation in a boost converter feeding a dc microgrid. Energies **16**(3), 1106 (2023). https://doi.org/10.3390/en16031106
5. T. Jing, A.S. Maklakov, in *Proceedings—2018 International Ural Conference on Green Energy, UralCon 2018*. A Review of Voltage Source Converters for Energy Applications (2018), pp. 275–281. https://doi.org/10.1109/URALCON.2018.8544364
6. H. Komurcugil, in *2009 35th Annual Conference of IEEE Industrial Electronics*. Nonlinear Control Strategy for Single-Phase PWM Current-Source Inverters (IEEE, 2009). https://doi.org/10.1109/iecon.2009.5415002
7. S. Kouro, J. Rodriguez, B. Wu, S. Bernet, M. Perez, Powering the future of industry: high-power adjustable speed drive topologies. IEEE Indus. Appl. Mag. **18**(4), 26–39 (2012) (Funding Information: The authors gratefully acknowledge support by CCTVal (FB0821) and Fondecyt 1110783). https://doi.org/10.1109/MIAS.2012.2192231
8. P. Liu, Z. Wang, S. Wei, Y. Bo, S. Pu, Recent developments of modulation and control for high-power current-source-converters fed electric machine systems. CES Trans. Electr. Mach. Syst. **4**, 215–226 (2020). https://doi.org/10.30941/CESTEMS.2020.00027

9. K. López-Rodríguez, W. Gil-González, A. Escobar-Mejía, Design and implementation of a PI-PBC to manage bidirectional power flow in the dab of an sst. Results Eng. **14**, 100437 (2022). https://doi.org/10.1016/j.rineng.2022.100437
10. O.D. Montoya, F.M. Serra, W. Gil-Gonzalez, E.M. Asensio, C.H. De Angelo, in *2021 IEEE International Autumn Meeting on Power, Electronics and Computing (ROPEC)*. Adaptive Control of a Single-Phase Grid-Forming for Feeding Unknown Resistive Loads. (IEEE, 2021). https://doi.org/10.1109/ropec53248.2021.9668136
11. O. Montoya, W. Gil-Gonzalez, A. Garces, F. Serra, J. Hernandez, in *2020 IEEE International Autumn Meeting on Power, Electronics and Computing (ROPEC)*. PI-PBC Approach for Voltage Regulation in Ćuk Converters with Adaptive Load Estimation (IEEE, 2020). https://doi.org/10.1109/ropec50909.2020.9258716
12. O.D. Montoya, F.M. Serra, G. Espinosa-Pérez, On the equivalence between pi-pbc and ioc designs: an application involving three-phase front-end converters. IEEE Trans. Circ. Syst. II Exp. Briefs **71**(1), 241–245 (2024). https://doi.org/10.1109/tcsii.2023.3299203
13. R.C. Montoya, PI passivity-based control: application to physical systems. Ph.D. thesis, Université Paris Saclay (COmUE) (2016)
14. W. Montoya, O.D. Riffo, S. Restrepo, C. Muñoz, W. Gil-González, O.D. Montoya, S. Riffo, C. Restrepo, J. Muñoz, A global tracking sensorless adaptive PI-PBC design for output voltage regulation in a boost converter feeding a dc microgrid. Energies **16**, 1106 (2023). https://doi.org/10.3390/EN16031106, https://www.mdpi.com/1996-1073/16/3/1106/htm and https://www.mdpi.com/1996-1073/16/3/1106
15. D.H. Tungadio, Y. Sun, Load frequency controllers considering renewable energy integration in power system. Energy Rep. **5**, 436–453 (2019). https://doi.org/10.1016/J.EGYR.2019.04.003
16. M. Tõnso, V. Kaparin, J. Belikov, J.: Port-hamiltonian framework in power systems domain: a survey. Energy Rep. **10**, 2918–2930 (2023). https://doi.org/10.1016/J.EGYR.2023.09.077

Chapter 19
The Energy Transition of a University Campus Through Photovoltaic and Battery Storage Systems

Sebastian Dittmann, Santiago Tosello, Carlos Meza, Hugo Sanchez, Micheal Klein, Sabine Thalmann, and Sabine Tischew

Abstract Anhalt University of Applied Sciences is striving to achieve carbon neutrality by 2035. This objective will be attained by primarily decreasing CO_2 emissions through the utilization of renewable energy sources, promoting eco-friendly transportation, and cultivating a verdant campus that fosters biodiversity. The University's sustainability plan emphasizes an interdisciplinary methodology that effectively combines various actions to realize this ambitious aim. However, this paper centers its attention on a crucial aspect: the examination of energy consumption, production, and technical viability. We will evaluate the real energy usage over the past six years, encompassing a sizeable rooftop photovoltaic (PV) system as well as an agrivoltaic system. This unique system combines agriculture, energy generation, and biodiversity measures, referred to Anhalt's AgriPVplus approach. Additionally, the paper will present an intricate technical analysis of distinct PV systems alternatives, including energy storage, across various locations on the campus. Various factors like orientation, technology, and design will be considered, with a focus on the economic viability of the PV systems and the restrictions imposed by the existing infrastructure. This will provide inputs for a roadmap to expand and integrate PV systems, accounting to the Campus's demand. The University faces limitations when it comes to injecting surplus energy into the grid. Consequently, a maximum threshold for increasing PV nominal power has been determined, considering the curtailed energy, which stands at approximately 600 kWp.

Keywords Energy transition · Photovoltaic · Renewable energy

S. Dittmann (✉) · S. Tosello · C. Meza · H. Sanchez · M. Klein · S. Thalmann · S. Tischew
Anhalt University of Applied Sciences, Köthen, Germany
e-mail: sebastian.dittmann@hs-anhalt.de

D. Rossit et al. (eds.), *Proceedings of the VII Ibero-American Congress of Smart Cities, ICSC-Cities 2024, 12-14 November, San Carlos, Costa Rica*, Springer Proceedings in Energy, https://doi.org/10.1007/978-981-96-4301-1_19

19.1 Introduction

Climate change is having and will have massive negative impact in people and nature across the earth. An urgent reduction of greenhouse emissions is needed as part of the strategy to avoid this situation. Renewable energies such as Photovoltaic play a key role in reducing greenhouse emissions and continue supporting our modern way of living.

To achieve the targets of the European Commission, as part of the European Green Deal, for 2030 [1], the energy transition (covering the energy demand as far as possible from renewable energies) will be supported among others by a growing expansion of Photovoltaic (PV) Systems [2, 3]. In case of Germany the expected installed PV power is approx. 400 GW [2, 3], currently installed PV capacity 92 GW [4]. However, public institutions, such as Universities, have faced a steady increase in electricity prices. Over the past five years, Anhalt University of Applied Sciences (HSA) has experienced an annual electricity cost increase of 20%. Given the escalating energy costs and the decreasing installation costs of PV systems, these systems are becoming economically viable for public institutions [5, 6].

This work simulates and analyzes the integration of photovoltaic (PV) systems into the university campus under realistic power demand and meteorological conditions. This analysis is essential for the campus, as it not only provides a roadmap for meeting electricity demand but also enables the creation of a holistic infrastructure. This infrastructure will serve as a dynamic platform for advancing knowledge through teaching, research, and operational excellence, further solidifying the campus's position as a hub of innovation and sustainability."

19.2 Background

Anhalt University of Applied Sciences (HSA) has three campuses located in Bernburg-Strenzfeld, Dessau-Roßlau, and Köthen, all in Saxony-Anhalt, Germany. The Bernburg-Strenzfeld campus, the focus of this study, excels in both academic achievements and quality of life. Student dormitories, laboratories, lecture halls, canteen, cafeteria, and sports facilities are situated within a green space spanning approximately 20 hectares. Surrounding the campus are 50 hectares of experimental fields and numerous project gardens and greenhouses, providing excellent conditions for teaching and applied research.

In 2021, HSA commenced the development of a sustainability strategy aimed at realizing the sustainability goals of the UN Agenda 2030 [1]. This strategy, developed through a collaborative process, identified eight key fields of action for the university [7–9]. These include, among others, sustainable mobility, sustainable research, and the sustainable operation of the three campuses to reduce CO_2 emissions. A crucial component of this strategy is the integration of photovoltaic (PV) systems to enhance self-sufficient and reduce dependency on costly energy. Public institutions, including

HSA, have faced a steady increase in electricity prices. Over the past five years, HSA has experienced an annual electricity cost increase of 20%. The current price of electricity stands at 39.5 cents per kWh (2023/24), up from 21.47 cents per kWh in 2020. Given the escalating energy costs and the decreasing installation costs of PV systems, these systems are becoming economically viable for public institutions.

However, the installation of additional PV systems at the Bernburg-Strenzfeld campus involves technical and legal considerations. This campus operates similarly to a small city or microgrid, as a significant portion of its 3,000 national and international students reside on campus. The campus layout and green areas are illustrated in Fig. 19.4. The university itself manages and operates the campus grid, which is connected to the public grid via a medium voltage (20 kV) connection point and uses an internal 3-phase 0.4 kV low voltage network. Heating of the buildings in winter and the supply of warm water is done by gas heating. As a public institution, the university is prohibited from feeding energy into the public grid. According to the Renewable Energy Sources Act (EEG) [10], PV system owners become energy suppliers and sell surplus energy to local grid operators. The EEG legislation originally provided a feed-in tariff (FIT) to promote renewable electricity generation. However, a publicly funded university cannot generate financial income through electricity sales. Any surplus energy produced by the campus PV system must either be stored or curtailed if not consumed.

Since June 2019, HSA has been operating a traditional rooftop PV system with a nominal power of 144.5 kWp. The PV system is composed of 516 mono-facial PV modules, each with a capacity of 280 Wp. The tilt angle is 20°, and the azimuth angle is -24° (south-south-east).

19.3 Energy Consumption Analyse

Despite a significant number of students living on campus and the constant demand generated by facilities such as cafeterias and greenhouses, the load profile and the amount of energy consumed are highly dependent on the academic year's schedule. The highest consumption occurs during the semester (from April to mid-July and from October to February), followed by a lower consumption during semester breaks (March and mid-July to September). However, the campus maintains a quite consistent base load throughout the year. The following section provides a detailed analysis of the energy consumption and the impact of the existing rooftop PV system. The analysis is the base of the design of the proposed PV systems and battery storage in terms of increasing the self-sufficient and avoiding energy feeding into the public grid.

19.3.1 Total Demand of the Campus

The analysis of the total campus demand is based on data from 2018 to 2023. Measurement data was taken at the main transition station, which connects the campus to the public grid. The data are recorded with a 15-min time resolution.

Table 19.1 shows the total energy consumption over the last six years, the year-to-year variation, and the variation of each year compared to the total consumption of 2018, while Fig. 19.1 shows the monthly energy consumption of each year.

The total energy consumption in 2018 and 2019 is around 1,650 MWh, showing only a slight variation of 1.55% in 2019 compared to the previous year. The monthly comparison in Fig. 19.1 shows no significant differences between individual months; the energy consumption has a lower deviation. However, in both years, there were

Table 19.1 Total energy consumption over the last six years and variation to the total consumption in 2018

Year	Total energy consumption [MWh]	Variation to 2018 (%)	Year-to-year variation (%)
2018	1.628	–	–
2019	1.653	1.6	1.6
2020	1.445	– 11.2	– 12.6
2021	1.370	– 15.9	– 5.2
2022	1.258	– 22.7	– 8.1
2023	1.152	– 29.2	– 8.4

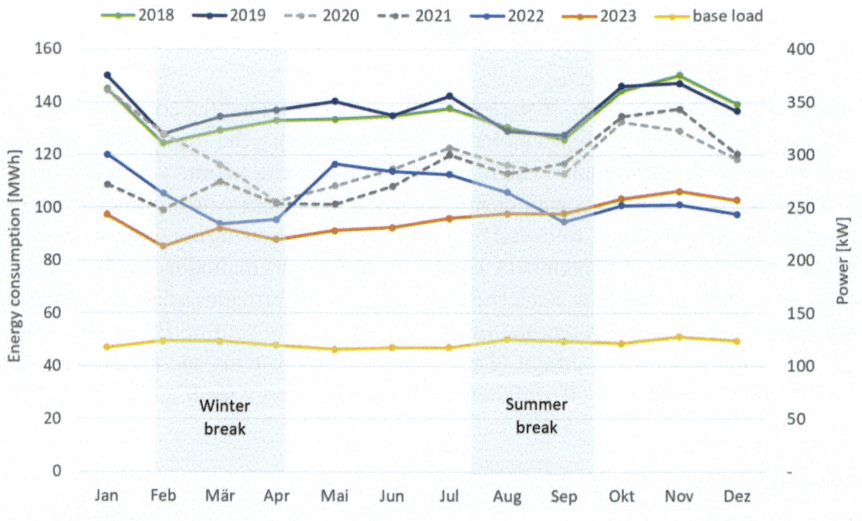

Fig. 19.1 Monthly energy consumption of six year from 2018 to 2023, green columns indicate the university winter and summer break. The yellow curve shows the average base power

three periods with significantly lower energy consumption, ranging from 9 to 15%. These are the winter and summer breaks and the Christmas vacation. During these periods, there are fewer people on campus since no lectures or lab activities are carried out. During the summer break and in December, most employees are on vacation.

Looking at 2020, 2021, and the beginning of 2022, represented by light gray dashed, dark gray dashed, and orange lines, respectively, it is clearly visible that the total and monthly energy consumption are significantly lower. There is a large drop of 20% from February 2020 to April 2020. This is due to the beginning of the COVID-19 pandemic. At that time, the university was closed, and thus, the energy demand decreased. Throughout 2020 and 2021, the university started teaching classes online, postponing all activities that could only be carried out in person, such as laboratories and exams, and many employees worked from home. Classroom lectures and laboratories were only possible again at the beginning of 2022.

However, analyzing the data from 2022 and 2023, it is notable that the energy consumption does not reach the levels of 2018 or 2019. This is probably because the university implemented several measures to save energy due to the increasing energy prices caused by the global crisis. Additionally, the work environment after the pandemic changed, and the flexibility to work from home led to fewer people on campus, and thus, a reduction in energy consumption.

Daily analysis

Figure 19.2 shows the load profile for 2019 with 15 min values (top), boxplot of power demand on weekdays (bottom left), power demand on weekends (bottom right). As already seen in the monthly data from Fig. 19.1 Winter, summer break, and Christmas vacation periods are clearly visible with a lower power demand. Power demand is also significantly different between weekdays (Monday to Friday) and weekends (Saturday and Sunday). Furthermore, the data show that the campus has a very stable base load of approximately 120 kW–140 kW; even in December, with the lowest demand, the base load does not drop below this value. Figure 19.2 (Bottom left and right) shows the hourly average power demand variation between weekdays and weekends, respectively. During weekdays, the energy demand rises from 6 am to a maximum at 12 pm and decreases until 6 pm. In contrast, on weekends the demand remains very constant, which describes the base load of the campus throughout the year. Since the highest energy production of south-facing PV systems occurs during mid-day in summer, and given that all produced energy must be consumed within the campus, understanding the variation in energy consumption and the approximate base load are key factors for the implementation of further PV systems. The following design of the proposed PV systems (Sect. 19.4) incorporates these considerations by exploring different orientations.

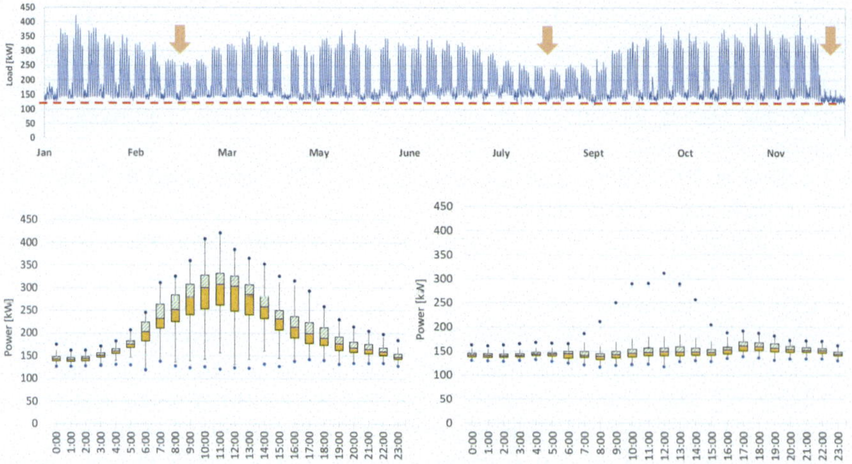

Fig. 19.2 Load profile for 2019 with 15 min values (top), boxplot of power demand on weekdays (bottom left), power demand on weekends (bottom right)

19.3.2 Analysis of the First PV System at the Campus

Since June 2019, HSA has been operating a traditional rooftop photovoltaic (PV) system on Building 20, with a nominal power of 144.5 kWp (Table 19.2). The PV system comprises 516 monofacial PV modules, each with a capacity of 280 Wp. The tilt angle is 20°, and the azimuth angle is − 24° (south-south-east). The system is grid-connected without an energy storage system or power regulation. Due to its orientation, the energy output exhibits a classical midday peak, with a maximum output power of approximately 130 kW. However, no energy has been fed into the public grid due to the relatively high and constant base load. Figure 19.3. (Left) compares the base load, the lowest measured values, for 2019 in comparison to 2018. Following the commissioning of the PV system in June 2019, the base load decreased during the summer months and increased in autumn and winter. The lowest value recorded in 2019 was 32 kW. The lowest base load values were observed during the summers of 2020–2023, with a minimum value ranging from 2 to 4 kW. However, these values were only briefly observed around midday on summer days with high solar irradiation.

The PV system generates an average of 132,000 kWh annually, contributing approximately 9–11% of the campus's total energy consumption.

Table 19.2 Design of the proposed PV systems

			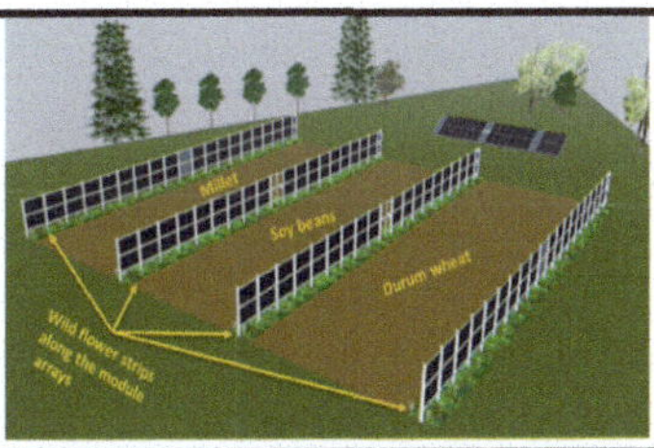
Location:	Building 20 *(Block 1)*	Location:	AgriPVplus *(Block 1)*
$P_{nominal}$:	144 kWp	$P_{nominal}$:	66 kWp + 15 kWp (south)
Orientation:	south-south-east 20° tilt	Orientation:	east-west 90° tilt; south 35°
Energy yield:	147.000 kWh/a	Energy yield:	81.000 kWh/a
	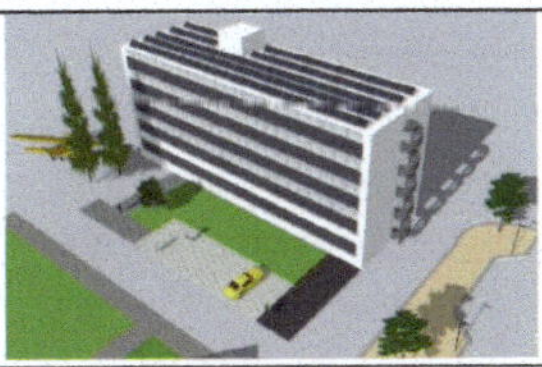		
Location:	Building 01 *(Block 3)*	Location:	Garparking *(Block 3)*
$P_{nominal}$:	46 kWp + 28 kWp (façade)	$P_{nominal}$:	260 kWp
Orientation:	south, 20° tilt; façade 90°	Orientation:	south, 35° tilt
Energy yield:	51.000 kWh/a roof 22.000 kWh/a façade	Energy yield:	265.000 kWh/a
	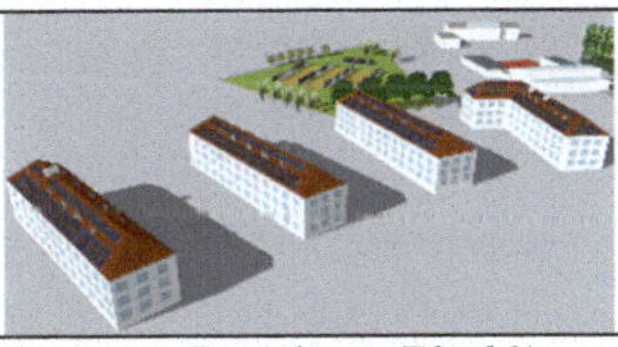		
Location:	Dormitory *(Block2)*	Location:	Building 29 *(Block 4)*
$P_{nominal}$:	235 kWp	$P_{nominal}$:	70 kWp
Orientation:	west 30° tilt	Orientation:	south 25° tilt
Energy yield:	265.300 kWh/a	Energy yield:	80.500 kWh/a

Location:	Building 06 and 07 *(Block 6)*	Location:	Mensa *(Block 5)*
$P_{nominal}$:	180 kWp	$P_{nominal}$:	146 kWp
Orientation:	east-west 40° tilt	Orientation:	east-west 40° tilt; south 30°
Energy yield:	160.000 kWh/a	Energy yield:	148.800 kWh/a

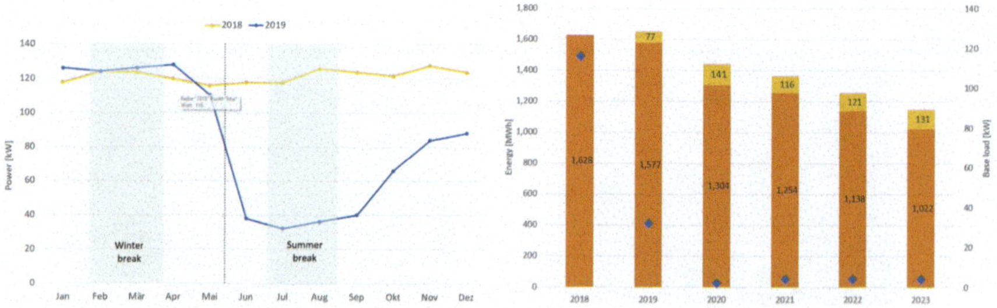

Fig. 19.3 Comparison of base load of 2018 and 2019 (left), Energy consumption and contribution of the first PV system at Building 20

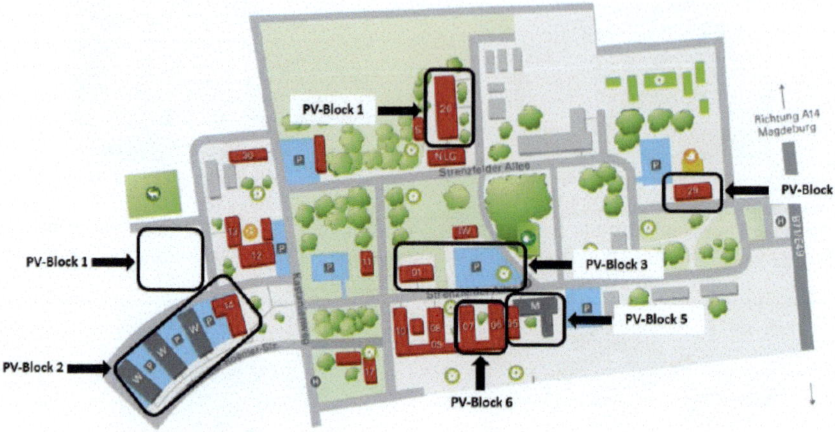

Fig. 19.4 Map of the campus Bernbug-Strenzfeld [4] with the identified buildings and open areas named PV-Blocks

19.4 Design and Simulation of Proposed PV System and Battery Storage

An extensive campus survey was conducted to identify potential locations and buildings where PV systems could be installed, including rooftops, façades, parking areas, and open spaces. The survey investigated possible shading of the PV systems, conducted a static analysis to determine whether roofs are suitable for PV installation, and considered options with different orientations (e.g., east–west or façades) to ensure energy production shifts throughout the day.

Figure 19.4 shows a map of the campus with the identified buildings and open areas labeled as PV Blocks. An individual design was created for all identified buildings and open areas. The PV systems were simulated using PVSyst with real measured meteorological data from the campus. Table 19.2 provides a 3D visualization of the proposed PV systems, along with their key parameters, and the simulated energy yields.

Table 19.3 Combination of investigated PV systems (PV-Blocks) and the selected battery capacity used in the simulation

Case	Combination of PV-blocks	Pn_total [kWp]	Battery capacity [kWh]							
1	Block 1	225	76	152	228	304	380	456	532	608
2	Block 1 + Block 4	294								
3	Block 1 + Block 2	460								
4	Block 1 + Block 3	559								
5	Block 1 + Block 2 + Block 3 + Block 4	864								
6	Block 1 + Block 2 + Block 3 (no parking) + Block 4	604								
7	Block 1 + Block 2 + Block 3 (no parking) + Block 4 + Block 5	750								
8	Block 1 + Block 2 + Block 3 (no parking) + Block 4 + Block 5 + Block 6	930								

To determine the maximum PV capacity that can be utilized on the campus, the PV Blocks were combined in various configurations to achieve different total power outputs and orientations, thereby investigating shifts in energy production to the morning and evening. This results in eight different cases/scenarios which have been investigated.

The second step of the simulation was to investigate a suitable battery capacity to store the surplus energy. Commercially available products were considered, with battery units starting from 76 kWh and are scalable to larger units. This study used battery capacities ranging from 76 to 1064 kWh, with incremental capacity steps of 76 kWh. Table 19.3 summarizes the simulation cases with the combined PV Blocks, including the total power and the battery capacities used.

19.5 Results and Discussion

This section will first describe the influence of additional PV systems on Building 20 (144 kWp) and then identify the optimal battery storage capacity to ensure that no energy is either curtailed or fed into the public grid.

19.5.1 Self-Sufficient and Curtailment of PV Energy

Figure 19.5 illustrates the increase in nominal installed PV capacity for the eight different scenarios, along with the self-sufficiency rate (i.e., the percentage of total loads covered by PV generation) shown by the gray bars, and the curtailment rate shown by the orange bars. In the case of Block 1 (existing PV system Building 20 + AgriPVplus system), the contribution to the total energy consumption can be increased to 15% without any curtailment. Because, the AgriPVplus system has a vertical east–west orientation, its energy contribution is higher in the mornings and afternoons and lower at midday compared to a south-facing PV system.

However, adding more PV systems, even with different orientations, leads to increased self-sufficiency but also raises the curtailment rate. By adding 604 kWp (case 6), the self-sufficiency rate reaches approximately 35%, while the curtailment rate is 25%. If the installed PV capacity is increased further, the self-sufficiency rate saturates at 38%, with any additional installed PV capacity increasing only the curtailment rate.

To meet the requirement that all produced energy must be consumed within the campus, the maximum power of additional PV systems should not exceed 81 kWp, bringing the total to 225 kWp. If more PV systems are to be installed with the goal of increasing the self-sufficiency rate, an appropriate storage system, such as battery storage, needs to be installed.

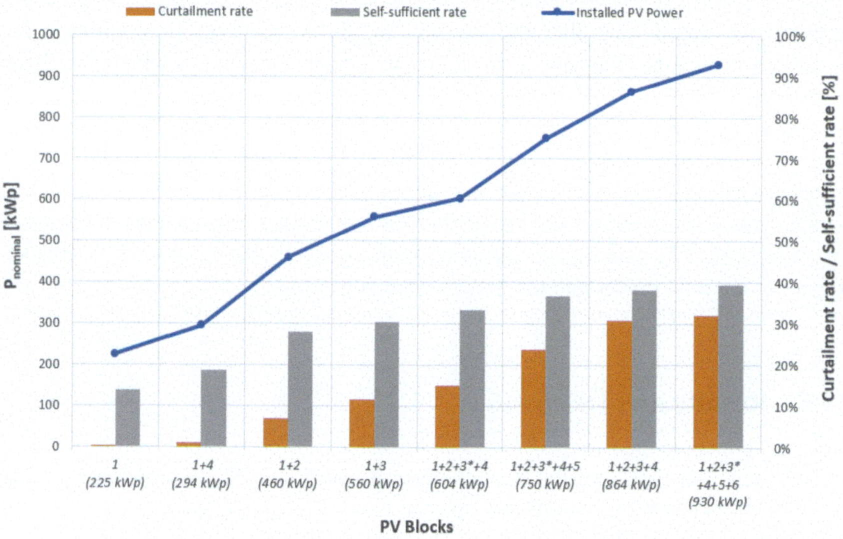

Fig. 19.5 Nominal installed PV capacity for the eight different scenarios together with the self-sufficient (i.e., percentage of the total loads covered by PV generation) (gray bars) and the curtailment rate (orange bars)

Daily analysis

Figure 19.6 shows a comparison of the load and PV generation for Case 7 (Block 1 + Block 2 + Block 3 (excluding parking) + Block 4), with a total installed PV capacity of 604 kWp. The graph presents data for a day in December and June (top left and right, respectively), as well as for a weekday and a weekend day (bottom left and right, respectively). The graphs clearly indicate that during winter, the generated energy will not exceed the base load, and thus, a storage system will not be needed. However, in spring, summer, and autumn, more energy needs to be either curtailed or stored.

The storage system should be designed to account for variations in energy demand, such as those between weekdays and weekends, and ensure that the battery capacities are appropriately sized to discharge overnight. Despite this, due to the campus load profile with its seasonal variations, achieving a 100% self-sufficient system will not be feasible. A balance needs to be found between storage, curtailment, and energy sourced from the grid.

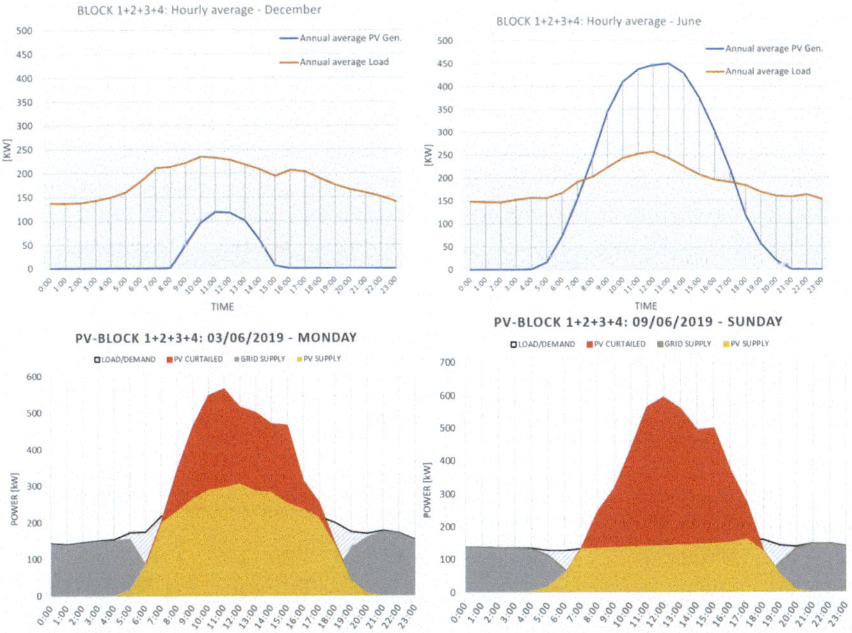

Fig. 19.6 Comparison of the load and the PV generation (case7, 604 kWp) for a day in December and in June (top left and right, respectively) and a weekday and a weekend day (bottom left and right, respectively). In red is the surplus energy which needs to be either curtailed or stored

19.5.2 *Storage System*

In total, 56 different configurations of Storage + PV systems have been simulated and analyzed. Table 19.4 presents the results for Case 7 (Block 1 + Block 2 + Block 3 (excluding parking) + Block 4) with a total PV capacity of 604 kWp. These PV systems collectively produce 650 MWh of energy annually. The PV system directly supplies 33.3% of the annual energy demand. However, without a storage system, approximately 96.4 MWh (14%) of generated energy would be curtailed annually. By integrating a storage capacity of 608 kWh, the curtailed energy can be reduced to 23.6 MWh (3.7%).

In every simulated configuration, we observed that reducing the curtailed energy from the PV system by increasing storage capacity requires significant effort. Additionally, as the storage capacity increases, the rate of decrease in curtailed PV energy diminishes. Figure 19.7 illustrates simulations for storage systems with capacities up to 1064 kWh.

All configurations exhibit similar behavior, which is more pronounced for smaller PV systems where the initial curtailed energy is lower compared to larger PV systems. The effort to achieve curtailed energy levels below a certain threshold (depending on the size of the PV system, $< 5\%$ for Blocks 1 + 2, $< 8\%$ for Blocks 1 + 3 + 4, $< 18\%$ for Blocks 1 + 2 + 3 + 4) is not justified when considering the investment relative to the saved energy. Analyzing all simulations, it is evident, that the PV systems on campus face significant challenges in supplying energy during winter months. Therefore, the effort required to save more energy during this period is substantially higher. Moreover, storage systems do not alleviate this issue, as most of the stored energy is generated during summer, with no ability to shift it to winter.

Table 19.4 Results of the simulation of case 7 (Block 1 + Block 2 + Block 3 (w/o parking) + Block 4) with a total PV capacity 604 kWp

	Storage capacity [kWh]								
	0	76	152	228	304	380	456	532	608
Self-consumption from battery [MWh]	0.0	15	26	35	43	49	55	59	63
Grid supply [MWh]	1102	1087	1076	1067	1059	1052	1047	1043	1039
Curtailment [MWh]	96	80	67	56	47	40	34	28	24
Battery losses [MWh]	0.0	1.8	3.3	4.7	5.8	6.9	7.8	8.6	9.3
Rate of self-sufficiency (PV + Storage)	33%	34%	35%	36%	36%	36%	37%	37%	37%
Self-consumption rate	85%	87%	89%	91%	92%	93%	94%	94%	95%
Storage supply rate	0.0%	0.9%	1.6%	2.2%	2.6%	3.0%	3.3%	3.6%	3.8%
Direct PV supply rate	33%	33%	33%	33%	33%	33%	33%	33%	33%
Curtailment (over PV generation)	15%	12%	10%	8.7%	7.3%	6.1%	5.2%	4.3%	3.7%

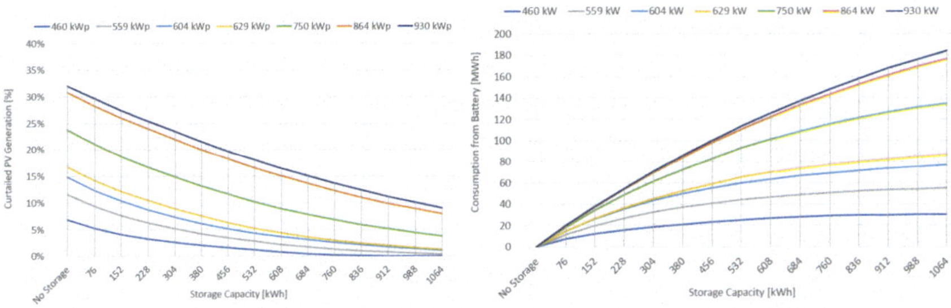

Fig. 19.7 PV blocks comparison of curtailed PV energy generation with different storage capacities (right) and comparison of battery self-consumption with different storage capacities (left)

19.6 Conclusion

The present analysis shows that the integration of renewable energy, namely photovoltaic (PV), into the University's campus is feasible. Considering the constraints and boundaries established by the grid, the necessity of including storage systems is justified only if the PV generators are sufficiently large. After examining several PV alternatives—varying in topology, technology, arrangement, and orientation—a saturation point was observed at around 600 kWp nominal PV power. Beyond this point, the rate of increase in added PV power and annual curtailed energy becomes nearly identical.

Further analysis incorporating battery storage revealed that for smaller PV systems, adding battery capacity did not substantially increase self-consumption. However, larger PV systems benefited more from battery integration, as excess energy could be stored and utilized effectively. Therefore, while small PV plants may experience energy curtailment, the current economic justification for adding battery systems to these systems is limited. Regarding the self-sufficiency rate (load coverage with own-generated energy), values up to 40% can be reached without battery storage, which are still economically feasible and attractive, albeit at the expense of the self-consumption rate (percentage of used self-generated energy), which reaches up to 69%. This means that although 40% of the demand could be covered by PV energy without battery storage, 31% of the generated PV energy would remain unused due to curtailment. When introducing battery storage, self-sufficiency can increase to 47%, with a self-consumption rate of up to 81%.

Based on these findings, HSA decided to install a second PV system with a different orientation to shift energy production to the morning and evening. This installation is an AgriPV system with vertically installed bifacial PV modules. This unique system integrates agriculture, energy generation, and biodiversity measures following Anhalt University's AgriPVplus approach [11]. The system comprises 144 vertically installed bifacial PV modules in an east–west orientation, each with a nominal power of 460 Wp, totaling 66.24 kWp. The distance between arrays is 14 m, allowing for the study of various types of crops (primarily drought-resistant crops such as millet, soybean, and durum wheat). Wildflower strips are planted under the

PV arrays (one meter on each side) to examine their impact on crops and ecosystem services [12]. The AgriPVplus system is currently under construction and is expected to be fully operational by the end of November 2024.

References

1. European Commission—2030 climate & energy framework, https://ec.europa.eu/clima/euaction/climate-strategies-targets/2030-climate-energy-framework_en. Accessed 15 February (2022)
2. H. Wirth, Current facts about photovoltaics in Germany, Fraunhofer ISE, 23.02.2020
3. V. Quaschning, For a sustainable electricity transition: increase in annual PV expansion to 15 GW/a sector coupling through the energy transition, HTW Berlin, (2016)
4. Fraunhofer ISE, www.energy-charts.info. Accessed 20.09.2024
5. J. Lee et al., Economic feasibility of campus-wide photovoltaic systems in new England. Renew. Energy **99**, 452–464 (2016)
6. C.H. Kost et al., Levelized cost of electricity renewable energy technologies, Fraunhofer ISE, (2024)
7. HSA sustainable strategy, https://www.hs-anhalt.de/en/sustainability/strategy.html. Accessed 20.09.2024
8. M. Dave, Z. Gou, et al. Greening Universities Toolkit 2.0, Transforming Universities into green and sustainable campuses: a toolkit for implementers, New South Wales: United Nations Environment Programme, (2014)
9. H.T. Fachrudin, Green campus concept based on architect perspective, in IOP Conference Series: Materials Science and Engineering, Medan, (2019)
10. Act on the Development of Renewable Energy Sources (Renewable Energy Sources Act—RES Act 2014) (PDF). Retrieved 28 July 2016. Unofficial translation of the RES Act of the version in force as of 1 August, (2014)
11. H. Sánchez Ortiz, S. Dittmann, S. Tosello, C. Meza, S. Dullau, M.H. Meyer, P. Scholz, S. Tischew, R. Gottschalg, Novel measurement concept for AgriPVplus systems—a triple use approach, WCPEC-8, (2022)
12. M.H. Meyer, S. Dullau, P. Scholz, M.A. Meyer, S. Tischew, Bee-friendly native seed mixtures for the greening of solar parks. Land **12**(6), 1265 (2023). https://doi.org/10.3390/land12061265

Part IV
Control Strategies for Smart Grid

Chapter 20
Personal Microgrid Concept According with the Puerto Rico Microgrid Rule—Comparative Analysis

Omar F. Rodriguez-Martinez, Yuly V. Garcia, Jan L. Diaz, Fabio Andrade, and Agustín A. Irizarry-Rivera

Abstract This study evaluates the feasibility of implementing microgrids in Puerto Rico in accordance with the regulatory requirements stipulated in the Public Renewable Energy Policy (PREB). Two different case studies were conducted to assess the feasibility of meeting the necessary conditions imposed to renewable microgrids of produce at least 75% of their energy annually. In addition, the study focuses on a specific photovoltaic case, which demonstrates the availability of enough rooftop space on residential houses for the deployment of the proposed Photovoltaic system. While transient voltage fluctuations were observed during simulations, the study indicates that these variations remain within the prescribed 5% boundary, conforming to standard regulations. These findings emphasize the potential for microgrids to meet PREB regulations and highlight the importance of region-specific considerations for broader applicability.

Keywords Microgrid · PV system · Regulation · Wind turbine

20.1 Introduction

In 2017, Hurricane Maria made landfall in Puerto Rico, twenty five percent (25%) of the electric transmission towers and distribution poles were severely damaged rendering the electric network useless [1]. The severity of this event was so great that, according to the official government report, 95% of the users had their electric service restored by march 21, 2018, six months after the hurricane hit the Island [2, 3].

In 2018 the Puerto Rico Energy Bureau (PREB) approved and published the "Regulation on the Development of Microgrids", (PREB microgrids rule). This rule was developed to " establish the legal and regulatory framework necessary to foster the development of the microgrid system in Puerto Rico". The rule is a direct result of an investigation in which case studies have been developed for selected renewable energy microgrids(MGs) described in the PREB microgrid Rule.

O. F. Rodriguez-Martinez · Y. V. Garcia · J. L. Diaz · F. Andrade (✉) · A. A. Irizarry-Rivera
University of Puerto Rico, Mayaguez, PR, USA
e-mail: Fabio.Andrade@upr.edu

D. Rossit et al. (eds.), *Proceedings of the VII Ibero-American Congress of Smart Cities, ICSC-Cities 2024, 12-14 November, San Carlos, Costa Rica*, Springer Proceedings in Energy, https://doi.org/10.1007/978-981-96-4301-1_20

In this paper the results obtained from the implementation of two personal renewable microgrids that meet the requirements of the PREB microgrid Rule are reported. In order to achieve the two working MGs that appease the rule's constraints, two requirements need to be met. The first one was be a personal MG, which means that the MG needs to have a maximum of two (2) owners. The second one is divided in two parts, the first one was be a renewable microgrid, which means that the principal source of energy must be either solar Photovoltaic (PV), wind, geothermal energy, renewable biomass, or biomass gas combustion. The second part is that, to meet the definition of renewable MG seventy-five percent (75%) of the energy produced by the MG in a period of 12 months must be delivered from the renewable generators within the MG.

The rest of this paper is organized as follows. Section 20.2 describes the models selected to represent the different components of each MG. Section 20.3 summarize the parametrization, considerations and hardware used for the different scenarios under study. Section 20.4 shows results. Finally, Sect. 20.5 present the general conclusions of this paper.

20.2 Photovoltaic, Wind and Distribution System Modeling

According to the Microgrid regulation requirements, two personal renewable MGs were selected and simulated in this work. Therefore, a PV and wind turbine personal system were selected to analyzed the performance of both energy resources. From literature review [4–11], the simplified single diode (SSD) or R_s model, the single-diode model (SDM) and the two-diode model to represent PV were identified. The SDM was chosen to conduct this study, since it demonstrates enhanced precision representing the I-V curve's characteristics during significant temperature variations in contrast to the SSD model. Based on that, the PV system was modeled using (20.1) that describes the behavior of a single-diode PV current panel.

$$I_{pv} = I_{ph} - I_o \left[e^{\frac{q(V_{pv}+I_{pv}R_S)}{N_S kTA}} - 1 \right] - \frac{V_{pv+I_{pv}R_s}}{R_{sh}} \tag{20.1}$$

To represent the wind turbine, a simplification of the model proposed by Teodorescu et al. in [11] was implemented. This model described a full power back-to-back converter to control the power injection of the wind turbine into the grid. Although this configuration improves the energy injection profile, all the dynamics of the wind turbine disappear when the generated power is converted to DC. Taking into account the loss of dynamics due to the inclusion of the power converters, the wind turbine was modeled as a function block present in the Simulink library containing the polynomial regression that describe the relationship between wind speed and the active power produced by the turbine. This behavior was obtained from the data sheet provided by the manufacturer of the selected wind turbine [12]. The polynomial regression describing the turbine energy injection is presented in (20.2).

Table 20.1 Parameters description

	Parameters	Personal source
Solar PV	Lines distribution system	4.16 kV feeder
	Transformers	Typical manufacturer data
	Demand Profile Loads	Residential (5 profiles)
	Irradiance generation	Coto Laurel's Farm
	Simulated period	One year
	Installed capacity	8.6 kW
	Simulation results	Annual energy production
Wind turbine	Lines distribution system	4.16 kV feeder
	Transformers	Typical manufacturer data
	Demand profile Loads	Residential + commercial
	Wind speed	Soderstrom
	Simulation period	One year
	Wind turbine data	Nozzi TN535 (10kW)
	Simulation results	Annual energy production

$$P_{wind} = 0.0018x^5 - 0.0375x^4 + 0.3049x^3 - 1.1561x^2 + 2.0965x - 1.4655 \tag{20.2}$$

To model the distribution system, a 4.16kV feeder from a town in southern Puerto Rico was used as reference. The main grid was modeled as an ideal three-phase source . The transformers were modeled using the classic model from the simscape library for the single- and three-phase transformer. All the required parameters were obtained by analyzing the data collected at the aforementioned feeder. The single- and three-phase distribution/transmission lines were modeled using the branch blocks available in the simscape library. The series branch represents series connection of resistor, inductor, and capacitor elements, depending on the required combinations. It is important to mention that all lines were modeled with branches, since for short lines this model provides an accurate approximation.

To model the system's loads, five (5) residential load profiles were obtained from a South community of Puerto Rico. Having these five residential curves of one day, a weekly profile curve for a residential load was created using these different behaviors to represent a specific day of normal week in Puerto Rico. For the connection of the generation model with the distribution system, the inverter was modeled as a continuous ideal AC current source.

The implemented models consider irradiance, and temperature variations as inputs for the PV system and wind speed for the Wind turbine. The other inputs and information required to carry out the different scenarios are shown in Table 20.1.

20.3 Study Cases

Once the models were chosen, they were implemented using MATLAB/Simulink, a simulation platform suitable for modeling microgrids. Two personal MG scenarios have been selected in this study, a photovoltaic and wind MGs scenarios. PREB microgrid standard, compliance is evaluated over one year, thus 12 months were simulated. To reduce the simulation time, one minute in the time scale used in the simulation is represented by three (3) seconds. This is scaling is suitable for this analysis due to the fast dynamics of the system that makes the simulation reach the steady state very quickly.

Despite the scaling, performing a year of simulation still requires significant simulation time and computational resources. For this reason, the OP5707, a parallel processing platform of OPAL-RT technologies, was used to perform the simulations. To make the connection between MATLAB/Simulink and the OP5707, the "RT-LAB" software, the laptop application used to control the OP5707, was used. To compile and load the simulation on the OP5707, it is necessary to divide the system into two subsystems. One subsystem contains the entire electrical system, including the distribution system, generators and loads. The second subsystem contains the console system that allows controlling the variables and displaying the results [13].

20.3.1 Demand Profile Models

Six (6) different load profiles were used to model the systems loads; five residential and one commercial. The five (5) residential load profiles were collected from a Southern community in Puerto Rico. Using one-day residential consumption of each of the five measured loads, a weekly profile behavior for a residential load was created using these different behaviors to represent a specific day of normal week in Puerto Rico. The commercial load profile was obtained from measurements of a commercial building in Colombia.[1] For the commercial load, only one-day consumption behavior was found. This means that the energy demand over the year for this load remains constant. Analyzing the behavior of the Colombia commercial curve, it could be compared to the behavior of Puerto Rico's largest malls. Considering this, it is possible to assume that the consumption behavior of a mall tends to be similar every day of the year without losing generality.

[1] Commercial load data obtained from utility via private correspondence.

20.3.2 PV Scenario

The PREB microgrid rule establishes that, in order to be considered a personal microgrid, the microgrid must supply a maximum of two users. Therefore, Fig. 20.1 shows the implemented microgrid considering the distribution feeder, PREPA represented by the block "Generation and Transmission System", the residential loads and the PV microgrid. As can be seen in Fig. 20.1, the PV microgrid is connected only to the residential load connected to the Phase A of the system. Therefore, it is important to mention that in order to have a balanced systems, the three residential loads use the same demand profile.

To implement the SDM model, manufactured data from the standard test conditions (STC) and the Nominal Operating Cell Temperature (NOCT) of a 280 Watts panel were used. Also, the design characteristics of the installed capacity of the MG is shown in Table 20.2. The PV design was based on the evaluated household's maximum daily energy demand. Using this value, the number of solar panels is calculated by considering the typical peak sun hours under normal conditions in Puerto Rico, the power rating of each panel, and the inverter efficiency.

20.3.3 Wind Scenario

Similar to the PV scenario, the wind microgrid (MG) is designed to supply the energy demand of a single user. Given the selected wind turbines, the wind MG was integrated into the commercial load, as shown in Fig. 20.2. The residential load follows the same demand profile used in the PV scenario. Wind speed data was

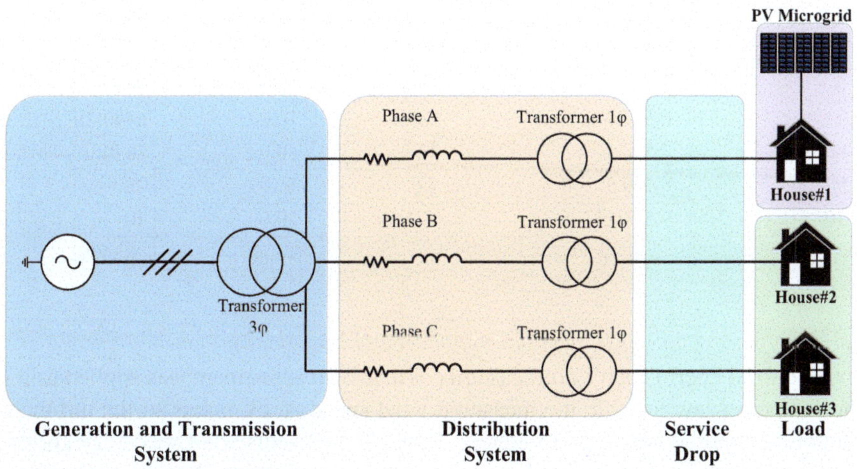

Fig. 20.1 PV personal Microgrid

Table 20.2 Personal Microgrid design

Residential demand (kWh/day)	18
% Load supplied by PV	100
Sun peak hours (h)	4
Single panel power (kW)	0.28
DC-to-AC derate factor	0.96
PV area (m^2)	1.68
PV system capacity (kW)	4.76
Quantity of solar panels	17
PV energy (kWh/day)	19.0

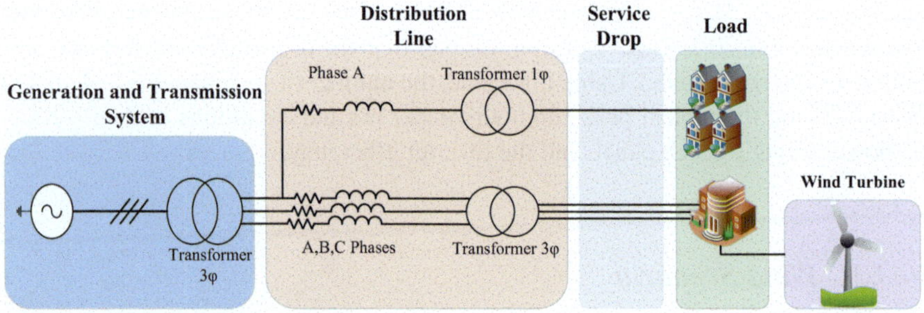

Fig. 20.2 Wind turbine personal microgrid

Table 20.3 Expected energy production

Wind speed (m/s)	Wind speed range (m/s)	Hours at wind speed	kW	kWh
3	2.5–3.5	1238	0.54	675
4	3.5–4.5	1732	1.75	3024
5	4.5–5.5	1954	3.81	7447
6	5.5–6.5	1354	6.97	9435
7	6.5–7.5	649	10.00	6490
8	7.5–8.5	167	10.00	1670
9	8.5–9.5	22	10.00	220
10	9.5–10.5	1	10.00	10
			Total	28,970.2

sourced from Puerto Rico's northeastern coast, and an adjustment was applied using the 1/7 power law to scale the measured wind speed at 13 meters to the turbine's minimum required height of 15 meters. Table 20.3 presents wind speed categories, the hours per year at each speed, and the estimated energy production of the wind turbine, calculated using wind speed data and the turbine's power equation. In a similar manner to the PV design, the wind turbine was selected based on the building's maximum daily energy demand, wind speeds and their frequency.

20.4 Results

Carrying out the different scenarios described in the past section, the following information was obtained and processed in order to see if the compliance of the PREB regulation was achieved.

20.4.1 PV Scenario

One-year simulation was performed to evaluate the annual performance of the MG. Figure 20.3 shows irradiance of the second day of April and December, which are the months with the highest and lowest irradiance values. Using the irradiance data set, the energy per month produced and demanded from the PV MG and the load is depicted in Fig. 20.4. From this figure it can be seen that the MG produces more energy than is required by one house in each month. In the whole year the MG produced 8.62 MWh of energy and the load demand 4.58 MWh for one house. Having these amount of energy produced, the system could supply a 88% of the energy demand of a second house with the same energy consumption. Another option is to sell the excess of energy produced to PREPA using net-metering. Also, due to the impossibility of sell the energy to any other entity different from PREPA, the owner/s of the MG could think in install a battery storage system in order to store remaining energy produced by the MG to optimize its operation and become resilient against natural disasters as hurricanes, earthquakes, etc.

Considering that an average house in Puerto Rico is about 152 m^2 [14] and assuming that just the 60% of the rooftop area can be used to accommodate the PV system, this means that 91.2 m^2 of area are available for PV panels. Each panel requires 1.68 m^2 and a total of 17 panels needs to be installed, that means a total area of 28.56 m^2 to deploy the system. Allocating space for corridors to avoid shade and provide maintenance this result will be multiply by 1.5 for a total estimated area of 42.84 m^2. Thus approximately 47% of the average rooftop surface area of a house in Puerto Rico is needed to generate enough energy to meet the load.

Because of the variability and excess of energy produced by the microgrid, it is important to analyse the behavior of the system's voltage. Figure 20.5 shows the distribution voltage profile of the residential load with the MG connected. Voltage variations at the load can be attributed to two primary factors: the increase in voltage levels occurs when the active power injected by the microgrid exceeds the load demand, while low voltage levels indicate voltage drops across the single-phase distribution lines. Although there is variability in the voltage behavior, the values remains into the boundaries permitted by the standard.

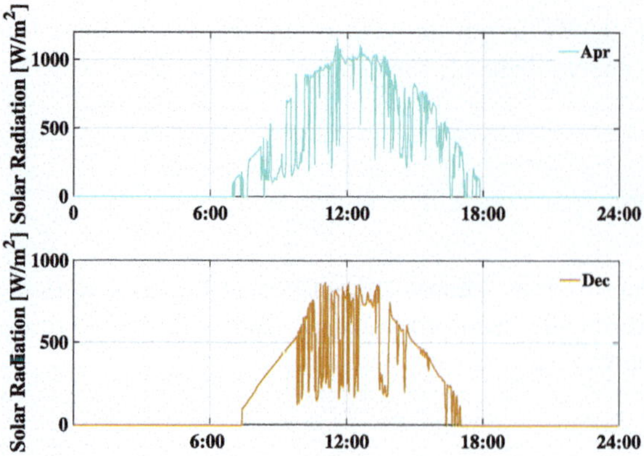

Fig. 20.3 Solar radiation

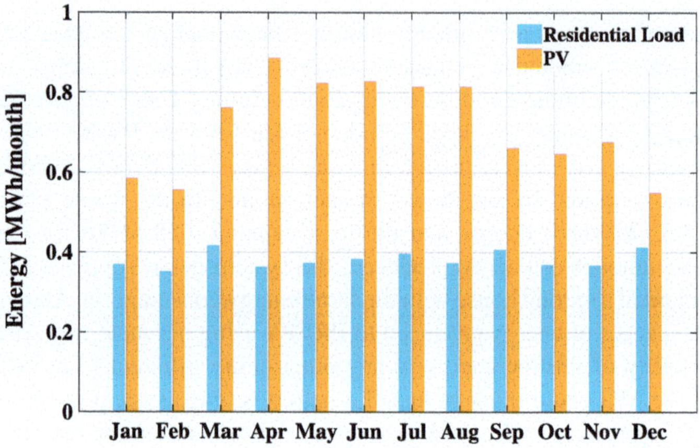

Fig. 20.4 Monthly MG energy production versus energy demand

20.4.2 Wind Scenario

Figure 20.6 shows the behavior of the wind speed during one day of the months with higher and lower wind speeds, May and February respectively. Using the wind speed data, Fig. 20.7 shows the amount of energy produced by the MG against the energy demanded by the commercial load. Looking these results, it can be seen that different from the previous case, the MG produces more energy than the required by the load just in some months. That means that PREPA needs to supply the energy shortage. Despite of that, the MG produces 27.61 MWh of energy of the 35.05 MWh demanded by the load in the whole year. This represent a 78.8% of the energy consumed in the year, value that exceed the 75% required by the PREB regulation.

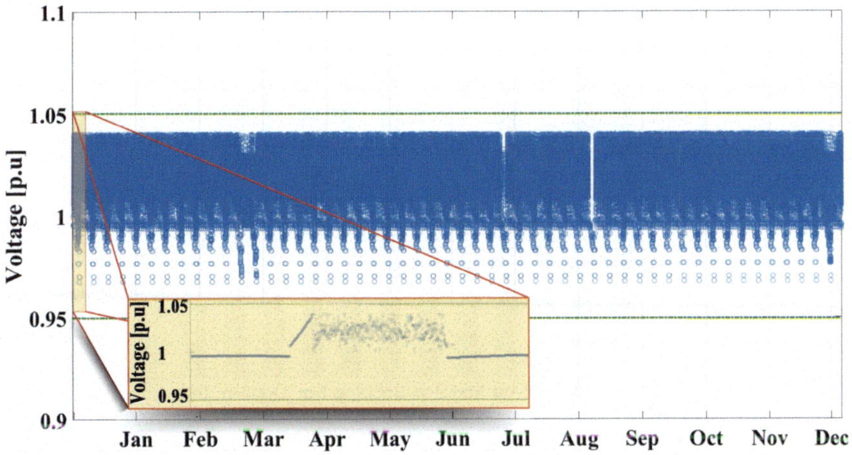

Fig. 20.5 Annual voltage profile in per unit

Fig. 20.6 Wind speed

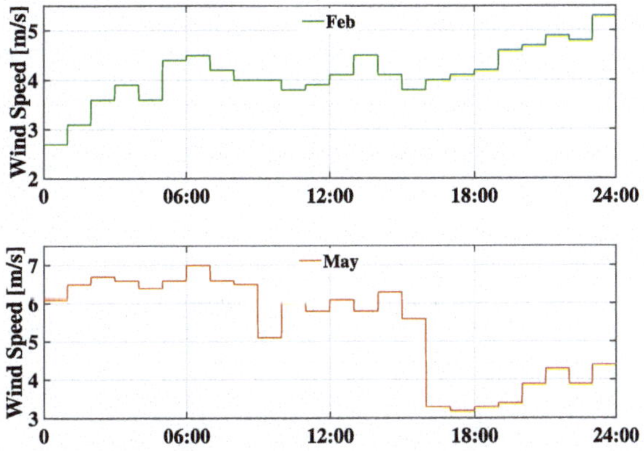

Looking for the distribution voltage behavior due to the implementation of the MG, it can be seen in Fig. 20.8 that both profiles remains under the boundaries permitted by the standard. Also, different from the PV case, the residential load has lower variability and do not surpass the 1 p.u value.

20.5 Conclusions

The case studies demonstrate that it is feasible to implement MGs that meet the requirements of the PREB regulation in Puerto Rico. In both cases, the requirement of at least 75% of energy production in a year period for renewable MGs is achievable. Also, for the PV case,it was demonstrated that there is enough rooftop space available

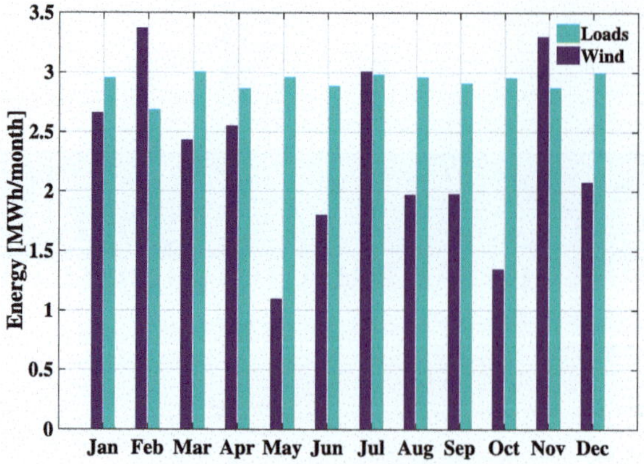

Fig. 20.7 Monthly MG energy production versus energy demand

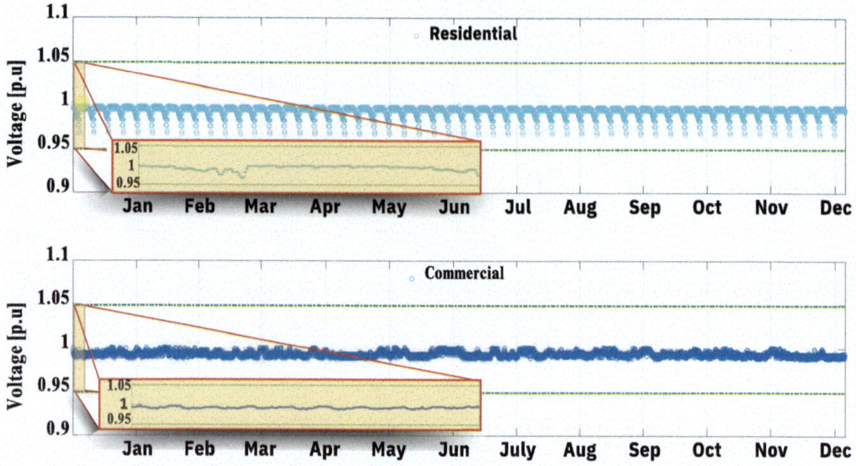

Fig. 20.8 Residential and commercial annual voltage profile in per unit

in the residential houses for the deployment of the proposed MG, assuming a 60% of the total space available for a average house. Furthermore, despite voltage variations were found in both cases along the simulation, it can be seen that voltage remains into the boundaries of 5% permitted by the standard [15]. An important aspect to consider in this case, its that the compliance of the energy production requirement was achieved using weather data from specific areas of Puerto Rico. So it is important to consider the availability and behavior of the resources in other areas in order to generalize this findings for the complete island [16–19].

Acknowledgements The Authors acknowledge, and thank, Windmar LLC, specially Mr. Víctor González, for sharing valuable data on the operation of its Coto Laurel photovoltaic farm located in Ponce, Puerto Rico. The authors acknowledge the support provided by the Thematic Network

723RT0150 "Red para la integración a gran escala de energías renovables en sistemas eléctricos (RIBIERSE-CYTED)" financed by the call for Thematic Networks of the CYTED (Ibero-American Program of Science and Technology for Develop-ment) for 2022.

References

1. A. Irizarry-Rivera, K. Montano-Martinez, S.I. Alzate-Drada, F. Andrade, A case study of residential electric service resiliency through renewable energy following hurricane maria (2018)
2. S.-E. Chen, M.A. Pando, A.A. Irizarry, Y. Baez-Rivera, W. Tang, Y. Ng, in *Advanced Geotechnical and Structural Engineering in the Design and Performance of Sustainable Civil Infrastructures: Proceedings of the 6th GeoChina International Conference on Civil & Transportation Infrastructures: From Engineering to Smart & Green Life Cycle Solutions–Nanchang, China, 2021*. Resiliency of Power Grid Infrastructure Under Extreme Hazards-Observations and Lessons Learned from Hurricane Maria in Puerto Rico (Springer, 2021), pp. 1–17
3. A. Kwasinski, F. Andrade, M.J. Castro-Sitiriche, E. O'Neill-Carrillo, Hurricane maria effects on puerto rico electric power infrastructure. IEEE Power Energy Technol. Syst. J. **6**(1), 85–94 (2019)
4. S. Bal, A. Anurag, B.C. Babu, in *Annual IEEE India Conference (INDICON)*. Comparative Analysis of Mathematical Modeling of Photo-Voltaic (pv) Array (IEEE, 2012), pp. 269–274
5. B.C. Babu, S. Gurjar, A novel simplified two-diode model of photovoltaic (pv) module. IEEE J. Photovol. **4**(4), 1156–1161 (2014)
6. Y.T. Tan, D.S. Kirschen, N. Jenkins, A model of pv generation suitable for stability analysis. IEEE Trans. Energy Convers. **19**(4), 748–755 (2004)
7. S. Gupta, H. Tiwari, M. Fozdar, V. Chandna, in *Asia-Pacific Power and Energy Engineering Conference*. Development of a Two Diode Model for Photovoltaic Modules Suitable for Use in Simulation Studies (IEEE, 2012), pp. 1–4
8. S. Bader, X. Ma, B. Oelmann, A comparison of one-and two-diode model parameters at indoor illumination levels. IEEE Access **8**, 172057–172064 (2020)
9. C.-T. Sah, R.N. Noyce, W. Shockley, Carrier generation and recombination in pn junctions and pn junction characteristics. Proceedings of the IRE **45**(9), 1228–1243 (1957)
10. F. Blaabjerg, R. Teodorescu, M. Liserre, A.V. Timbus, Overview of control and grid synchronization for distributed power generation systems. IEEE Transactions on industrial electronics **53**(5), 1398–1409 (2006)
11. R. Teodorescu, M. Liserre, P. Rodriguez, *Grid Converters for Photovoltaic and Wind Power Systems* (Wiley, 2011)
12. A. Musso, D. Spago, G. Soraperra, U. Spallinger, G. Cimatti, M. Rialti, Tn-11-47 tn 535 general description
13. Flagship real-time digital simulator / simulation tools / op5707xg. https://www.opal-rt.com/simulator-platform-op5707/
14. A.A. Irizarry, J.A. Colucci, E. O'Neill, *Chapter 5 Solar Photovoltaics* (2014), pp. 1–96. https://www.uprm.edu/aret/docs/Ch_5_PV_systems.pdf
15. N.F.P. Association, *NFPA 70: National Electrical Code* (National Fire Protection Assoc, 2007)
16. J.F. Patarroyo-Montenegro, J.D. Vasquez-Plaza, O.F. Rodriguez-Martinez, Y.V. Garcia, F. Andrade, Comparative and cost analysis of a novel predictive power ramp rate control method: a case study in a pv power plant in puerto rico. Appl. Sci. **11**(13), 5766 (2021)
17. O.D. Garzon, A. Hauman-Rivera, M.P. Gomez, J.L. Diaz, Y.V. Garcia, F. Andrade, A.C.L. Hernández, A.A. Irizarry-Rivera, in *IEEE PES Innovative Smart Grid Technologies Latin America (ISGT-LA)*. Integration and Assessment of Photovoltaic Systems in Puerto Rican Communities (IEEE, 2023), pp. 375–379

18. R. Calloquispe-Huallpa, A. Huaman-Rivera, A.F. Ordoñez-Benavides, Y.V. Garcia-Garcia, F. Andrade-Rengifo, E.E. Aponte-Bezares, A. Irizarry-Rivera, in *IEEE PES Innovative Smart Grid Technologies Latin America (ISGT-LA)*. A Comparison Between Genetic Algorithm and Particle Swarm Optimization for Economic Dispatch in a Microgrid (IEEE, 2023), pp. 415–419
19. Y.V. Garcia, O.F. Rodriguez-Martinez, A.A. Irizarry-Rivera, F. Andrade, in *IEEE PES Innovative Smart Grid Technologies Latin America (ISGT-LA)*. Cooperative Microgrids: Technical Feasibility of the Preb's Microgrid Rule (IEEE, 2023), pp. 350–354

Chapter 21
IDA-PBC Controller Design for Grid-Connected Quasi-Z-Source Three-Phase Inverter

Federico M. Serra, Francisco D. Esteban, Fernando A. Mezzano, Carlos R. Baier, Eduardo E. Espinoza, and Oscar D. Montoya

Abstract This paper presents a new passivity-based control strategy using interconnection and damping assignment for a three-phase quasi-impedance source inverter connected to the grid through an inductive filter. The designed controller regulates the DC bus voltage to the required value while simultaneously regulating the active and reactive power injected into the electrical grid. As a nonlinear stabilizing control technique, it achieves good dynamic performance and ensures system stability in response to changes in the reference voltage and in the power injected into the grid. Simulation results are presented to validate the performance of the proposed control strategy.

Keywords Quasi-Z-source inverter · IDA-PBC controller · Averaged model · Three phase inverter · Electrical grid

21.1 Introduction

In recent decades, the demand for electrical energy has increased considerably, along with the emission of greenhouse gases [1]. Consequently, numerous alternatives have been proposed, with the main objective of generating electrical energy from

F. M. Serra (✉) · F. D. Esteban · F. A. Mezzano
Laboratorio de Control Automático (LCA), Facultad de Ingeniería y Ciencias Agropecuarias, Universidad Nacional de San Luis, CONICET, Villa Mercedes, San Luis, Argentina
e-mail: serrafederico@gmail.com

C. R. Baier
Departamento de Ingeniería Eléctrica, Universidad de Talca, Curicó, Chile

E. E. Espinoza
Departamento de Ingeniería Eléctrica, Facultad de Ingeniería, Universidad Católica de la Santísima Concepción, Talca, Chile

O. D. Montoya
Facultad de Ingeniería, Universidad Distrital Francisco José de Caldas, Laboratorio Inteligente de Energía, Bogotá, Colombia

D. Rossit et al. (eds.), *Proceedings of the VII Ibero-American Congress of Smart Cities, ICSC-Cities 2024, 12-14 November, San Carlos, Costa Rica*, Springer Proceedings in Energy, https://doi.org/10.1007/978-981-96-4301-1_21

renewable resources, thereby helping to meet the growing energy demand and reduce emissions of gases that contribute to climate change [2].

In this context, Distributed Generation Systems (DGS) are gaining greater importance, as they are primarily based on renewable energy sources such as photovoltaic, wind, and fuel cells, among others. These systems offer an efficient and sustainable alternative to centralized generation systems, enabling greater integration of renewable energy sources with the power grid and thereby reducing the transmission and distribution losses associated with the current electrical system [3].

However, since DGS can operate in direct current (DC) or alternating current (AC), the use of power electronic converters is necessary to achieve an efficient and safe connection between DGS and the electrical grid. These converters allow the interconnection of DC systems with AC ones, or vice versa, as well as DC systems with different voltage levels, and AC systems with varying voltage levels, frequencies, or number of phases [4].

Particularly, when it is necessary to inject the power generated by a DC DGS (photovoltaic generation system) into the electrical grid, a DC-AC converter (inverter) must be employed, which can be either single-phase or three-phase depending on the type of grid to which it is connected. Therefore, control strategies must be developed to maximize both its efficiency and reliability, and ensure stable operation. In [5], a control strategy based on proportional integral resonant (PI-R) controllers for a three-phase inverter with an inductive-capacitive-inductive (LCL) filter that injects the power generated by a photovoltaic panel array into the electrical grid is proposed. The controller is designed to regulate the DC bus voltage and inject sinusoidal currents during conditions of grid imbalance. However, due to the nonlinear dynamics of such systems, linear controllers must be designed using the linearized model of the converter around a specific operating point, which limits the performance and stability of the system under changes in operating conditions, such as variations in injected power or in the DC bus voltage of the converter.

Consequently, the literature also proposes nonlinear control strategies aimed at ensuring stability and good dynamic performance of the system across its entire operating range. In [6], a nonlinear current-limiting controller for a three-phase inverter connected to the grid via an LCL filter is proposed. The controller consists of two internal control loops for current and voltage, and an external power control loop, which performs as grid support. The designed controller allows for the regulation of voltage and active and reactive power, respectively, with a current limit that ensures system stability in closed loop. Whereas, [7] proposes a Terminal Sliding Mode Control (TSMC) strategy for a three-phase inverter with an LCL filter connected to the grid. The designed controller includes a control loop for regulating the DC bus voltage and a loop for controlling the grid currents, and ensures stability and efficiency even in the presence of disturbances and uncertainties in the converter's parameters.

However, when there is a significant difference between the AC grid voltage and the DC stage voltage, an additional DC-DC conversion stage is necessary to boost the DC voltage from the DGS to the level required by the inverter's DC bus.

To achieve this, multi-stage topologies can be employed, such as isolated or non-isolated DC-DC converters connected to the inverter's DC bus, or single-stage DC-AC topologies [8]. Non-isolated multi-stage DC-DC topologies are a good option for photovoltaic inverters connected to the grid; however, they have higher losses, are less reliable, and are more costly compared to single-stage DC-AC topologies [9, 10].

Regarding single-stage DC-AC converters, new topologies have been developed in recent decades, such as the Impedance Source Inverter (ZSI), initially presented in 2002 [11]. This topology allows for both reduction and elevation of the input DC voltage in a single stage, achieving high efficiency, reliability, robustness, and low cost, although its control presents a significant challenge [12]. A variant of the ZSI is the quasi-Impedance Source Inverter (qZSI), which has improved several aspects compared to the ZSI, such as the voltage across the semiconductors and the smoothness of the current drawn from the DC source [13]. Due to the increasing popularity of this topology, numerous studies have been conducted in recent decades proposing various control approaches for grid-connected qZSI topologies, aimed at regulating the DC bus voltage and the power injected into the grid. In [14], a cascade control strategy based on proportional-integral (PI) controllers for a three-phase qZSI connected to the grid through an LCL filter is proposed. The design objective is to regulate the DC bus voltage and the active and reactive currents injected to the grid. The controller is designed using the linearized Generalized State-Space Averaged model (GSSA), resulting in a control strategy that is simple to implement, with good dynamic performance and accurate reference tracking. However, since the qZSI results in a nonlinear system, variations in its operating point can deteriorate the performance and cause stability issues in the closed-loop system. To address this, the literature also explores nonlinear control approaches, such as in [15], which proposes an Energy Shaping Control (ESC) method for a three-phase qZSI connected to the grid via an L filter. The controller is designed to regulate the DC bus voltage and the power injected into the grid, resulting in a control strategy with good dynamic performance and stability, though it has high computational load and requires measuring the voltage across both capacitors in the quasi-impedance source network.

This paper proposes a new control strategy based on passivity, using interconnection and damping assignment (IDA-PBC), for a three-phase qZSI connected to the grid via an L filter. The controller is designed to regulate the DC bus voltage and the active and reactive power injected into the grid. By employing a passivity-based design technique, the controller achieves global stability and exhibits excellent dynamic performance in response to changes in the DC bus voltage reference and the currents injected into the grid. Because the controller is designed using the *dq* model of the converter, and due to a proper selection of the assigned interconnection matrix, a simple-to-implement controller is obtained, requiring only the measurement of the voltage across one of the DC stage capacitors, the currents and the voltages of the AC grid. The performance of the proposed control strategy is validated through simulation results.

This work is organized as follows: Sect. 21.2 derives the mathematical model of the converter, Sect. 21.3 presents the proposed control strategy, Sect. 21.4 shows the obtained simulation results, and Sect. 21.5 provides the conclusions.

21.2 Mathematical modeling

To design a control strategy that meets the established performance objectives, the mathematical model describing the dynamic behavior of the converter shown in Fig. 21.1 must be obtained. Initially, the switched model of the qZSI will be derived, followed by its averaged model in dq coordinates, and finally, the pH model will be obtained, which will be used in the next section for the control strategy design.

21.2.1 Switched Model

The topology under study in the present work is shown in Fig. 21.1. It consists of a quasi-impedance network on the DC side and a three-phase inverter connected to the electrical grid via an L filter. The quasi-impedance network allows the input voltage V_{in} to be increased to the level required by the inverter on the DC bus (v_{pn}), and is composed by capacitors C_1 and C_2, inductors L_1 and L_2, and a diode D. Meanwhile, the inverter is made up of six transistors (Q_a, Q_b, Q_c, $\overline{Q}_a$, $\overline{Q}_b$, and $\overline{Q}_c$), and the grid inductive filter L_g.

To achieve an increase in the DC bus voltage, the qZSI operates with two switching modes. The first is known as Shoot-through (see Fig. 21.2a), and is achieved by simultaneously closing two transistors in the same branch of the converter shown in Fig. 21.1. The second mode (Non-Shoot-through) (see Fig. 21.2b) corresponds to the conventional pulse-width modulation (PWM) operation mode of the inverter.

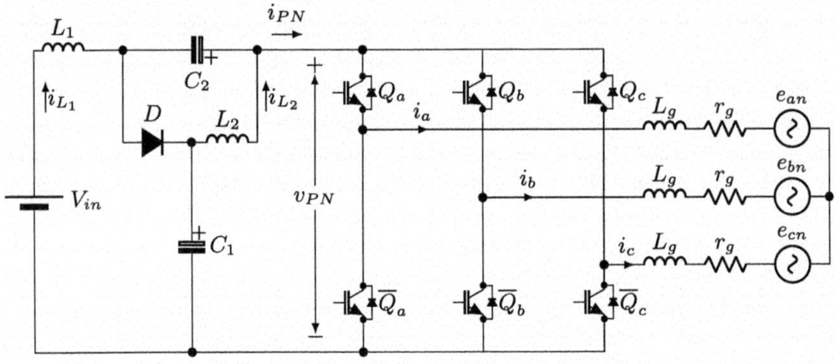

Fig. 21.1 Topological diagram of the three-phase qZSI connected to the grid

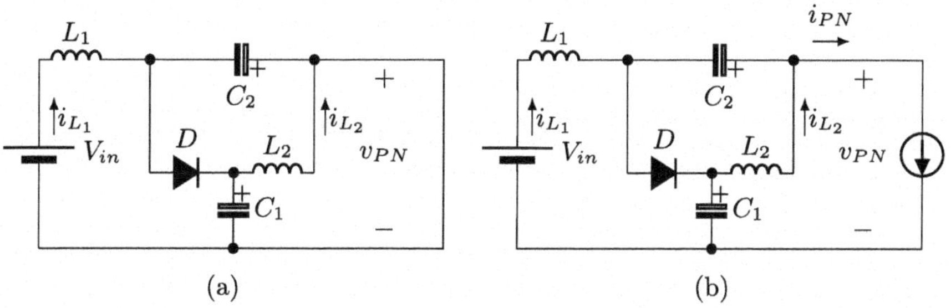

Fig. 21.2 Equivalent circuit of the qZSI in **a**) shoot-through state ($S_T = 1$) and **b** non-shoot-through state ($S_T = 0$)

Therefore, if two switching variables are defined, (S_T for the Shoot-through mode, and S_k for the PWM mode),

$$S_T = \begin{cases} 1 \text{ for the Shoot-through state} \\ 0 \text{ for the Non-Shoot-through state} \end{cases} \tag{21.1}$$

$$S_k = \begin{cases} 1 \text{ when transitor k is on} \\ 0 \text{ when transitor k is off} \end{cases} \tag{21.2}$$

where $S_k \in \{0; 1\}$, with $k = a, b, c$.

Next, by performing a power balance between the DC and AC stages, the relationship between the current i_{pn} and the three grid currents (i_a, i_b, and i_c) is obtained, which results in,

$$S_T' i_{pn} = S_a i_a + S_b i_b + S_c i_c. \tag{21.3}$$

Therefore, from the circuit in Fig. 21.1 for the switching states shown in Fig. 21.2 given by (21.1) and (21.2), considering that the transistors switch instantaneously and are lossless, and taking into account (21.3), the switched model of the qZSI is obtained as,

$$L_1 \dot{i}_{L_1} = - r_1 i_{L_1} - S_T' v_{C_1} + S_T v_{C_2} + V_{in}, \tag{21.4}$$

$$L_2 \dot{i}_{L_2} = - r_2 i_{L_2} + S_T v_{C_1} - S_T' v_{C_2}, \tag{21.5}$$

$$C_1 \dot{v}_{C_1} = S_T' i_{L_1} - (S_a i_a + S_b i_b + S_c i_c) - S_T i_{L_2}, \tag{21.6}$$

$$C_2 \dot{v}_{C_2} = S_T' i_{L_2} - (S_a i_a + S_b i_b + S_c i_c) - S_T i_{L_1}, \tag{21.7}$$

$$L_g \dot{i}_a = - r_g i_a - e_{an} + \left(\frac{2S_a}{3} - \frac{S_b}{3} - \frac{S_c}{3} \right) v_{PN}, \tag{21.8}$$

$$L_g \dot{i}_b = - r_g i_b - e_{bn} + \left(\frac{2S_b}{3} - \frac{S_a}{3} - \frac{S_c}{3} \right) v_{PN}, \tag{21.9}$$

$$L_g \dot{i}_c = -r_g i_c - e_{cn} + \left(\frac{2S_c}{3} - \frac{S_a}{3} - \frac{S_b}{3}\right) v_{PN}. \tag{21.10}$$

where $S_T^{'} = 1 - S_T$, r_1 and r_2 are the internal resistances of inductors L_1 and L_2, and r_g is the internal resistance of the grid filter inductors.

21.2.2 Averaged Model

To design the control strategy proposed in this work, the switched variables of the model given by (21.4)–(21.10) must be averaged.

Initially, it is assumed that the voltage v_{pn} can be expressed as,

$$v_{pn} = v_{C_1} + v_{C_2}. \tag{21.11}$$

Thus, after averaging the model (21.4)–(21.10), substituting (21.11), and applying an amplitude-invariant dq transformation, the averaged model of the qZSI converter in dq coordinates is obtained as,

$$L_1 \dot{i}_{L_1} = -r_1 i_{L_1} - (1-D) v_{C_1} + D v_{C_2} + V_{in}, \tag{21.12}$$

$$L_2 \dot{i}_{L_2} = -r_2 i_{L_2} + D v_{C_1} - (1-D) v_{C_2}, \tag{21.13}$$

$$C_1 \dot{v}_{C_1} = (1-D) i_{L_1} - m_d i_d - m_q i_q - D i_{L_2}, \tag{21.14}$$

$$C_2 \dot{v}_{C_2} = (1-D) i_{L_2} - m_d i_d - m_q i_q - D i_{L_1}, \tag{21.15}$$

$$L_g \dot{i}_d = -r_g i_d + \omega_g L_g i_q + m_d (v_{C_1} + v_{C_2}) - e_d, \tag{21.16}$$

$$L_g \dot{i}_q = -r_g i_q - \omega_g L_g i_d + m_q (v_{C_1} + v_{C_2}) - e_q. \tag{21.17}$$

where i_d and i_q are the d and q components of the grid currents (i_a, i_b, and i_c), m_d and m_q are the modulation indexes of the switching signals in dq coordinates, and D is the modulation index of the Shoot-Through function, representing the time while the inverter remains in this state relative to a switching period, i.e., $D = \frac{T_{ST}}{T_s}$, where T_{ST} is the duration of the Shoot-Through state and T_s is the switching period, which can be expressed as $T_s = 1/f_s$, where f_s is the switching frequency.

21.2.3 pH Model for Grid-Connected qZSI

Next, the dq model of the three-phase qZSI connected to the grid, given by (21.12)–(21.17), will be represented in its pH form, given by,

$$\dot{\mathbf{x}} = [\mathbf{J}(\mathbf{u}) - \mathbf{R}(\mathbf{x})] \frac{\partial H}{\partial \mathbf{x}} + \boldsymbol{\zeta}. \tag{21.18}$$

where $\mathbf{x}$ is the state vector, $\mathbf{J}(\mathbf{u})$ is the interconnection matrix (antisymmetric), $\mathbf{R}(\mathbf{x})$ is the damping matrix (symmetric and positive semi-definite), H is the system's energy function, and $\boldsymbol{\zeta}$ is a vector of external sources, which are given by,

$$\mathbf{x} = \begin{bmatrix} L_1 i_{L_1} \\ L_2 i_{L_2} \\ C_1 v_{C_1} \\ C_2 v_{C_2} \\ L_g i_d \\ L_g i_q \end{bmatrix}, \ \mathbf{R}(\mathbf{x}) = \begin{bmatrix} r_1 & 0 & 0 & 0 & 0 & 0 \\ 0 & r_2 & 0 & 0 & 0 & 0 \\ 0 & 0 & 0 & 0 & 0 & 0 \\ 0 & 0 & 0 & 0 & 0 & 0 \\ 0 & 0 & 0 & 0 & r_g & 0 \\ 0 & 0 & 0 & 0 & 0 & r_g \end{bmatrix}, \ \boldsymbol{\zeta} = \begin{bmatrix} V_{in} & 0 & 0 & 0 & -e_d & -e_q \end{bmatrix}^T,$$

$$\mathbf{J}(\mathbf{u}) = \begin{bmatrix} 0 & 0 & -D' & D & 0 & 0 \\ 0 & 0 & D & -D' & 0 & 0 \\ D' & -D & 0 & 0 & -m_d & -m_q \\ -D & D' & 0 & 0 & -m_d & -m_q \\ 0 & 0 & m_d & m_d & 0 & \omega_g L_g \\ 0 & 0 & m_q & m_q & -\omega_g L_g & 0 \end{bmatrix}. \tag{21.19}$$

where $D' = (1 - D)$.

21.3 Control Strategy

In the following section, the proposed control strategy in this work is presented, with design objectives focused on regulating the DC bus voltage (v_{pn}) and the active and reactive powers (p and q) to a specified reference value.

The proposed control strategy is based on IDA-PBC [16–18]. This technique involves assigning the desired closed-loop dynamics by setting the matrices $\mathbf{R}_a(\mathbf{x})$, $\mathbf{J}_a(\mathbf{x})$, and the energy functions $H_a(\mathbf{x})$, and solving the expression (21.20),

$$[\mathbf{J}(\mathbf{u}) - \mathbf{R}(\mathbf{x})]\frac{\partial H_a}{\partial \mathbf{x}} = -[\mathbf{J}_a(\mathbf{u}) - \mathbf{R}_a(\mathbf{x})]\frac{\partial H_d}{\partial \mathbf{x}} + \boldsymbol{\zeta}, \tag{21.20}$$

where,

$$\frac{\partial H_a}{\partial \mathbf{x}} = \begin{bmatrix} -i_{L_1}^* & -i_{L_2}^* & -v_{C_1}^* & -v_{C_2}^* & -i_d^* & -i_q^* \end{bmatrix}^T, \ \frac{\partial H_d}{\partial \mathbf{x}} = \begin{bmatrix} \epsilon_1 & \epsilon_2 & \epsilon_3 & \epsilon_4 & \epsilon_5 & \epsilon_6 \end{bmatrix}^T,$$

$$\mathbf{J}_a(\mathbf{x}) - \mathbf{R}_a(\mathbf{x}) = \begin{bmatrix} 0 & 0 & -D' & D & 0 & 0 \\ 0 & 0 & D & -D' & 0 & 0 \\ D' & -D & R_1 & 0 & -m_d & -m_q \\ -D & D' & 0 & 0 & -m_d & -m_q \\ 0 & 0 & m_d & m_d & R_2 & -\omega_g L_g \\ 0 & 0 & m_q & m_q & \omega_g L_g & R_3 \end{bmatrix}, \tag{21.21}$$

with R_1, R_2, and R_3 being the damping constants assigned through $\mathbf{R}_a(\mathbf{x})$ to achieve the desired dynamic response, while $\epsilon_1 = i_{L_1} - i^*_{L_1}$, $\epsilon_2 = i_{L_2} - i^*_{L_2}$, $\epsilon_3 = v_{C_1} - v^*_{C_1}$, $\epsilon_4 = v_{C_2} - v^*_{C_2}$, $\epsilon_5 = i_d - i^*_d$, and $\epsilon_6 = i_q - i^*_q$ represent the error functions.

The reference values are obtained by evaluating the model given by (21.12)–(21.17) in steady state, considering that the inductor resistances are negligible ($r_1 = r_2 = r_g \simeq 0$), resulting,

$$
\begin{aligned}
D^* &= \frac{1}{2}\left(1 - \frac{V_{in}}{v^*_{pn}}\right), \quad v^*_{C_1} = \left(\frac{1-D^*}{1-2D^*}\right)V_{in}, \quad v^*_{C_2} = \left(\frac{D^*}{1-2D^*}\right)V_{in}, \\
m^*_d &= \frac{1}{v^*_{C_1} + v^*_{C_2}}\left(-\omega_g L_g i^*_q + e_d\right), \quad m^*_q = \frac{1}{v^*_{C_1} + v^*_{C_2}}\left(\omega_g L_g i^*_d + e_q\right), \\
i^*_{L_1} &= \frac{1}{1-2D^*}\left(m^*_d i^*_d + m^*_q i^*_q\right), \quad i^*_{L_2} = \frac{1}{D^*}\left((1-D^*)i^*_{L_1} - m^*_d i^*_d - m^*_q i^*_q\right).
\end{aligned}
\tag{21.22}
$$

Given that the active and reactive power in dq coordinates are expressed as,

$$p^* = e_d i^*_d + e_q i^*_q, \tag{21.23}$$

$$q^* = e_q i^*_d - e_d i^*_q, \tag{21.24}$$

The references (i^*_d and i^*_q) can be obtained from (21.23)–(21.24) considering that the grid voltages are balanced ($e_q = 0$),

$$i^*_d = \frac{p^*}{e_d}, \quad i^*_q = -\frac{q^*}{e_d}. \tag{21.25}$$

where p^* and q^* denote the active and reactive power references that need to be injected into the grid.

On the other hand, from (21.11) and knowing that $v_{C_2} = v_{C_1} - V_{in}$, it can be obtained,

$$v_{C_1} = \frac{v_{pn} + V_{in}}{2}. \tag{21.26}$$

Therefore, by using (21.26), and substituting (21.19) and (21.21) into (21.20), and taking the third, fifth, and sixth rows, the control actions are obtained,

$$D = \frac{1}{i^*_{L_1} + i^*_{L_2}}\left(i^*_{L_1} - R_1\epsilon_3 - m_d i^*_d - m_q i^*_q\right), \tag{21.27}$$

$$m_d = \frac{1}{2v^*_{C_1} - V_{in}}\left(-\omega_g L_g i^*_q - R_2\epsilon_5 + r_g i^*_d + e_d\right), \tag{21.28}$$

$$m_q = \frac{1}{2v^*_{C_1} - V_{in}}\left(\omega_g L_g i^*_d - R_3\epsilon_6 + r_g i^*_q + e_q\right). \tag{21.29}$$

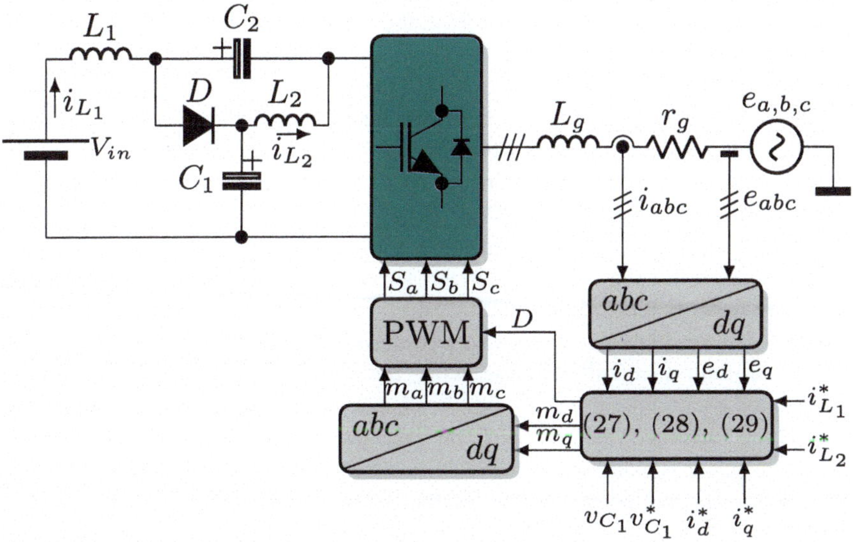

Fig. 21.3 Proposed control scheme

Figure 21.3 shows the control scheme proposed in this work, which consists of a coordinate transformation block from a, b, c to dq coordinates, from which the currents i_d and i_q and the grid voltages e_d and e_q are obtained. These are used to derive the control actions given by (21.27)–(21.29), which are then applied to a coordinate transformation block from dq to a, b, c coordinates to obtain the modulation indices m_a, m_b, and m_c, from which the PWM signals (S_a, S_b, and S_c) are generated to drive the transistors of the three-phase inverter. Note that the control laws only require the measurement of grid currents and voltages, and the voltage v_{C_1}.

21.4 Simulation Results

To validate the performance of the proposed control strategy, two simulation tests of 100ms duration were conducted on the converter shown in Fig. 21.1 using the control strategy depicted in Fig. 21.3, and employing Matlab Simulink with a simulation step of $T = 0.1\mu$s. The parameters of the converter used are listed in Table 21.1, while

Table 21.1 Converter parameters

V_{in}	v_{pn}^*	$L_{1,2}$	$r_{1,2}$	$C_{1,2}$	L_g	r_g	f_s
250 V	400 V	0.25 mH	0.5 Ω	460 μF	2 mH	0.6 Ω	20 kHz

the controller parameters were set to $R_1 = 0.02$, $R_2 = 0.3$ and $R_3 = 0.3$, aiming to achieve a settling time of 2ms for the voltage v_{pn} and 20ms for the grid currents. In both tests, an effective phase voltage of 110V was used.

The initial test begins with a 400 V reference for the DC bus voltage, a 1 kW active power reference, and 0 VAr for the reactive power reference. At time $t = 20$ ms, the active power reference is increased to 1.5 kW. Finally, at $t = 60$ ms, the reactive power is increased to 250 VAr.

In Fig. 21.4, the voltages, currents, and powers in the grid are shown. It can be observed that, following the change in the active power reference, the grid currents reach their new reference value within 20 ms (see Fig. 21.4b and c), with a response that shows no oscillations. The active power also reaches its new reference value after 20 ms with a smooth response, while the reactive power exhibits a transient with a small oscillation in response to the change in the active power reference but returns to its reference value within 20 ms, as shown in Fig. 21.4d. When the change in reactive power is made at $t = 60$ ms, the grid currents again reach their reference value, presenting a response without oscillations or overshoot (see Fig. 21.4b and c). The active power shows a small transient but stabilizes at its reference value after

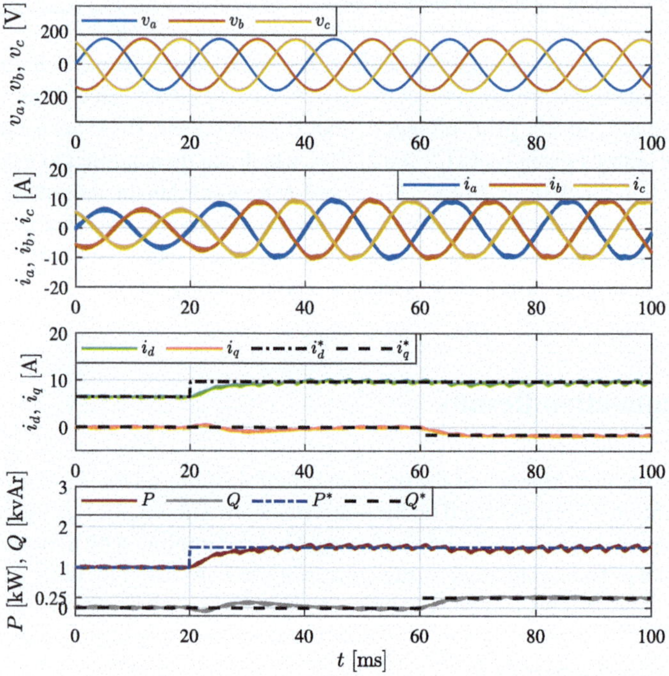

Fig. 21.4 AC voltages and currents for the qZSI during two changes in the power references. **a** Grid voltages in *abc* coordinates. **b** Grid currents in *abc* coordinates. **c** Grid currents in *dq* coordinates. **d** Active and reactive power

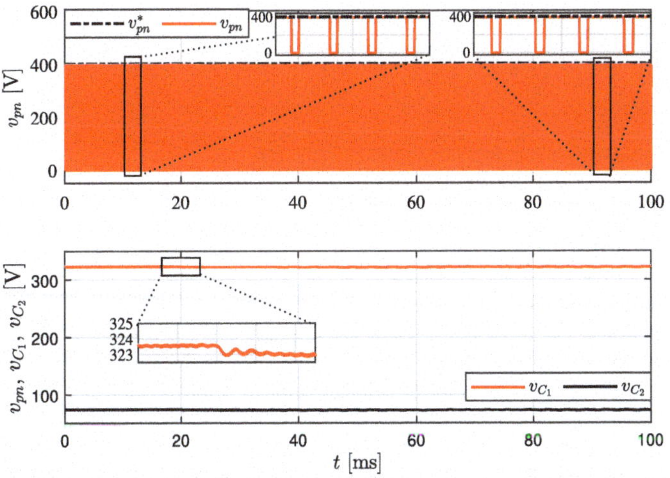

Fig. 21.5 DC-side voltages for the qZSI in response to a change in power reference. Top: DC bus voltage. Bottom: Voltage across both capacitors

20 ms, and the reactive power exhibits a response without oscillations or overshoot, stabilizing at its new reference value within the required settling time (see Fig. 21.4d).

In Fig. 21.5, the DC bus voltage (v_{pn}) and the voltages across both capacitors are shown. It can be observed that, before the first reference change, the voltage v_{pn} (see Fig. 21.5 (top)) is stable at its reference value. After the change in active power, it exhibits a small transient but returns to its reference value with a slight steady-state error, similar to the response observed following the change in reactive power. In Fig. 21.5 (bottom), the voltages across both capacitors of the quasi-impedance network are shown. Like the DC bus voltage, these voltages exhibit a transient response without jitter and a small steady-state error after the changes in active and reactive power references, respectively.

The second test involves a change in the DC bus voltage reference from 400 V to 420 V at $t = 40$ ms, while keeping the active and reactive power references at 1 kW and 0k VAr, respectively.

In Fig. 21.6b, it can be seen that following the change in the DC bus voltage reference, the currents exhibit a transient response with overshoot, and they stabilize back to their reference value approximately after 20 ms. Additionally, the dq-coordinate components of the grid currents also show a transient response with overshoot and slight oscillation, although they return to their reference values (see Fig. 21.6c). Furthermore, Fig. 21.6d shows the active and reactive powers injected into the grid. Both exhibit a transient response with overshoot and a small oscillation due to the change in the voltage reference, but both stabilize back to their reference values.

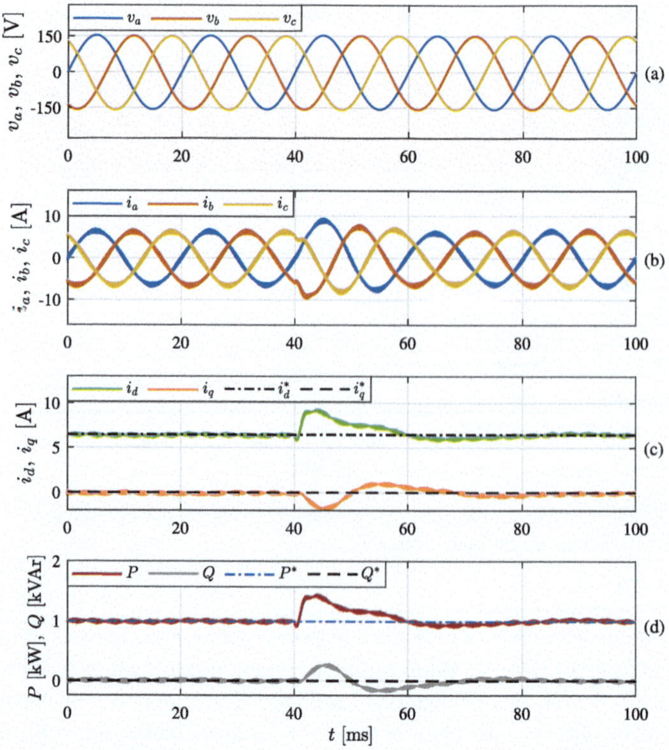

Fig. 21.6 AC side voltages and currents for the qZSI in response to a change in the DC bus voltage reference. **a** Grid voltages in *abc* coordinates. **b** Grid currents in *abc* coordinates. **c** Grid currents in *dq* coordinates. **d** Active and reactive power

Finally, in Fig. 21.7, the DC bus voltage and the voltage across both capacitors of the quasi-impedance network are shown. It can be observed that following the change in reference, the DC bus voltage (see Fig. 21.7 (top)) exhibits a transient response with overshoot and slight oscillation, and stabilizes at its new reference value after 2 ms. Similarly, the voltages across both capacitors also show a transient response with overshoot (see Fig. 21.7 (bottom)).

21.5 Conclusion

This paper proposes the design of a new control strategy for a three-phase qZSI connected to the grid through an inductive filter. The system is modeled in dq coordinates, which simplifies the controller design. Given the nonlinear nature of the system, a passivity-based controller is proposed using interconnection and damping assignment, achieving a globally stable controller with good dynamic performance

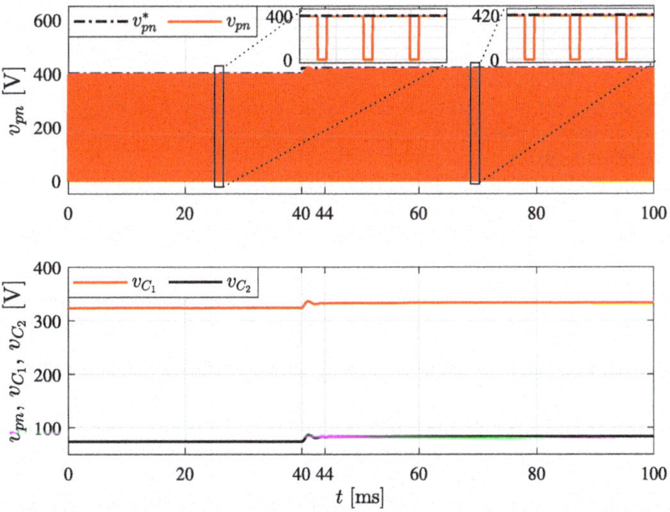

Fig. 21.7 DC side voltages for the qZSI in response to a change in the DC bus voltage reference. Top: DC bus voltage. Bottom: Voltage across both capacitors

in response to changes in the power reference injected to the grid, and in the DC bus voltage. By appropriately selecting the assigned interconnection matrix, the measurement of currents and the voltage across one of the capacitors in the DC stage is avoided. The control strategy performance was validated through simulation results, proving that the DC voltage, as well as the AC currents and powers, exhibit transient behavior without overshoot during changes in the active power reference and small oscillations when the DC bus voltage reference is altered, although in both scenarios, they eventually return to their reference values.

Acknowledgements The authors acknowledge the support provided by the Thematic Network 723RT0150 “Red para la integración a gran escala de energías renovables en sistemas eléctricos (RIBIERSE-CYTED)” financed by the call for Thematic Networks of the CYTED (Ibero-American Program of Science and Technology for Development) for 2022.

Disclosure of Interests The authors have no competing interests to declare that are relevant to the content of this article.

References

1. A. Rahman, O. Farrok, M.M. Haque, Environmental impact of renewable energy source based electrical power plants: solar, wind, hydroelectric, biomass, geothermal, tidal, ocean, and osmotic. Renew. Sustain. Energy Rev. **161**, 112279 (2022)
2. Q. Hassan, S. Algburi, A.Z. Sameen, H.M. Salman, M. Jaszczur, A review of hybrid renewable energy systems: Solar and wind-powered solutions: Challenges, opportunities, and policy implications. Results Eng. **20**, 101621 (12 2023)

3. R. Ufa, Y. Malkova, V. Rudnik, M. Andreev, V. Borisov, A review on distributed generation impacts on electric power system. Int. J. Hydrogen Energy **47**, 20347–20361 (2022)
4. M.E.T.S. Junior, L.C.G. Freitas, Power electronics for modern sustainable power systems: distributed generation, microgrids and smart grids–a review. Sustainability **14**, 3597 (2022)
5. M. Alharbi, Control approach of grid-connected PV inverter under unbalanced grid conditions. Processes **12**, 212 (2024)
6. S. Dedeoglu, G.C. Konstantopoulos, *Three-Phase Grid-Connected Inverters Equipped with Nonlinear Current-Limiting Control* (IEEE, 2018), pp. 38–43
7. Y. Feng, B. Liu, F. Han, *Terminal Sliding Mode Control Approach of DC/AC Voltage Source Inverters in Balanced 3-Phase 3-Wire Grid* (IEEE 2023), pp. 1–5
8. D. Mathew, R.C. Naidu, A review on single-phase boost inverter technology for low power grid integrated solar PV applications. Ain Shams Eng. J. **15**, 102365 (2024)
9. A. Colmenar-Santos, A.R. Linares-Mena, J.F. Velázquez, D. Borge-Diez, Energy-efficient three-phase bidirectional converter for grid-connected storage applications. Energy Convers. Manage. **127**, 599–611 (2016)
10. H. Heydari-doostabad, M. Pourmahdi, M. Jafarian, A. Keane, T. O'Donnell, Three-switch common ground step-down and step-up single-stage grid-connected PV inverter. IEEE Trans. Power Electron. **37**, 7577–7589 (2022)
11. F. Peng, *Z-Source Inverter* (IEEE), pp. 775–781
12. E.P.P. Soares-Ramos, L.D. Oliveira-Assis, R. Sarrias-Mena, P. Garcia-Trivino, C.A. Garcia-Vazquez, L.M. Fernandez-Ramirez, Averaged dynamic modeling and control of a quasi-Z-source inverter for wind power applications. IEEE Access **9**, 114348–114358 (2021)
13. C.R. Baier, F.A. Villarroel, M.A. Torres, M.A. Perez, J.C. Hernandez, E.E. Espinosa, A predictive control scheme for a single-phase grid-supporting quasi-Z-source inverter and its integration with a frequency support strategy. IEEE Access **11**, 5337–5351 (2023)
14. K. Mokhtari, S. Shokri-Kojori, M. Aliyari-Shoorehdeli, A new generalized state-space averaged model, control design and stability analysis for three phase grid-connected quasi-Z-Source inverters. Int. J. Electr. Power Energy Sys. **158**, 109932 (2024)
15. D. Ma, Z. Cai, R. Wang, Q. Sun, P. Wang, P.: Energy shaping controller design of three-phase quasi-Z-source inverter for grid-tie. IET Power Electron. **13**, 3601–3612 (12 2020)
16. F.M. Serra, C.H.D. Angelo, D.G. Forchetti, Interconnection and damping assignment control of a three-phase front end converter. Int. J. Electrical Power and Energy Systems 60, 317–324 (2014). https://doi.org/10.1016/j.ijepes.2014.03.033
17. S. Barman, S. Samanta, J.P. Mishra, P. Roy, B.K. Roy, Design and Implementation of an IDA-PBC for a grid connected inverter used in a photovoltaic system. IFAC-PapersOnLine **51**, 680–685 (2018)
18. F.M. Serra, C.H. De Angelo, Ida-pbc controller design for grid connected front end converters under non-ideal grid conditions. Electric Power Syst. Res. **142**, 12–19 (2017). https://www.sciencedirect.com/science/article/pii/S0378779616303455

Chapter 22
Optimization of a Fixed-Focus Scheffler Dish with Dual-Axis Tracking System for Solar Concentration in Northwestern Mexico

S. Islas-Pereda, N. Velázquez-Limón, J. A. Aguilar-Jiménez, J. Ríos-Arriola, J. D. Gámez-Avilez, and N. A. Pérez de Alva

Abstract This study focuses on optimizing a fixed-focus Scheffler dish solar concentrator with a dual-axis tracking system, specifically designed for the northwestern region of Mexico. The objective is to improve the performance of the system and operating hours by using an opto-geometric analysis and a mathematical model that links key design parameters, such as the interception angle and dish diameter, to the overall efficiency of the system. Through simulations and empirical evaluations, the impact of the solar declination angle and hour angle on solar tracking and concentration was analyzed. The dual-axis tracking system implemented allows precise adjustment to the seasonal sun and daily variations, thereby maximizing energy capture throughout the year. The results indicate that, in the Mexicali region, a system with a 60° interception angle achieves the best annual performance, capturing up to 1008.4 kWh and operating for over 3300 h per year. The conclusions highlight the importance of an optimized design that balances cosine losses and operating hours, maximizing system efficiency in a region with high solar irradiance, and emphasizing its potential application in the local industry and thermal energy generation.

Keywords Fixed focus · Scheffler dish · Solar concentration · Dual-axis tracking system

S. Islas-Pereda (✉) · N. Velázquez-Limón · J. A. Aguilar-Jiménez · J. Ríos-Arriola · J. D. Gámez-Avilez · N. A. P. de Alva
Centro de Estudios de Las Energías Renovables, Instituto de Ingeniería, Universidad Autónoma de Baja California, Mexicali, BC, México
e-mail: islass@uabc.edu.mx

D. Rossit et al. (eds.), *Proceedings of the VII Ibero-American Congress of Smart Cities, ICSC-Cities 2024, 12-14 November, San Carlos, Costa Rica*, Springer Proceedings in Energy, https://doi.org/10.1007/978-981-96-4301-1_22

Nomenclatures

Symbol Description

P	Focal distance, meters
ξ_w	Hourly tracking angle of the tracking system, degrees
ξ_d	Solar altitude tracking angle of the tracking system, degrees
θ_{int}	Interception angle of the dish, degrees
D_{dish}	Diameter of the dish, meters
D_{AP}	Aperture area diameter, meters
ω	Hour angle, degrees
h	Time in 24-h format, hours
day	Julian day
δ	Declination of the Earth, degrees
α_{sun}	Solar altitude, degrees
ψ_{sun}	Azimuth angle, degrees
α_{mec}	Target angle for solar altitude tracking, degrees
ψ_{mec}	Target angle for azimuth tracking, degrees
$\%A_{ef}$	Effective area, m^2
$Disc_{ratio}$	Radius ratio
W_n	Solar image diameter, meters
C_C	Concentration ratio
ε	Solar subtended angle, degrees
Z	Spatial dimension for altitude, meters
Y	Spatial dimension North–South, meters
X	Spatial dimension East–West, meters

22.1 Introduction

Solar energy in thermal systems is increasingly essential within renewable energy, significantly reducing environmental impacts linked to fossil fuel use. Particularly, solar thermal energy offers a feasible alternative to conventional fuels in electricity generation and food processing, sectors critical for rural communities distant from power grids, where energy scarcity worsens poverty conditions.

Solar thermal technologies fall into two primary categories: non-concentrating and concentrating. Non-concentrating technologies, such as evacuated tube heaters and flat plate collectors, are generally used for water heating and do not require precise tracking systems. Conversely, concentrating technologies including parabolic troughs, compound parabolic concentrators (CPC), parabolic dishes, and Scheffler dishes—depend on solar tracking systems with one or two axes to achieve optimal performance, reaching temperatures above 200 °C. This temperature range enables diverse applications from small-scale cooking and heating to large-scale electricity generation, water desalination, and refrigeration.

Numerous initiatives leverage Scheffler-type dishes for solar concentration. For electricity generation, Dish-Stirling systems are particularly promising. Kleih [1] demonstrated a Dish-Stirling system at the Centre for Solar Energy and Hydrogen Research in Stuttgart with power outputs between 5 and 25 kW, which were affected by solar radiation variations and tracking issues. In Thailand, Snidvongs [2] installed a 10 kW plant under challenging conditions with high humidity and 500 W/m^2 average irradiance. At the regional level, projects like that by Alarcón et al. [3] in Colombia employ optimized parabolic dishes to supply electricity or gas to rural areas with restricted access, showcasing the potential of solar concentrators to enhance living conditions in underserved areas.

The global energy deficit has spurred the development of technologies like solar cookers. Balzar and Stumpf [4] created a solar cooker using vacuum tubes and heat pipes, achieving oil temperatures of up to 252 °C. Similarly, John [5] developed a solar cooker in Africa based on a parabolic dish, reducing fossil fuel dependence and combating deforestation. These devices demonstrated high efficiency, as shown by boiling 3 kg of water, underscoring the value of parabolic dishes and tracking systems for effective operation.

In addition to cooking, solar water heating is gaining attention across various sectors. Arulkumaran [6] conducted experiments with a parabolic dish to generate steam, reaching temperatures up to 215 °C for rural needs. Ladan Mohammed [7] designed a domestic water heating system that achieved up to 100 °C, incorporating a linear actuator for autonomous tracking. Srithar [8] later created an autonomous triple-basin desalination system that reached 85 °C and produced a distillation rate of 16.94 kg/m^2, evidencing solar efficacy of concentration for water desalination.

Further Scheffler dish applications exist in tracking and non-tracking systems. Ruelas et al. [9] developed ray tracing software to optimize a Scheffler concentrator for a Stirling engine, enhancing heat transfer with elliptical geometries. Rapp and Schwartz [10] constructed a 2 m^2 Scheffler dish without tracking to lower costs and examine thermal storage feasibility. Later, Dafle [11] developed a 16 m^2 mono-axial system for water heating, while Munir et al. [12] created an 8 m^2 fixed-focus Scheffler system with a 90° interception angle and solar declination tracking, maximizing solar concentration yearly.

Various authors recognize the need for Scheffler dishes with fixed focal points and annual positional adjustments to maintain optimal solar concentration, a subject addressed in this mathematical and operational study to optimize system performance.

22.2 System Description

22.2.1 Study Location

The present study was conducted in the northwestern region of Mexico, specifically in the city of Mexicali, Baja California. This area has a hot desert climate (BWh) according to the Köppen climate classification, extremely hot summers and mild winters, making Mexicali a favorable location for the implementation of solar concentration technologies. The geographical of the city location, with a latitude of 32.66° N and a longitude of 115.47° W, allows it to receive high levels of solar radiation throughout most of the year. According to historical records, the average solar energy in the region ranges between 5 and 6 $kWh/m^2/day$, peaking during the summer months. These conditions make Mexicali an ideal site for the installation of solar concentration systems.

22.2.2 Physical and Operational Description of the Scheffler Solar Concentration System

The mechanical limits of the movement axes in the tracking subsystem of a fixed-focus Scheffler dish system were evaluated to ensure its proper functioning. The system includes two main movement axes that allow the aperture plane to be oriented perpendicular to the incident solar rays, maximizing concentration and preventing the solar image from becoming defocused.

The first axis, denoted as ξ_D, is responsible for adjusting the solar declination or the altitude of the sun, compensating for variations in the solar trajectory throughout the year. This angle is zero when the interception angle θ_{int} is perpendicular to the installation surface. Its magnitude varies, as shown in Fig. 1a, between $-90°$ and 90°, depending on the season and the position of the sun. The second axis, ξ_w, is parallel to the interception axis θ_{int} and follows the solar hour angle ω, allowing solar tracking in its daily path, as illustrated in Fig. 1b.

When the system is oriented along the North–South axis, it follows the trajectory of the sun to focus the radiation on a fixed focal point. This design ensures that the solar image is consistently projected onto a fixed position. However, the movement of the system depends on both the apparent trajectory of the sun, which varies according to the latitude of the site φ, and the mechanical limits of the system relative to the interception angle θ_{int}. The following mathematical analysis describes the relationship between these parameters for the optimal design of the tracking system.

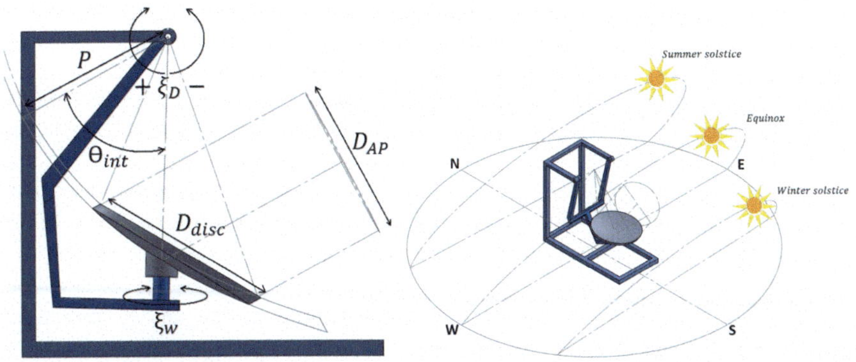

Fig. 22.1 Disc-Scheffler fixed focus concentrator system: **a** Season-hour tracking axis, **b** system on *N-S* axis

22.3 Mathematical Model

The mathematical model was developed with the objective of identifying and correcting potential inconsistencies in the movements of the tracking axes of the fixed-focus Scheffler dish system. To achieve this, equations were formulated to describe the solar altitude and azimuth angle as functions of various parameters. These parameters allow for the evaluation and analysis of the of the system behavior throughout the year, taking into account the specific geographical location. This model aims to provide a more precise understanding of how the angles and positions of the dish vary to maximize solar capture, thereby optimizing the overall performance of the tracking system during the entire annual cycle. This model, used by Duffie and Beckmann [13], allows for the estimation of the trajectory of the sun throughout the day and year for application in solar concentration systems.

22.3.1 Opto-Geometric Parameters

In selecting the interception angle (θ_{int}), it is crucial to account for cosine losses, which are calculated by the following ratio:

$$\% A_{ef} = \frac{D_{AP}}{D_{disc}} \tag{22.1}$$

where the effective area ($\%A_{ef}$) estimates the proportion of material used in the construction of the dish relative to the aperture area. This ratio takes a value of 1 when the plane of the dish is completely parallel to the aperture plane and 0 when the two planes are orthogonal.

22.3.2 Movement of the Tracking Axes

As explained previously in the system description section, the system features two movement axes: the declination axis (solar altitude) and the hour angle axis. These axes allow for adjusting the position of the Scheffler dish to maximize solar radiation capture throughout the day. The system is modeled within three-dimensional space in the *xyz* octant, with the objective of identifying the relationship between the mechanical angles of the axes and the angles that describe the trajectory of the sun throughout the day and year.

In the *YZ* plane (Fig. 22.2), the trajectory of the aperture area can be represented by a straight line in the form $z = my + b$. In this case, Z_0 corresponds to the minimum position when $\omega = 0$, projected onto Y_0. On the other hand, Z_m is obtained when $\omega = \pm 90^\circ$, projected onto Y_m, while Z_1 is generated when $\omega = \pm 180^\circ$. From these positions, the following equations are derived as functions of the declination angle (ξ_D) and the interception angle (θ_{int}).

$$Z_0 = \sin(\theta_{int} - 90 - \xi_D) \tag{22.2}$$

$$Y_0 = \cos(\theta_{int} - 90 - \xi_D) \tag{22.3}$$

$$Z_1 = \sin(\theta_{int} - 90 + \xi_D) \tag{22.4}$$

$$Y_1 = \cos(\theta_{int} - 90 + \xi_D) \tag{22.5}$$

$$Y_m = \frac{Z_m - b}{m} \tag{22.6}$$

The solar image projected onto the *YZ* plane is crucial for precisely adjusting the tracking system. By projecting these equations onto the hour angle axis (ω), it is possible to determine how the orientation of the Scheffler dish varies throughout the day. In this way, the concentration of radiation on the fixed receiver is optimized, ensuring that solar energy is efficiently harnessed under different solar conditions.

The resulting value of Y_m is crucial for the analysis of the azimuthal angle, thus requiring detailed attention. Given that the radius in this model is unitary, the solar altitude is determined solely by the magnitude of Z. Solving the equation for Z in terms of its projections onto Y and using the trigonometric relationship with the hour angle (ω), the mechanical solar altitude (α_{mec}) can be expressed as follows:

$$\alpha_{mec}(\pm\omega) = \pm \arcsin([Y1 + Y_m(1 + \cos \xi_w)]m + b) \tag{22.7}$$

When analyzing the *XY* plane to determine the azimuthal angle, it can be observed that the circular motion described by the hour angle axis (ω) is projected as an

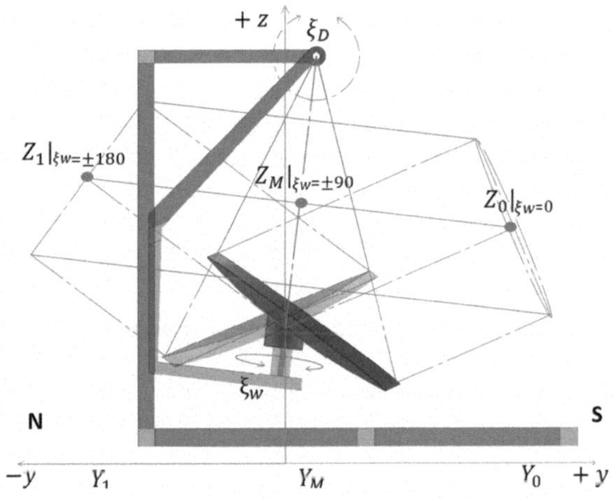

Fig. 22.2 Plane *YZ* of the Disc-Scheffler system

ellipse whenever $\xi_D \neq 0$. This motion will only be circular when $\xi_D = 0$, meaning that no additional adjustments are required to compensate for solar declination. This elliptical projection introduces additional considerations in the design of the system, which must be taken into account to ensure optimal solar radiation concentration and system efficiency throughout the year.

22.4 Methodology

Due to the complexity of the tracking mechanism, the mathematical model cannot be solved using simple algebraic methods. Therefore, to determine the angles ξ_D and ξ_w based on the values of α_{sun} and ψ_{sun}, numerical methods were employed. These methods ensure that $\alpha_{sun} = \alpha_{mec}$ and $\psi_{sun} = \psi_{mec}$, guaranteeing accurate solar tracking.

It is important to note that this study utilized a system with a focal length of 1 m, and all the tested disks had a diameter of 1 m. However, the aperture area of each disk varies depending on the interception angle (θ_{int}). To further clarify how variable sweeps and explorations were conducted, the resulting mathematical model is illustrated in the flowchart shown in Fig. 22.3.

22.5 Results and Discussion

In the city of Mexicali, Baja California, Mexico, solar irradiance values reach up to 905 W/m^2 between May and July, making this region an excellent location for photovoltaic and solar thermal applications. However, as shown in Fig. 4a, the most frequent irradiance values in this area range around 600 W/m^2. This solar resource

Fig. 22.3 Diagram flow of parameters determination

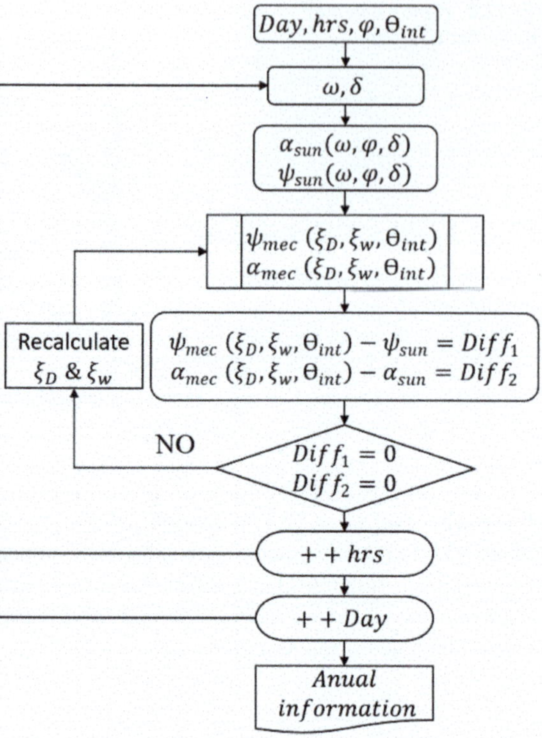

profile was derived from meteorological data provided by the Autonomous University of Baja California, Mexicali Campus. These data were used to implement the optimization methodology for the fixed-focus Scheffler dish concentrator, tailoring its design to the specific conditions of this region.

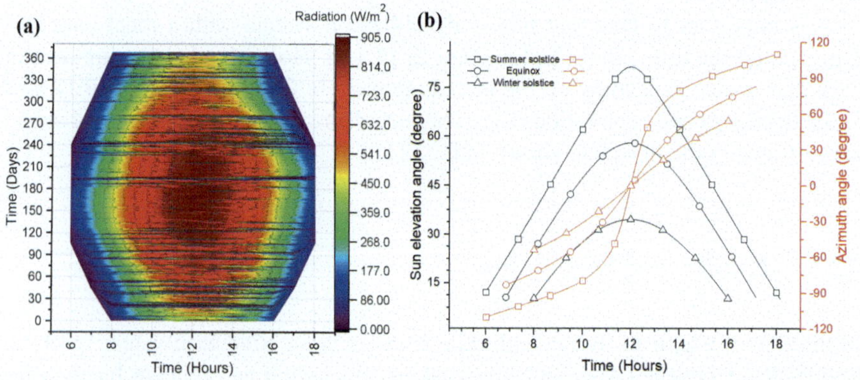

Fig. 22.4 Location description parameter: **a** Direct irradiation. **b** Angles of solar trajectory in Mexicali B.C

Since the tracking mechanism is intrinsically linked to the movement required to keep the solar image focused on the fixed receiver, it is essential to analyze the apparent trajectories of the Sun in the region of interest. Solar altitude and azimuthal angles will dictate the key parameters for the tracking design of the mechanism. Figure 4b presents the angles of the apparent trajectory of the Sun for the region of interest. Mexicali, located at approximately 32.62° N latitude, has a maximum solar elevation of 81.45°, which occurs during the summer solstice.

In contrast, the minimum solar elevation during the winter solstice is 34.45°, reached at 12:00 solar time. Additionally, at noon, the average solar elevation is approximately 58°, which corresponds to the complementary angle of the latitude of the site.

The analysis of the azimuthal angle shows that, in winter, the Sun rises in the southeast and sets in the southwest, providing days with no more than 8 h of sunlight. In summer, however, the Sun rises in the northeast and sets in the northwest, resulting in days with more than 14 h of daylight. It is important to note that this study was limited to analyzing solar trajectories with a minimum elevation of 10° to ensure the accuracy of the tracking system design.

22.5.1 *Equinox: Behavior of the Fixed-Focus System*

As previously mentioned, the tracking mechanism is governed strictly by the angular movements of ξ_D and ξ_w, ensuring that the normal line to the Scheffler aperture of the dish is directed towards the solar angles α_{sun} and ψ_{sun}, thereby achieving an incidence angle close to or equal to 0°. Figure 22.5 shows the tracking angles of the system and their movement during the equinox, based on the selected dish cut (depending on interception angles of 45°, 60°, 75°, or 90°).

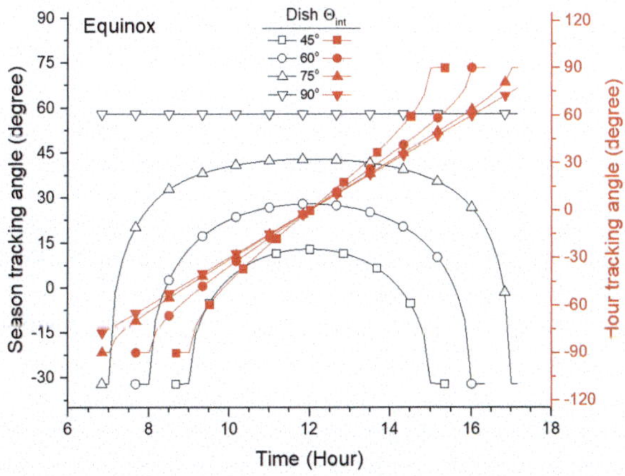

Fig. 22.5 Season and hour tracking movement in Equinox

Fig. 22.6 Season and hour tracking angular speed in Equinox

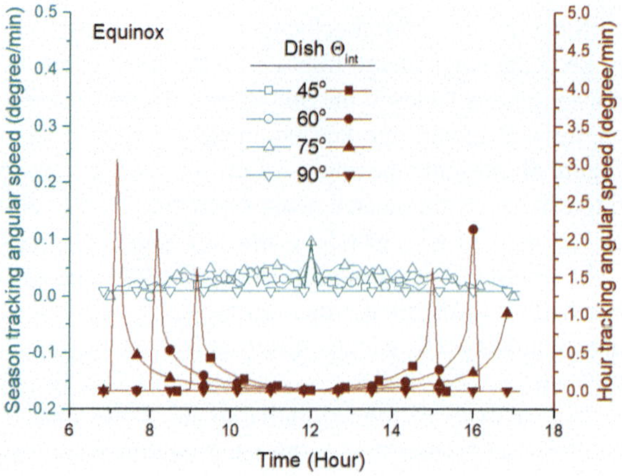

It is observed that for the dish with a 90° interception angle, the seasonal tracking angle (ξ_D) remains fixed at 58°. This occurs because, at this time of the year, the tracking mechanism reflects the orthogonality of the movement axes due to the 90° interception angle. Conversely, for the other dishes shown in the Fig. 22.5, a dependency between ξ_D and ξ_w is evident, indicating that orthogonality between the movement axes is not achieved in any possible trajectory.

On the other hand, not only is it important to analyze the possible trajectories of this tracking mechanism, but also to study the angular velocities of ξ_D and ξ_w, as shown in Fig. 22.6. The analysis of these angular velocities allows us to assess whether the selection of transmissions and/or motors for solar tracking is appropriate. It is observed that, for seasonal tracking, velocities greater than 0.1° per minute are not required. Furthermore, the maximum velocity occurs at 12:00 solar time.

In contrast, hourly tracking (ξ_w) requires speeds of up to 3° per minute, which occur at sunrise and sunset. This is because when $\xi w = \pm 90^{\circ}$, the maximum solar altitude that can be achieved is $\alpha_{mec} = 90^{\circ} - \theta_{int}$, potentially generating singularities or points of non-convergence in the system.

22.5.2 *Summer Solstice: Behavior of the Fixed-Focus System*

As shown in Fig. 22.7, during the summer solstice, the apparent trajectory of the sun reaches its highest altitude of the year, setting the upper limit for the declination axis angle (ξ_D) of the system. Similarly, it is observed that with an interception angle (θ_{int}) of 45°, fewer hours of full operation are obtained approximately 6 useful hours, assuming clear skies. This occurs because, during this time of year, the Sun does not rise exactly in the East nor set exactly in the West, which requires the declination axis angle to oscillate at values different from 58° for a system with $\theta_{int} = 90^{\circ}$, allowing the system to operate for a total of 11 h.

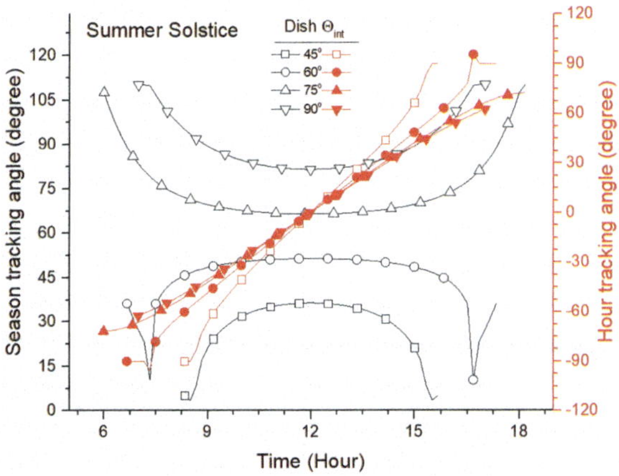

Fig. 22.7 Season and hour tracking movement in summer solstice

This example clearly demonstrates the necessity of coordinated movement between both axes of the tracking system, as the vector established between the object and the Sun, with components in α_{sun} and ψ_{sun}, must be followed closely.

The system with an interception angle of 45° provides approximately 7 h of operation, despite the fact that there are more than 14 h of daylight in the summer. Therefore, the importance of correctly selecting the Scheffler dish to be implemented is reiterated. It is also important to note that there is a trade-off between the selected interception angle and the effectiveness of the reflective surface. In other words, system performance is not only determined by the hours of operation but also by how much energy is effectively intercepted and reflected towards the fixed-focus receiver area.

In Fig. 22.8, the angular velocities of the seasonal and hourly tracking axes are presented. It can be observed that around 12:00, the angular velocities of the seasonal tracking axis (ξ_D) do not exceed 0.12° per minute, a behavior very similar to that observed during the equinox. However, during sunrise and sunset, the velocities reach more than 0.3° per minute. Given this, it is recommended to reduce the operational time of the system by a few minutes each day to avoid these high-velocity positions. Doing so would decrease the mechanical requirements of the motors and transmissions, optimizing the performance of the system and reducing component wear.

22.5.3 *Winter Solstice: Behavior of the Fixed-Focus System*

After analyzing the behavior of the system during the equinox and summer solstice, this section examines how the tracking mechanism for a fixed focal point encounters difficulties when operating with $\psi_{sun} \geq 90^{\circ}$ and $\psi_{sun} \leq -90^{\circ}$. This is particularly

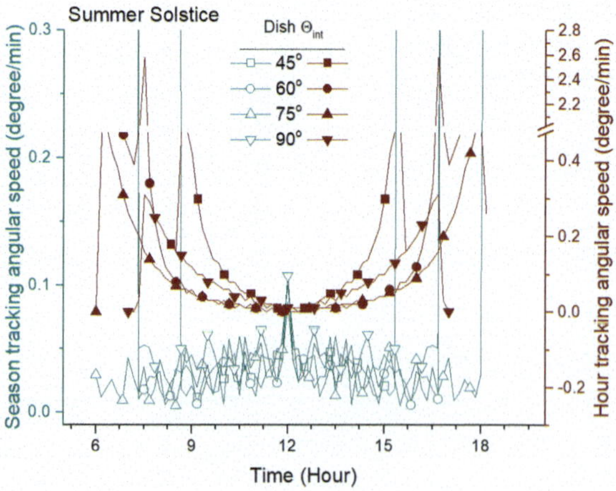

Fig. 22.8 Season and hour tracking angular speed in summer solstice

relevant during the winter solstice, when the system requires smoother movements to maintain proper tracking.

Figure 22.9 presents the angular movements of the tracking axes for the winter solstice. It can be observed that the operation time is similar across the four systems evaluated, indicating that the performance of the system relies heavily on the effectiveness of the selected interception angle. The effectiveness as a function of the interception angle is discussed in the following section.

Based on Fig. 22.10 and the information discussed in the previous sections, it can be observed that angular velocities of the seasonal tracking axis ($\xi_D \leq 0.12°$ per minute) are typical for the fixed-focus Scheffler dish solar concentrator system. However, for the hourly tracking axis, velocities exceed 0.3° per minute in a system with a dish having an interception angle (θ_{int}) of 45°.

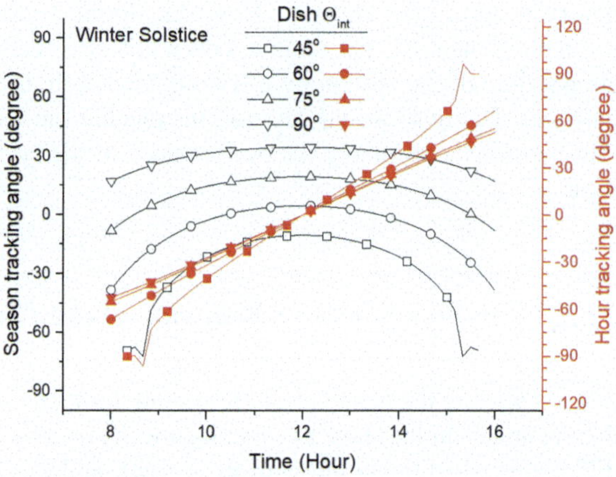

Fig. 22.9 Season and hour tracking movement in winter solstice

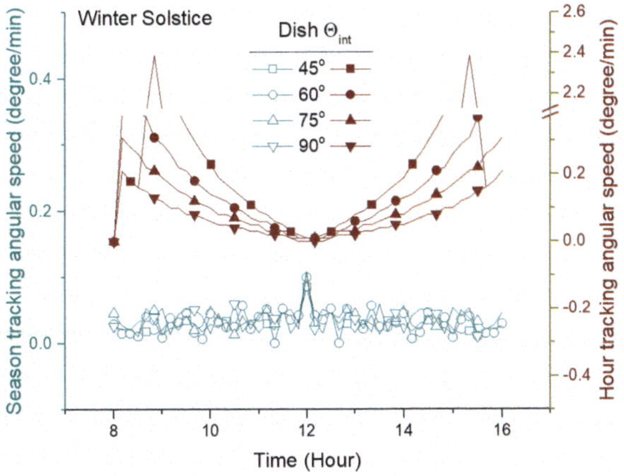

Fig. 22.10 Season and hour tracking angular speed in winter solstice

22.5.4 Performance and Optimization

Based on prior knowledge of the aperture area behavior in heliostat systems, and by analyzing the surface of the Scheffler dish as a set of small differential planes, it is possible to determine the factor related to cosine losses in each section of the reflective surface. Figure 11a shows the profile associated with cosine losses for a Scheffler dish with a 60° interception angle. The x and y variables correspond to the coordinates of the matrix paraboloid, revealing that the center of this dish is located 1.15 m from the origin.

It is important to note that the closer the selected dish is to the origin of the matrix paraboloid, the better the factor related to cosine losses. This highlights the trade-off involved in selecting the interception angle. As the dish moves further from the origin in the x and y coordinates, the ratio between the aperture area and the required reflective surface decreases significantly, affecting the efficiency of the system.

Figure 11b presents the operating hours of the various systems analyzed in relation to the season of the year, as well as the effectiveness of the Scheffler dish concerning cosine losses. This study reveals an inverse relationship between efficiency and operating hours, suggesting that there may be an optimal θ_{int} value that achieves the best performance in the region of Mexicali, Baja California, Mexico.

Furthermore, it is shown that systems with interception angles of 30°, 40°, and 45° maintain relatively stable operating hours throughout the year, without significant variation. In contrast, systems $\theta_{int} > 50^{\circ}$ begin to show dependence on the time of year, with a greater variability in their operating time. This highlights the importance of selecting the appropriate interception angle to optimize the performance of the system based on the climatic of the region and solar conditions.

Since the number of operating hours and the cosine loss factor suggest the optimal θ_{int} value for selecting the system with the best performance, additional information is required to fine-tune this value. Therefore, a study was conducted to evaluate the previously mentioned systems, considering their interaction with the irradiance of the site throughout the year.

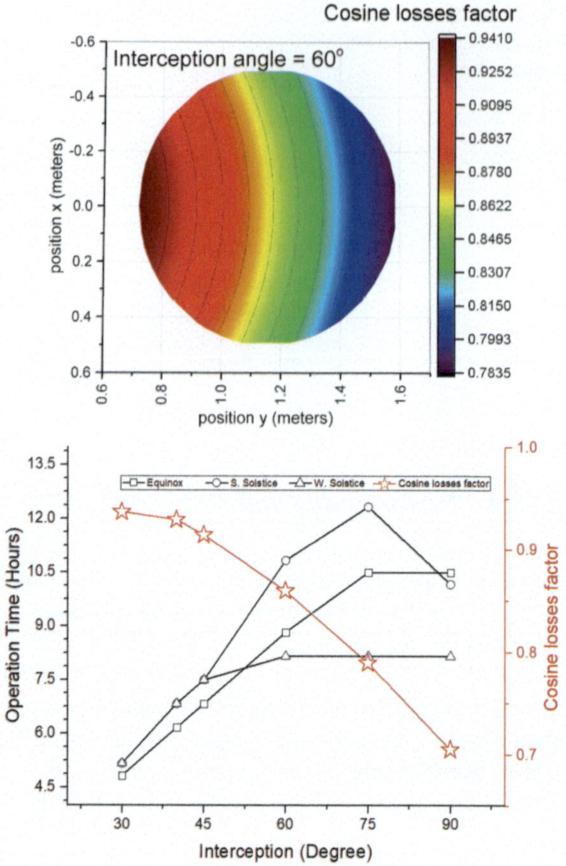

Fig. 22.11 Analysis of operation time by each season and cosine losses factor

The combination of operating hours, disk efficiency, and site irradiance determined that the dish with a 60° interception angle offers the best performance, as shown in Fig. 22.12. This system can intercept 1,008.4 kWh annually, with 3,333 h of operation. In other words, a dish with a one-meter diameter and a one-meter focal length can intercept an annual average of 2.76 kWh per day, with an aperture area of 0.68 m^2.

Furthermore, it was observed that this system performed the best during both the summer solstice and the equinox. However, during the winter solstice, the dish with $\theta_{int} = 45^{\circ}$ showed the highest performance.

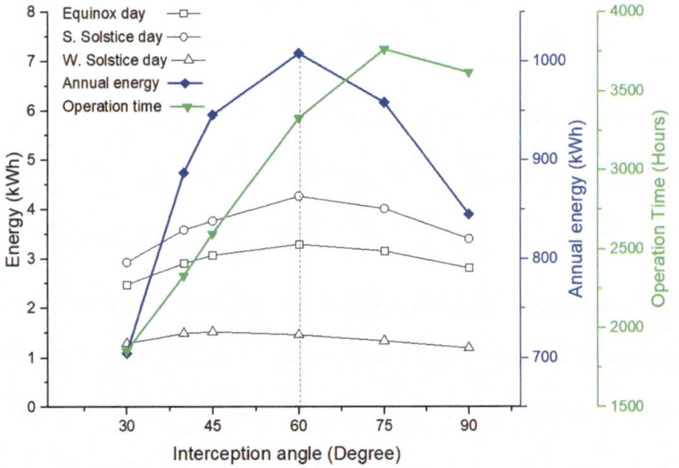

Fig. 22.12 System annual and season performance

22.6 Conclusions

The analysis and optimization of a fixed-focus Scheffler dish with a dual-axis tracking system in this study demonstrate that adapting solar concentrators to specific climate conditions can greatly improve performance. A mathematical model was developed to identify optimal parameters, such as interception angle and dish diameter, which are essential for maximizing solar energy capture, minimizing cosine losses, and ensuring accurate solar tracking throughout the year.

The dual-axis tracking system, designed to adjust both solar declination and hour angles, allows the Scheffler dish to remain focused on a fixed receiver across all seasons. This adaptability is especially beneficial in regions like Mexicali, where solar altitude varies considerably with the seasons. The results show that a 60° interception angle provides optimal performance, capturing over 1,000 kWh annually and achieving more than 3,300 operational hours per year. Efficiency is enhanced by balancing operating hours with the reflective surface effectiveness, thereby reducing solar ray misalignment losses. The study also emphasizes that selecting the correct interception angle is crucial to optimizing the system. A smaller angle may limit operating hours, while a larger angle may reduce the effective concentration of solar radiation on the receiver. Thus, optimal design requires balancing energy capture with minimized losses.

Acknowledgements The authors would like to thank the Engineering Institute of the Autonomous University of Baja California for the support received to carry out this work in the best possible way.

Disclosure of Interests The authors declare that they have no conflict of interest.

References

1. J. Kleih, Dish-stirling test facility. Solar Energy Materials **24**(1–4), 231–237 (1991). https://doi.org/10.1016/0165-1633(91)90063-Q
2. S. Snidvongs, The structure and foundation design for small solar thermal dish stirling 10 kW power plant for Thailand Softland and poor isolation nature, in Solar Energy, ASME, pp. 729–735. (2005). https://doi.org/10.1115/ISEC2005-76017
3. J.A. Alarcón, J.E. Hortúa, A.G. Lopez, Design and construction of a solar collector parabolic dish for rural zones in Colombia. TECCIENCIA **7**(14), 14–22 (2013). https://doi.org/10.18180/tecciencia.2013.14.2
4. A. Balzar, P. Stumpf, S. Eckhoff, H. Ackermann, M. Grupp, A solar cooker using vacuum-tube collectors with integrated heat pipes. Sol. Energy **58**(1–3), 63–68 (1996). https://doi.org/10.1016/0038-092X(96)00024-2
5. M. John, A solar parabolic dish cooker for rural African setting, 2015. Accessed: Sep. 18, 2018. [Online]. Available: www.iosrjen.org
6. M. Arulkumaran, W. Christraj, W. Christraj, Experimental analysis of non-tracking solar parabolic dish concentrating system for steam generation. Eng. J. **16**(2), 53–60 (2012). https://doi.org/10.4186/ej.2012.16.2.53
7. I. Ladan Mohammed, Design and development of a parabolic dish solar water heater, Vol. 2, pp. 822–830, Accessed: Sep. 18, 2018. [Online]. Available: www.ijera.com
8. K. Srithar, T. Rajaseenivasan, N. Karthik, M. Periyannan, M. Gowtham, Stand alone triple basin solar desalination system with cover cooling and parabolic dish concentrator. Renew. Energy **90**, 157–165 (2016). https://doi.org/10.1016/J.RENENE.2015.12.063
9. J. Ruelas, J. Palomares, G. Pando, Absorber design for a Scheffler-type solar concentrator. Appl. Energy **154**, 35–39 (2015). https://doi.org/10.1016/j.apenergy.2015.04.107
10. J. Rapp, P. Schwartz, Construction and improvement of a scheffler reflector and thermal storage device, 2010. Accessed: Aug. 22, 2018. [Online]. Available: https://digitalcommons.calpoly.edu/cgi/viewcontent.cgi?article=1023&context=physsp
11. V.R. Dafle, Design, development & performance evaluation of concentrating monoaxial scheffler technology for water heating and low temperature industrial steam application. Accessed: Aug. 22, 2018. [Online]. Available: www.ijera.com
12. A. Munir, O. Hensel, W. Scheffler, Design principle and calculations of a Scheffler fixed focus concentrator for medium temperature applications. Sol. Energy **84**(8), 1490–1502 (2010). https://doi.org/10.1016/j.solener.2010.05.011
13. J.A. Duffie, W.A. Beckman, Solar engineering of thermal processes. Wiley, (2013)

Chapter 23
Technical and Economic Analysis of an Off-Grid Microgrid with Solar PV and Battery Storage: A Case Study for Rural Communities in Mexico

J. A. Aguilar-Jiménez, S. Islas-Pereda, N. Velázquez-Limón, J. Ríos-Arriola, C. A. Cásares-De la Torre, J. A. Suástegui-Macías, J. D. Gámez-Avilez, and N. A. Pérez de Alva

Abstract This study presents a technical and economic analysis of an off-grid microgrid system based on photovoltaic energy and battery storage, designed to meet the energy needs of the rural community of Ejido Delicias in Baja California, Mexico. The analysis focuses on the impact of varying photovoltaic capacity and storage on key performance indicators, including solar contribution, auxiliary energy use, energy losses, and the Levelized Cost of Energy (LCOE). The results indicate that while increasing photovoltaic capacity reduces auxiliary energy consumption, it requires a significant initial investment. The optimal configuration identified in this study consists of a 200 kW photovoltaic system and 450 kWh of battery storage, achieving an LCOE of 230 USD/MWh and covering 97.4% of the community's annual energy demand. In comparison, extending the electrical grid would result in an LCOE of 244.43 USD/MWh over a 25-year analysis period, but it would limit the community's energy independence. The findings suggest that a photovoltaic microgrid with battery storage is a viable and competitive solution for electrifying remote communities, offering both energy security and economic efficiency. The study highlights the importance of optimizing photovoltaic generation and storage to maximize solar energy utilization while minimizing auxiliary energy consumption and operational costs.

Keywords Microgrid · Renewable energy · Solar energy · Isolated community

---

J. A. Aguilar-Jiménez (✉) · S. Islas-Pereda · N. Velázquez-Limón · J. Ríos-Arriola · C. A. Cásares-De la Torre · J. D. Gámez-Avilez · N. A. Pérez de Alva
Centro de Estudios de las Energías Renovables, Instituto de Ingeniería, Universidad Autónoma de Baja California, Mexicali, Baja California, México
e-mail: aguilarj30@uabc.edu.mx

J. A. Suástegui-Macías
Facultad de Ingeniería, Universidad Autónoma de Baja California, Mexicali, Baja California, México

D. Rossit et al. (eds.), *Proceedings of the VII Ibero-American Congress of Smart Cities, ICSC-Cities 2024, 12-14 November, San Carlos, Costa Rica*, Springer Proceedings in Energy, https://doi.org/10.1007/978-981-96-4301-1_23

23.1 Introduction

Access to electricity in remote and isolated communities poses significant challenges, especially in regions without connection to the national power grid [1]. In this context, off-grid microgrids based on solar photovoltaic (PV) energy have emerged as a viable solution to provide clean, reliable, and sustainable electricity in these areas. These microgrids enable the integration of solar technologies with energy storage systems, reducing dependence on fossil fuels and promoting greater energy independence.

In recent years, the decrease in the initial investment costs of PV systems and battery storage technologies has been a key factor in the viability of solar microgrids in isolated communities. This reduction in costs has improved the competitiveness of solar energy compared to conventional solutions, such as diesel generators, enabling a wider implementation of these technologies in small- and medium-scale projects [2]. This development has opened new opportunities for rural communities to adopt renewable sources as an economically viable and sustainable alternative to meet their energy needs.

The analysis of the Levelized Cost of Energy (LCOE) has been widely used in previous studies to assess the profitability of PV systems in off-grid microgrids. Various studies have shown that, when considering factors such as system lifespan, operation and maintenance costs, and storage capacity, the LCOE of solar energy has proven to be competitive compared to other technologies. For example, Graça Gomes et al. [3] proposed an optimization model to determine the most cost-efficient capacity for off-grid microgrids based on renewable energies with energy storage. In a study applied to the regions of Corvo and the Azores archipelago, Portugal, they achieved an LCOE of 210 €/MWh, which was lower than the diesel-only alternative.

Similarly, Tarife et al. [4] analyzed the LCOE of off-grid microgrids for rural agricultural communities in Southern Philippines, which included systems of hydropower, photovoltaic energy, diesel generation, and battery storage. They found that the optimal size for the microgrid consisted of 250 kW of photovoltaic power, 100 kWh of energy storage, and a 13 kW diesel generator, resulting in an LCOE of 179.5 USD/MWh.

Aguilar-Jiménez et al. [5] compared the use of AGM (absorbed glass mat) batteries and lithium batteries in off-grid microgrids. In the case study of a community in Bigene, Guinea-Bissau, the economic analysis revealed that the LCOE could reach up to 162 USD/MWh using lithium batteries. It is important to note that, in this microgrid, with a PV capacity below 700 kW and a storage system of less than 2580 kWh, there was no significant difference in the LCOE between AGM and lithium storage technologies. This suggests that AGM technology could be preferred due to its lower initial investment cost compared to lithium batteries.

A significant number of studies have explored the technical and economic feasibility of implementing microgrids based on renewable energy, comparing them with other alternatives such as the use of fossil fuels or the extension of the power grid [6–8]. However, in most of these studies, the LCOE is strongly influenced by the specific context of the project, including factors such as the number of users, energy consumption profile, renewable energy potential, technology and installation costs, as well as the customs and habits of the community. These elements can vary greatly from one case to another, making it necessary to carry out optimization studies to determine the optimal capacity of the microgrid according to these specific conditions.

In this study, we present a technical and economic feasibility analysis of an off-grid microgrid for the community of Ejido Delicias, located in Baja California, Mexico. This community is isolated from the national power grid and experiences high levels of social marginalization due to the lack of basic services, including electricity. To address this issue, we propose the design and sizing of a hybrid system composed of solar photovoltaic energy, auxiliary generation through diesel, and energy storage in batteries. The goal is to determine the optimal capacity of the microgrid using the LCOE as the main economic indicator. Additionally, the extension of the power grid to the community is considered as an alternative electrification solution, and both options are compared in terms of their LCOE.

23.2 System Description

Figure 23.1 shows the schematic diagram of the systems analyzed in this study. First, the technical and economic feasibility of installing an off-grid microgrid to meet the electrical energy needs of a community is examined. The microgrid consists of a photovoltaic system for generating renewable electrical energy, an inverter/controller to regulate the energy supply to the community, and a battery bank to store excess photovoltaic energy, which can then be used during periods of low or no solar radiation (such as cloudy days or nighttime). Additionally, a diesel-based auxiliary generator is considered to meet demand when the photovoltaic resource or stored energy is insufficient. The electricity generated by the microgrid is delivered to users via a medium-voltage distribution network.

To economically compare the microgrid with other electrification options, the extension of the power grid is also considered as an alternative for bringing electrical service to the community. This involves installing transmission lines from the nearest point of the existing power grid to supply centralized electricity to the users.

23.2.1 Study Community Description

To analyze the technical and economic feasibility of the proposed systems, Ejido Delicias, a community located in the municipality of San Felipe, Baja California, Mexico (30.6882, − 114.7103), was chosen as the case study. This community is isolated from the national power grid, approximately 28 km from the nearest interconnection point. Currently, the community relies on a 75 kVA diesel generator to

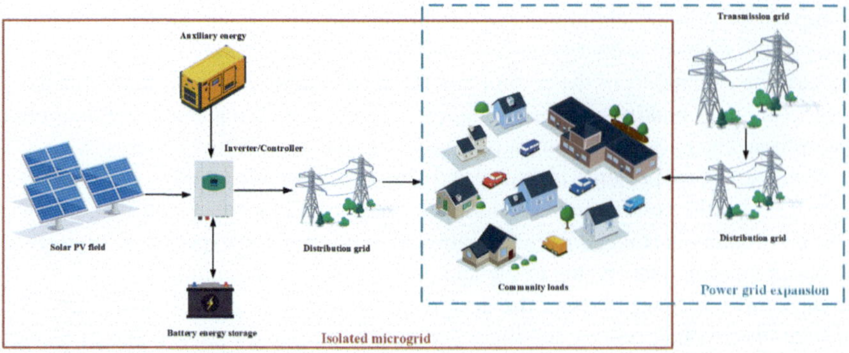

Fig. 23.1 Schematic diagram of the systems analyzed: off-grid microgrid with renewable energies and grid extension

supply electricity to its 25 resident families. However, due to the high cost of fuel, this generator is only used for two hours per day, severely limiting access to electricity.

The climate conditions in Ejido Delicias are extreme, with temperatures exceeding 40 °C in summer and dropping to as low as 5 °C in winter, making access to electricity essential for thermal comfort. In addition to thermal comfort, electricity is also critical for food and medicine preservation, a vital function given the community's remoteness from other populated areas. The nearest city with access to electricity is San Felipe, approximately 46 km away.

According to the National Solar Radiation Database [9], Ejido Delicias receives an average daily global horizontal solar radiation of 6.14 kWh/m^2, with an average wind speed of 3.4 m/s and an annual average temperature of 24.2 °C. These conditions make the community a prime location for the implementation of renewable energy technologies, particularly photovoltaic systems.

23.3 Methodology

The microgrid simulation was performed using the System Advisor Model (SAM) and Transient System Simulation (TRNSYS) software. SAM was used to estimate photovoltaic energy production due to its precise energy estimation model and its photovoltaic technology databases, which facilitated the assessment of photovoltaic generation under the specific climate conditions of Ejido Delicias. On the other hand, TRNSYS was used to link photovoltaic production with the community's energy demand, utilizing an external file generated in SAM. In addition, the integration of energy storage with batteries and auxiliary energy supply was studied using TRNSYS. Different scenarios with various installed capacities of photovoltaic technology and energy storage were analyzed to assess their effect on the need for auxiliary energy.

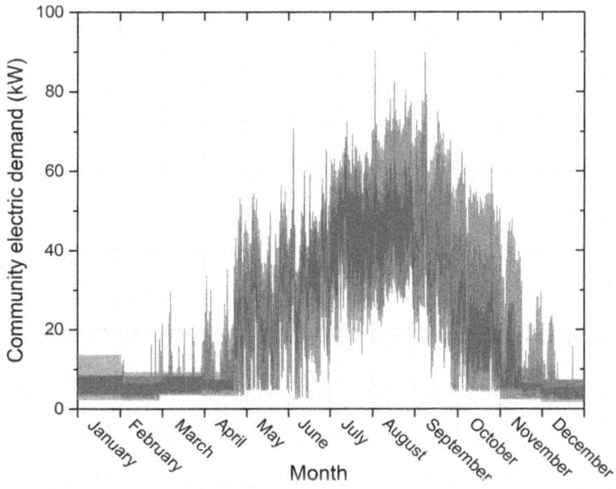

Fig. 23.2 Estimated electricity demand profile of the 25-family community of Ejido Delicias

23.3.1 Estimation of the Electrical Energy Demand of Ejido Delicias Community

Since the community of Ejido Delicias does not have a continuous electricity supply, as previously mentioned, there is no real energy demand profile available as a reference. For this reason, an estimated energy demand profile was created for the domestic users in the community, considering electrical loads required for food preservation, thermal comfort, and entertainment. The annual electrical demand profile for the 25 families residing in the community is shown in Fig. 23.2.

This demand profile is heavily influenced by the use of air conditioning (A/C) during the spring and summer months, as the region's high temperatures make space cooling a critical necessity. As a result, the community's electrical energy demand is significantly elevated during a large portion of the year.

23.3.2 Estimation of the Electrical Energy Demand of the Community Due to Air Conditioning

To estimate the energy demand from A/C, the TRNSYS software was used with Type56 to simulate the thermal load of a typical local house throughout the year. A 8.16×6.17 meter house was considered, with concrete walls and a wooden roof, including insulation in the roof using 2-inch thick polystyrene. This resulted in an annual thermal load of 17,117 kWh per house, assuming a comfort temperature of 25 °C.

A coefficient of performance (COP) of 3.2 was assumed, which is a common value for minisplit A/C units. This led to an estimated electrical energy demand of 5349 kWh per user annually for A/C. For the 25 families in the community, the total annual energy demand was estimated at 176.5 MWh, with 75.7% of the total (133.7 MWh) used for A/C.

23.3.3 Simulation Process

To perform the operational simulation of the microgrid system, the SAM and TRNSYS software were used, as mentioned earlier. SAM was utilized to estimate photovoltaic production under the climatic conditions of the study region. Subsequently, the hourly photovoltaic production estimates were exported to TRNSYS through an external .csv file and used as input data via Type9. Additionally, the hourly electrical demand profile of the 25 families was used in the simulation as an input variable, also through Type9.

When photovoltaic production was insufficient to meet the demand and no stored energy was available in the batteries, the auxiliary generator supplied the energy shortfall. The simulator performed energy balances to determine when the demand should be met by photovoltaic energy, batteries, or the auxiliary generator. It also assessed whether there were any photovoltaic energy surpluses that could be stored in the battery bank.

At the end of the simulation, data on photovoltaic electricity production, energy supplied by the batteries, energy provided by the auxiliary generator, and energy losses were recorded. Figure 23.3 shows the user interface developed to simulate the microgrid in TRNSYS.

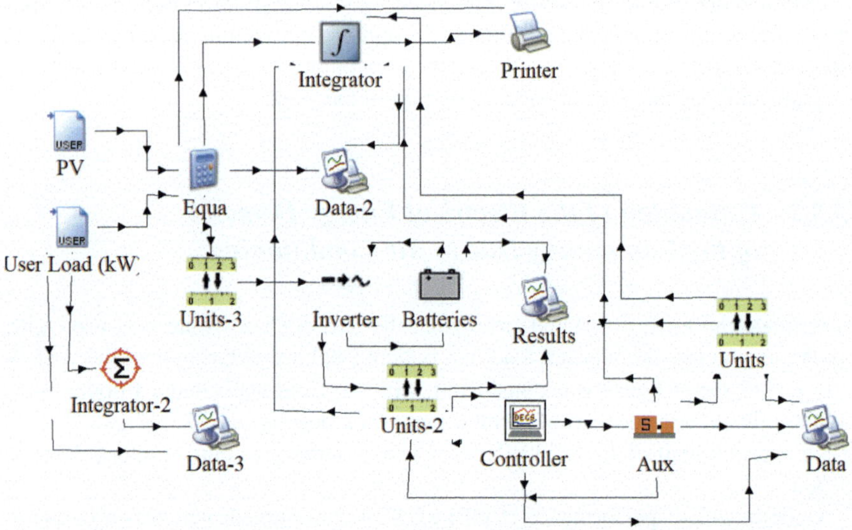

Fig. 23.3 User interface of the simulator developed in TRNSYS

23.3.4 Energy Indicators

To quantify energy generation from different technologies and conduct a technical and economic comparison, various indicators were used, including energy generation and losses, auxiliary energy requirements, and the contribution of solar energy to the demand. These indicators were based on variations in the installed capacity of photovoltaic generation and battery energy storage.

Solar fraction

In certain scenarios, not all energy generated by photovoltaic modules is used by the demand or users, as the solar energy generation period does not always align with energy needs. This can result in energy losses, especially when there is insufficient storage capacity or inefficient demand management.

The solar fraction is an indicator used in photovoltaic systems to measure the proportion of the total energy demand that is covered by solar energy. Specifically, the solar fraction indicates how much of the energy demand is directly met by solar energy rather than conventional sources such as diesel generators or the power grid. A value of 1 indicates that the entire energy demand was covered by solar energy, while values less than 1 indicate contributions from other energy sources, typically the auxiliary system. The solar fraction is calculated according to Eq. 23.1:

$$f_{sol} = \frac{E_{load} - E_{aux}}{E_{load}} \tag{23.1}$$

where f_{sol} corresponds to the solar fraction, E_{load} is the annual energy consumed by the community and E_{aux} is the annual auxiliary energy supplied.

Auxiliary energy

In cases where solar energy capture is insufficient and/or there is not enough stored energy in the battery system, the electricity demand in the off-grid microgrid must be met by an auxiliary system, providing the required amount of energy. Typically, this auxiliary system is based on a diesel generator or another backup source. It ensures that the electricity supply to the users is maintained continuously.

To calculate the auxiliary energy required to meet the electricity demand that is not covered by solar energy or stored energy, Eq. 23.2 is used:

$$E_{aux} = \sum_{i=0}^{8760} \mathrm{P}_{load}(i) - \mathrm{P}_{pv}(i) \tag{23.2}$$

where $\mathrm{P}_{load}(i)$ is the power, in watts (W), of the community's demand at hour (i) of the simulation, and $\mathrm{P}_{pv}(i)$ is the photovoltaic power, in W, generated by the system at hour (i) of the simulation. If the value of E_{aux} is positive, it indicates that there is not enough solar or stored energy, requiring the auxiliary system to supply the needed energy. If the resulting value is negative, it means there is sufficient stored energy in the photovoltaic system, or the photovoltaic production is enough to meet the demand, and no auxiliary energy is required.

The result of this equation represents the annual auxiliary energy needed to satisfy the microgrid's electricity demand.

23.3.5 *Economic Analysis*

The LCOE was used to perform an economic comparison of different photovoltaic power and energy storage capacities. LCOE is a key indicator used to assess the economic competitiveness of various energy generation technologies. It represents the total cost of generating one megawatt-hour (MWh) of electricity over the system's lifetime.

LCOE is calculated by dividing the sum of the initial investment costs, operation and maintenance costs, and replacement costs by the total energy generated by the system over its lifetime, as shown in Eq. 23.3:

$$LCOE = \frac{\sum_{t=1}^{t=25}(CAPEX_t + OPEX_t + REPEX_t)}{\sum_{t=1}^{t=25} E_{load}} \tag{23.3}$$

where CAPEX represents the initial investment costs, OPEX are the operational and maintenance costs in year *t,* and REPEX are the replacement costs in year *t*.

For off-grid microgrids, LCOE is particularly relevant as it not only includes the costs of the photovoltaic system but also those of energy storage, which is crucial to ensure continuity of supply in the absence of a grid connection. A low LCOE indicates that the system is economically viable and competitive compared to other energy alternatives, such as diesel generators, whereas a higher LCOE may suggest the need to optimize the design or reduce costs in key system components. Table 23.1 shows the investment costs, operating expenses, as well as inflation rates, discount rates and other values considered in the economic analysis.

23.4 Results

This section presents the energy and economic results of the proposed system based on the installed capacity of photovoltaic generation and battery energy storage. For the purposes of this study, variations in photovoltaic capacity were considered within a range of 25 kW to 275 kW, both without battery energy storage and with energy storage capacities of up to 600 kWh. The effect of these variations on solar energy contribution to the community's energy demand, auxiliary energy usage, energy losses, solar fraction, and finally the LCOE, was analyzed to determine the optimal photovoltaic generation-to-energy storage ratio.

Table 23.1 Costs and rates considered in the economic study[1]

Cost of photovoltaic system	1000 USD/kW
Cost of storage system	286 USD/kWh
Balance of plant cost of storage system	28.6 USD/kWh
Auxiliary system capacity (diesel)	100 kW
Auxiliary system cost	100 USD/kW
Inverter replacement cost	58 USD/kW
Year replacement of inverters and batteries	13 years
Annual maintenance cost	10 USD/kW
Diesel cost	1.31 USD/liter
Annual energy consumption growth rate	0.25%
Inflation rate	4.70%
Discount rate	5%
Annual power degradation of photovoltaic modules	0.50%
System lifetime	25 years

23.4.1 Solar Contribution to Demand

The effect of varying photovoltaic installed capacity and battery energy storage capacity on solar contribution can be seen in Fig. 23.4. With a photovoltaic capacity in the range of 25 kW, there is no increase in solar contribution when using energy storage, as the entire photovoltaic generation is used to meet the community's demand, leaving no surplus to store for later use. In these scenarios, the energy shortfall is supplied by the auxiliary diesel generator.

As photovoltaic capacity increases, energy surpluses are generated, which can be stored in batteries for use during periods of low or no photovoltaic generation (cloudy days or nighttime), resulting in a greater solar contribution to the demand. With a photovoltaic capacity of 100 kW, the optimal storage capacity is 200 kWh, as a higher storage capacity would not increase the solar contribution. However, with a photovoltaic capacity of 200 kW, a battery bank with a maximum capacity of 500 kWh would be required to achieve the highest solar contribution.

The maximum solar contribution observed within the study range was 172 MWh, which corresponds to 97.4% of the total annual demand of the study community (176.5 MWh).

[1] The costs of photovoltaic and storage technologies were determined based on the costs of local distributors. The cost of diesel was based on the official cost in the case study region. 10% of the cost of the storage system was considered as BoP.

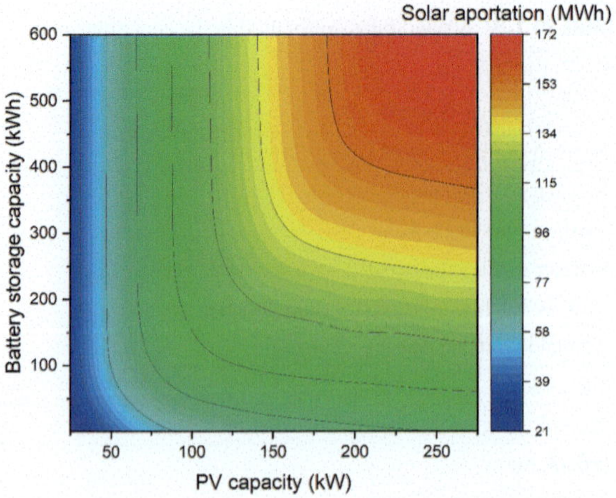

Fig. 23.4 Microgrid solar contribution to demand as a function of installed photovoltaic capacity and battery storage capacity

23.4.2 *Auxiliary Energy to Meet the Demand*

As mentioned in the system description, the microgrid includes an auxiliary energy supply system for periods when solar energy or stored energy in the batteries is unavailable. This auxiliary energy is provided by diesel generators. Figure 23.5 shows the annual amount of auxiliary energy required as a function of the installed photovoltaic capacity and the battery storage capacity.

To achieve a significant reduction in auxiliary energy consumption, a photovoltaic capacity substantially higher than the community's maximum demand (88.7 kW) is required. This is necessary both to meet daytime demand and to generate enough energy for nighttime consumption using a storage system that accommodates the photovoltaic generation surplus. With 100 kW of installed photovoltaic capacity

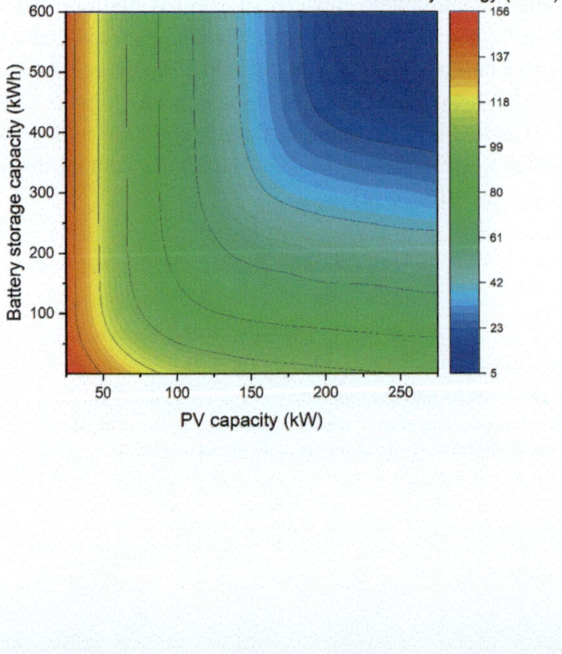

Fig. 23.5 Auxiliary energy required to meet demand as a function of installed photovoltaic capacity and battery storage capacity

and a 150 kWh battery bank, auxiliary energy consumption is 80.5 MWh per year. Increasing the battery storage capacity to 300 kWh reduces auxiliary energy consumption to 70.8 MWh, a 12% reduction.

As battery storage capacity increases but photovoltaic power remains constant, the percentage reduction in auxiliary energy decreases because insufficient energy is generated for storage and nighttime consumption. Conversely, increasing photovoltaic capacity results in a greater contribution to the microgrid's demand, but this increase is not significant unless accompanied by an increase in storage capacity. The lowest levels of auxiliary energy, around 5 MWh, were achieved with 275 kW of photovoltaic capacity and 600 kWh of storage.

23.4.3 Inverter and Regulator Energy Loss

Figure 23.6 shows the amount of energy wasted by the microgrid due to the conversion between direct current (DC) and alternating current (AC) in the inverter and regulator. As mentioned in the methodology section, an inverter efficiency of 0.96 and a regulator efficiency of 0.95 were considered, based on conservative values representative of currently available technologies. These losses in photovoltaic system-generated energy are directly related to the installed photovoltaic capacity and, to a lesser extent, to the battery storage capacity.

As photovoltaic capacity increases, there is a notable rise in energy losses, primarily due to the greater amount of DC photovoltaic energy being converted to AC through the inverter. Increasing storage capacity causes a slight increase in energy losses, resulting from the use of the current regulator and its efficiency. However, when storage capacity reaches its maximum, energy losses stabilize because the

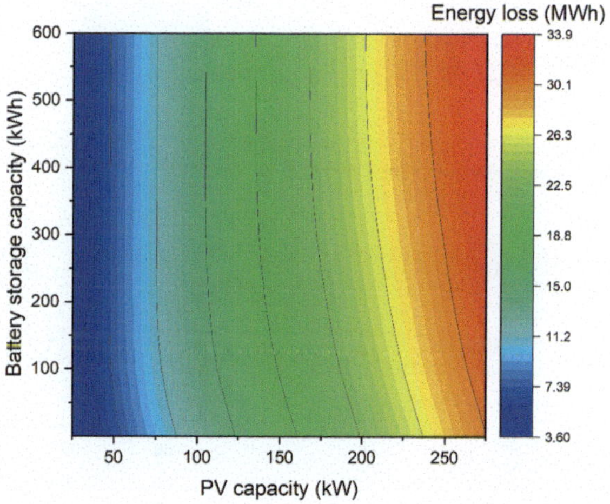

Fig. 23.6 Energy loss due to AC/DC current conversion in inverter and controller

photovoltaic production is no longer sufficient to store more energy in the batteries, which halts further losses associated with the regulator.

With the maximum photovoltaic and battery storage capacity studied, energy losses reached up to 33.9 MWh, equivalent to 19.2% of the annual energy consumption of the study community. Therefore, conversion and regulation efficiencies can be considered a critical factor in the microgrid's operation.

23.4.4 Solar Fraction

Figure 23.7 shows the solar fraction as a function of photovoltaic capacity and energy storage. This indicator allows for the determination of the degree of contribution from solar generation to the community's electricity demand. High solar fraction values indicate a low requirement for auxiliary energy, but they may also represent a high initial investment due to the photovoltaic and storage capacities needed to achieve such values.

It can be observed that for photovoltaic capacities below 50 kW, increasing battery energy storage capacity does not result in an increase in the solar fraction. This is primarily because nearly all the photovoltaic energy produced is used to meet the microgrid users' demand, leaving little to be stored for later use. However, as photovoltaic capacity increases, expanding the battery bank does lead to an increase in the solar fraction. Solar fraction values close to 0.6 can be achieved with photovoltaic capacities greater than 100 kW and storage systems smaller than 200 kWh.

The highest estimated solar fraction was 0.97, achieved with a photovoltaic capacity of 300 kW and a storage bank of 600 kWh.

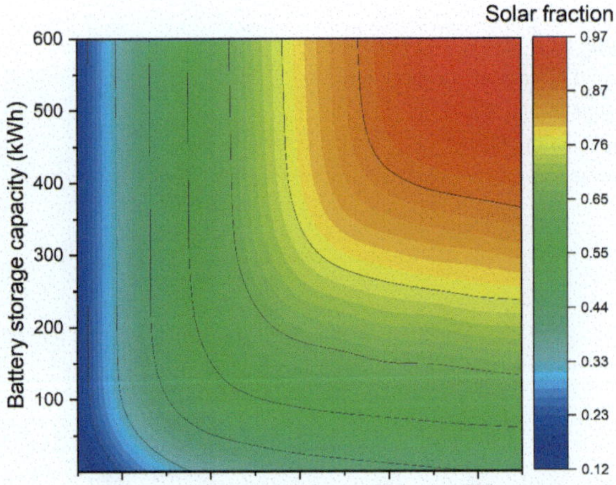

Fig. 23.7 Solar fraction as a function of installed photovoltaic capacity and battery energy storage capacity

23.4.5 LCOE of the Microgrid with Battery Storage

Although the solar fraction helps determine how solar generation-storage ratios impact the community's energy demand, such as reducing auxiliary energy consumption, other indicators, especially economic ones, are required to establish the appropriate photovoltaic and storage capacity for the community's context. The LCOE shows the investment, operation, and maintenance costs over the project's lifetime, relative to the useful energy generated. Therefore, it is a relevant comparative indicator for decision-making in these types of investments.

Figure 23.8 shows the effect of installed PV capacity and battery energy storage on the LCOE. The highest LCOE value, at 447 USD/MWh, was observed with a photovoltaic capacity of 25 kW and 600 kWh of storage. This is due to the large initial investment required for batteries, as well as the high operational costs from using auxiliary energy. As the photovoltaic capacity increases, the LCOE tends to decrease because the cost of photovoltaic technology is not as high as the operational cost of auxiliary energy.

However, unlike the solar fraction indicator, the lowest LCOE is not achieved with the highest photovoltaic and storage capacity studied. The lowest LCOE value, at 230 USD/MWh, was achieved with a photovoltaic capacity of 200 kW and 450 kWh of battery storage. When these capacities are reduced, the LCOE tends to increase due to higher auxiliary energy consumption, which has a high cost. Increasing photovoltaic generation and storage capacities reduces auxiliary energy consumption but significantly raises the initial investment due to the cost of photovoltaic modules and batteries, without a proportional reduction in auxiliary energy consumption.

Nevertheless, values close to 250 USD/MWh can be achieved with systems of 150 kW photovoltaic and 250 kWh storage, representing a significant reduction in the initial investment.

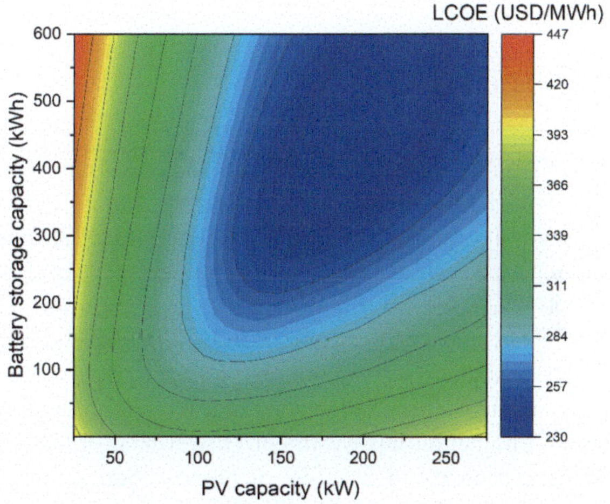

Fig. 23.8 LCOE of the microgrid as a function of installed photovoltaic capacity and battery energy storage capacity

Table 23.2 Energy and economic considerations for economic analysis of grid extension

Distance to last point of power grid	28 km
Cost per kilometer of power grid [10]	$18,919 USD/km
Annual energy consumption per user	7458 kWh
Average energy price with operator [10]	0.065 USD/kWh
Community energy consumption in 25 years	4539 MWh
Total cost of energy in 25 years	$579,772 USD
LCOE	244.43 USD/MWh

23.4.6 LCOE Considering Expansion of the Electric Grid

Another option for electrifying rural areas is extending the power grid to deliver centralized energy. This option usually involves significant initial investment costs, depending on the community's context, such as the distance to the nearest grid point, the number of users, and the region's topography. Additionally, when the grid is extended, users will have to pay for the amount of electricity consumed according to the operator's rate.

To compare the previously analyzed microgrid option with the grid extension option, Table 23.2 presents the technical and economic considerations involved in the comparison.

Considering that the community is located 28 km away from the nearest grid connection point, and taking into account the costs per kilometer for extending the grid and the electricity prices provided by the sole electricity operator in Mexico, it has been calculated that extending the grid would result in an LCOE of 244.43 USD/MWh over the 25-year analysis period. This represents an LCOE 6.2% higher than the most economical PV-based microgrid option, which resulted in 230 USD/MWh.

23.5 Conclusions

This research presented a technical and economic analysis of an off-grid microgrid based on photovoltaic energy and battery storage, designed to meet the energy needs of the rural community of Ejido Delicias in Baja California, Mexico. Through detailed simulations, various scenarios of photovoltaic capacity and storage were evaluated to optimize solar contribution to demand, minimize auxiliary energy consumption, and reduce energy losses associated with current conversion and storage.

The results show that although a larger photovoltaic capacity increases the solar fraction and reduces reliance on auxiliary sources such as diesel generators, it also entails a higher initial investment. However, an optimal operating point was identified, where a capacity of 200 kW of photovoltaic generation and 450 kWh of storage achieves an LCOE of 230 USD/MWh, effectively balancing the initial investment and operational costs. This scenario also enabled solar energy to cover 97.4% of the community's annual energy demand.

Furthermore, when comparing the microgrid option with extending the electrical grid, it was determined that the grid extension would result in an LCOE of 244.43 USD/MWh over 25 years. While this is a comparable economic alternative, it comes with the disadvantage of relying on the operator's infrastructure and rates, which could limit the community's energy independence.

In conclusion, the photovoltaic microgrid with battery storage represents a technically and economically viable solution for providing clean and reliable energy to isolated communities like Ejido Delicias. Although the initial investment can be substantial, the reduction in auxiliary energy use and the low LCOE achieved demonstrate that this technology can be competitive compared to other alternatives, such as grid extension. The study also highlights the importance of optimizing generation and storage capacities to maximize solar energy utilization, reduce energy losses, and ensure the economic viability of the project.

Acknowledgements The authors would like to thank the Engineering Institute of the Autonomous University of Baja California for the support received to carry out this work in the best possible way.

Disclosure of Interests The authors declare that they have no conflict of interest.

References

1. J.A. Aguilar-Jiménez, N. Velázquez, A. Acuña, R. Cota, E. González, L. González, R. López, S. Islas, Techno-economic analysis of a hybrid PV-CSP system with thermal energy storage applied to isolated microgrids. Sol. Energy **174**, 55–65 (2018). https://doi.org/10.1016/J.SOLENER.2018.08.078
2. J.A. Aguilar-Jiménez, N. Velázquez, R. Beltrán, L. Hernández-Callejo, R. López-Zavala, E. González-San Pedro, Potential for thermal water desalination using microgrid and solar thermal field energy surpluses in an isolated community, in *Smart Cities. ICSC-CITIES 2019*, ed by N. Sergio, H.C. Luis (Springer, Cham, 2020), pp. 162–175. https://doi.org/10.1007/978-3-030-38889-8_13
3. J. Graça Gomes, H.J. Xu, Q. Yang, C.Y. Zhao, An optimization study on a typical renewable microgrid energy system with energy storage. Energy **234**, 121210 (2021). https://doi.org/10.1016/J.ENERGY.2021.121210
4. R. Tarife, Y. Nakanishi, Y. Chen, Y. Zhou, N. Estoperez, A. Tahud, Optimization of hybrid renewable energy microgrid for rural agricultural area in southern Philippines. Energies (Basel) **15** (2022). https://doi.org/10.3390/en15062251
5. J.A. Aguilar-Jiménez, L. Hernández-Callejo, J.A. Suástegui Macías, V. Alonso Gómez, A. García-Álvaro, R. Maján-Navalón, L.J. Obregón, Energy and economic analysis of renewable energy-based isolated microgrids with AGM and lithium battery energy storage: case study Bigene, Guinea-Bissau. Urban Sci. **7**, (2023). https://doi.org/10.3390/urbansci7020066
6. Z. ul A. Qureshi, S.A.A. Kazmi, S. Mushtaq, M. Anwar, An integrated assessment framework of renewable based Microgrid deployment for remote isolated area electrification across different climatic zones and future grid extensions. Sustain. Cities Soc. **101**, (2024). https://doi.org/10.1016/j.scs.2023.105069

7. E.A. Esparcia, M.T. Castro, C.M.F. Odulio, J.D. Ocon, A stochastic techno-economic comparison of generation-integrated long duration flywheel, lithium-ion battery, and lead-acid battery energy storage technologies for isolated microgrid applications. J Energy Storage. **52**, 104681 (2022). https://doi.org/10.1016/J.EST.2022.104681
8. R.A. Abdelsalam, Y.S. Abdallah, N.M. Shabar, O.A. Afifi, M.H. Fouda, A.A. El-Deib, Energy management and techno-economic optimization of an isolated hybrid AC/DC microgrid with green hydrogen storage system, in *IEEE Conference on Power Electronics and Renewable Energy, CPERE 2023* (2023). https://doi.org/10.1109/CPERE56564.2023.10119587
9. NSRDB, https://nsrdb.nrel.gov/data-viewer. Last accessed 17 Sept 2024
10. Comisión Federal de Electricidad: Esquema tarifario vigente. https://app.cfe.mx/Aplicaciones/CCFE/Tarifas/TarifasCRECasa/Casa.aspx

Part V
Other Innovative Developments for Smart Cities

Chapter 24
Comparison of Charging Infrastructure for Electric Vehicles in Multi-unit Residential Buildings in Spain

Ester Garcia Blàzquez, Roberto Villafáfila-Robles, and Marina Codina Escolar

Abstract The increasing adoption of Electric Vehicles (EV) and the transition towards sustainable transportation have necessitated the development of robust EV charging infrastructure, particularly in Multi-Unit Residential Buildings (MURB) within the context of cities. This transition poses unique challenges due to the complex dynamics of shared residential spaces and the need for regulatory compliance. This study explores the techno-economic aspects of implementing EV charging systems in MURB, analyzing both collective and individualized metering approaches. Through a case study, the energy demands and state of charge patterns are evaluated, alongside the infrastructure's capacity to support these needs. A comprehensive cost-benefit analysis is conducted, assessing the impact on the user's bill and the consequences of considering different charging schemes. The findings highlight the importance of aligning building power management strategies with regulatory guidelines, particularly ITC-BT-52, to optimize energy use and ensure compliance. The results demonstrate that strategic implementation of electric vehicle charging infrastructure not only supports the energy transition but also offers significant cost savings and operational efficiencies for MURBs.

Keywords Electric vehicle · Multi-unit residential buildings · Charging infrastructure · Techno-economic analysis

Abbreviations

ANFAC	Asociación Española de Fabricantes de Automóviles y Camiones
CNMC	Comisión Nacional de los Mercados y la Competencia
CP	Charging point

https://www.citcea.upc.edu/.

E. Garcia Blàzquez (✉) · R. Villafáfila-Robles · M. C. Escolar
Centre d'Innovació Tecnològica en Convertidors Estàtics i Accionaments (CITCEA), Departament d'enginyeria elèctrica, Universitat Politècnica de Catalunya (UPC), Barcelona, Spain
e-mail: ester.garcia.blazquez@estudiantat.upc.edu

D. Rossit et al. (eds.), *Proceedings of the VII Ibero-American Congress of Smart Cities, ICSC-Cities 2024, 12-14 November, San Carlos, Costa Rica*, Springer Proceedings in Energy, https://doi.org/10.1007/978-981-96-4301-1_24

EU	European Union
EV	Electric vehicle
LMS	Load management system
MFL	Main feeder line
MURB	Multi-unit residential buildings
PNIEC	Plan Nacional Integrado de Energía y Clima
PSM	Protection system of the MFL
SoC	State of charge

24.1 Introduction

Electric mobility is increasingly recognized as a crucial element in the future of sustainable transportation, primarily due to its potential to significantly reduce environmental impact. Currently, road transport is responsible for approximately 15% of global CO_2 emissions, a figure that notably surpasses the 2.5% contributed by the aviation sector [1]. EVs, with their zero tailpipe emissions, offer a promising solution to reduce greenhouse gas emissions and combat climate change.

In line with the European Green Deal, which aims for climate neutrality by 2050, electric mobility is a cornerstone of the "Fit for 55" initiative [2]. However, disparities among European Union (EU) countries highlight varying progress in EV adoption and charging infrastructure development. In Spain, the Asociación Española de Fabricantes de Automóviles y Camiones (ANFAC) rates the country below the EU average in both electromobility and charging infrastructure [3].

To bridge this gap, Spain's government launched the Plan Nacional Integrado de Energía y Clima (PNIEC) in 2021, targeting a 20% emission reduction by 2030 and aspiring to have 5 million EVs on the road by the same year. Despite these ambitions, current infrastructure falls short, with only 330,000 EVs and 30,000 public Charging Points (CP) as of 2023 [4, 5].

Several factors hinder the growth of EV usage in Spain. High purchase costs, underdeveloped infrastructure, and range anxiety-consumers' fear of limited EV range-are significant barriers. Effective policy instruments, such as subsidies, incentives, and regulations, are crucial for driving large-scale EV adoption. The Plan Moves III, introduced by the Spanish government, aims to incentivize EV purchases and promote low-emission zones [6].

The distribution and type of CPs also present challenges. 57% of charging stations are urban and predominantly chargers below 22 kW. Although long-distance travel calls for interurban fast chargers, modern cities are the center of development for charging infrastructure. The increased number of EVs will impact electricity consumption, raising concerns about grid capacity and stability. Over-reliance on fast chargers could lead to peak demand issues and affect power quality [7]. For this reason, placing chargers below 22 kW would be a recommended strategy. The placement should be in locations where the user is bound to spend a lot of time, such as the household during nighttime or the office during the work day.

Given that 65% of Spain's population resides in Multi-Unit Residential Buildings (MURB) in cities, addressing the unique challenges of installing charging infrastructure in these settings is critical [8]. While previous research has focused on fast-charging infrastructure, the challenges of MURB in Spain remain underexplored [9–11].

Recent studies highlight the unique challenges of implementing electric vehicle (EV) charging infrastructure in multi-unit residential buildings (MURBs), emphasizing the need for effective policy and technical solutions to support sustainable urban mobility. Lopez-Behar et al. [12] mapped critical feedback mechanisms in MURBs and proposed policy interventions to address barriers in shared residential spaces, while a similar study by the same authors (2019) further explores regulatory and operational considerations for sustainable EV infrastructure in urban settings [12, 13]. Additionally, Ramsebner et al. [14] examined load management strategies, providing insights into efficient energy distribution for battery electric vehicle (BEV) charging in multi-apartment buildings. Together, these studies underscore the need for tailored approaches to address both technical constraints and policy gaps in MURB EV infrastructure [14].

This work seeks to address the complexities of introducing EV charging infrastructure in multi-unit residential settings specifically within the context of Spain. It proposes analysing both the technical and financial aspects of existing charging infrastructure in MURBs within the framework of the ITC-BT-52.

24.2 Techno-Economic Framework

In 2014, the Spanish Ministry of Industry, Energy and Tourism enacted ITC-BT-52, "Instalaciones con fines especiales. Infraestructura para la recarga de vehículos eléctricos," which addresses the technical aspects of installing EV charging infrastructure [15]. This regulation supplements the existing Reglamento de Baja Tensión, establishing a foundation for safe EV charging across various scenarios and facilitating standardized installation schemes to streamline implementation. The framework sets standards for new MURBs and guides the adaptation of existing buildings.

The primary challenge in developing charging infrastructure in MURBs is selecting the appropriate installation scheme. ITC-BT-52 outlines four schemes; however, Scheme 4, designed for single-family houses, is excluded from this analysis. The installation schemes are categorized into collective and individualized schemes. Collective schemes offer the advantage of centralized load management, preventing grid overload. In contrast, individual schemes, while initially lacking centralized management, provide easier billing and expense control for users.

Scheme 1 employs a collective metering system, where a central meter measures the total electricity consumption across all charging stations. This setup includes a secondary, user-specific meter at each charging station, intended to facilitate individual billing. However, a key challenge lies in the accurate allocation of costs among users, particularly since the use of secondary meters for billing purposes depends on user consent to share their consumption data.

Scheme 2 involves an individual installation with a common main meter for the household and the charging station. The joined meter allows for simple billing and management of consumption but poses challenges regarding installation. Installation complexities appear due to the distance and voltage drop from the meter to the garage. The feasibility of running individual branches for each user may be limited. Another factor to take into account is that the usage of this scheme is reserved for household owners in the building; people who want to rent a parking spot cannot use it.

Scheme 3 is an individual installation with a main meter specific for each charging station, separate from the household meter and the parking meter. Its advantages are similar to those of scheme 2 in terms of individualized billing. However, it also entails the highest installation cost per user due to the need for a separate main meter for each station and increased administrative complexity with multiple meters per user.

Understanding the implications of each installation scheme is crucial for users and the community. Table 24.1 collects the main advantages and disadvantages of each scheme to simplify the decision-making process.

According to ITC-BT-52, new MURBs must include a pre-installation. For existing buildings without pre-installation, individual owners can undertake pre-installation without community permission, provided it does not alter communal elements. This provision has both technical and financial implications, as individual investments may be required, and unilateral decisions can impact community well-being. The pre-installation provision allows each user to choose their preferred scheme, potentially resulting in mixed installation types within a single parking lot. This diversity increases the complexity of load management and requires careful planning to ensure efficiency and safety.

A first analysis of the technical framework reveals the system's complexity, encompassing both technical challenges and social implications. Additionally, the decisions of users of private transportation are highly influenced by economic factors, considering the purchasing price of the vehicle, the price of electricity, and the cost of installing or adapting charging infrastructure. This work focuses on the effect in the final electricity bill. The choice of installation scheme for EV charging in MURBs significantly impacts the electricity bill of the user. Each scheme (collective or individualized) dictates how consumption is metered and billed. Collective schemes, such as Scheme 1, centralize metering, which can complicate cost allocation among users. Individual schemes, like Scheme 3 and 2, provide more precise billing for each charging station.

Electricity tariffs, which often vary by time of use, further influence costs. Users' ability to take advantage of cheaper rates during off-peak hours depends on the flexibility of their chosen scheme. Additionally, combining different schemes within the same building can pose challenges in managing overall electricity costs and ensuring equitable distribution of expenses among residents.

Table 24.1 Advantages and disadvantages of CP installation schemes

Scheme	Pros	Cons
1	• Simplified management	• Increased administrative complexity due to shared usage
	• Easy to implement	• Potential overload
	• Scalability	• Complex billing per user
	• Cost efficiency of shared infrastructure	• High initial installation cost
2	• Individualized billing	• Does not allow individualization of charging station consumption
	• Integrated management	• May be insufficient if electric vehicle consumption is high
	• Integration with existing household electrical systems simplifies installation	• Complex installation
	• Low initial cost	• Difficulty in controlling individual expenses
3	• Individualized billing	• Increased administrative complexity
	• Enables individual management and control of consumption	• Requires the most infrastructure
	• Easy to control individual expenses	• Space requirements
	• Highest initial cost	
	• Complex maintenance	

24.3 Methodology for Analyzing EV Charging Infrastructure in MURBs

The analysis methodology aims to characterize charging point installation in MURBs by conducting the following analysis:

- Assessment of the current situation of the charging infrastructure and the electric capacity of the building.
- Financial analysis of the impact on the electricity bill of the consumer.
- Comparison of cost variations for unified choices of installation schemes.

In order to study the MURB, inputs in Fig. 24.1 are defined. These are divided into 3 groups considering their nature. General data for the characterization of the building configuration is defined to understand the technical limitations and basic characteristics of the MURB such as the ampacity of the Main Feeder Line (MFL). The characteristics of each EV and the CP used are defined. Moreover, to understand the charging needs of the EV, owner habits are considered through daily usage and time at which the user plugs in the EV. Finally, basic economic data to understand household consumption is included.

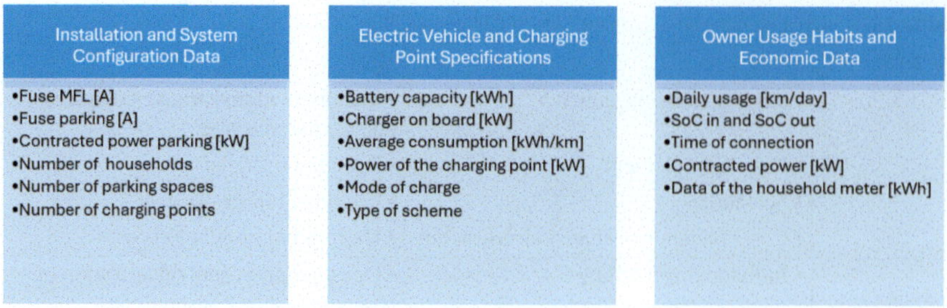

Fig. 24.1 Classification of the inputs for the analysis

The methodology facilitates a comparison between the building's electrical installation capacity and its current consumption, including the EV load. This comparison helps identify whether further development of the charging infrastructure is feasible. Figure 24.2 defines the principal outputs of the program in a simplified manner. For each EV, different outputs are computed. The evolution of the State of Charge (SoC) computes how the charge of the vehicle changes per hour considering the user needs and the capacity of the charger. This derives into a weekly evolution, considering the average daily consumption of the vehicle and the time of connection. The evolution of the SoC allows to compute the charging power used per hour per day.

Although different charging strategies can be considered, a first analysis is done with the "Arrive and Charge" strategy. This implies that the users connect the EV when arriving at the MURB and this is charged at full power until the desired or maximum SoC. This consideration defines the worst-case scenario.

From the economic standpoint, the total electric bill for the EV user is computed and the part of the bill associated with EV consumption is identified. To compare the advantages and disadvantages of the different schemes, the bill considers three possibilities: all schemes are 1, 2, or 3.

The following assumptions have been made for the first stage of the technical analysis:

- Daily mobility: starts to charge on Sunday (Sunday to Thursday).
- Vehicles charge at full power.
- Owners connect the vehicles every day to achieve the desired SoC.
- Charging strategy: arrive and charge.
- All vehicles use Mode of Charge 3 allowing for load management.

This initial analysis focuses on typical weekday mobility, assuming that users drive similar distances each day and plug in their vehicles at consistent times. Weekends are excluded, as weekend travel often involves irregular patterns and less predictable charging behavior, with users likely charging away from home.

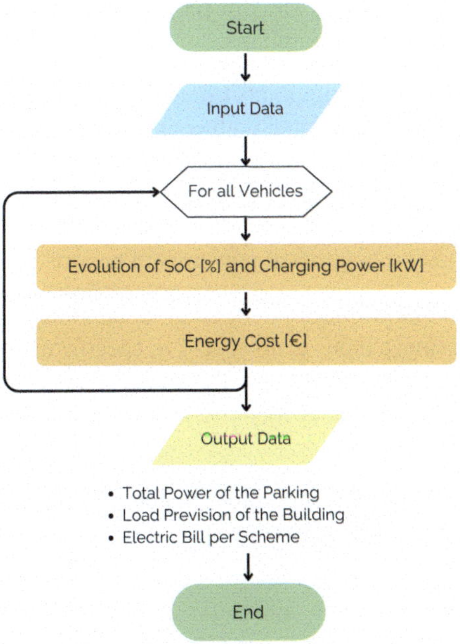

Fig. 24.2 Flow diagram of the methodology for the analysis

The power used for the analysis of the electric system capacity is the load of the entire building considering the load prevision defined in ITC-BT-52 for collective schemes.

$$P_{Building} = P_1 + P_2 + P_3 + P_4 + f_{PSM} \cdot P_5 \tag{24.1}$$

where:

- P_1: Power load corresponding to the group of households. Household consumption is assumed to be of basic electrification (5.75 kW/25 A) defined in ITC-BT-10. The simultaneity factor is chosen considering the number of households in the building.
- P_2: Power load corresponding to common services. Although usually given by square meters, an approximation is used of 1.4 by the number of households.
- P_3: Power load corresponding to offices and commercial establishments.
- P_4: Power load corresponding to the garage, different from charging of the EVs. Parking consumption, aside from EVs charging is computed considering a standard parking space measures 2.3 m x 5 m and a factor of 1.2 for the free space.
- P_5: Power load of the EVs consumption.
- f_{PSM}: Simultaneity factor of EV charging. If the simultaneity factor is unknown, it is assumed that the Protection System of the MFL (PSM) is installed and therefore, the value of f is 0.3.

For each EV, the energy and power cost is computed considering the standing charge defined by the Comisión Nacional de los Mercados y la Competencia

(CNMC) [16]. The bill for the three types of schemes is computed considering the following assumptions:

- All charging points use the same scheme.
- Costs associated with contracted power in the parking lot will be divided equally among all charging point owners.
- Costs associated with the energy consumption of the charging points in Scheme 1 will be paid by the owner of the EV proportionally to their consumption.
- Same contracted power in all periods.

The billing system used to distribute energy costs among users of Scheme 1 assumes there is no access to specific energy consumption per period. The energy cost is computed with the main meter lecture. As the total energy consumed per CP is known, the energy cost is distributed proportionally.

24.4 Case Study

The case study considers a simple MURB without commercial establishments that includes a garage with CPs installed. A general definition of the configuration of the building and the characteristics of the electric installation is done in Table 24.2.

Each CP user has a set of inputs. The charger on board of the EV is considered to be 7.2 kW in all cases. All CPs are defined to be AC charging point and single-phase with a power of 3.7 kW. The mode of charge is 3, specific for EV charging. To be able to obtain an average daily consumption of energy of the EV, an average consumption of the EV per kilometer is defined as 0.14 kWh/km within the urban area. Finally it is considered that all EV users plug in the vehicle to obtain a 100 % SoC.

In Table 24.3, the specific data for each EV is defined using information about the car models and approximate values to simulate a real-case scenario. The chosen car models are among the most widely adopted EVs globally and cover a diverse range of purchase prices and battery capacities. To be able to compute the total electricity bill, an average reading of the household meter across all time periods is used. The contracted power for the household consumption is defined at a value of 4.6 kW. However, when considering Scheme 2, the contracted power amounts to 6.9 kW, to accommodate the household and EV load. For the case considering scheme 2, it is assmued that the owners of the CPs are residents in the MURB.

Regarding the cost data, two tariffs are considered. The tariff 2.0TD has the same price periods all year, consequently, if the consumption is the same each month, also the bill. However, for tariff 3.0TD there is discrimination among months. The analysis will consider the worst case scenario with higher prices, that is, considering the periods for January.

Table 24.2 Installation and system configuration data of the MURB

Installation and system configuration data	Value
Fuse MFL [A]	200
Fuse parking [A]	160
Contracted power parking [kW]	25
Number of households	8
Number of parking spaces	16
Number of charging points	4

Table 24.3 Specific CP data per user

Data per user	Charging point			
	1	2	3	4
Name	Nissan Leaf	Nissan Townstar	BMW iX1	BMW i3
Battery capacity [kWh]	62	45	64.8	42
Daily usage [km/day]	27	60	60	30
Initial SoC [%]	40	40	50	50
Time of connection [h]	20:00	18:00	21:00	20:00
Household meter [kWh]	30	38	27	35

24.5 Results

The initial analysis involves modeling the evolution of the SoC to determine the maximum simultaneous load due to the consumption of the EVs. As depicted in Fig. 24.3, users typically connect their EVs on Sunday with a low SoC due to increased weekend mobility. Consequently, the period with the highest number of EVs connected simultaneously and for the longest duration occurs from Sunday night to Monday morning. During the following days, the daily average mobility generates a consistent SoC profile. This pattern underscores the importance of considering weekend travel behavior in the design of EV charging infrastructure.

Examining the power consumption in relation to the SoC profile reveals a peak on Sunday and Monday evenings, correlating with when all EV users connect their vehicles in the afternoon, leading to nighttime peaks. Figure 24.4 illustrates the daily power profile of EV consumption, highlighting these peaks. These values will determine the adequacy of the installation of the building and if change is needed. In order to smooth out consumption if necessary, a Load Management System (LMS) should be used to reallocate consumption or reduce charging power per vehicle.

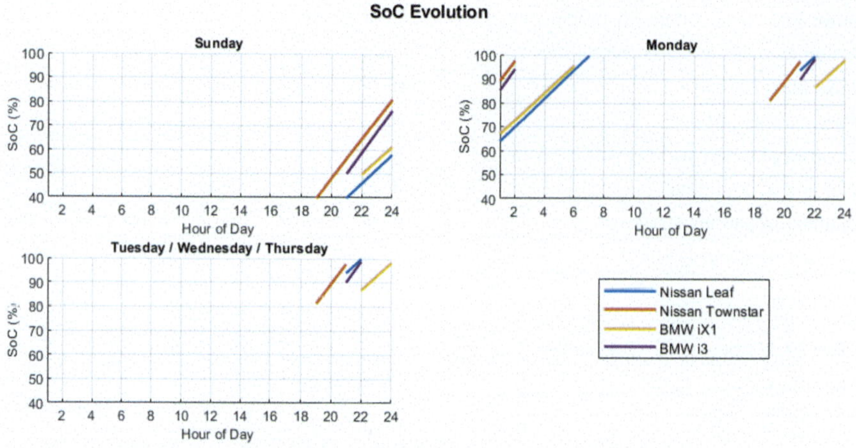

Fig. 24.3 SoC evolution for EVs

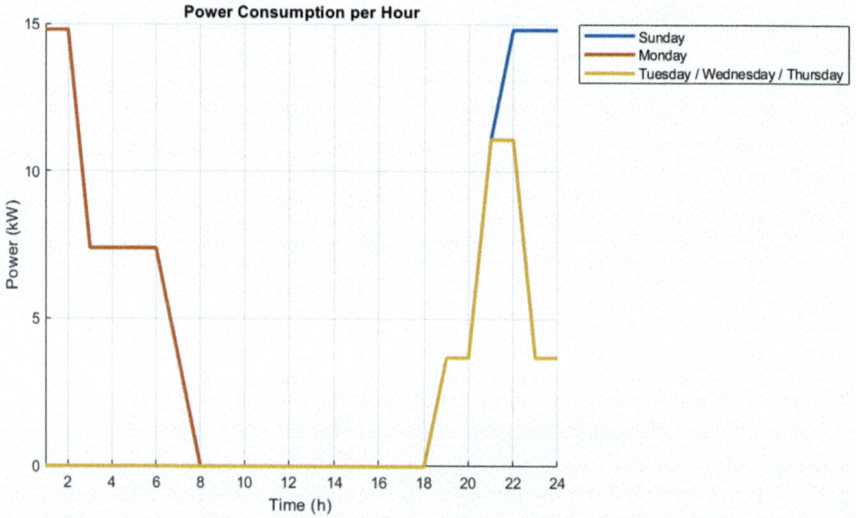

Fig. 24.4 Power consumption profiles of EV charging stations

In the context of the building's electrical system, it is evident that the existing protection system is adequately dimensioned to handle the building's consumption, staying within the MFL fuse capacity. Figure 24.5 indicates that, provided other consumptions remain constant, there is technical feasibility for installing additional EV chargers.

A comparative economic analysis was conducted assuming all users select the same charging scheme. The differences among EVs can be explained by differences in mobility patterns and user habits. These include connection time and daily usage

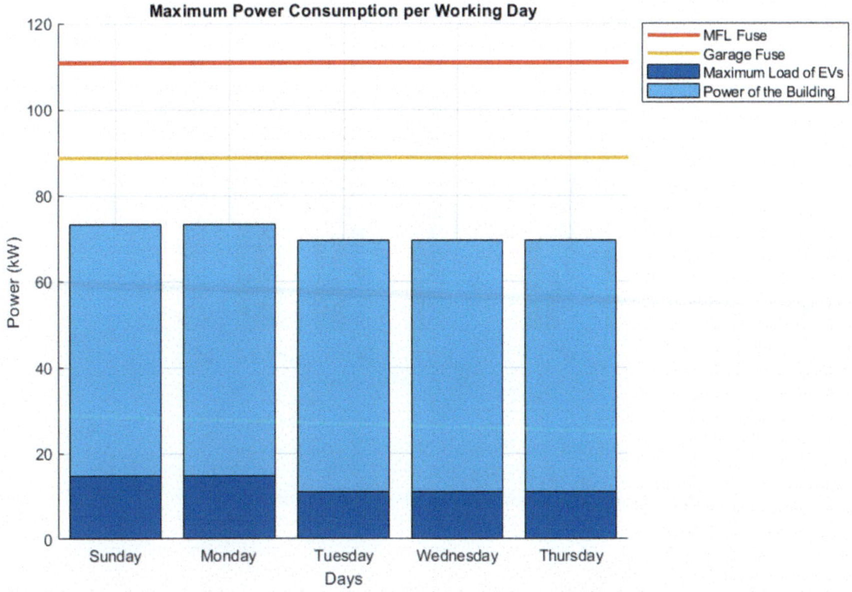

Fig. 24.5 Daily maximum power of the building and EV consumption

among others. The analysis indicates that Scheme 1 is the most expensive in most cases, while Scheme 2 incurs the least additional cost. This is attributed to the shared contracted power between household consumption and EV charging in Scheme 2, eliminating the need for an additional meter. An exception is found in Fig. 24.6 when analysing the Nissan Townstar user. Scheme 3 is more expensive than Scheme 1 in this case. The user benefits from the collective billing as it charges in more expensive hours than the other vehicles. This highlights the importance of the billing strategy in collective schemes and the effects it may have on the individual bills of the users.

Figure 24.7 provides a detailed breakdown of the EV consumption bill for the Nissan Leaf user. This represents the part of the user bill consequence of EV charging. The power cost is a fixed value determined by the contract with the company, based on the maximum power consumed simultaneously. In contrast, the energy cost fluctuates directly with consumption, varying according to the charging power used over time.

The high power cost in Scheme 1 arises from the need for a contracted power exceeding 15 kW in the garage to support all EVs charging simultaneously. This individual cost would decrease with an increase in the number of consumers, as the power cost is evenly distributed. In terms of energy costs, Schemes 2 and 3 are equivalent. Scheme 1 has lower energy costs because energy is charged based on the share per EV user of the total EV consumption without considering charging periods. Vehicles charging during expensive hours benefit more from this scheme. Other billing strategies could be considered for Scheme 1.

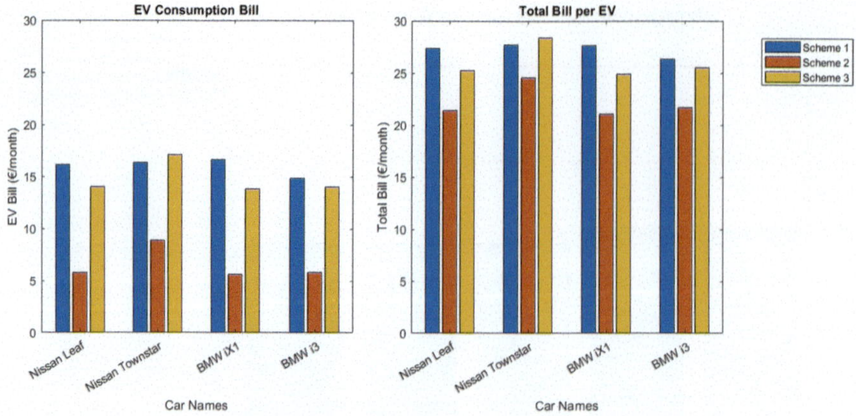

Fig. 24.6 Total electricity bill and specific costs of the EV consumption

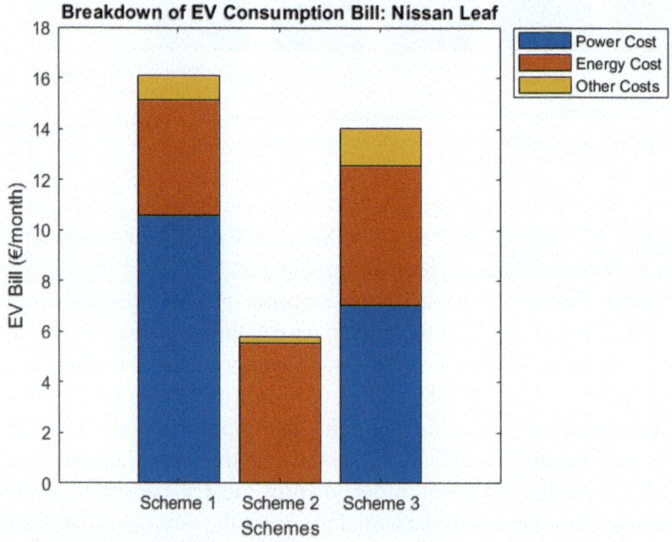

Fig. 24.7 Breakdown of the EV consumption bill for the Nissan Leaf user

Increasing the number of installed charging points will only impact consumers if Scheme 1 is universally chosen. Individual schemes will remain unaffected. Another consideration for Scheme 1 is that the total contracted power of the building - considering that of each meter- can be lowered through using LMS and distributing efficiently power among the vehicles.

While Scheme 2 appears to be the most cost-effective, it would be valuable to consider the potential regulatory and logistical challenges of implementing such power scheme. The need for an individual branch from the household meter to the garage could pose technical limitations and an increase in initial investment.

24.6 Conclusions

Electric mobility is poised to become the future of transportation. Although ownership of EVs is increasing, charging infrastructure is falling behind. This being one of the main topics limiting the widespread of electromobility, there is a need for investing on its development. Focusing on slow charging within urban areas, MURBs represent one of the main areas of interest. The study investigates the integration of CPs for EVs in MURBs in Spain, focusing on both economic and technical aspects. It evaluates different installation schemes for CPs, assessing their impact on electricity billing and the overall feasibility of widespread EV adoption in urban settings.

An economic analysis reveals that while Scheme 2 appears the most cost-effective, its implementation may face regulatory and logistical challenges. The need for individual branches from household meters to the garage can impose technical limitations and increase initial investment. Conversely, individual schemes like Schemes 2 and 3 allow users to manage their charging independently, providing more precise billing and potentially reducing bill costs.

The centralized approach of Scheme 1 introduces complexities in energy cost distribution, necessitating transparent reporting and oversight mechanisms to ensure fair billing practices. This scheme is most effective when adopted by a large portion of the community, thereby reducing power costs per user.

From a technical perspective, the case study presents the relevance of EV consumption when analysing the capacity of the electric system of the building. This highlights the importance of LMSs to smooth out the power profile. On the same note, future challenges include analysing the consequences of installing more CPs. The increase in load could imply the need for modifications in the system. Extensions to the system may require redimensioning wiring and protection devices, incurring additional administrative and economic consequences. LMSs could avoid this up to a certain point by controlling load peaks. The analysis should also consider the possibility of combining different schemes within the same building to optimize cost-effectiveness and operational efficiency.

Future studies should address these challenges, incorporating flexible installation schemes and considering other charging strategies that encourage off-peak charging and enhance the financial viability of EV ownership in MURBs. Additionally, long-term considerations should include ongoing operational costs and their impact on the community's energy consumption patterns. Finally, more realistic simulations should incorporate detailed user habits and behaviors, ensuring that solutions are aligned with the actual usage patterns of EV owners.

The economic implications of CP installation schemes are particularly significant in cities with diverse financial capabilities among residents. The choice of scheme impacts not only individual electricity costs but also the overall affordability and appeal of EV ownership, which influences adoption rates. By considering the technical and economic aspects of each scheme, stakeholders can make informed decisions that balance individual needs with broader goals, such as reducing costs, enhancing system efficiency, and fostering sustainable urban transportation.

References

1. H. Ritchie, Cars, planes, trains: where do CO_2 emissions from transport come from? [Online]. Available: https://ourworldindata.org/co2-emissions-from-transport (2020)
2. Consejo Europeo y Consejo de la Unión Europea, Objetivo 55. Available: https://www.consilium.europa.eu/es/policies/green-deal/fit-for-55/
3. Asociación Española de Fabricantes de Automóviles y Camiones (ANFAC), Barómetro de la electromovilidad (2024). [In Spanish]
4. Gobierno de España, *Plan Nacional Integrado de Energía y Clima (PNIEC) 2021–2030* (Ministerio para la Transición Ecológica y el Reto Demográfico, Ene, 2020). ([In Spanish])
5. ECODES, Plan de despliegue de puntos de recarga eléctrica en España (2021). [In Spanish]
6. Asociación Española de Fabricantes de Automóviles y Camiones (ANFAC), Vehículo electrificado—Informe anual 2023 (2023). [In Spanish]
7. L. Noel, G.Z. de Rubens, B.K. Sovacool, J. Kester, Fear and loathing of electric vehicles: the reactionary rhetoric of range anxiety. Energy Res. Soc. Sci. **48**, 96–107 (2019). [Online]. Available: https://doi.org/10.1016/j.erss.2018.10.001
8. European Commission [Eurostat], House or flat: where do you live? [Press release] (2020). https://ec.europa.eu/eurostat/en/web/products-eurostat-news/-/DDN-20200513-1?inheritRedirect=true&redirect=/eurostat/en/news/whats-new
9. F. Pardo-Bosch, P. Pujadas, C. Morton, C. Cervera, Sustainable deployment of an electric vehicle public charging infrastructure network from a city business model perspective. Sustain. Cities. Soc. **71**, 102957 (2021). Available: https://doi.org/10.1016/j.scs.2021.102957
10. A. Saldarini, S.M. Miraftabzadeh, M. Brenna, M. Longo, Strategic approach for electric vehicle charging infrastructure for efficient mobility along highways: a real case study in Spain. Vehicles **5**(3), 761–779 (2023). Available: https://www.mdpi.com/2624-8921/5/3/42
11. A. Colmenar-Santos, C. De Palacio, D. Borge-Diez, O. Monzón-Alejandro, Planning minimum interurban fast charging infrastructure for electric vehicles: methodology and application to Spain. Energies **7**(3), 1207–1229 (2014). Available: https://www.mdpi.com/1996-1073/7/3/1207
12. I. Lopez-Behar, M. Tran, T. Froese, J.R. Mayaud, O.E. Herrera, W. Merida, Charging infrastructure for electric vehicles in multi-unit residential buildings: mapping feedbacks and policy recommendations. Energy Policy **126**, 444–451 (2019). Available: https://www.sciencedirect.com/science/article/pii/S0301421518306840
13. D. Lopez Behar, Installation of charging infrastructure for electric vehicles in multi-unit residential buildings in British Columbia, Master's Thesis, University of British Columbia (2017). Available: https://doi.org/10.14288/1.0361163
14. J. Ramsebner, A. Hiesl, R. Haas, Efficient load management for BEV charging infrastructure in multi-apartment buildings. Energies **13**(22), 1–15 (2020). Available: https://doi.org/10.3390/en13225927
15. Ministry of Industry, Energy and Tourism, Instalaciones con fines especiales. Infraestructura para la recarga de vehículos eléctricos [Instructional Technical Complementary (ITC) BT 52]. Boletín Oficial del Estado **301** 101732–101744 (2014)
16. Comisión Nacional de los Mercados y la Competencia. (2023, December 21). Resolución de 21 de diciembre de 2023, de la Comisión Nacional de los Mercados y la Competencia, por la que se establecen los valores de los peajes de acceso a las redes de transporte y distribución de electricidad de aplicación a partir del 1 de enero de 2024 [Resolution of the National Commission for Markets and Competition, of December 21, 2023, establishing the values of the tolls for access to electricity transmission and distribution networks applicable from January 1, 2024]. Boletín Oficial del Estado, BOE-A-2023-26251, 171491-171500

Chapter 25
Water Level Control in Small-Scale Hydroelectric Plants Using Differential Flatness

Alexis Castelan-Perez, Francisco Beltran-Carbajal, Ivan Rivas-Cambero, and Jesus C. Hernandez

Abstract The forecasted increase in electricity demand presents challenges due to the desire for renewable generation with minimal environmental impact. In this context, small-scale hydroelectric plants represent a viable alternative to traditional generation methods. This paper presents the design of a water level controller based on differential flatness, applied to a small-scale hydroelectric plant, with the aim of using water efficiently. The dynamic model of the system resembles that of a two-tank system, which exhibits the property of differential flatness. Leveraging this property, its representation in terms of differential flatness is developed, facilitating the controller's design. Two case studies are presented through computer simulation, in which the system must follow reference trajectories and is subject to operational constraints, with the aim of validating the proposed control scheme. In the first case study, the regulation of the water level in the head tank is addressed, while in the second, the objective is to regulate the water level in the surge tank.

Keywords Differential flatness · Water level control · Small scale hydroelectric plants

A. Castelan-Perez · I. Rivas-Cambero
Departamento de Posgrado, Universidad Politécnica de Tulancingo, Hidalgo, Mexico
e-mail: alexis.castelan1294@upt.edu.mx

I. Rivas-Cambero
e-mail: ivan.rivas@upt.edu.mx

F. Beltran-Carbajal (✉)
Departamento de Energía, Universidad Autónoma Metropolitana, Unidad Azcapotzalco, Mexico City, Mexico
e-mail: fbeltran@azc.uam.mx

J. C. Hernandez
Departamento de Ingeniería Eléctrica, Universidad de Jaén, Andalucía, Spain
e-mail: jcasa@ujaen.es

D. Rossit et al. (eds.), *Proceedings of the VII Ibero-American Congress of Smart Cities, ICSC-Cities 2024, 12-14 November, San Carlos, Costa Rica*, Springer Proceedings in Energy, https://doi.org/10.1007/978-981-96-4301-1_25

25.1 Introduction

The demand for electrical energy has been growing in recent years, presenting challenges in both generation and distribution. In response, many countries are increasingly turning to renewable sources in an effort to reduce the use of fossil fuels [28]. Among the main sources of renewable energy are solar, wind, and hydroelectric power, each with its own advantages and challenges.

Hydroelectric generation offers the advantage of allowing planned operation due to the capacity of dams to ensure a constant water flow. However, large-scale hydropower generation has significant disadvantages. The construction of dams can severely alter ecosystems, change water quality, and affect local flora and fauna [23].

Hydroelectric power plants are the main source of generation in more than thirty countries and account for approximately one-fifth of the world's electricity consumption [27]. This is because hydroelectric technology has been developed and refined over the past century, although the use of water's kinetic energy is a much older practice. Hydroelectric plants can be classified according to their generation capacity. Smaller capacity plants, known as small-scale hydroelectric plants, have a generation capacity ranging from 10 to 100 kW [5]. The advantages of small-scale hydroelectric plants include their low implementation cost and minimal environmental impact, provided they are properly planned [17, 22]. Currently, there are approximately 80,000 small-scale hydroelectric plants in operation or under construction worldwide, with China and Europe leading their development through private investments and rural electrification programs [8].

Various works address the challenge of regulating water levels in reservoirs to optimize hydroelectric power generation. Proper regulation not only enhances energy production, particularly during periods of flooding or drought, but also ensures a minimum flow necessary for the survival of aquatic wildlife [11, 14]. The control techniques used for water level regulation are varied. PI and PID controllers are widely used for this purpose, as they have proven effective in practical applications [6, 15]. Likewise, model predictive control (MPC) has been successfully implemented due to its ability to optimize water management and allow rapid reactions to variations in production plans [13, 31]. Additionally, model-free control for hydroelectric plants effectively maintains water level setpoints under severe operating conditions, improving energy generation performance [16]. As observed, water level control in hydroelectric plants is an active area of research, significantly contributing to the stability and operational efficiency of these systems.

Differential flatness is a structural property observed in various systems, enabling trajectory tracking to be simplified without solving complex differential equations [10, 18]. This property facilitates the parametrization of a system's state variables and control inputs in terms of a flat output and a finite number of its time derivatives [24]. Controllers leveraging differential flatness have been successfully implemented across diverse applications, significantly enhancing both robustness and trajectory tracking performance. In robotics, differential flatness-based controllers have demonstrated notable advantages. For example, [7] showcases their ability to improve robustness and trajectory tracking in robotic exoskeletons designed for rehabilitation. Similarly, [26] highlights a controller for mobile robotic manipulators based on differential flatness, demonstrating the property's adaptability in robotic systems. In mechanical systems, differential flatness has been applied for various purposes. For instance, [1] employs this property in a trajectory control and vibration suppression scheme, achieving robust and efficient performance under uncertainties and external disturbances. Likewise, [4] combines differential flatness with algebraic parameter identification techniques to analyze and control mechanical systems with multiple degrees of freedom. The control of electrical machines has also benefited from differential flatness. In [3], a differential flatness-based controller is used for position control in a permanent magnet DC motor, yielding precise and reliable results. Additionally, [2] proposes a trajectory tracking controller for permanent magnet synchronous motors, effectively managing various uncertainties. Applications in power electronics further illustrate the versatility of differential flatness-based controllers. For instance, [25] presents a controller for a modular multilevel converter, achieving high power performance even in the presence of noise and parametric disturbances. In energy management for electromobility, [29] introduces a differential flatness-based control algorithm to optimize hybrid battery/supercapacitor energy sources. Similarly, in water level regulation for coupled tank systems, [12] combines differential flatness with parameter estimation to achieve precise and efficient water level control. These diverse examples underscore the versatility and effectiveness of differential flatness-based controllers in enhancing performance and robustness across a wide range of systems and applications.

The present manuscript is organized as follows: Sect. 25.2 addresses the dynamic modeling of the small-scale hydroelectric plant, as well as its representation in differential flatness and the design of the water level controller. Section 25.3 presents the computer simulations conducted to validate the proposed control scheme, considering two case studies. Finally, the simulation results are analyzed, and conclusions from the developed work are drawn.

25.2 Differential Flatness-Based Water Level Control

25.2.1 System Dynamic Model

The dynamic modeling of small-scale hydroelectric plants, which have a head tank and a surge tank, can be done by comparing them to a two-tank system [9]. Figure 25.1 shows a diagram representing this system. The system modeling is explained in detail in [20], and the equations describing its dynamics are as follows:

$$\dot{h}_1 = \frac{1}{A_1}\left[q_{in} - \frac{1}{R}(h_1 - h_2)\right], \tag{25.1}$$

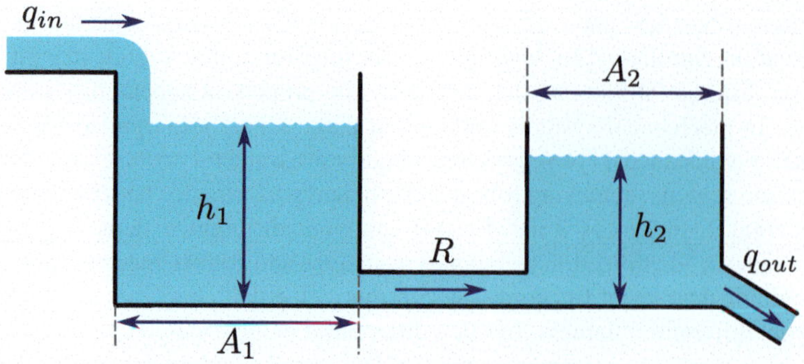

Fig. 25.1 Representation of the small-scale hydroelectric plant as a two-tank system

$$\dot{h}_2 = \frac{1}{A_2}\left[\frac{1}{R}(h_1 - h_2) - q_{out}\right], \tag{25.2}$$

where h_1 is the water level in the head tank, h_2 is the water level in the surge tank, R is the hydraulic resistance, A_1 is the horizontal area of the head tank, and A_2 is the horizontal area of the surge tank. q_{in} is the inflow rate from a river, and q_{out} is the outflow rate sent to the turbine.

The regulation of water levels in two-tank systems has been widely studied in the literature, generating a substantial knowledge base on the subject. For instance, in [19], water level regulation is addressed using neural networks, and its performance is compared with a PID controller. In another study, [21] presents the design and experimental validation of a nonlinear controller for a coupled two-tank system, demonstrating its effectiveness in liquid level tracking. Similarly, the control algorithm proposed in [30], based on multiparametric programming, effectively regulates liquid levels in two tanks, minimizing control effort and adapting to systems with dynamic variations or parameter uncertainties.

25.2.2 *Differential Flatness*

The two-tank system is considered to be differentially flat, with the flat output being the water level in the head tank $y = h_1$. As a flat system, it can be parameterized in terms of the flat output and a finite number of its time derivatives:

$$\begin{aligned} y &= h_1, \\ h_2 &= A_1 R\dot{y} - Rq_{in} + y, \\ q_{out} &= -A_1 A_2 R\ddot{y} - (A_2 + A_1)\dot{y} + A_2 R\dot{q}_{in} + q_{in}. \end{aligned} \tag{25.3}$$

The system can be represented in state-space using the system parameterization as:

$$\dot{z}_1 = z_2,$$
$$\dot{z}_2 = -\left(\frac{1}{A_1 R} + \frac{1}{A_2 R}\right) z_2 + \frac{1}{A_1}\dot{q}_{in} + \frac{1}{A_1 A_2 R} q_{in} - \frac{1}{A_1 A_2 R} q_{out},$$

where $z_1 = y$ and $z_2 = \dot{y}$, which are components of the state vector $z = [z_1, \ z_2]^T$. Based on the system parameterization shown in Eq. (25.3), it is possible to plan reference trajectories for the other system variables:

$$\begin{aligned} y^\star &= h_1^\star, \\ h_2^\star &= A_1 R \dot{y}^\star - R q_{in} + y^\star, \\ q_{out}^\star &= -A_1 A_2 R \ddot{y}^\star - (A_2 + A_1)\dot{y}^\star + A_2 R \dot{q}_{in} + q_{in}, \end{aligned} \tag{25.4}$$

where the notation "*" indicates the reference trajectory of the variable. From the third expression of Eq. (25.3), the following PD action controller based on differential flatness is obtained for tracking the reference trajectories of the flat output y:

$$\begin{aligned} q_{out} &= -A_1 A_2 R u - (A_2 + A_1)\dot{y} + A_2 R \dot{q}_{in} + q_{in}, \\ u &= \ddot{y}^\star - k_d \dot{e} - k_p e. \end{aligned} \tag{25.5}$$

The tracking error is defined as $e = y - y^\star$. The controller gains are selected using a second-order reference system for the closed-loop error dynamics:

$$\ddot{e} + 2\omega_n \zeta \dot{e} + \omega_n^2 e = 0,$$

so that $k_d = 2\omega_n \zeta$ and $k_p = \omega_n^2$. Thus, as long as $\omega_n > 0$ and $\zeta > 0$, the system stability is ensured.

25.3 Simulation

To validate the differential flatness-based controller, two case studies are proposed. In the first, the goal is to regulate the water level in the head tank, h_1. In the second case study, the goal is to regulate the water level in the surge tank, h_2. In both cases, the system must follow reference trajectories generated using Bézier polynomials for water level regulation. The system is subject to operational constraints similar to those presented in [13]:

- The flow rate to the turbine must be greater than $0.4\,m^3/s$ to avoid affecting the local fauna.
- The water level in both tanks must always be between 6 and 7.5 m.

The system parameters, obtained from [6], are $A_1 = 92\,m^2$, $A_2 = 5\,m^2$, and $R = 0.1$ s m^{-2}. The simulation is performed in MATLAB using the fourth-order Runge-Kutta method with an integration step of 0.01 s to solve the differential equations.

25.3.1 Case Study 1

In this case study, the objective is to track trajectories for regulating the water level in the head tank h_1, which is the flat output of the system. The following trajectory, based on Bézier polynomials, is proposed:

$$y^{\star} = \begin{cases} \alpha_0, & \text{for } 0 \leq t < t_1 \\ \alpha_0 + (\alpha_1 - \alpha_0) \sum_{k=1}^{3} r_k \left(\frac{t-t_1}{t_2-t_1}\right)^{2+k}, & \text{for } t_1 \leq t < t_2 \\ \alpha_1, & \text{for } t_2 \leq t < t_3 \\ \alpha_1 + (\alpha_2 - \alpha_1) \sum_{k=1}^{3} r_k \left(\frac{t-t_3}{t_4-t_3}\right)^{2+k}, & \text{for } t_3 \leq t < t_4 \\ \alpha_2, & \text{for } t \geq t_4 \end{cases}, \tag{25.6}$$

where $\alpha_0 = 7.2\,\text{m}$, $\alpha_1 = 7.5\,\text{m}$, $\alpha_2 = 7\,\text{m}$, $t_1 = 0\,\text{s}$, $t_2 = 150\,\text{s}$, $t_3 = 300\,\text{s}$, $t_4 = 450\,\text{s}$, $r_1 = 10$, $r_2 = -15$, $r_3 = 6$. Similarly, the inflow rate q_{in} is proposed as a Bézier polynomial:

$$q_{in} = \begin{cases} \beta_0, & \text{for } 0 \leq t < t_1 \\ \beta_0 + (\beta_1 - \beta_0) \sum_{k=1}^{3} r_k \left(\frac{t-t_1}{t_2-t_1}\right)^{2+k}, & \text{for } t_1 \leq t < t_2 \\ \beta_1, & \text{for } t \geq t_2 \end{cases}, \tag{25.7}$$

where $\beta_0 = 0.9\,\text{m}^3/\text{s}$, $\beta_1 = 1.5\,\text{m}^3/\text{s}$, $t_1 = 0\,\text{s}$, $t_2 = 400\,\text{s}$. The simulation runs for 500 s, with initial conditions of $h_1(0) = 7.1\,\text{m}$ and $h_2(0) = 7\,\text{m}$.

Figure 25.2 shows a graphical representation of the proposed controller for this case study. The reference trajectory for the flat output and the inflow rate, as shown in Eqs. (25.6) and (25.7), respectively, provide the necessary information for regulating the water level. According to Eq. (25.5), it is not necessary to know the state of the variable h_2.

Figure 25.3 shows the simulation result for the variable h_1. Since the initial condition of the system is lower than the reference, it takes about 25 s to reach the reference,

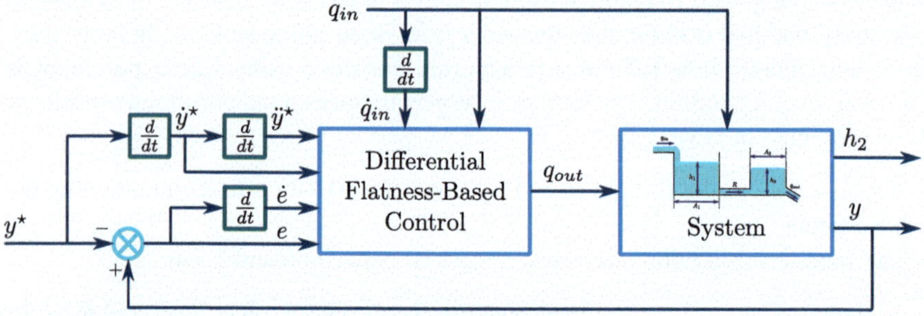

Fig. 25.2 Differential flatness-based control diagram for case study 1

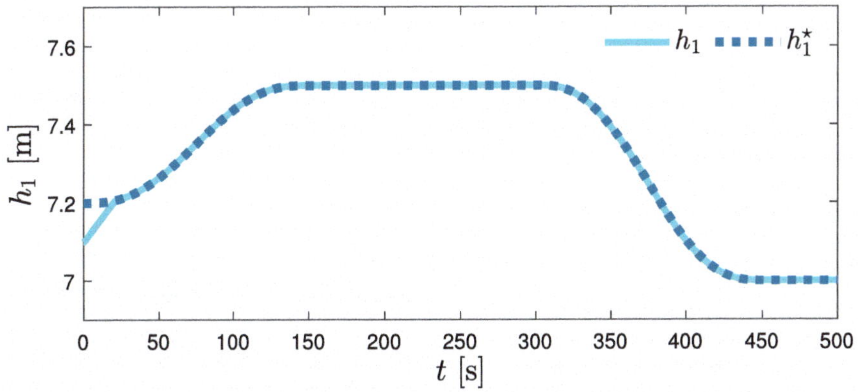

Fig. 25.3 Simulation result for h_1 in case study 1

depending on the inflow rate. From this point, the designed control closely follows the reference trajectory described in Eq. (25.6), both in the constant and varying segments.

The reference trajectories for the other variables, parameterized as a function of the flat output as shown in Eq. (25.4), allow us to predict the dynamics of all variables, not just the flat output. Thus, Fig. 25.4 shows the behavior of h_2 relative to the reference trajectory and the conditions established for this case study. It is observed that the reference trajectory also differs from the initial condition of the system for h_2. However, like h_1, it reaches the reference thanks to the inflow rate.

Observing the reference trajectory of the outflow rate q_{out} and comparing it with the simulation signal, as shown in Fig. 25.5, reveals an initial discrepancy. This occurs because the system's initial conditions are lower than the proposed trajectories, causing the control to minimize outflow to quickly reach the reference. At this point, operational constraints come into play: ideally, additional inflow would be preferable, but this would disrupt normal dam operation. The constraints require a minimum flow of 0.4 m^3/s to the turbine, as seen in Fig. 25.5.

Finally, Fig. 25.6 shows the proposed inflow rate q_{in} for this case study. This has high variability considering the type of system being addressed in this work, but it provides valuable information for validating the proposed controller. Observing Fig. 25.3 in the last 50 s, it is evident that the level of h_1 remains constant. During this time interval, the inflow rate q_{in} is also constant.

This means that to maintain a constant level, as much water must leave the dam as enters the system, which is perfectly observed in Figs. 25.5 and 25.6.

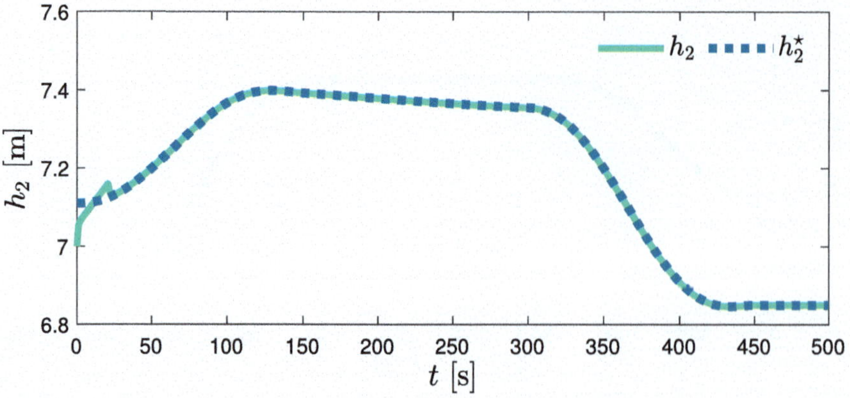

Fig. 25.4 Simulation result for h_2 in case study 1

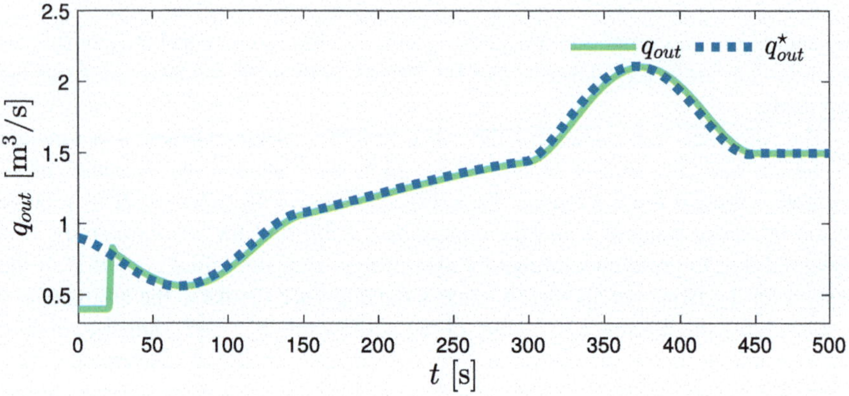

Fig. 25.5 Simulation result for q_{out} in case study 1

25.3.2 Case Study 2

In this second case study, the objective is to follow trajectories for regulating the water level in the surge tank, h_2. This differs from the previous case as here it is necessary to control a non-flat variable. The proposed approach to tackle this problem is to include a trajectory generator for the flat output, which, from a reference trajectory of h_2, can generate a trajectory for y. Figure 25.7 shows the incorporation of this element into the control diagram.

One option for generating the reference trajectory $y^\star$ can be derived from the dynamic model of the system. Taking Eq. (25.1), considering that $y^\star = h_1^\star$ and substituting h_2 with $h_2^\star$, we obtain:

$$\dot{y}^\star = \frac{1}{A_1}\left[q_{in} - \frac{1}{R}\left(y^\star - h_2^\star\right)\right]. \tag{25.8}$$

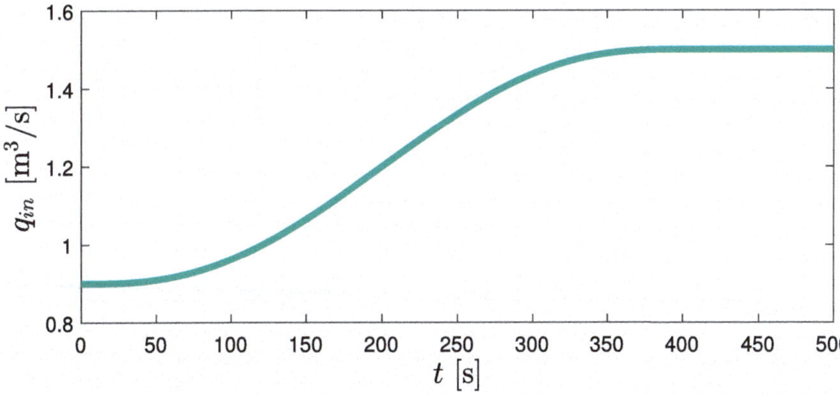

Fig. 25.6 Simulation result for q_{in} in case study 1

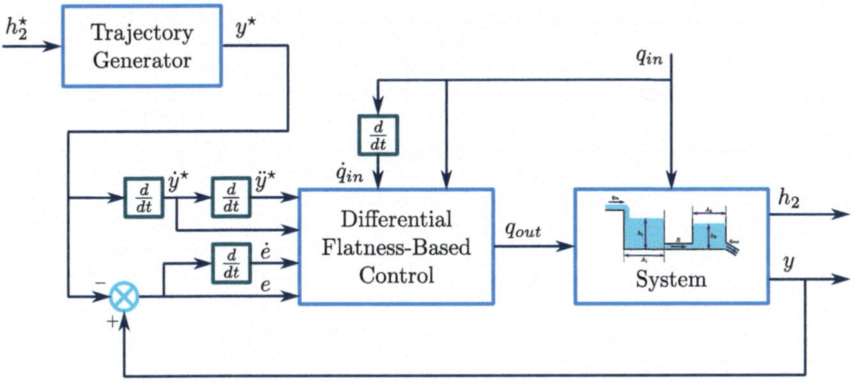

Fig. 25.7 Differential flatness-based control diagram for case study 2

By setting the initial condition of $y^\star$ and the reference trajectory $h_2^\star$, the differential equation can be solved to obtain the trajectory for the flat output y, with which the water level h_2 can be regulated.

The proposed reference trajectory $h_2^\star$ for this case study is based on Bézier polynomials:

$$h_2^\star = \begin{cases} \gamma_0, & \text{for } 0 \le t < t_1 \\ \gamma_0 + (\gamma_1 - \gamma_0) \sum_{k=1}^{3} r_k \left(\frac{t-t_1}{t_2-t_1}\right)^{2+k}, & \text{for } t_1 \le t < t_2 \\ \gamma_1, & \text{for } t_2 \le t < t_3 \\ \gamma_1 + (\gamma_2 - \gamma_1) \sum_{k=1}^{3} r_k \left(\frac{t-t_3}{t_4-t_3}\right)^{2+k}, & \text{for } t_3 \le t < t_4 \\ \gamma_2, & \text{for } t \ge t_4 \end{cases}, \tag{25.9}$$

where $\gamma_0 = 6.3\,\mathrm{m}$, $\gamma_1 = 7\,\mathrm{m}$, $\gamma_2 = 7.3\,\mathrm{m}$, $t_1 = 0\,\mathrm{s}$, $t_2 = 100\,\mathrm{s}$, $t_3 = 200\,\mathrm{s}$, and $t_4 = 350\,\mathrm{s}$. For this case, a slightly more complex inlet flow q_{in} is proposed to observe the controller's behavior in various scenarios. The following Bézier polynomial is proposed for this purpose:

$$q_{in} = \begin{cases} \sigma_0, & \text{for } 0 \le t < t_1 \\ \sigma_0 + (\sigma_1 - \sigma_0) \sum_{k=1}^{3} r_k \left(\frac{t-t_1}{t_2-t_1}\right)^{2+k}, & \text{for } t_1 \le t < t_2 \\ \sigma_1 + (\sigma_2 - \alpha_1) \sum_{k=1}^{3} r_k \left(\frac{t-t_2}{t_3-t_2}\right)^{2+k}, & \text{for } t_2 \le t < t_3 \\ \sigma_2, & \text{for } t \ge t_3 \end{cases}, \tag{25.10}$$

where $\sigma_0 = 1.6\,\mathrm{m^3/s}$, $\sigma_1 = 0.85\,\mathrm{m^3/s}$, $\sigma_2 = 1.2\,\mathrm{m^3/s}$, $t_1 = 0\,\mathrm{s}$, $t_2 = 250\,\mathrm{s}$, $t_3 = 400\,\mathrm{s}$. The simulation is performed over a period of 500 s, with initial conditions $h_1(0) = 6.2\,\mathrm{m}$, $h_2(0) = 6.2\,\mathrm{m}$, and $y^\star(0) = 6.2\,\mathrm{m}$.

By setting the initial condition of h_1 and the reference trajectory $y^\star$ to the same value, an almost perfect trajectory tracking in h_1 is achieved, as shown in Fig. 25.8. It is important to note that the trajectory generation is done by numerically solving Eq. (25.8).

In the case of h_2, as seen in Fig. 25.9, there is a noticeable difference between the variable and its reference $h_2^\star$ in the initial moments. This is because the initial condition of the system is lower than that presented in the reference trajectory of Eq. (25.9). Therefore, it is necessary for the inlet flow to increase the water level to start following the reference trajectory.

Figure 25.10 shows the outlet flow to the turbine q_{out}. Initially, the outlet flow is at a minimum because it is intended for the water level h_2 to reach the reference trajectory. Once reached, it starts to follow the trajectory obtained from the system parametrization. It is apparent that there are sections where the trajectory and the simulation result are not exactly the same, which may be due to small differences

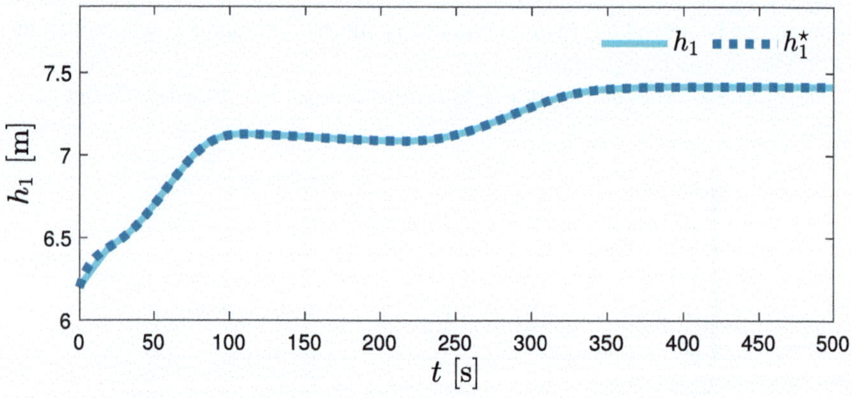

Fig. 25.8 Simulation result for h_1 in case study 2

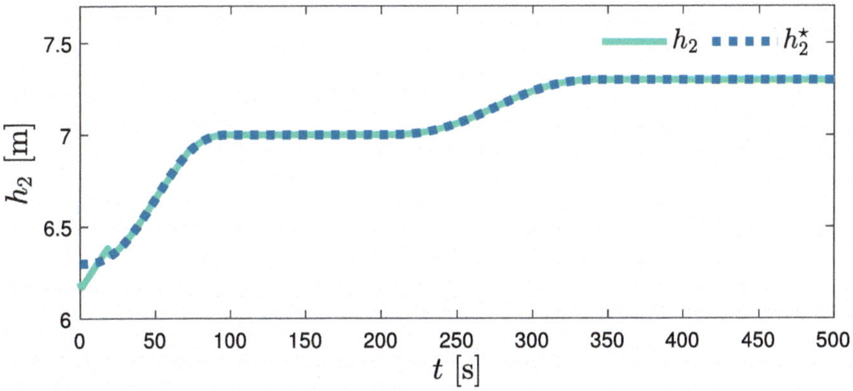

Fig. 25.9 Simulation result for h_2 in case study 2

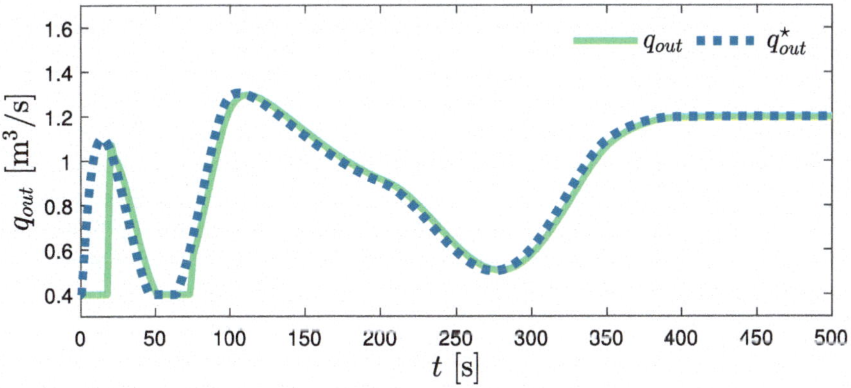

Fig. 25.10 Simulation result for q_{out} in case study 2

between the reference trajectory $y^\star$ and the flat output y, as well as their temporal derivatives.

Finally, Fig. 25.11 shows the inlet flow to the system for the second case study, represented in Eq. (25.10). Initially, there is a descending water input that starts at 1.6 m^3/s and ends at 0.85 m^3/s. This water input was proposed to verify that, even in situations where the inlet flow decreases, the developed control is capable of following the established reference trajectory. After that, a faster increase in flow was incorporated to observe the system's behavior.

As in the first case study, a constant reference level was established at the end, aiming to observe whether, in this situation, the inflow of water is the same as the outflow. This can be observed by comparing the graphs in Figs. 25.10 and 25.11.

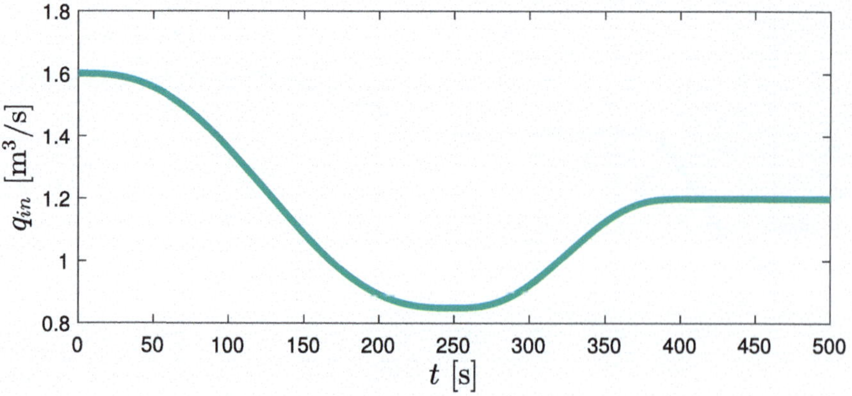

Fig. 25.11 Simulation result for q_{in} in case study 2

25.4 Conclusions

This paper has presented the design of a water level controller based on differential flatness, applied to a small-scale hydroelectric plant, aiming to use water efficiently. The dynamic model of the system resembles a two-tank system with the property of differential flatness. Utilizing this property, its representation in terms of differential flatness is developed, facilitating the controller's design. Two case studies are presented through computer simulations, where the system follows reference trajectories and is subject to operational constraints to validate the proposed control scheme. In the first case study, the regulation of the head tank water level is addressed, while in the second, the goal is to regulate the water level in the surge tank.

The results from the simulations demonstrate that the differential flatness-based controller effectively regulates water levels in both tanks, adhering to the set constraints. The controller ensures the system's operational efficiency, proving its applicability in small-scale hydroelectric plants. This work contributes to the ongoing research in the field of water level control in hydroelectric plants, highlighting the potential for differential flatness-based controllers to enhance the performance and robustness of these systems. Future work could explore the implementation of this control approach in real-world settings and investigate its adaptability to different types of hydroelectric systems and varying operational conditions.

Acknowledgements The authors of this article thank the Consejo Nacional de Humanidades, Ciencias y Tecnologías (CONAHCYT), support provided for developing this work. In addition, the authors acknowledge the support provided by the Thematic Network 723RT0150 "Red para la integración a gran escala de energías renovables en sistemas eléctricos (RIBIERSE-CYTED)" financed by the call for Thematic Networks of the CYTED (Ibero-American Program of Science and for Development) for 2022.

References

1. F. Beltran-Carbajal, G. Silva-Navarro, Output feedback dynamic control for trajectory tracking and vibration suppression. Appl. Math. Model. **79**, 793–808 (2020). https://doi.org/10.1016/j.apm.2019.11.004
2. F. Beltran-Carbajal, A. Favela-Contreras, J.L. Hernandez-Avila, O. Olvera-Tapia, D. Sotelo, C. Sotelo, Dynamic output feedback control for desired motion tracking on synchronous motors. Int. Trans. Electr. Energy Syst. **30**(3), e12260 (2020). https://doi.org/10.1002/2050-7038.12260
3. F. Beltran-Carbajal, A. Favela-Contreras, A. Valderrabano-Gonzalez, J.C. Rosas-Caro, Output feedback control for robust tracking of position trajectories for dc electric motors. Electr. Power Syst. Res. **107**, 183–189 (2014). https://doi.org/10.1016/j.epsr.2013.09.020
4. F. Beltran-Carbajal, G. Silva-Navarro, Generalized nonlinear stiffness identification on controlled mechanical vibrating systems. Asian J. Control. **21**(3), 1281–1292 (2019). https://doi.org/10.1002/asjc.1807
5. M.H. Bollen, F. Hassan, *Integration of Distributed Generation in the Power System*. (Wiley, 2011)
6. D. Borkowski, Water level control of a small hydropower plant with a surge tank. J. Hydraul. Res. **55**(2), 284–291 (2017). https://doi.org/10.1080/00221686.2016.1252800
7. B. Brahmi, T. Ahmed, I. El Bojairami, A.A.Z. Swapnil, M. Assad-Uz-Zaman, K. Schultz, E. McGonigle, M.H. Rahman, Flatness based control of a novel smart exoskeleton robot. IEEE/ASME Trans. Mechatron. **27**(2), 974–984 (2021). https://doi.org/10.1109/TMECH.2021.3076956
8. T.B. Couto, J.D. Olden, Global proliferation of small hydropower plants-science and policy. Front. Ecol. Environ. **16**(2), 91–100 (2018). https://doi.org/10.1002/fee.1746
9. S. Endo, M. Konishi, H. Imabayasi, Water level control of small-scale hydroelectric power plant by deadbeat control method, in *2000 26th Annual Conference of the IEEE Industrial Electronics Society. IECON 2000. 2000 IEEE International Conference on Industrial Electronics, Control and Instrumentation. 21st Century Technologies*, vol. 2 (2000), pp. 1123–1128. https://doi.org/10.1109/IECON.2000.972280
10. M. Fliess, J. Lévine, P. Martin, P. Rouchon, Flatness and defect of non-linear systems: introductory theory and examples. Int. J. Control. **61**(6), 1327–1361 (1995). https://doi.org/10.1080/00207179508921959
11. S. Heimerl, N. Schwiersch, Dynamic water-level regulation at run-of-river hydropower plants to increase efficiency and generation. Water **13**(21), 2983 (2021). https://doi.org/10.3390/w13212983
12. C. Huang, B. Du, Differentially flatness active disturbance rejection control approach via algebraic parameter identification to double tank problem, in *2016 35th Chinese Control Conference (CCC)* (IEEE, 2016), pp. 2108–2113. https://doi.org/10.1109/ChiCC.2016.7553678
13. C. Jeong, B. Furenes, R. Sharma, Implementation of simplified sequential stochastic model predictive control for operation of hydropower system under uncertainty. Comput. Chem. Eng. **179**, 108409 (2023). https://doi.org/10.1016/j.compchemeng.2023.108409
14. C. Jeong, R. Sharma, Multistage model predictive control with simplified method on scenario ensembles of uncertainty for hjartdøla hydropower system, in *2023 IEEE Conference on Control Technology and Applications (CCTA)* (IEEE, 2023), pp 733–740. https://doi.org/10.1109/CCTA54093.2023.10252685
15. O.F. Jiménez, M.H. Chaudhry, Water-level control in hydropower plants. J. Energy Eng. **118**(3), 180–193 (1992). https://doi.org/10.1061/(ASCE)0733-9402(1992)118:3(180)
16. C. Join, G. Robert, M. Fliess, Model-free based water level control for hydroelectric power plants. IFAC Proc. Vol. **43**(1), 134–139 (2010). https://doi.org/10.3182/20100329-3-PT-3006.00026
17. T.S. Kishore, E.R. Patro, V. Harish, A.T. Haghighi, A comprehensive study on the recent progress and trends in development of small hydropower projects. Energies **14**(10), 2882 (2021). https://doi.org/10.3390/en14102882

18. J. Levine, *Analysis and Control of Nonlinear Systems: a Flatness-Based Approach* (Springer Science & Business Media, 2009)
19. M. Majstorovic, I. Nikolic, J. Radovic, G. Kvascev, Neural network control approach for a two-tank system, in *2008 9th Symposium on Neural Network Applications in Electrical Engineering* (IEEE, 2008), pp 215–218. https://doi.org/10.1109/NEUREL.2008.4685619
20. K. Ogata, Ingeniería de control moderna. Pearson educación (2010)
21. H. Pan, H. Wong, V. Kapila, M.S. de Queiroz, Experimental validation of a nonlinear backstepping liquid level controller for a state coupled two tank system. Control. Eng. Pract. **13**(1), 27–40 (2005). https://doi.org/10.1016/j.conengprac.2003.12.019
22. M. Rotilio, C. Marchionni, P. De Berardinis, The small-scale hydropower plants in sites of environmental value: an Italian case study. Sustainability **9**(12), 2211 (2017). https://doi.org/10.3390/su9122211
23. S.N. Silva, J.Á.D.Castillo, An approach of the hydropower: advantages and impacts. a review. J. Energy Res. Rev. **8**(1), 10–20 (2021). https://doi.org/10.9734/jenrr/2021/v8i130201
24. H. Sira-Ramirez, S.K. Agrawal, *Differentially Flat Systems* (CRC Press, 2004)
25. P.B. Steckler, J.Y. Gauthier, X. Lin-Shi, F. Wallart, Differential flatness-based, full-order nonlinear control of a modular multilevel converter (mmc). IEEE Trans. Control. Syst. Technol. **30**(2), 547–557 (2021). https://doi.org/10.1109/TCST.2021.3067887
26. C.P. Tang, P.T. Miller, V.N. Krovi, J.C. Ryu, S.K. Agrawal, Differential-flatness-based planning and control of a wheeled mobile manipulator-theory and experiment. IEEE/ASME Trans. Mechatron. **16**(4), 768–773 (2010). https://doi.org/10.1109/TMECH.2010.2066282
27. J.G. Velasco, Energías renovables. Editorial Reverte (2010)
28. L.C. Voumik, M.A. Islam, S. Ray, N.Y. Mohamed Yusop, A.R. Ridzuan, CO_2 emissions from renewable and non-renewable electricity generation sources in the g7 countries: static and dynamic panel assessment. Energies **16**(3), 1044 (2023). https://doi.org/10.3390/en16031044
29. B. Yodwong, P. Thounthong, D. Guilbert, N. Bizon, Differential flatness-based cascade energy/current control of battery/supercapacitor hybrid source for modern e-vehicle applications. Mathematics **8**(5), 704 (2020). https://doi.org/10.3390/math8050704
30. A. Zakaria, E. Mid, M. Mohamed, M. Hussin, A. Shaari, E. Ruslan, D. Hadi, M. Masri, Control algorithm for two-tank system using multiparametric programming. J. Phys. Conf. Ser. **2641**, 012028. IOP Publishing (2023). https://doi.org/10.1088/1742-6596/2641/1/012028
31. S.M. Zanoli, C. Pepe, G. Astolfi, F. Luzi, Reservoir advanced process control for hydroelectric power production. Processes **11**(2), 300 (2023). https://doi.org/10.3390/pr11020300

Chapter 26
Smart Cities Strategies: Data-Driven Urban Energy Transition Using QGIS Heat Maps to Foster Photovoltaic Micro-generation in Manta, Ecuador

Marcos A. Ponce-Jara, Milton Moreano-Alvarado, Catalina Rus-Casas, and Daniel Segovia Muñoz

Abstract This paper presents a data-driven approach to promote photovoltaic (PV) micro-generation systems in the city of Manta, Ecuador, by using QGIS-generated heat maps to identify high-demand areas and assess energy consumption trends citywide. The study not only identifies high-energy consumption areas but also characterizes the city's overall energy profile, providing a comprehensive overview that informs tailored strategies for promoting PV micro-generation. By examining energy demand across various socio-economic zones—including low-income, middle-income, and high-end residential sectors—the research highlights specific areas where targeted interventions can accelerate the adoption of renewable energy. For instance, QGIS analysis revealed that middle-income regions consume between 300 and 1500 kWh/month, while high-end areas exhibit consumption levels exceeding 3350 kWh/month, making them ideal candidates for PV systems. The study emphasizes the use of QGIS, a free and open-source tool, as an accessible platform for developing countries like Ecuador to conduct detailed geographic analyses at minimal cost. Special consideration is given to tourism-driven regions, where seasonal energy demand fluctuations call for more efficient and resilient energy solutions. The paper also explores policy recommendations, such as community-based solar initiatives and government-led financial incentives, to foster PV adoption across

M. A. Ponce-Jara (✉) · M. Moreano-Alvarado · D. S. Muñoz
Faculty of Engineering, Universidad Laica Eloy Alfaro de Manabí, Manta, Ecuador
e-mail: marcos.ponce@uleam.edu.ec

M. Moreano-Alvarado
e-mail: milton.moreano@uleam.edu.ec

D. S. Muñoz
e-mail: eddison.segovia@uleam.edu.ec

C. Rus-Casas
Electronic Engineering and Automatic Department, University of Jaén Las Lagunillas Campus, Jaén, Spain
e-mail: crus@ujaen.es

D. Rossit et al. (eds.), *Proceedings of the VII Ibero-American Congress of Smart Cities, ICSC-Cities 2024, 12-14 November, San Carlos, Costa Rica*, Springer Proceedings in Energy, https://doi.org/10.1007/978-981-96-4301-1_26

different socio-economic groups. By integrating geographic information systems (GIS) with renewable energy policies, this research provides valuable insights into Ecuador's energy transition and supports the development of sustainable, low-carbon smart cities.

Keywords PV micro-generation · GIS-based energy planning · Smart cities

26.1 Introduction

Electricity is a cornerstone of modern society, driving industries, homes, and digital infrastructure essential for economic growth and development. Global electricity demand has emerged in recent decades, fueled by rapid industrialization, urbanization, and growing technological reliance [1, 2]. While this growth has fueled economic progress, it has also led to higher greenhouse gas emissions from fossil fuel-based power generation. With growing concerns over climate change, there is widespread consensus that transitioning to renewable energy is essential to mitigate environmental impacts [3]. Among these, Non-Conventional Renewable Energy (NCRE) technologies, particularly rooftop photovoltaic (PV) systems, have gained popularity. These systems offer benefits such as reduced transmission losses, improved energy security, and greater resilience to power outages [4], making them essential to the global shift toward a sustainable energy future. In fact, NCRE significantly driven by solar PV and wind technologies, are transforming the energy markets by increasingly contributing to the global energy mix [5]. Solar PV systems are vital for meeting rising electricity demand due to their declining costs and scalability. Entities managing power systems are focusing on implementing PV generation near end-user connection nodes, reducing system expansion costs, increasing energy injections from PV generators, lowering consumer bills, and cutting distribution costs in the wholesale market [6, 7].

Despite extensive efforts over past decades in South American countries to advance solar PV energy, the introduction of new energy policies and structural reforms in regulatory frameworks have not fully achieved the anticipated results in promoting widespread deployment of NCREs. Only Brazil, Chile, Peru, and Uruguay have notably succeeded, emerging as regional leaders in solar energy capacity [8]. Most of the studies in this field have focused on assessing and adapting energy policies to meet the specific needs of individuals countries, aiming to enhance the adoption and integration of renewable energy sources. These studies emphasize the importance of tailored policy frameworks that consider local socio-economic, environmental, and technological factors to effectively promote solar energy development [8–11]. Additionally, Geographic Information Systems (GIS) have proven invaluable, enabling researchers and policymakers to identify and prioritize areas with high potential for PV generation. GIS-based analyses incorporate various environmental factors (such as solar radiation levels, land availability, and proximity to existing infrastructure) to optimize the siting of PV systems and maximize their efficiency impact [12–14].

This dual approach, combining policy refinement with advanced spatial analysis, not only accelerates the transition to sustainable energy systems but also supports the development of smart cities, which rely on intelligent data-driven solutions to enhance urban living and sustainability.

However, despite established energy policy frameworks and solar radiation studies in many South American countries, the broad adoption of solar energy might still be impeded by suboptimal planning and design. To ensure the successful deployment of PV systems, it is imperative to engage in systematic planning and design that carefully consider a broad range of factors and constraints, including technical, economical, and social aspects. Neglecting such comprehensive considerations could significantly curtail the potential for extensive solar energy integration across the region. To address these challenges, this study proposes a comprehensive approach that considers multiple dimensions and suggests alternatives for the integration of grid-connected solar systems. Such an approach, when applied to cities with high energy demand, could serve as a model for the targeted introduction of solar photovoltaic energy. In the context of Ecuador, this strategy could play a critical role in mitigating the country's energy challenges, particularly those exacerbated by severe regional droughts [15]. The objective of this paper is twofold: (a) to stratify residential electricity users based on consumption levels by analyzing data provided by the regional electricity distribution company through heat maps generated using Quantum GIS (QGIS) software; and (b) to identify residential areas with the highest energy consumption in the city of Manta and its surrounding districts. The article is organized as follows. Section 26.2 outlines Ecuador's energy crisis and reviews relevant studies. Section 26.3 details data sources and the approach for creating heat maps. Section 26.4 presents the data analysis results on energy consumption in Manta. Section 26.5 discusses strategies to promote PV micro-generation in key residential and tourism areas. Finally, Sect. 26.6 presents the conclusion.

26.2 Ecuador's Energy Challenges and Sustainable Solutions

This section analyzes the challenges of Ecuador's energy sector and the need for renewable energy solutions. The problem description highlights key issues, while the related work reviews studies on solar PV systems using GIS in Ecuador, laying the foundation for data-driven strategies in the country's energy transition and smart city initiatives. The section also considers Manta as a case study, analyzing its urban structure and the implications for energy efficiency and smart city initiatives.

26.2.1 Problem Description

Historically, Ecuador's electric sector has been characterized by significant governmental oversight across Generation, Transmission, and Distribution. The 2008 Constitution designated electricity as a strategic resource, granting the state full control over its regulation, decision-making, and provision through public enterprises. While the Constitution allows for the delegation of activities to private, mixed, or community-based entities, non-public involvement has remained limited, primarily in small-scale generation projects, including renewable energy [16]. Since its enactment, the contribution of the non-public sector to the national electrical system has been minimal.

Ecuador's reliance on hydroelectric and thermoelectric generation has severely constrained the system's ability to adapt to growing demand and climate change-related challenges. The current energy crisis, exacerbated by a severe drought, has significantly reduced water levels in reservoirs feeding hydroelectric plants. This, combined with the limited operational capacity of thermoelectric facilities, has placed additional strain on the national power system. Delays in the construction of new generation projects, technical issues in hydroelectric plants, and reduced electricity imports from Colombia (which is also facing drought) have further contributed to the crisis, resulting in widespread outages and economic losses exceeding USD 500 million since January 2024 [17]. In response, the government enacted the Organic Law of Energy Competitiveness (LOCE) in January 2024, which allows private companies and foreign state-owned enterprises to participate in electricity generation, including renewable energy projects, and energy transmission when public or mixed enterprises are unable to do so. This shift opens opportunities for data-driven strategies to promote renewable energy deployment.

26.2.2 Related Work

In recent years, several studies have explored the integration of solar photovoltaic (PV) systems in Ecuador, emphasizing the use of Geographic Information Systems (GIS) for site selection and resource assessment. These investigations, which include detailed analyses of rooftop potential in cities like Santa Isabel [18], and multi-city studies across Quito, Guayaquil, Cuenca, and the Galapagos Islands [19], underscore the significant energy generation potential from urban rooftop installations, particularly in regions with favorable climatic conditions and existing infrastructure. The approach taken in this study complements these efforts by adding a detailed analysis of electricity consumption data provided by the energy distribution company CNEL Manabí to residential users, utilizing Quantum GIS (QGIS), a free, open-source platform [20, 21]. Using QGIS heat maps in Manta allows to identify critical areas of electricity demand.

26.2.3 *The City of Manta and Its Physical-Spatial Organization*

Manta exemplifies a diffuse urban model common in Latin America, with unsustainable development marked by uncontrolled expansion and low-density settlements. This sprawl complicates efforts to transform it into a smart city, making infrastructure deployment (such as sensors, IoT devices, and connectivity networks) more challenging and costly [22, 23]. Energy distribution also tends to be less efficient in these cities, making the establishment of smart grids and renewable energy solutions more resource-intensive [24]. Moreover, fostering community engagement and ensuring the widespread adoption of smart city technologies can be more difficult, as the physical distance between residents can hinder the development of cohesive community driven initiatives. Figure 26.1 illustrates the boundaries and divisions of the Manta district.

Manta, a medium-sized city in Ecuador, grew rapidly after the construction of its deepwater dock in 1960 while maintaining its traditional orthogonal layout. By the 1980s, widespread "urban invasions" (unauthorized land occupations) led to disorganized peripheral expansion. In general, the development of many neighborhoods has been shaped by land ownership patterns, where properties are re-parcelled and integrated into the existing road network [25].

During the 1990s, Manta was predominantly composed of working-class neighborhoods, with certain areas featuring modest housing (1990–1997 expansion, see Fig. 26.2). The tallest buildings were primarily concentrated in the city center and

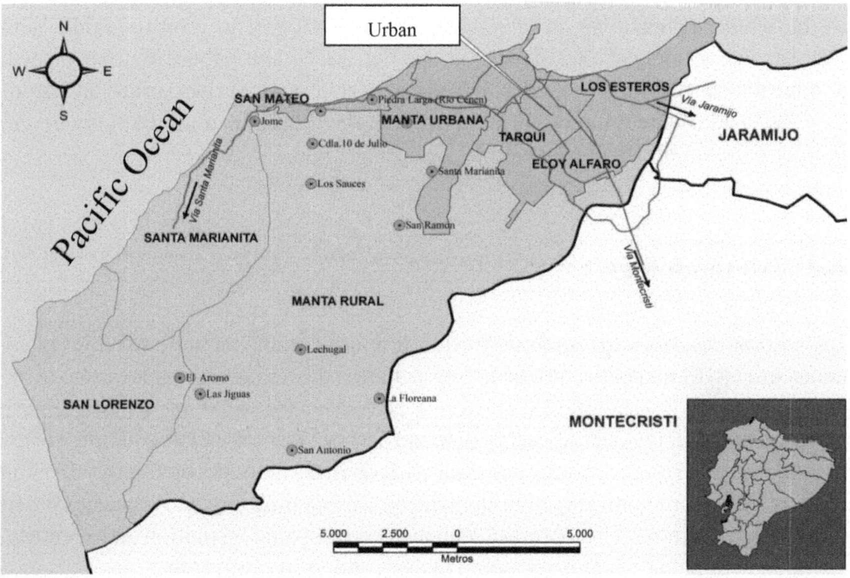

Fig. 26.1 Manta and its spatial organization [23]

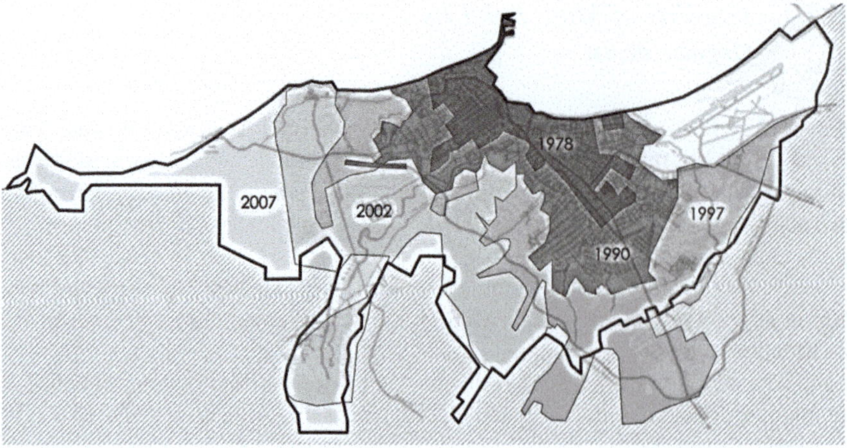

Fig. 26.2 Growth of the city of Manta in recent decades and socioeconomic distribution [23]

near the port, where much of the city's commercial activity took place [25]. According to [23], in recent decades, the city's growth has been marked by the development of neighborhoods around the perimeter access road, "Via Circumvalación", which provide access to previously inaccessible areas. This expansion, coupled with the rise of condominiums and gate close communities, has led to the creation of large residential developments characterized by low population density. Condominium construction has been concentrated in two key areas: the southwest coastal region, where high-end residential complexes are located (2007 expansion, see Fig. 26.2), and the northern area along the ring road, where developments cater to middle- and lower-income residents (2002 expansion, see Fig. 26.2). This pattern of urban growth has contributed to a fragmented urban landscape, reinforcing urban sprawl and deepening the spatial separation of social classes. Figure 26.2 illustrates the growth and urban orientation of Manta since 1978.

26.3 Methodology

This section describes the analysis for the proposed characterization. While many studies use QGIS to analyze urban energy patterns, this research stands out by using actual electricity consumption data from CNEL Manabí, allowing a more precise characterization of residential energy demand across Manta's socio-economic zones. By integrating verified consumption data with QGIS heat maps, this study offers an empirical basis for identifying high-demand areas, enabling targeted strategies based on real usage patterns. This approach enhances both spatial resolution and accuracy, providing actionable insights for policymakers and utility providers.

26.3.1 Description of the Proposed Approach and Data Sources

The proposed approach leverages electricity consumption data provided by CNEL-EP Manabí to develop a detailed characterization of residential energy usage in Manta. This data, which spans from 2016 to 2020, reveals a 24.9% increase in electricity demand and a 20.5% rise in the number of residential users over this period. These figures highlight the growing energy needs in the city and underscore the urgency for implementing sustainable solutions like PV micro-generation. By analyzing this data through GIS tools, we can visualize consumption patterns and identify areas with the highest demand, offering a pathway to optimizing energy distribution and supporting renewable energy integration.

A critical element in the generation of heat maps for this analysis is the use of a color scale to differentiate consumption levels across residential areas. The color gradient typically ranges from cooler tones (e.g., blue) to represent low electricity consumption to warmer tones (e.g., red) for areas with higher consumption. This visual tool enables quick identification of high-demand regions, which are prime targets for interventions aimed at improving energy efficiency and promoting renewable energy sources. The scaling of the heat map is fine-tuned to reflect electricity consumption brackets as defined by Ecuador's electricity tariff system, with consumption levels categorized into 14 distinct brackets [26]. These consumption ranges help contextualize the data within the pricing structure, showing the correlation between electricity usage and cost, particularly in areas with elevated consumption.

The methodology proposed for generating heat maps to track electricity consumption in residential areas using QGIS software is comprised of five distinct phases (Fig. 26.3). Each phase of this methodology is systematically outlined and described in detail below.

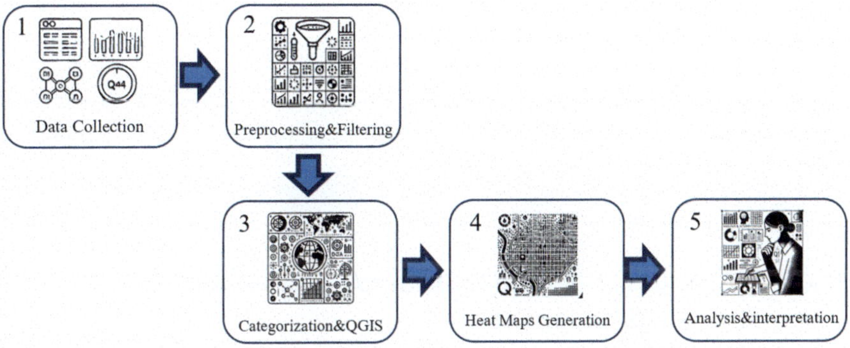

Fig. 26.3 Methodology workflow for generating energy consumption heat maps in QGIS

1. **Data Collection**: in this initial phase, electricity consumption data is gathered from relevant sources, such as utility companies. The accuracy and comprehensiveness of this data are required to ensure reliable outcomes in subsequent phases.
2. **Data Preprocessing and Filtering**: the collected data undergoes preprocessing to clean and standardize it for analysis. This includes removing any inconsistencies or errors and converting the data into a format suitable for GIS processing. Then, the dataset is filtered to isolate residential electricity consumption. Geographic coordinates (x, y) are added to the filtered data to enable the identification of specific locations within the study area.
3. **Data Categorization and import into QGIS**: the filtered data is then categorized according to predefined electricity consumption ranges, typically stablished by electricity companies. This categorization allows for the differentiation of consumption levels across various households. This categorized data is imported into QGIS software, where it is organized into layers for visualization. Proper structuring of the data within QGIS is essential for accurate and meaningful map generation.
4. **Heat Map Generation**: QGIS tools are utilized to generate heat maps based on the imported data. This phase involves adjusting visualization parameters such as color gradients and point density to effectively highlight areas of high and low electricity consumption.
5. **Analysis and Interpretation**: in the final phase, the generated heat maps are analyzed to identify patterns and regions with highest electricity consumption. The insights gained from this analysis are used to inform strategic planning and policymaking, particularly in the context of energy efficiency and renewable energy.

26.4 Results and Analysis

This section presents the findings from the data analysis, focusing on energy consumption patterns in Manta city. The results highlight key areas of high demand and the potential for photovoltaic micro-generation, contributing to the city's transition towards smart city initiatives. A detailed interpretation of these findings is provided to inform strategic urban energy solution.

26.4.1 QGIS-Based Heat Map Analysis for Residential Energy Consumption

To create a reliable heat map, the average annual electricity consumption data for 2020 in the residential sector was utilized. The geographical coordinates of each user were integrated, and the data was categorized into the 14 consumption ranges established in the tariff schedule for the Ecuadorian electricity sector [26]. This classification allows for a detailed spatial analysis of energy demand within the city.

Using QGIS software, a density raster was generated from a point vector layer representing each user. The process involves applying a kernel density estimation to calculate the intensity of electricity consumption in specific areas. Each kernel is assigned a color gradient, where darker or warmer colors indicate higher energy consumption levels. The visualization enables a clear spatial interpretation of consumption patterns, highlighting areas with higher demand, which are crucial for identifying potential zones for targeted energy efficiency measures or renewable energy interventions, such as PV micro-generation. Figure 26.4 presents the heat map generated using QGIS software, highlighting electricity consumption patterns across different areas. Zones 1 and 2 represent the urban regions, while Zone 3 covers the rural areas.

As seen in Figure 26.4, energy consumption varies significantly across different parts of the city and surrounding areas. The map illustrates that regions with low population density and lower-income neighborhoods generally exhibit basic consumption, up to 300 kWh/month (depicted in blue and categorized as a small consumer [27]). These areas are typically lower-income or sparsely populated, where energy use is limited to basic household needs. Meanwhile, low-to-middle-income regions show moderate energy consumption, ranging from 300 to 1500 kWh/month (depicted in light green and categorized as a medium consumer [27]).

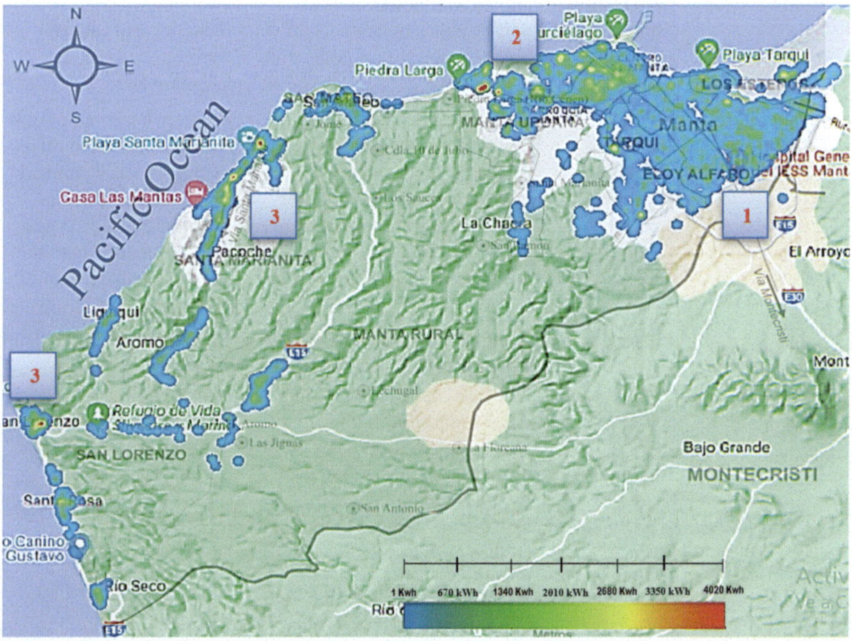

Fig. 26.4 QGIS-based heat map of Manta city

Higher consumption levels are categorized beyond 1500 kWh/month in Ecuador and represent the high-income population. For a more detailed analysis, we have divided this range into two segments to better distinguish between middle and high-end consumers. The transition from green to light orange represents consumption between 1500 and 3350 kWh/month, typically found in areas populated by middle and high-class residents with higher economic capacity and more household appliances, highlight areas with substantial energy use, likely driven by access to additional comforts and luxury amenities. Finally, high-end residential developments and certain rural areas show notably higher consumption levels, depicted in red, with figures ranging from 3350 to 4020 kWh/month. These areas include affluent residential neighborhoods and tourist-heavy regions, where large properties, luxury amenities, and seasonal tourism drive up electricity consumption. A detailed analysis of these data is presented in Table 26.1.

Table 26.1 Analysis for residential energy consumption

Zone	Range of consumption (kWh/month)	Interpretation of results	Suggestions for fostering solar PV
1. Manta (1990 to 2002 growth expansion). Central and northern areas	Up to 300 kWh (Blue) 300 to 1500 kWh (Blue to Light green)	Moderate energy consumption due to lower population density and limited economic capacity, reflecting essential household energy use. Some spots of yellow zones indicate slightly higher usage from middle-class residents	Introduce small-scale solar PV systems to meet essential household needs, focusing on affordability and reducing grid reliance for moderate-income households
2. Manta (2002 and 2007 growth expansion). Southwest coastal region	1500 to 2680 kWh (Light green to yellow) 3350 to 4020 kWh (Orange/Red)	High energy consumption in high-end residential areas, driven by luxury amenities such as pools and air conditioning. Some spots of orange-red zones highlight affluent households with energy-intensive lifestyles	Promote large-scale solar PV installations to offset the high energy demand in luxury homes, with potential for high returns on investment due to the energy-intensive nature of the area
3. Santa Marianita and San Lorenzo. Rural areas	350 to 1340 kWh (Blue to Light green) 1340 to 2680 kWh (Light green to Yellow) up to 4020 kWh (Red)	A bluish base of residents with low consumption is observed. However, exceptionally high energy consumption in tourist-driven rural areas, with second residences often owned by affluent individuals. Seasonal peaks due to tourism further elevate demand	Encourage solar PV adoption for second homes and tourism facilities to reduce peak consumption, especially during high-demand tourist seasons

26.5 Discussion

The findings of this study, based on the heat map analysis, reveal two distinct energy consumption patterns across different areas of Manta: lower-income neighborhoods with moderate energy use and high-end residential zones with significant higher energy consumption. These patterns suggest the need for two tailored approaches to foster the adoption of PV micro-generation systems that align with the economic capacity and energy needs of these areas.

26.5.1 Targeting Low-Income Population with Accessible PV Systems

In low-income households, where energy consumption typically is up to 300 kWh, promoting the adoption of PV systems requires carefully crafted policies to overcome the financial and logistical barriers that often limit access to renewable energy. In these communities, the primary hurdle is the initial cost of installation, which can be prohibitive for households that rely on limited financial resources. Addressing this challenge requires a multi-pronged approach that combines government support, community engagement, and innovative financing mechanisms.

1. **Government subsidies and financial incentives**: Government subsidies can lower the upfront cost of PV systems, making them more accessible to low-income households. Ecuador could implement a tired subsidy model, offering greater support based on income and energy consumption. These subsidies could cover a significant portion of the installation costs, ensuring that households in underserved areas can participate in the energy transition. Additionally, tax exemptions for renewable energy investments could be expanded to include both homeowners and local community-based organizations, further reducing financial barriers to entry.
2. **Educational campaigns and technical support**: Educational campaigns should raise awareness in low-consumption areas about the long-term savings and environmental benefits of PV systems. Collaborating with local NGOs and universities, the government can organize workshops to inform residents about solar energy advantages. Providing technical support and training will also help ensure the sustainability of these projects.
3. **Policy alignment with national social and energy goals**: Ecuador can integrate PV micro-generation into its national strategy to address energy poverty and promote sustainability. By offering targeted subsidies and financing, affordable PV systems in low-income neighborhoods the government can address energy poverty while promoting environmental sustainability and energy security.

26.5.2 Fostering PV Adoption in Middle and High-End Residential Zones

Middle-income and high-end residential areas, identified in the 1500 to 4020 kWh/month range (depicted in yellow, orange, and red), present unique opportunity for promoting PV micro-generation systems. These households have the financial capacity to invest in larger PV installations, which can help reduce their reliance on the national grid while offering substantial cost savings and greater energy independence. To promote PV adoption in these areas, the following strategies could effectively accelerate the deployment of solar energy:

1. **Tax incentives and subsidized financing programs**: In these areas, tailored policies should include tax incentives such as property tax deductions and income tax exemptions for renewable energy investments. These incentives would reduce the financial burden of installing PV systems. Additionally, subsidized financing programs, offering low-interest loans, can make larger, high-capacity systems more accessible, helping homeowners reduce electricity costs over the long term. A tiered incentive model could be designed where subsidies decrease as income or energy consumption increases, ensuring that wealthier residents contribute proportionately while still encouraging significant investment in renewable energy.
2. **Community-based solar initiatives**: In affluent gated communities, community-based solar initiatives offer a cost-effective model for PV adoption. By pooling resources, residents can invest in a shared PV system for communal properties, reducing individual installation costs while benefiting from solar energy. This approach powers shared amenities, fosters community engagement, and scales renewable energy adoption in high-consumption areas efficiently.

26.5.3 Targeting Tourist-Driven Areas

The rural areas of Santa Marianita and San Lorenzo, although characterized by lower population density, exhibit high levels of energy consumption, primarily driven by seasonal tourism and second-home ownership. Targeting these areas with tailored incentives for PV installations would not only stabilize energy consumption during tourist influxes but also reduce the strain on Ecuador's national grid.

1. **Tourism-specific incentives for PV installations**: To encourage the adoption of PV systems in tourism-heavy regions, the government could offer tourism-specific incentives. These incentives could take the form of subsidies or tax breaks aimed at property owners and businesses that invest in PV systems for holiday homes, resorts, and hotels. By reducing the financial burden on property owners, the government can ensure that these high-consumption areas transition toward more sustainable energy use.

2. **Promoting sustainable development**: The adoption of PV systems in Santa Marianita and San Lorenzo supports Ecuador's goals for sustainable tourism development. By incentivizing solar energy in tourist-heavy areas, the government can lower the tourism industry's carbon footprint and enhance energy reliability. Distributed PV systems can stabilize energy demand during peak seasons, reducing strain on the national grid and contributing to Ecuador's renewable energy transition.

26.6 Conclusions

In this study, we have showcased the potential of QGIS-based heat maps as a powerful tool for visualizing energy consumption patterns across intermediate cities like Manta and promoting PV micro-generation systems. QGIS, being a free and open-source software, offers a practical and accessible solution for developing countries, enabling them to harness geographic information systems (GIS) without financial barriers. By analyzing residential electricity demand, we identified key zones that could benefit from targeted solar interventions, aligning with smart city strategies aimed at improving urban energy efficiency and sustainability. For instance, low-income areas exhibited basic consumption levels up to 300 kWh/month, while middle-income regions ranged from 300 to 1500 kWh/month. High-end residential zones, including affluent neighborhoods and tourism-driven areas, displayed consumption exceeding 3350 kWh/month, making them prime targets for PV interventions.

The proposed strategies for low-income households, middle- and high-income residential areas, provide practical pathways for fostering PV adoption across different socio-economic groups. By addressing the specific energy needs of these diverse populations, including tourism-driven regions, these strategies support Ecuador's broader energy transition goals. This work contributes to the growing body of knowledge on sustainable urban energy planning and demonstrates how integrating geographic information systems (GIS) with renewable energy policies can accelerate the development of resilient, low-carbon cities.

Acknowledgements The projects oriented towards the ecological transition and the digital transition. (Grant No. TED2021-131137B-I00 ("Aportación a la Transición Ecológica en el sector Industrial a través del Autoconsumo Fotovoltaico") and the support provided by the Thematic Network 723RT0150 "Red para la integración a gran escala de energías renovables en sistemas eléctricos (RIBIERSE-CYTED)" financed by the call for Thematic Networks of the CYTED (Ibero-American Program of Science and Technology for Development) for 2022. In addition, the authors also acknowledge the support of the "Universidad Laica Eloy Alfaro de Manabí (Proyecto de Investigación: Observatorio de Energía y Sostenibilidad)" and the Escuela de Doctorado from the UNED (Universidad Nacional de Educación a Distancia), which has supported part of this work as a part of the accreditation for the Ph.D. program (Milton Moreano).

References

1. UN—United Nations, *World Urbanization Prospects The 2018 Revision* (2018). Accessed: Aug 22, 2024 (Online). Available: https://population.un.org/wup/Publications/Files/WUP2018-Report.pdf
2. IEA—International Energy Agency, *World Energy Outlook 2020* (2020) (Online). Available: www.iea.org/weo
3. IRENA—International Renewable Energy Agency, *World energy transitions outlook 2023: 1.5°C pathway* (International Renewable Energy Agency IRENA, 2023). Accessed: Aug 22, 2024 (Online). Available: https://www.irena.org/-/media/Files/IRENA/Agency/Publication/2023/Jun/IRENA_World_energy_transitions_outlook_summary_2023.pdf
4. B. Parida, S. Iniyan, R. Goic, *A Review of Solar Photovoltaic Technologies* (2011). https://doi.org/10.1016/j.rser.2010.11.032
5. IEA—International Energy Agency, *Renewables 2023 Analysis and Forecast to 2028*. Accessed: Aug 22, 2024 (Online). Available: https://iea.blob.core.windows.net/assets/96d66a8b-d502-476b-ba94-54ffda84cf72/Renewables_2023.pdf
6. IRENA—International Renewable Energy Agency, *Innovation Landscape for a Renewable-Powered Future: Solutions to Integrate Variable Renewables* (2019). Accessed: Aug 22, 2024 (Online). Available: https://www.irena.org/-/media/Files/IRENA/Agency/Publication/2019/Feb/IRENA_Innovation_Landscape_2019_report.pdf
7. DOE—Department of Energy U.S., *The Potential Benefits of Distributed Generation and the Rate-Related Issues That May Impede Its Expansion Report Pursuant to Section 1817 of the Energy Policy Act of 2005 Background*. Accessed: Aug 23, 2024 (Online). Available: https://www.energy.gov/oe/articles/potential-benefits-distributed-generation-and-rate-related-issues-may-impede-its
8. M.A. Ponce-Jara, M. Pelaez-Samaniego, M. Moreano-Alvarado, C. Velásquez-Figueroa, M. Castro, An assessment on energy policies and challenges to promote solar PV in South America: the Ecuadorian case, in *Ecuador Perspectives of the Past, Present and Future: A Multi-Criteria Approach to Social Evolution* (Nova Science Publisher, 2021)
9. IRENA, *Policy Brief Renewable Energy in Latin America 2015 : An Overview of Policies*. Accessed: Jan 31, 2018 (Online). Available: https://www.irena.org/-/media/Files/IRENA/Agency/Publication/2015/IRENA_RE_Latin_America_Policies/IRENA_RE_Latin_America_Policies_2015.pdf
10. L. Cadavid, C.J. Franco, Impact of devaluation on grid parity for residential solar generation in Latin America. IEEE Lat. Am. Trans. **15**(11), 2097–2102 (2017). https://doi.org/10.1109/TLA.2017.8070414
11. IRENA, *International Renewable Energy Agency. Analisis del Mercado de energias Renovables: America Latina*, (Abu Dabi, Emiratos Árabes Unidos, 2016) (Online). Available: http://www.irena.org/DocumentDownloads/Publications/IRENA_Market_Analysis_Latin_America_summary_ES_2016.pdf
12. M. Jahangiri, O. Nematollahi, H. Saghaei, A. Haghani, Using GIS maps to assessing wind energy in Asian countries: finding the high potential countries and examining their current status and outlook. Iranica J. Energy Environ. **15**(4), 428–443 (2024). https://doi.org/10.5829/ijee.2024.15.04.10
13. A.A.A. Gassar, S.H. Cha, Review of geographic information systems-based rooftop solar photovoltaic potential estimation approaches at urban scales. Appl. Energy **291** (2021). https://doi.org/10.1016/j.apenergy.2021.116817
14. Y. Choi, J. Suh, S.M. Kim, GIS-based solar radiation mapping, site evaluation, and potential assessment: a review, May 01, 2019 (MDPI AG, Switzerland, 2019). https://doi.org/10.3390/app9091960
15. M.A. Ponce-Jara, I. Pazmino, Á. Moreira-Espinoza, A. Gunsha-Morales, C. Rus-Casas, Assessment of single-axis solar tracking system efficiency in equatorial regions: a case study of Manta, Ecuador. Energies (Basel) **17**(16), 3946 (2024). https://doi.org/10.3390/en17163946

16. CE—Constitución del Ecuador, articulo 314 y 315. 2015. Accessed: Sept 03, 2024 (Online). Available: https://www.finanzas.gob.ec/wp-content/uploads/downloads/2015/03/constitucion_de_bolsillo.pdf
17. Gestión Digital, *Cortes de Luz ya dejan 576 millones de dólares en perdidas para Ecuador*. Accessed: Sept 03, 2024 (Online). Available: https://revistagestion.ec/noticias/cortes-de-luz-ya-dejan-576-millones-en-perdidas-para-ecuador/
18. A.H. Idrovo Macancela, M.V. Velecela Zhindón, *Evaluación Basada en GIS del Potencial Solar Fotovoltaico en Techos de Edificios de Áreas Urbanas: Caso de Estudio Santa Isabel-Ecuador* (Universidad Politécnica Salesiana, Cuenca, 2024). Accessed: Aug 25, 2024 (Online). Available: https://dspace.ups.edu.ec/bitstream/123456789/28008/1/UPS-CT011441.pdf
19. M. Tapia, L. Ramos, D. Heinemann, E. Zondervan, Power to the city: assessing the rooftop solar photovoltaic potential in multiple cities of Ecuador. Phys. Sci. Rev. **8**(9), 2285–2319 (2023). https://doi.org/10.1515/psr-2020-0061
20. QGIS, *Open Source Geographic Information System*. Accessed: Aug 25, 2024 (Online). Available: https://www.qgis.org/
21. QGIS Desktop. Accessed: Aug 25, 2024 (Online). Available: https://www.osgeo.org/projects/qgis/
22. M. Ponce-Jara, M. Moreano Alvarado, J.I. Gutiérrez Lino, Ciudades compactas y difusas: el crecimiento de Manta y su relación con la movilidad humana," in *Análisis Corporativo, Desarrollo y Finanzas* (Editorial Martinéz, Durango, Méjico, 2016), ch. IV, p. 465
23. M. Ponce-Jara, M. Moreano-Alvarado, W. Zamora, K.I. Velez, Impacto ambiental del transporte público de Manta: una propuesta de sostenibilidad, in *ParroquiaTarqui* (MAWIL, Ed., Manta, 2022). https://doi.org/10.26820/978-9942-622-14-3
24. S. Borlase, *Smart Grids—Infrastructure, Technology and Solutions* (2013)
25. L.J. Sáinz, R. del Caz Enjuto, M.C. Solorzano, F. Jové, J. Orozco, Transformación de las ciudades latinoamericanas y europeas de tamaño intermedio. Estudio comparativo entre Manta, Ecuador y Valladolid, España, in *Seminario Internacional de Investigación en Urbanismo. "V Seminario Internacional de Investigación en Urbanismo, Barcelona-Buenos Aires, junio* **2013**(1) (2013)
26. ARCONEL, *Agencia de regulación y control de electricidad. Pliego tarifario del sector eléctrico Ecuatoriano*. Accessed: Sep 06, 2021 (Online). Available: https://www.controlrecursosyenergia.gob.ec/wp-content/uploads/downloads/2021/06/Anexo_1_pliego_tarifario_spee_2021.pdf
27. Guillermo Lazo Mendoza, *Reglamento general de la ley orgánica de eficiencia energética*. Ecuador (2021). Accessed: Sep 16, 2024 (Online). Available: https://www.geoenergia.gob.ec/wp-content/uploads/downloads/2022/11/25_Reglamento_Ley_Organica_Eficiencia_Energetica.pdf

Chapter 27
A Mixed Combinatorial Optimization Model for the Río Negro Hydroelectric Complex

Claudio Risso, Sergio Nesmachnow, Rodrigo Porteiro, J. Mario Vignolo, Emiliano Sierra, and Mario Ibarburu

Abstract The management of power generation systems requires the optimized coordination of resources and investments across timeframes ranging from days to decades. This paper introduces a mixed-integer optimization model (MIP) for the short-term operation of the Río Negro Hydroelectric Complex, a crucial asset in Uruguay's efforts to achieve energy sovereignty by primarily relying on wind, solar, and hydroelectric power sources. The model addresses the challenge of balancing fluctuating renewable energy supply with hydroelectric resources while ensuring cost-effective dispatch and system reliability. The experimental results demonstrate the accuracy with which a MIP approximation can model an extremely nonlinear problem.

Keywords Energy optimization · Hydrothermal dispatch · Mixed integer linear programming

---

C. Risso (✉) · S. Nesmachnow · J. Mario Vignolo
Facultad de Ingeniería, Universidad de la República (UdelaR), Montevideo, Uruguay
e-mail: crisso@fing.edu.uy

S. Nesmachnow
e-mail: sergion@fing.edu.uy

J. Mario Vignolo
e-mail: jesus@fing.edu.uy

R. Porteiro · E. Sierra · M. Ibarburu
UTE, Administración Nacional de Usinas y Trasmisiones Eléctricas, Montevideo, Uruguay
e-mail: rporteiro@ute.com.uy

E. Sierra
e-mail: esierra@ute.com.uy

M. Ibarburu
e-mail: mibarburu@ute.com.uy

D. Rossit et al. (eds.), *Proceedings of the VII Ibero-American Congress of Smart Cities, ICSC-Cities 2024, 12-14 November, San Carlos, Costa Rica*, Springer Proceedings in Energy, https://doi.org/10.1007/978-981-96-4301-1_27

27.1 Introduction

Managing the generation power system involves optimizing over different time horizons spanning from days to several decades. The complexity of this management arises from the large number of components and decisions involved, and from the uncertainty in key parameters. Technological advancements and environmental protection initiatives can influence the management of an electrical system. The issue is explored through an analysis of the Uruguayan electricity system, which has become a global reference after achieving energy sovereignty, relying heavily on renewable energy sources to meet its electricity demand [1, 3]. Uruguay has demonstrated remarkable capacity in integrating these energy sources into its electrical matrix, achieving a balance between efficiency and sustainability. This paper presents a mixed-integer optimization model (MIP) for energy production in the Río Negro Hydroelectric complex, which has been a cornerstone in the successful renewal of Uruguay's electrical system. This complex is bound to play a key role in adapting the system to future challenges, serving as a central element in the country's strategy to maintain its leadership in the use of clean energy and efficient planning of the electrical system.

27.1.1 Integration of Non-conventional Renewable Energy and Its Impact on the Energy System

Between 2017 and 2020, Uruguay successfully supplied over 95% of its energy demand through renewable generation, with minimal dependence on fossil fuels [12]. This achievement results from the country's strategic approach in balancing its existing controllable hydroelectric capacity with new investments in non-conventional renewable energy sources.

In 2021 and 2022, Uruguay continued to meet its domestic energy needs predominantly with renewables [15]; however, approximately 20% of its electricity production during this period came from thermal fossil sources. This shift was largely due to an unusual drought in Southeastern South America, which severely affected major hydroelectric plants in the region. Despite these conditions, Uruguay exported electricity to Argentina and Brazil (see Fig. 27.1) to help meeting their energy demands, substantially increasing its production beyond internal consumption.

In 2023, the drought severely impacted the Uruguayan hydroelectric generation, which on average accounts for over 40% of the country's energy supply. To address this shortfall, Uruguay increased its reliance on thermal generation to meet domestic demand, while also importing energy, primarily from Brazil. This notable shift in energy exchange conditions is reflected in the contrast between Figs. 27.1 and 27.2. The inter-annual variability in water inflows is one of the main features of the Uruguayan power system [11]. Moreover, hourly power supply from existing wind and solar power plants is highly volatile as in most power systems. In this context

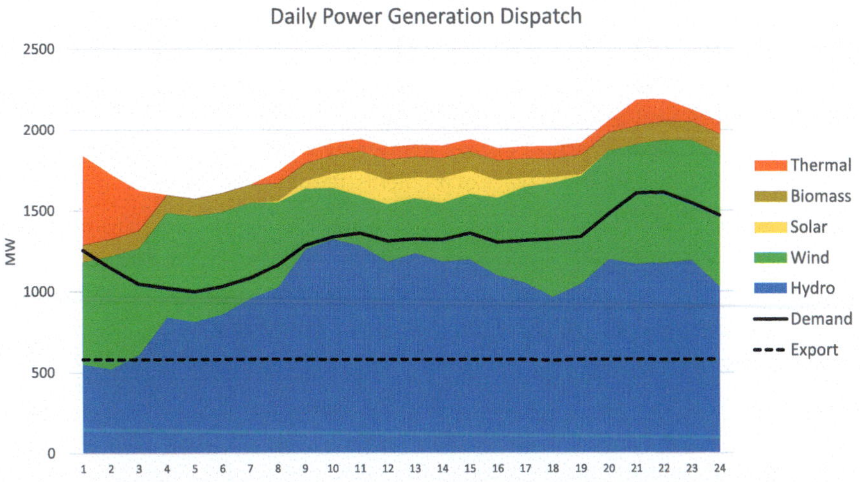

Fig. 27.1 Daily dispatch with an abundance of renewables and export (October 8, 2023)

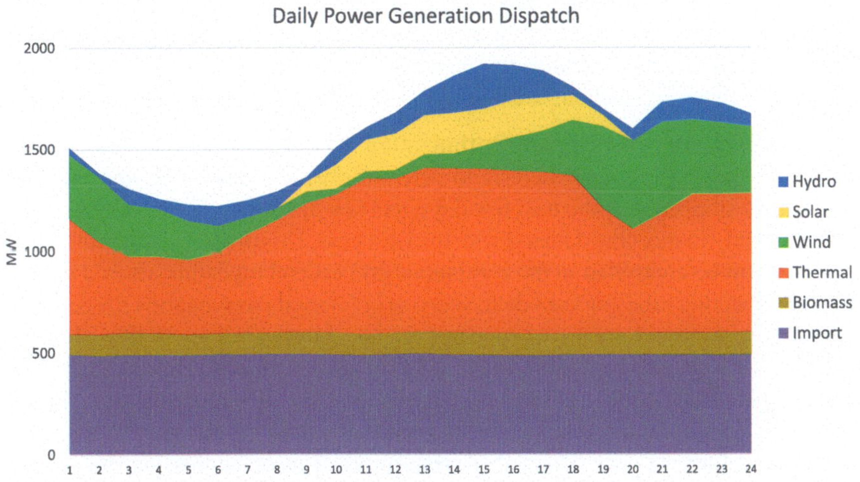

Fig. 27.2 Daily dispatch with renewable shortages and high imports (March 8, 2023)

both the managing of water resources stored in hydro power plants reservoirs and power exchanges with neighboring systems (Argentine and Brazil) are of paramount importance for the system.

Thermal plants and power imports must fill the gap as hydro decreases, introducing two sources of inter temporal restrictions: (i) the startup time of the largest and most efficient thermal units which requires up to 24 h to achieve full power; (ii) the conditions of power trade with Brazil, based upon a portfolio of short-term offers. Each offer has a term during which energy can be bought, a limited activation window before unaccepted offers expire, and once accepted a minimum amount of energy to be purchased. Offers' terms often extend longer than a week. Hence, Uruguay's short-term planning horizon has been extended to incorporate periods of up to fifteen days, to account for those restrictions. The perspective of rapid demand growth in emerging sectors such as electric vehicles [10] and hydrogen production [2] could introduce greater flexibility into energy management but also add complexity to planning models. We outline the types of problems, the case study, and the hypotheses on which this work is based.

27.1.2 Frameworks for Electric System Operation and Planning

Electric system management is divided into different time horizons, with problems being solved sequentially in reverse order from how they are presented here.

Continuous-Time Grid Control: It is the continuous-time grid control, where real-time production is coordinated among a predefined set of generating units to meet demand at every moment, ensuring that network frequency, active/reactive power flow, and other physical parameters remain within limits [8]. This is primarily an electrical engineering challenge and lies outside the scope of this paper.

Unit Commitment and Dispatch Optimization: It is the short-term operation planning problem. The objective here is to determine which generating units will be started up or stopped over the next few days, and how much power each plant should produce hourly to meet the demand with the least cost. This process defines the set of units that will be controlled in the continuous grid control model mentioned earlier. Dispatch planning relies on cost data, short-term forecasts or scenarios for demand, water inflows to reservoirs, hourly power from uncontrollable sources like wind and solar, cross-border energy trading with neighboring countries and scheduled maintenance of generating units [5]. Typically, the dispatch horizon spans between 48 and 480 h, divided into hourly intervals, and is updated multiple times per day to adjust for the latest forecast data. This paper primarily focuses on addressing this type of short-term problem. To optimize the short-term operation problem, it is necessary an approximation of the Bellman value function at the end of the short-term horizon [7], which gives the least expected cost of longer-term operation thereafter.

Medium- and Long-Term Planning Models: This level involves medium- and long-term operation planning models. In many power systems, hydro power plants have large reservoirs which when full can store water to maintain the plant production for weeks, months or years without using the water inflows. Medium- and long-term operation planning models, which cover horizons ranging from a few weeks to several years, aim to assign value to strategic assets, such as the water stored in reservoirs [6, 14]. These values are transmitted to short-term models by means of Bellman value function approximations.

Investment Decision Models: Finally, very long-term models simulate the electric system over several decades, adding to the random variables aforementioned, others like fossil fuel prices, and technology costs. These models evaluate the economic outcomes of system operations over extended periods and are out of the scope of this work.

UTE has developed a set of systems to manage the previous problems, including MOP [9], a model for the short and medium term operation and Mingo, to optimize generation investment. This work is part of a joint venture between UTE and UdelaR, exploring the use of MIP models as an alternative approach for addressing short-term operations. The remainder of the document focuses exclusively on the unit commitment and dispatch problem, treating the grid as a single node and excluding complexities such as international energy transactions.

27.1.3 Case Study: Simplification of Uruguay's Electric System

While the specifics of the thermal units are not addressed in detail, the hydroelectric facilities are thoroughly examined, as the primary subject of this study. Figure 27.3 illustrates the layout of the four dams that comprise Uruguay's hydroelectric complex. Complementarily, Table 27.1 shows its main parameters. For dispatch planning purposes, the future cost function (Bellman values) is an input derived from medium and long-term planning models.

Before 2005, electricity demand was met using a combination of hydroelectric plants and controllable thermal units. As shown in Table 27.1, starting from its maximum water level and operating at full capacity, the Rincón del Bonete power plant has an estimated depletion period of six months, assuming no natural inflows to replenish the reservoir (i.e., no hydrological inputs). The Bonete reservoir, notable for its size, was once the largest artificial lake in the world for several years following its inauguration in 1945. Together with the two downstream hydroelectric plants on the same river: Baygorria and Palmar, it constitutes Uruguay's primary energy storage system. The ability to function as a large energy accumulator has enabled the effective integration of non-conventional renewable energy sources in Uruguay, potentially reducing the reliance on batteries to manage the intermittency of solar and wind energy. If they generate more energy than expected, hydroelectric production is scaled back, allowing reservoirs to store the surplus energy indirectly through water reserves. Conversely, when renewable generation falls short of forecasts, hydroelectricity compensates.

In 2005, Uruguay embarked on a substantial investment in non-conventional renewable energy sources, among which, biomass (mainly pulp mills) stands out due to their controllability and operation, which closely resemble traditional thermal plants but without the reliance on fossil fuels. Simultaneously and because of country's favorable conditions, Uruguay saw a considerable expansion of wind farms,

Fig. 27.3 Map of Uruguay with the 4 hydroelectric plants of the system

Table 27.1 Technical parameters of hydroelectric plants in Uruguay

Name of the Unity	Maximum Power (MW)	Days to empty	Source of the inflows
Bonete	149	182	Río Negro
Baygorria	111	3	Bonete [+8h] and others
Palmar	343	13	Baygorria [+16h] and others
Salto Grande	1890	15	Río Uruguay

reaching a combined installed capacity of 1500 MW, and more recently solar energy, whose installed capacity has reached 250 MW. The fossil fuel-based thermal generation capacity currently stands at 1154 MW. Around 36% of Uruguay's installed capacity comes from non-dispatchable sources, whose intermittence presents significant challenges for system dispatch planning.

Prior to 2005, Uruguay's electricity generation was predominantly controllable. Within a 72 h planning horizon, both hydrological inflows and national electricity demand could be forecasted with high accuracy, and short-term planning systems were inherently deterministic. With the integration of wind, solar, and biomass, these sources are prioritized, supplying electricity as long as it does not exceed demand. Controllable components then meet the residual demand: the gap between total demand and non-conventional renewable generation.

The volatility of renewables shifts to this residual demand, becoming the key variable in dispatch models. This rapid growth of non-conventional energy sources has driven a significant redesign of short- and medium-term planning systems. Short-term planning currently uses Stochastic Dynamic Programming [4], crucial to Uruguay's energy transition. However, new challenges arise, including managing flexible demands, complex thermal unit commitments, and the increasing role of international energy exchanges. Large amounts of wind and solar capacity increase the level of uncertainty of the problem and the need of methods to take it into account [16]. One of the ways to address this problem is through the use of a Mixed-Integer Stochastic Programming (MISP) approach. A notable example of this is the case of [13], which serves as a direct reference for such implementations. MISPs are well suited to deal with a small power system like the Uruguayan one. Conversely, the non-linearities of hydroelectric plants, especially in the Río Negro basin, complicate linear optimization models, as lake states affect production. This paper presents a MIP model to optimize dispatch costs for the Río Negro hydroelectric complex, accounting for system failures: a fine for not being able to fulfill demand. For sake of simplicity, the model assumes known residual demand and hydrological contributions over the next fifteen days.

27.2 Production Model for the Río Negro

The Río Negro Hydroelectric Complex houses three of the four hydro-generation units of Uruguay. It is of utter importance the fact that these three electric dams are in tandem in the following sequence: Bonete, Baygorria and Palmar. The outflows from Bonete are integrated into the stock of Baygorria's lake after an 8-hour delay, while the outflows from Baygorria reach Palmar's lake after 16 h. The actual production function of a hydroelectric unit is complex and depends on countless variables. A reference production function of a hydroelectric unit is

$$P = \rho g \eta \cdot Q^{trb} \cdot (h^{lk} - h^{rv}). \tag{27.1}$$

The formulation in Eq. (27.1) relies on the efficiency of turbines and the potential energy at the height of the reservoir's lake, where P corresponds to the generated power [MW], Q^{trb} to the water flowing downstream through the turbines [m^3/s], h^{lk} is the height of the top of the lake and h^{rv} is the level of the river after the dam, both expressed in [m]. $(h^{lk} - h^{rv})$ is the head of the dam. Remaining parameters are constants, namely: the Earth's gravitational acceleration g [m/s^2]; the density of water ρ [kg/m^3]; the energy conversion efficiency of turbines η [dimensionless]. In this model, the state variable h^{lk} solely depends on the volume of water at the reservoir. The height of Bonete and Palmar lakes can be adjusted with degree two polynomials at a relative error lower than 1%. Baygorria is operated as a run-of-the-river hydroelectric generator, so that water coming from upstream is either turbinated or spilled at that moment.

Regarding control variables, besides turbinated water Q^{trb}, another source of water discharges to account for is the spilling Q^{spl}. At times, either for security reasons or simply for profitability, water must be released from the dam without generating power. To fully account water discharges Q^{wds}, both components are included, since $Q^{wds} = Q^{trb} + Q^{spl}$. Water discharges raise the level downstream (i.e., h^{rv}), which, according to (27.1), decreases the productivity of water passing through the turbines. This level is also affected by the height of the next lake downstream, if one exists. Therefore, the river height downstream Palmar depends only on its water discharges, Baygorria's river h^{rv}_{Bay} depends on its own discharges Q^{wds}_{Bay} and on the height h^{lk}_{Pal} of Palmar's lake:

$$h^{rv}_{Bay} = F_R(h^{lk}_{Pal}, Q^{wds}_{Bay}). \tag{27.2}$$

Baygorria's lake is constant in height (run-of-the-river unit), so Bonete's river depends only on its discharges Q^{wds}_{Bon}. Downstream rivers height can be approximated by linear functions or degree two polynomial. Such interdependencies end up linking production variables of some units with state variables of others. The Río Negro Complex must be optimized as a whole.

27.2.1 Formulation and Parameters Adjustment

To adjust all these functions, UTE provided us with the complete hourly time series for the Río Negro, covering production, control, and state variables, as well as natural inflows, over a ten-year period from January 1st, 2010, to December 31th, 2019. The final adjustment and formulation are in the polynomial of Eq. (27.3), which renames control and state variables for sake of simplicity.

$$\begin{aligned} P(x_{1h}, y_{1h}, v_{1h}, x_{2h}, y_{2h}, x_{3h}, y_{3h}, v_{3h}) = \sum_{t=1}^{T} & p_1^{(1)} x_{1h,t} + p_1^{(2)} x_{1h,t} v_{1h,t} - p_1^{(3)} x_{1h,t} v_{1h,t}^2 \\ & - p_1^{(4)} x_{1h,t}^2 - p_1^{(4)} x_{1h,t} y_{1h,t} + p_2^{(1)} x_{2h,t} - p_2^{(2)} x_{2h,t}^2 - p_2^{(2)} x_{2h,t} y_{2h,t} - p_2^{(3)} x_{2h,t} v_{3h,t} \\ & + p_2^{(4)} x_{2h,t} v_{3h,t}^2 + p_3^{(1)} x_{3h,t} + p_3^{(2)} x_{3h,t} v_{3h,t} - p_3^{(3)} x_{3h,t} v_{3h,t}^2 - p_3^{(4)} x_{3h,t}^2 \\ & - p_3^{(4)} x_{3h,t} y_{3h,t} + p_3^{(5)} x_{3h,t}^3 + p_3^{(5)} x_{3h,t} y_{3h,t}{}^2 + 2 p_3^{(5)} x_{3h,t}{}^2 y_{3h,t}, \end{aligned} \tag{27.3}$$

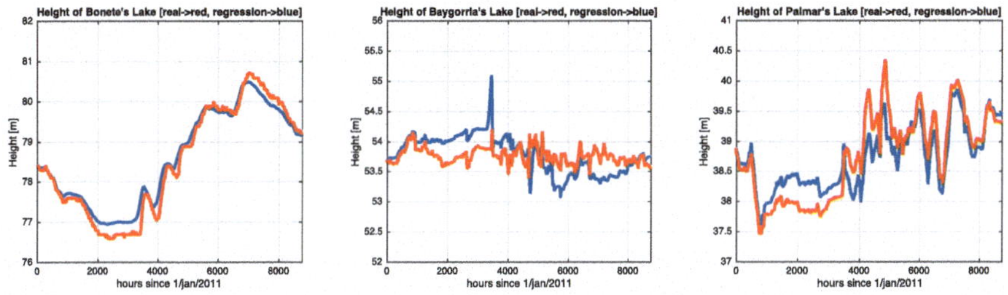

Fig. 27.4 Records and estimations for the height (over sea level) at each lake over the Río Negro along 2011

Subscripts $1h$, $2h$ and $3h$ allude to hydraulic units, respectively Bonete, Baygorria and Palmar. Constants $p_i^{(j)}\geq 0$ are the coefficients for the production of each unit $1\leq i\leq 3$. Variable names x, y and v respectively refer to turbinated flow [m^3/s], spilling [m^3/s] and lake's volume [m^3] at each unit. It is assumed that variables are discretized into time-slots of equal length, of one hour in this case, along a time-horizon of T hours. While the result of the sum in Eq. (27.3) accounts for the whole energy [MWh] produced during the period, each addend captures the energy production at its time slot. To assess the quality of some regressions, Fig. 27.4 presents the real-world data (in red) for the height of each lake along a reference year (2011), compared with the result of regressions (in blue).

Considering that some input-data (e.g., natural inflows) have sampling errors, and other variables affecting the result (e.g., wind speed and direction) were disregarded for lack of data, the overall result is quite remarkable. After computing the differences between the model in equation Eq. (27.3) and the actual power series during 2011, is noted that: (i) 79.5% of the samples have an absolute error below 6 MW, which is 1% of the combined capacity of the Río Negro's power plant; (ii) this figure increases to 92.9% when the absolute error is below 12 MW (i.e., 2% of the plant); and (iii) the regression was based on data from the entire ten-year period, during which several turbines were offline during days due to scheduled maintenance or malfunctions, affecting the nominal production.

In summary, we conclude that the non-linear reference model in Eq. (27.3) is highly accurate overall.

27.2.2 Medium-Term Optimization Reference

The reference production model in Eq. (27.3) is a differentiable function, but the Hessian matrices of these polynomials are generally not definite (i.e. neither convex nor concave) on a slot-by-slot basis, which poses a significant challenge for conventional optimization techniques. The first step in addressing the problem is to find solutions that allow optimizing energy production in a long-term reference context of one year. The goal is to get insights of an ideal operation for a specific context. The year 2011 was selected as the reference because it had the lowest water inflows to the Río Negro basin during the ten-year period. This required near-optimal use of water reserves, providing insights to simplify the general model. The non-linear problem addressed in this section is

$$\begin{cases} \max\limits_{X,Y,V} P(X,Y,V) \\ d_t \geq P_t(X_t, Y_t, V_t), \ (i) \\ (X, Y, V) \in F. \qquad (ii) \end{cases} \tag{27.4}$$

where $P(X, Y, V) = \sum_{t=1}^{T} P_t(x_{1h,t}, y_{1h,t}, v_{1h,t}, x_{2h,t}, y_{2h,t}, x_{3h,t}, y_{3h,t}, v_{3h,t})$, being $P_t(X_t, Y_t, V_t)$ the addends/terms in the sum of Eq. (27.3). For the system to be operational, the flows of water through the turbines (X), the spillways (Y), as well as the water volume in the reservoirs (V), must stay within technical limits, expressed in Eq. (27.4)$-(ii)$ in the form of a set F. Additionally, at each time step t, the power P_t generated and supplied to the grid cannot exceed the demand d_t of that period (Eq. (27.4)$-(i)$). The volume of water in each reservoir at any time must always comply a mass balance equation (elaborated at Sect. 27.3). To find reference solutions, a representative period is established, which inherits the initial and final levels of each lake in 2011, as well as their inflows during the period, respectively sketched in Figs. 27.4 and 27.5.

Afterwards, a reference demand profile was needed. The available ten years of hydraulic data only overlaps the early stages of Uruguay's renewable energy expansion. To create a representative demand scenario (scenario A), the 2018 residual demand was used, as wind power was fully deployed by then, and its average inflows closely matched the ten-year dataset. Additionally, another synthetic demand was crafted (demand scenario B), in this case after copying the shape of the recorded production at the Río Negro during 2011. The second instance was scaled to match the accumulated energy of the demand in the previous scenario. To determine optimized solutions for both scenarios, an iterative optimization algorithm employing successive approximations was used. We remark here how the spilling was managed for demand instances A and B. Figure 27.6 sketches the result for the three solutions (red, blue and yellow curves) of better quality (i.e., energy production) resulting from

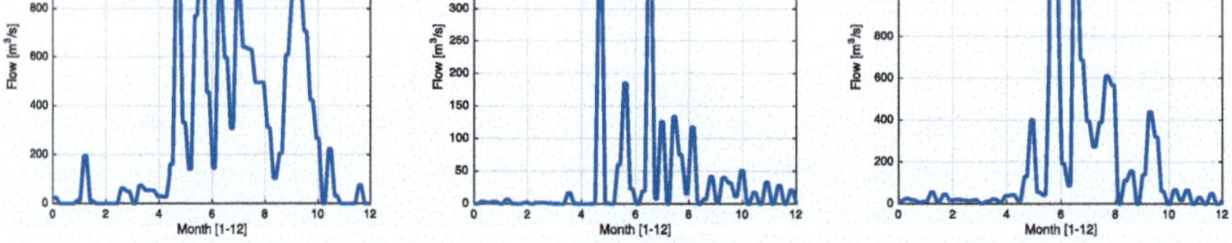

Fig. 27.5 Natural Inflows to the lake of every hydroelectric unit over the Río Negro [2011]

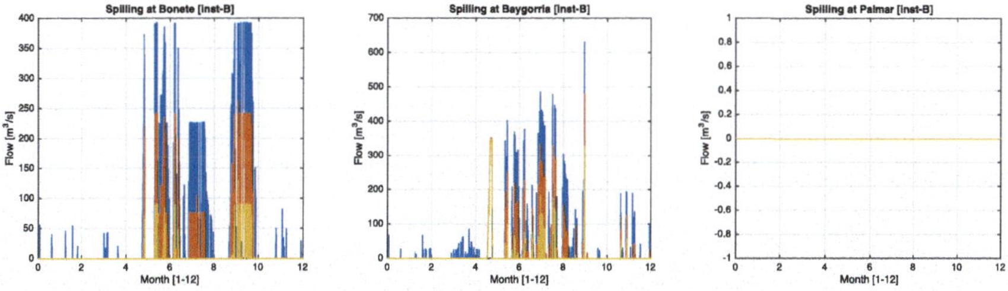

Fig. 27.6 Spilling graphic for the three better solutions found for demand instance B

the iterative algorithm previously mentioned. The results are for instance B, since it is slightly more extreme in terms of spilling. Note that the maximum flow spilled at Bonete is under 400 [m^3/s] at any time for the three solutions, a figure remarkably low, below the 10% of its technical limit. The relative gap is similar regarding Baygorria.

Regarding Palmar, no spillages are recorded in any solution across both instances. This behavior is the cornerstone to simplify the model.

27.3 MIP Approach for Río Negro's Complex

The section introduces a MIP model to optimize the Unit Commitment and Dispatch at the Río Negro Hydroelectric Complex. In Sect. 27.2.1, optimized electricity production solutions were identified under the hydrological conditions of a dry year (in terms of natural influxes). From these, six high-quality solutions were selected, all using the same inflow series. Three of these solutions were based on an annual demand scenario derived from the actual residual demand of 2018 (scenario A), while the other three adapted the production pattern from the 2011 energy series (scenario B). A year is composed of 24 consecutive fifteen-day periods, plus a remainder with a few days more. To obtain references of efficiently managed production at Río Negro during drought conditions, the six one-year-long solutions were chopped into fifteen-day periods. This produced a reference set of 144 fifteen-day control periods. Addends P_t in Eq. (27.3) are polynomials, each consisting of 18 monomials. However, since $y_{3h,t}$ is zero for every t across all 144 solutions (as spilling at Palmar is nil), the monomials $p_3^{(4)} x_{3h,t} y_{3h,t}$, $p_3^{(5)} x_{3h,t} {y_{3h,t}}^2$ and $2 p_3^{(5)} {x_{3h,t}}^2 y_{3h,t}$ can be disregarded. Additionally, across all 144 solutions, we checked that the combined relative influence of terms: $-p_1^{(4)} x_{1h,t} y_{1h,t}$, $-p_2^{(2)} x_{2h,t} y_{2h,t}$ and $p_3^{(5)} x_{3h,t}^3$ was negligible.

Therefore, Eq. (27.3), which provides a high-quality approximation of Río Negro's production under any weather conditions, can be simplified to the twelve monomials in Eq. (27.5) when the problem is considered within a sliding window during a mid-term dry weather season.

$$\hat{P}(x_{1h}, v_{1h}, x_{2h}, x_{3h}, v_{3h}) = \sum_{t=1}^{T} \hat{p}_1^{(1)} x_{1h,t} + \hat{p}_1^{(2)} x_{1h,t} v_{1h,t} - \hat{p}_1^{(3)} x_{1h,t} v_{1h,t}^2 - \hat{p}_1^{(4)} x_{1h,t}^2 + \hat{p}_2^{(1)} x_{2h,t}$$
$$- \hat{p}_2^{(2)} x_{2h,t}^2 - \hat{p}_2^{(3)} x_{2h,t} v_{3h,t} + \hat{p}_2^{(4)} x_{2h,t} v_{3h,t}^2 + \hat{p}_3^{(1)} x_{3h,t} + \hat{p}_3^{(2)} x_{3h,t} v_{3h,t} - \hat{p}_3^{(3)} x_{3h,t} v_{3h,t}^2 - \hat{p}_3^{(4)} x_{3h,t}^2 \tag{27.5}$$

Note that constants $p_i^{(j)}$ in Eq. (27.3) are renamed to $\hat{p}_i^{(j)}$ in Eq. (27.5). Their values were slightly adjusted to obtain a better match between the new/simplified model and the former over the six solutions found. For simplicity, the optimization problem discussed in this article includes only the opportunity cost of water at Bonete and Palmar reservoirs (i.e., Bellman costs coming from longer term planning models), as well as the penalty (i.e., cost of failure) for each MWh of unmet dispatch at every hour. This is expressed as shown in Eq. (27.6).

$$\begin{cases} \min\limits_{x_{ih}, y_{ih}, v_{ih}} \; ca_{1h}(v_{1h,0} - v_{1h,T}) + ca_{3h}(v_{3h,0} - v_{3h,T}) + CF \sum_{t=1}^{T} (d_t - g_{h,t}) & \\ d_t \geq g_{h,t}, & (i) \\ g_{h,t} = g_{1h,t} + g_{2h,t} + g_{3h,t}, & (ii) \\ g_{1h,t} = Pt\hat{B}on(x_{1h,t}, y_{1h,t}, v_{1h,t}), & (iii) \\ g_{2h,t} = Pt\hat{B}ay(x_{2h,t}, y_{2h,t}, v_{3h,t}), & (iv) \\ g_{3h,t} = Pt\hat{P}al(x_{3h,t}, y_{3h,t}, v_{3h,t}), & (v) \\ v_{1h,t} = v_{1h,t-1} + 3600(a_{1h,t} - x_{1h,t} - y_{1h,t}), & (vi) \\ x_{2h,t} + y_{2h,t} = a_{2h,t} + x_{1h,t-8} + y_{1h,t-8}, & (vii) \\ v_{3h,t} = v_{3h,t-1} + 3600(a_{3h,t} - x_{3h,t} - y_{3h,t} + x_{2h,t-16} + y_{2h,t-16}), & (viii) \\ (x_{ih}, y_{ih}, v_{ih}) \in (X, Y, V), & (ix) \end{cases} \tag{27.6}$$

As it was elaborated in Sect. 27.1.2, this model assumes the existence of Bellman costs to estimate the cost of opportunity of the water. In the objective function of (27.6), ca_{1h} and ca_{3h} (expressed in [USD/m^3]) respectively quantify those costs for Bonete and Palmar. The end-to-end variation of volume at reservoirs (i.e., $(v_{1h,0} - v_{1h,T})$ and $(v_{3h,0} - v_{3h,T})$) times its opportunity-cost are two of the costs accounted; the third is the total cost of failure.

In terms of constraints, Eq. (27.6)$-(i)$ ensures that the generation $g_{h,t}$ does not exceed the demand d_t at any time, which is essentially the same as Eq. (27.4)$-(i)$. Equations (27.6)$-(ii)$ to (v) represent that the total hydroelectric energy $g_{h,t}$ at any time-slot t, is the sum of the production from Bonete ($g_{1h,t}$), Baygorria ($g_{2h,t}$) and Palmar ($g_{3h,t}$).

Equations (27.6)$-(vi)$ to $(viii)$ enforce mass balance at each reservoir. Rain in remote areas takes days to reach the reservoir and impact storage. With permanent collection of rainfall data across the basin, natural inflows can be treated as deterministic for dispatch planning purposes. Therefore, is assumed that the inflows to the reservoirs are given: $a_{1h,t}$ (for Bonente), $a_{2h,t}$ (for Baygorria) and $a_{3h,t}$ (for Palmar). Equation (27.6)$-(vi)$ states that the water volume in Bonete's reservoir at time $t+1$ must equal the volume at time t, plus the natural inflows during that period, minus the discharges. The net inflow per second $(a_{1h,t} - x_{1h,t} - y_{1h,t})$—expressed in [m^3/s]– must be multiplied by the number of seconds in the time-slot, which in this case is 3600 (one hour). Since Baygorria is a run-of-the-river hydroelectric station, inflows and outflows must always match. Additionally, the balance at Baygorria must account for the water discharged from Bonete eight hours earlier, as indicated in the last column of Table 27.1 and captured in Eq. (27.6)$-(vii)$. The balance at Palmar's reservoir is similar (Eq. (27.6)$-(viii)$). Finally, Eq. (27.6)$-(ix)$ incorporates the technical limits of the units within the complex. The formulation in Eq. (27.6) is nonlinear, because even being simpler, the production addends at (27.5) are not linear yet. The terms in Eq. (27.5) can be regrouped to identify each unit, as in Eq. (27.7).

$$\begin{cases} Pt\hat{B}on = (\hat{p}_1^{(1)} x_{1h} - \hat{p}_1^{(4)} x_{1h}^2) + (\hat{p}_1^{(2)} v_{1h} - \hat{p}_1^{(3)} v_{1h}^2) x_{1h} \\ Pt\hat{B}ay = (\hat{p}_2^{(1)} x_{2h} - \hat{p}_2^{(2)} x_{2h}^2) - (\hat{p}_2^{(3)} v_{3h} - \hat{p}_2^{(4)} v_{3h}^2) x_{2h} \\ Pt\hat{P}al = (\hat{p}_3^{(1)} x_{3h} - \hat{p}_3^{(4)} x_{3h}^2) + (\hat{p}_3^{(2)} v_{3h} - \hat{p}_3^{(3)} v_{3h}^2) x_{3h} \end{cases} \quad (27.7)$$

These expressions complete the formulation in Eq. (27.6). Observe that the first term in the three production functions are analogous, and correspond to degree two polynomials $(\hat{p}x_{ih,t} - \hat{q}x_{ih,t}^2)$ in x_{ih}, all of which are concave. A concave function within a maximization process can easily and efficiently be captured by means of tangents. That idea is used to capture $(\hat{p}_3^{(1)} x_{3h} - \hat{p}_3^{(4)} x_{3h}^2)$ for Palmar with three tangents: one at the minimum technical (0); one at at the maximum technical; the other exactly in the middle. Hence, finding a good approach for that term is equivalent to maximizing a variable z_{3h}, while keeping it below these three tangents at any time:

$$\begin{cases} z_{3h,t} \leq \hat{r}_3^{(1)} x_{3h,t}, \\ z_{3h,t} \leq \hat{r}_3^{(2)} x_{3h,t} + \hat{s}_3^{(2)}, \\ z_{3h,t} \leq \hat{r}_3^{(3)} x_{3h,t} + \hat{s}_3^{(3)} \end{cases} \quad (27.8)$$

The solution was similar for Bonete (z_{1h}) and Baygorria (z_{2h}), but even simpler, as only two tangent lines were needed to achieve the desired accuracy. Up to this point, only continuous variables were used. Boolean variables are now introduced to account for the expressions $(\hat{p}_1^{(2)} v_{1h} - \hat{p}_1^{(3)} v_{1h}^2) x_{1h}$ and $(\hat{p}_3^{(2)} v_{3h} - \hat{p}_3^{(3)} v_{3h}^2) x_{3h}$ of (27.7). These terms respectively swell the efficiency of Bonete and Palmar, by incorporating the additional energy given from the extra height as the reservoir volume expands. Next, the focus is set upon how the issue was addressed at Bonete, with a similar approach applied to Palmar.

The approach quantizes Bonete's reservoir levels and introduces variables to capture additional efficiency. These variables activate or deactivate sequentially as the reservoir volume crosses certain thresholds, which are relative to the initial volume $v_{1h,0}$. Due to Bonete's large size, volume changes over a fifteen-day period are expected to be minimal.

After analyzing the volume series of Bonete ($v_{1h,t}$) across our reference set of 144 solutions, the conclusion is that three levels were sufficient to achieve the desired accuracy. Let's denote the three levels into which Bonete's volume is quantized—after the initial volume $v_{1h,0}$ for a given fifteen-day short-term instance—

as: $\overline{V1h1}$, $\overline{V1h2}$ and $\overline{V1h3}$. The approximation to $g_{1h,t}$ follows the replacement of Eq. (27.4)–(iii) with the following set of constraints

$$\begin{cases} g_{1h,t} = z_{1h,t} + \sum_{i=1}^{3} \theta_{1h,t}^{(i)}, & (i) \\ z_{1h,t} \leq \hat{r}_1^{(1)} x_{1h,t}, & (ii) \\ z_{1h,t} \leq \hat{r}_1^{(2)} x_{1h,t} + \hat{s}_1^{(2)} & (iii) \end{cases} . \tag{27.9}$$

The $z_{1h,t}$ addend approximates the concave portion, as shown for Palmar in Eq. (27.8), thought this case only uses two tangents. The variables $\theta_{1h,t}^{(i)}$ correspond to the additional production due to the extra height. To work properly, these variables require constraints as in Eq. (27.10).

$$\begin{cases} 0 \leq \theta_{1h,t}^{(1)} \leq (\hat{p}_1^{(2)} \overline{V1h1} - \hat{p}_1^{(3)} \overline{V1h1}^2) x_{1h,t}, & (i) \\ 0 \leq \theta_{1h,t}^{(2)} \leq (\hat{p}_1^{(2)} \Delta\overline{V1h2} - \hat{p}_1^{(3)} \Delta\overline{V1h2}^2) x_{1h,t}, & (ii) \\ 0 \leq \theta_{1h,t}^{(2)} \leq 680 (\hat{p}_1^{(2)} \Delta\overline{V1h2} - \hat{p}_1^{(3)} \Delta\overline{V1h2}^2) \varphi_{1h,t}^{(1)}, & (iii) \\ 0 \leq \theta_{1h,t}^{(3)} \leq (\hat{p}_1^{(2)} \Delta\overline{V1h3} - \hat{p}_1^{(3)} \Delta\overline{V1h3}^2) x_{1h,t}, & (iv) \\ 0 \leq \theta_{1h,t}^{(3)} \leq 680 (\hat{p}_1^{(2)} \Delta\overline{V1h3} - \hat{p}_1^{(3)} \Delta\overline{V1h3}^2) \varphi_{1h,t}^{(2)} & (v) \end{cases} \tag{27.10}$$

This equation introduces constants derived from $\overline{V1h1}$ to $\overline{V1h3}$, which are: $\Delta\overline{V1h2} = (\overline{V1h2} - \overline{V1h1})$, $\Delta\overline{V1h2}^2 = (\overline{V1h2}^2 - \overline{V1h1}^2)$, $\Delta\overline{V1h3} = (\overline{V1h3} - \overline{V1h2})$ and $\Delta\overline{V1h3}^2 = (\overline{V1h3}^2 - \overline{V1h2}^2)$. Equation (27.10)–$(i)$, which is linear, allows $\theta_{1h,t}^{(1)}$ to take on a value similar to $(\hat{p}_1^{(2)} v_{1h} - \hat{p}_1^{(3)} v_{1h}^2) x_{1h}$ in Eq. (27.7). These values align precisely when $v_{1h,t} = \overline{V1h1}$. For now, keep in mind that $\varphi_{1h,t}^{(1)} = 1$ only if $v_{1h,t} \geq \overline{V1h2}$, and $\varphi_{1h,t}^{(2)} = 1$ only if $v_{1h,t} \geq \overline{V1h3}$. If $\varphi_{1h,t}^{(1)} = 0$, Eq. (27.10)–(iii) forces $\theta_{1h,t}^{(2)} = 0$. The deactivation of $\varphi_{1h,t}^{(2)}$ similarly affects $\theta_{1h,t}^{(3)}$ due to Eq. (27.10)–(v). Conversely, since 680 [m^3/s] is the maximum technical turbinated flow for Bonete, Eq. (27.10)–(iii) deactivates when $\varphi_{1h,t}^{(1)} = 1$, and as $x_{1h,t}$ changes, the sum of $\theta_{1h,t}^{(1)} + \theta_{1h,t}^{(2)}$ approximates $(\hat{p}_1^{(2)} v_{1h} - \hat{p}_1^{(3)} v_{1h}^2) x_{1h}$ when the reservoir volume is around $\overline{V1h2}$. A similar trigger effect occurs when $\varphi_{1h,t}^{(2)} = 1$, which also implies that $\varphi_{1h,t}^{(1)} = 1$. In this case, $\sum_{i=1}^{3} \theta_{1h,t}^{(i)}$ approximates $(\hat{p}_1^{(2)} v_{1h} - \hat{p}_1^{(3)} v_{1h}^2) x_{1h}$ when the reservoir volume is near $\overline{V1h3}$. Equation (27.11) shows how $\varphi_{1h,t}^{(i)}$ variables can be forced to zero according on thresholds.

$$\begin{cases} \varphi_{1h,t}^{(1)} \leq 1 - \dfrac{\overline{V1h2} - v_{1h,t}}{M1}, \\ \varphi_{1h,t}^{(2)} \leq 1 - \dfrac{\overline{V1h3} - v_{1h,t}}{M1} \end{cases} \tag{27.11}$$

Note that, with a properly chosen value of $M1$, $\varphi_{1h,t}^{(1)}$ cannot be activated if $v_{1h,t} < \overline{V1h2}$, as it is a boolean variable constrained to a value less than 1. The same idea goes for $\varphi_{1h,t}^{(2)}$. The strategy for addressing the problem at Palmar is similar. The main difference lies in Palmar's faster dynamics and smaller reservoir, which allows its lake level to fluctuate more significantly than Bonete's. To account for this, five reservoir levels were used for Palmar instead of three.

The only remaining unresolved term is $-(\hat{p}_2^{(3)} v_{3h} - \hat{p}_2^{(4)} v_{3h}^2)x_{2h}$, which impacts Baygorria's efficiency as the level of Palmar's lake rises. Experimental evaluations—both on the 144 prior solutions and on the results of test models prior to this one– have shown that Palmar's power and reservoir size dominate the optimization outcome. It is Palmar's reservoir management that influences Baygorria, not the other way around. Therefore, the subproblem was handled through a simple linear regression of the expression. The regression is performed for each specific instance before running the problem, using the five designated level thresholds $\overline{V3h1}$ to $\overline{V3h5}$ for Palmar in that instance.

27.4 Experimental Evaluation

The evaluation of the MIP model was conducted using a test-set of 204 instances. These instances were grouped into 6 subgroups, each consisting of 34 instances. The subgroups are fifteen-day periods sampled over the three reference good-quality solutions of scenario A as defined in Sect. 27.2.2, meaning that the profile of residual demands match actual data from 2018. The first half of the instances replicates the residual demand, while the second half increases it by a factor of 1.5 and adds a sustained demand of an additional 50 MW, as in Table 27.2.

Note that sampling among different solutions allow exploring changes in the initial conditions of reservoirs of Bonete and Palmar, while natural inflows do the proper with the series of natural inflows sketched in Fig. 27.5. The economical parameters ca_{1h}, ca_{3h} and CF derive from the actual hydro-production during 2011 for a reference cost of energy of 80 [USD/MWh]. Bellman values ca_{1h} and ca_{3h} are equal at instances 3 and 6, whereas other instances shift toward either Bonete or Palmar. The

Table 27.2 Complementary parameters to complete the test-instances data-set

Subgroup	ca_{1h}[USD/m^3]	ca_{3h}[USD/m^3]	CF[USD/MWh]	Residual demand
1	0.0033	0.0063	3200	d_t as in 2018
2	0.0063	0.0033	3200	d_t as in 2018
3	0.0048	0.0048	3200	d_t as in 2018
4	0.0033	0.0063	1600	50 MW + $1.5 \times d_t$
5	0.0063	0.0033	1600	50 MW + $1.5 \times d_t$
6	0.0048	0.0048	1600	50 MW + $1.5 \times d_t$

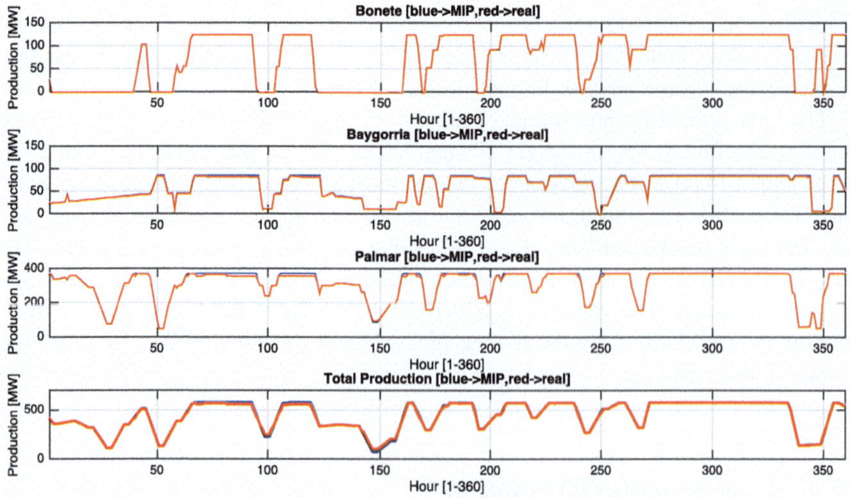

Fig. 27.7 Real [(27.3) in red] vs MIP approximated production: $g_{1h,t}$, $g_{2h,t}$, $g_{3h,t}$ and $g_{h,t}$

cost of failure for the last three subgroups is 20 times the reference energy cost, and it is doubled for the first half.

Instances were solved by means of the software `IBM(R) ILOG(R) CPLEX(R) Interactive Optimizer 12.6.3.0`, running on an `HP ProLiant DL385 G7`, with `AMD Opteron` processor 6172, 24 cores at 2.1GHz and 64GB of RAM.

All solutions reported here reached optimality, with relative differences between the cumulated reference production function in (27.3) and its equivalent $\sum_{t=1}^{T} g_{h,t}$ of the MIP's approximation (eq-(27.6)–(ii)) ranging from -0.012% to 0.046%, with average -0.001%, which is negligible. When we move from total energy to power (i.e., slot-by-slot energy) the difference in [MWh] ranges from 0.92 to 32.05, with a mean of 12.52. As a reference, consider that the combined installed plant of Bonete, Baygorria and Palmar is 603 MW. The maximum deviation in power occurs for the 10th instance of subgroup 4 in Table 27.2, and its corresponding curves are shown in Fig. 27.7.

Computation times ranged from under 1 to 1210 s, with an average of 121 s. Notably, in most cases, the majority of the time was spent on fine-tuning the solution to reach optimality. We remark that in over 95% of the 204 instances solved, a feasible solution with a proven duality gap (i.e., an error estimate) of less than 1% was achieved within the first 2 s of runtime.

27.5 Conclusions and Future Work

Merging conventional renewable energy sources with newer ones allows for the best use of both, but poses challenges in operations. MIP and MISP models are ideal for addressing parts of the problem and are better suited to adapt to emerging challenges. This is not the case for legacy hydroelectric complexes such as the one in Río Negro. This paper presents a MIP model for such a complex that effectively tackles the subproblem. This research team is actively working to incorporate additional components into the model to develop a more comprehensive and realistic Unit Commitment and Dispatch Optimization system.

Acknowledgements This work was supported by UTE (Administración Nacional de Usinas y Trasmisiones Eléctricas, Uruguay) and PEDECIBA (Programa de Desarrollo de las Ciencias Básicas, Uruguay).

References

1. R. Bertoni, P. Messina, The "wind revolution" in Uruguay and the role of the public sector in guiding energy transitions, in *Energy Transitions in Latin America: The Tough Route to Sustainable Development* (Springer, 2023), pp. 195–216
2. B. Bouzas, E. Teliz, V. Díaz, Green hydrogen production in Uruguay: a techno-economic approach. Int. J. Chem. React. Eng. **22**(7), 783–795 (2024)
3. G. Casaravilla, R. Chaer, Energy transition of Uruguay, in *IAEE Energy Forum, Fourth Quarter* (International Association for Energy Economics, 2021), pp. 11–14
4. M.P. Cristobal, L.F. Escudero, J.F. Monge, On stochastic dynamic programming for solving large-scale planning problems under uncertainty. Comput. Oper. Res. **36**(8), 2418–2428 (2009)
5. A.L. Diniz, T.M. Souza, Short-term hydrothermal dispatch with river-level and routing constraints. IEEE Trans. Power Syst. **29**(5), 2427–2435 (2014)
6. R. Ferrero, J. Rivera, S. Shahidehpour, A dynamic programming two-stage algorithm for long-term hydrothermal scheduling of multireservoir systems. IEEE Trans. Power Syst. **13**(4), 1534–1540 (1998)
7. J.J. Hargreaves, B.F. Hobbs, Commitment and dispatch with uncertain wind generation by dynamic programming. IEEE Trans. Sustain. Energy **3**(4), 724–734 (2012)
8. R. Khatami, M. Parvania, C. Chen, S.S. Guggilam, S.V. Dhople, Dynamics-aware continuous-time economic dispatch: a solution for optimal frequency regulation, in *HICSS* (2020), pp. 1–10
9. MOP: Modelo de operación desarrollado por UTE (2024), https://gitlab.com/utepladmm/MOP. Accessed 21 Sept 2024
10. D. Oroño, M. Vignolo, V. Echinope, Analysis of the penetration and impacts of hybrid and electric vehicles in the electrical system of Uruguay. In: *2015 IEEE PES Innovative Smart Grid Technologies Latin America (ISGT LATAM)* (IEEE, 2015), pp. 75–80
11. J. Patrone, A. Plat, G. Failache, Actualización de los criterios de operación durante crecidas normales y extraordinarias de los embalses en cadena del Río Negro (Uruguay). In: *Dam Maintenance and Rehabilitation* (Routledge, 2017), pp. 433–440
12. REN21: Renewables 2020 global status report. Tech. rep., Paris: REN21 Secretariat (2021), https://www.ren21.net/gsr-2020/

13. C. Risso, Benefits of demands control in a smart-grid to compensate the volatility of non-conventional energies. Revista Facultad de Ingeniería Universidad de Antioquia (93), 19–31 (Aug 2019). https://doi.org/10.17533/udea.redin.20190404, https://revistas.udea.edu.co/index.php/ingenieria/article/view/336255
14. V. Sherkat, R. Campo, K. Moslehi, E. Lo, Stochastic long-term hydrothermal optimization for a multireservoir system. IEEE Trans. Power Appar. Syst. **8**, 2040–2050 (1985)
15. U.S. Energy Information Administration: Uruguay-international-u.s. energy information administration (eia) (2024), https://www.eia.gov/international/data/country/URY/electricity/electricity-generation. Accessed 17 Sept 2024
16. W. Van Ackooij, I. Danti Lopez, A. Frangioni, F. Lacalandra, M. Tahanan, Large-scale unit commitment under uncertainty: an updated literature survey. Ann. Oper. Res. **271**(1), 11–85 (2018)

Chapter 28
Modeling Vehicular Flow Using an Approach Based on the Navier–Stokes Equations

Ernesto de la Cruz-Nicolás, Hugo Estrada-Esquivel, Alicia Martínez-Rebollar, Odette Pliego-Martínez, and Eddie Clemente

Abstract The study of vehicular flow is vital for traffic optimization and urban planning. Traditionally, traffic models have been based on specific theories of vehicular behavior, but this paper proposes an innovative approach that utilizes the Navier–Stokes equations, which are classical in fluid dynamics, to model the flow of vehicles on roadways. We adapt these equations to capture the dynamics of vehicular traffic, treating vehicle movement as fluid flow rather than simple motion on a road. The approach relies on the analogy between fluid dynamics and traffic behavior, applying concepts of mass and momentum conservation to describe how vehicles cluster and disperse on a roadway. The necessary adaptations for viscosity and pressure in the context of vehicular traffic are examined, and simulations are presented to illustrate how this model can offer new perspectives on congestion and traffic flow. This work

E. de la Cruz-Nicolás (✉) · H. Estrada-Esquivel · A. Martínez-Rebollar · O. Pliego-Martínez · E. Clemente
National Institute of Technology of México/CENIDET, Cuernavaca Morelos, México
e-mail: d21ce090@cenidet.tecnm.mx

H. Estrada-Esquivel
e-mail: hugo.ee@cenidet.tecnm.mx

A. Martínez-Rebollar
e-mail: alicia.mr@cenidet.tecnm.mx

O. Pliego-Martínez
e-mail: d21ce092@cenidet.tecnm.mx

E. Clemente
e-mail: eddie.ct@cenidet.tecnm.mx

E. de la Cruz-Nicolás
National Institute of Technology of México/Cuautla Institute of Technology, Cuautla, Morelos, México

O. Pliego-Martínez
National Institute of Technology of México/Milpa Alta Institute of Technology, México City, México

D. Rossit et al. (eds.), *Proceedings of the VII Ibero-American Congress of Smart Cities, ICSC-Cities 2024, 12-14 November, San Carlos, Costa Rica*, Springer Proceedings in Energy, https://doi.org/10.1007/978-981-96-4301-1_28

highlights the potential of the Navier–Stokes equations to provide a deeper understanding of traffic patterns and suggests new directions for future research in the optimization of vehicular flow and urban traffic management.

Keywords Vehicular flow · Fluid · Traffic congestion · Equations · Model

28.1 Introduction

Currently, effective vehicular traffic management faces a growing challenge due to the increasing number of vehicles in large cities. Traffic congestion not only affects the quality of urban life but also has significant economic and environmental repercussions [1]. Accurately understanding vehicular flow and having the ability to predict and manage traffic patterns is important for improving mobility and safety in urban environments [2].

Conventional methods for modeling traffic have employed various techniques, such as queueing theory, system dynamics, and both macroscopic and microscopic approaches that replicate vehicle behavior and their interactions. However, these methods often face limitations in capturing the complexity and variability of traffic behavior under different conditions, such as the impact of unforeseen events (accidents or adverse weather conditions), fluctuations in traffic volumes during peak and off-peak hours, and driver behavior in high congestion situations. These factors add levels of variability that traditional models find difficult to accurately represent, thereby limiting their predictive and analytical capabilities in dynamic environments [3].

The Navier–Stokes equations, fundamental in fluid dynamics, provide a robust mathematical foundation for describing the behavior of fluids in motion. These equations are based on the principles of mass, momentum, and energy conservation, and are widely applied in engineering to model phenomena such as the flow of air and water. Although some studies have proposed the analogy between fluid flow and vehicular traffic, their practical application has been limited [4, 5].

This article introduces an innovative approach by adapting the Navier–Stokes equations to model vehicular flow, considering traffic as a continuous fluid rather than a discrete series of vehicles. The premise is that vehicular movement can be described analogously to fluid flow, allowing for the application of fluid dynamics concepts and tools to traffic analysis. Adapting these equations to traffic involves treating vehicle density as a form of "viscosity" and the interactions between them as forces that affect the vehicular "fluid."

This study explores the necessary modifications to implement the Navier–Stokes equations in traffic modeling, adjusting the terms of viscosity and pressure to reflect the characteristics of traffic. Results are presented that demonstrate how this approach can provide a deeper understanding of traffic dynamics, including the formation of traffic jams and the influence of various factors on vehicular flow.

The main objective is to provide a theoretical and practical framework for applying the Navier–Stokes equations in the analysis of vehicular traffic, with the hope that this approach can enhance traffic planning and management in urban areas. By integrating advanced concepts of fluid dynamics into traffic modeling, the aim is to open new possibilities for research and the application of innovative techniques to optimize vehicular flow.

28.2 Related Works

Modeling vehicular traffic has been an intense field of research with various methodological approaches aimed at capturing the complexity of vehicular flow and improving traffic management in urban environments [6]. Among the most prominent traditional approaches are queueing theory-based models and system dynamics.

Queueing theory models have been widely used to analyze traffic at congestion points, such as intersections and toll booths [7, 8]. These models represent traffic as a series of queues of vehicles that must be processed, allowing for an evaluation of the efficiency and performance of transportation infrastructures [9]. The work in [10] provides an overview of how queueing models are applied to various situations in traffic management, including the analysis of waiting times at traffic signals and the capacity of intersections.

In contrast, system dynamics applies concepts of feedback and system dynamics to model traffic. This approach considers how variables such as vehicle density and speed interact and change over time [11]. The work in [12] introduces the application of system dynamics to traffic, providing a useful tool for understanding the evolution of traffic based on various parameters and feedback from the system.

Macroscopic traffic models, such as the Lighthill–Whitham–Richards (LWR) model, treat traffic in terms of aggregated variables like flow and density. These models have been fundamental for analyzing large transportation systems and have proven effective in capturing the overall behavior of traffic on roads and in urban areas [13]. In contrast, microscopic models simulate the behavior of each individual vehicle and their interactions, providing a higher level of detail but with greater computational complexity [14]. The research in [15] on the use of SUMO (Simulation of Urban Mobility) illustrates how microscopic models allow for detailed simulation of traffic and vehicle behavior in different scenarios.

Moreover, the application of fluid dynamics principles to vehicular traffic has emerged as a promising line of research. This approach seeks to draw analogies between various problems, such as the relationship between traffic and fluids, using concepts like viscosity to describe how interactions between vehicles affect overall flow [16]. Although this perspective has provided new insights into traffic modeling, its practical application has been limited thus far.

Fundamental to fluid dynamics, computational fluid dynamics (CFD) is widely applied in engineering to model fluid flow in various real-world problems, such as air and water [17]. The Navier–Stokes equations, based on principles of mass, momentum, and energy conservation, have extensive applications in predicting complex fluid phenomena [18]. Adapting these equations for modeling vehicular traffic represents an emerging area of research, offering the potential to provide a new dimension for understanding traffic dynamics.

In recent times, there has been an increase in the use of advanced techniques such as machine learning [19] and optimization to improve the accuracy of traffic modeling [20–22]. These techniques enable real-time predictions of traffic behavior and present new possibilities for urban traffic management and planning. Research analyzing and proposing the use of deep learning algorithms for traffic prediction illustrates how these tools can significantly enhance the ability to forecast and manage traffic in real time and in various contexts [23–26].

28.3 Methodology for Developing Vehicular Flow Modeling Using a Navier–Stokes Equations-Based Approach

The methodology for developing vehicular flow modeling adopts an approach based on the Navier–Stokes equations. This theoretical framework is chosen for its ability to accurately and precisely describe fluid behavior, which is essential for understanding traffic dynamics. Through a specific case study, the necessary data collection to support the model is addressed, followed by its formulation and the application of the Navier–Stokes equations to vehicular flow. Subsequently, the implementation of these equations is presented, culminating in a validation and adjustment process to ensure the model's effectiveness and accuracy. Each subsection provides a comprehensive view of the stages involved in model development, offering a systematic approach that allows for the optimization of vehicular flow simulation and analysis in real-world environments.

28.3.1 Case Study

First, the case study is defined, selecting Insurgentes Avenue in Mexico City as the setting for modeling vehicular flow. This avenue was chosen for being one of the most important access and transit routes in the capital, making it a representative environment for analyzing urban traffic dynamics. Modeling vehicular flow on Insurgentes, based on the Navier–Stokes equations, requires a detailed analysis of several key factors.

In this initial phase, the geometric characteristics of the stretch of the avenue are identified, including length, number of lanes, gradients, and structural elements that influence traffic flow. These characteristics are important as they help define the boundary conditions and physical constraints for the vehicular flow simulation. Additionally, typical traffic conditions are considered, such as average speed, vehicle density, and peak hours, providing a clear view of the circulation patterns that occur at different times of the day.

28.3.2 Data Collection

Data collection was conducted on Insurgentes Avenue in Mexico City to model vehicular flow using an approach based on the Navier–Stokes equations. The information obtained included precise data on traffic and real-time road incidents, collected through mapping applications (Here Maps), with records taken every 5 min over a 24-h period, covering the period from February to June 2024. A total of 1,440 records were collected.

Additionally, simulated data were generated for key variables such as friction, viscosity, and aerodynamic resistance, essential for modeling vehicular flow under real conditions. Friction between vehicles and the surface of the avenue, as well as viscosity, which represents internal resistance within traffic flow, were integrated into the Navier–Stokes equations to more accurately reflect vehicle interactions. Furthermore, the simulation scenarios included aerodynamic resistance and resistance due to slope, assessing their impact on speed and the overall behavior of traffic on the avenue. These models provide a deeper understanding of urban traffic dynamics, facilitating future interventions for controlling and optimizing vehicle flow. All these variables were incorporated as attributes in the set of 1,440 records collected.

Data simulation for variables such as friction, viscosity, and aerodynamic resistance was necessary due to the difficulty and high costs associated with directly collecting these data under real urban traffic conditions. Unlike traffic and incident data, which can be obtained in real-time through mapping applications and traffic sensors, measurements of friction, viscosity, and aerodynamic resistance require specialized equipment and controlled conditions to ensure accuracy, which is impractical on an active avenue like Insurgentes.

To represent friction, a normal distribution was used based on observations of resistance across different pavement types, allowing for the variability of surface conditions and the diversity of vehicles, such as cars and buses, to be reflected. Viscosity was modeled using a gamma distribution, suitable for simulating internal resistance within vehicular flow, as it captures how vehicle density and interaction affect traffic fluidity. Aerodynamic resistance, on the other hand, was represented by a log-normal distribution, based on studies of vehicle interactions with moving air resistance. This distribution is suitable for aerodynamic resistance data, which tend to be asymmetric and where extreme values in high- or low-speed scenarios are significant.

The use of these probabilistic distributions allows for approximating natural traffic variations without interfering with the avenue's flow or incurring significant direct measurement costs. This approach ensures that the model reflects complex interactions under realistic conditions, optimizing the accuracy of the analysis and facilitating future interventions in traffic management.

To determine viscosity in the context of vehicular flow on Insurgentes Avenue in Mexico City, specific adaptations were made to the classical approach of the Navier–Stokes equations, as there is no direct friction between vehicles as in a fluid. Insurgentes has multiple lanes in both directions, and not all travel at the same speed; this variation is modeled through a "vehicular viscosity" factor, reflecting speed differences between lanes and the effects of maneuvers like lane changes. Additionally, speed limits, which induce changes in density and average speed, act as external forces influencing vehicular "pressure" within the model. The edges of the avenue and surrounding infrastructure also affect flow, especially in the outer lanes, where they generate a resistance or deceleration effect similar to a wall in a fluid. Although there is no molecular friction, the model adapts the concept of viscosity to represent indirect interactions between vehicles, where a slower vehicle reduces the speed of those following it in high-density areas. This approach allows the application of mass and momentum conservation principles to describe the clustering and dispersal of vehicles within the flow, offering a new perspective on congestion and traffic on Insurgentes.

The following link provides access to a traffic dataset https://drive.google.com/file/d/1QaZe5D8WD2olC__zjj9zVkdbWRrneNce/view?usp=sharing, which contains a random sample of 100 records of traffic and incident data collected in real-time through Here Maps. It also includes simulated data for Insurgentes Avenue, with the purpose of offering a partial view of the data used.

28.3.3 Model Formulation

The formulation of the traffic flow model is based on the Navier–Stokes equations, adapted to apply the principles of fluid dynamics to the behavior of vehicular flow, rather than fluids. For this purpose, three stages were carried out, described as follows:

(a) Stage 1: Continuity Equation (Conservation of Mass): This is a fundamental principle in modeling dynamic systems. This equation ensures that mass—in this case, the number of vehicles—is conserved within a closed system or a segment of roadway. That is, as vehicles move along the roadway, the number of vehicles entering a segment must equal the amount exiting, unless there is a change in vehicle density.

Consequently, the Continuity Equation is formulated by considering the speed and density of traffic at different points on the roadway. Mathematically, it is defined under the principle that the temporal variation in vehicle density must correspond to the difference between the flow of vehicles entering and exiting. This implies that if the speed of vehicles changes or accumulation occurs in a specific area, the density will adjust proportionally, as illustrated in Eq. 28.1.

$$\frac{\partial \rho}{\partial t} + \frac{\partial (\rho v)}{\partial x} = 0 + \frac{\partial (\rho v)}{\partial x} \tag{28.1}$$

where:

$\rho v(x, t)$ is the vehicle density as a function of position. x and time tt,

$v(x, t)$ is the average speed of vehicles as a function of position and time,

$\frac{\partial \rho}{\partial t}$ represents the change in vehicle density with respect to time,

$\frac{\partial (\rho v)}{\partial x}$ is the rate of change of vehicle flow (ρv) with respect to position x.

The Continuity Equation ensures the model's coherence with the laws of mass conservation, providing a solid foundation for describing the movement and distribution of traffic on the roads.

(b) Stage 2 Adapted Navier–Stokes Equations: At this stage, the Navier–Stokes equations are adapted to model vehicular flow instead of fluids, capturing specific traffic factors such as vehicle density, average speed, and internal and external forces that influence its behavior. The adapted equation includes terms representing vehicle density (ρ), the velocity field (u), traffic pressure ($p_{traffic}$) which simulates the tendency to maintain a safe distance traffic viscosity ($\mu_{traffic}$), which describes the friction and resistance between vehicles, and external forces ($f_{traffic}$), such as traffic signals or road conditions. This model allows for the simulation of complex traffic phenomena, such as the formation and dispersion of congestion and the propagation of braking waves, facilitating the analysis and prediction of traffic patterns under various conditions, thereby aiding in the management and planning of transportation systems in real-time., as shown in Eq. 28.2.

$$\rho\left(\frac{\partial u}{\partial t} + u \cdot \nabla u\right) = -\nabla p_{traffic} + \mu_{traffic} \nabla^2 u + f_{traffic} \tag{28.2}$$

where:

u is the speed of the vehicles,

$\mathrm{p}_{\mathrm{traffic}}$ represents the traffic pressure (related to density),

$\mu_{traffic}$ is the viscosity of traffic (related to friction between vehicles),

$\mathrm{f}_{\mathrm{traffic}}$ are the external forces specific to traffic.

Equation 28.2 describes how traffic density, traffic pressure, resistance forces, and external forces interact to influence the behavior of vehicular flow. Just like in fluid flow, the balance between these factors determines the evolution of traffic over time and space.

(c) Stage 3 Friction and Resistance Equations: This stage involves integrating the concepts of friction and resistance into the Navier–Stokes equations of the model, which affects the movement of vehicles and their interactions on the roadway, thereby enabling an accurate simulation of traffic flow. The traffic model implements equations to describe how friction varies with traffic density; Eq. 28.3 relates friction and vehicle density:

$$\mathrm{F}_{\mathrm{friction}} = \mathrm{k}\rho^2 \mathrm{F}_{\mathrm{friction}} = \mathrm{k}\rho^2 \tag{28.3}$$

where:

k Represents the intensity of friction, depending on various factors such as road characteristics, vehicle type, traffic incidents, among others,

ρ^2 Indicates that friction increases non-linearly with vehicle density.

The equations that describe the total resistance a vehicle experiences while moving are formulated by considering all types of resistance, as shown in Eqs. 28.4, 28.5, 28.6, and 28.7:

$$\mathrm{R_{total}} = \mathrm{R_{aerodynamic}} + \mathrm{R_{slope}} + \mathrm{R_{friction}}\mathrm{R_{total}} = \mathrm{R_{aerodynamic}} + \mathrm{R_{slope}} + \mathrm{R_{friction}} \tag{28.4}$$

$$\mathrm{R}_{aerodynamic} = \frac{1}{2} C_d A \rho u^2 \tag{28.5}$$

where:

C_d is the drag coefficient,

A is the frontal area exposed to air,

ρ is the air density,

u is the vehicle's speed relative to the air.

$$R_{slope} = mg\sin(\theta) \tag{28.6}$$

where:

m is the mass of the vehicle,

g is the acceleration due to gravity (9.81 m/s2),

θ is the angle of the slope.

$$R_{friction} = \mu N \tag{28.7}$$

where:

μ is the coefficient of friction (depends on the type of soil or road),

N is the normal force.

28.3.4 Formulation of the Vehicle Flow Model Using Navier–Stokes Equations

By combining the equations presented in Sect. 3.3, it is possible to construct a model that describes traffic dynamics and allows for the evaluation of its performance under various conditions. This is reflected in Eq. 28.8.

$$\frac{\partial \rho}{\partial t} + \frac{\partial(\rho v)}{\partial x} + \left(\frac{1}{\rho}\left(\rho\left(\frac{\partial u}{\partial t} + u \cdot \nabla u\right) + \nabla p_{traffic} - \mu_{traffic} \nabla^2 \cdot u - k\rho^2\right)\right) = 0 + \frac{\partial(\rho v)}{\partial x} \tag{28.8}$$

The first part $\frac{\partial \rho}{\partial t} + \frac{\partial(\rho v)}{\partial x} + \frac{\partial(\rho v)}{\partial x}$ represents the conservation of mass, while the second part encapsulates the Navier–Stokes equation, which includes the forces acting on the traffic flow, balancing the dynamics of speed, pressure, and resistance. By implementing the previously described equations, a general model of vehicular flow is developed based on a Navier–Stokes approach. This method allows for the description of vehicular traffic behavior as a fluid, facilitating the representation and analysis of traffic dynamics under various conditions.

By employing the Navier–Stokes equations, both the interactions between vehicles and the effects of density, speed, and external forces such as aerodynamic resistance, friction, and slopes in the road infrastructure can be accurately modeled. This model captures the highly dynamic nature of vehicular flow, enabling more realistic simulations of traffic conditions, optimizing transit management, and assessing the impact of different scenarios, such as variations in speed or changes in infrastructure. Additionally, this approach is particularly useful for identifying bottlenecks and areas of congestion, as it considers the forces that affect the collective behavior of traffic, resulting in more precise and efficient modeling.

28.3.5 Implementation of Equations

In this section, the process of implementing the key equations that constitute the proposed vehicular flow model is detailed, using the Simulation of Urban Mobility (SUMO) tool for simulation. This approach allows for the reproduction and analysis of traffic behavior in urban environments under various conditions, integrating the fundamental equations that govern vehicular dynamics. First, the implementation of the Continuity Equation is addressed, ensuring the conservation of vehicular density over time and space. Subsequently, the implementation of the Adapted Navier–Stokes Equations is detailed to represent traffic behavior instead of fluids. Finally, the implementation of Friction and Resistance is explained, as these elements model the forces influencing traffic dynamics, such as aerodynamic resistance and interaction with the road surface.

(a) Implementation of the Continuity Equation: The implementation of the Continuity Equation is performed using numerical methods such as the finite difference scheme to approximate the partial derivatives. Traffic behavior on a road is simulated by dividing the road into segments and calculating the evolution of vehicle density and speed in each segment over time. The implementation of the Continuity Equation in the basic simulation involved the following steps:

- Initial vehicle density in each road segment: In this step, the initial traffic density was defined to represent localized congestion.
- Average speed of vehicles in each segment: In this step, a linear relationship between speed and density was used, reflecting that vehicles move more slowly as density increases.

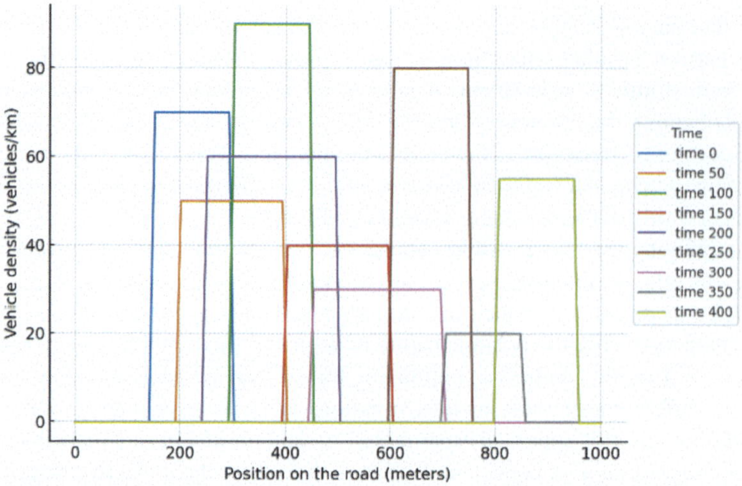

Fig. 28.1 Traffic simulation using the continuity equation (in meters)

- Time step and spatial step: In this step, the evolution of density along the road was analyzed at different time steps, allowing observation of how congestion changes over time.
- Use of the finite difference scheme to approximate the derivatives of density and flow along the road and over time: In this step, the density was updated at each time step, using finite differences to approximate the derivatives in the continuity equation.

The Fig. 28.1 shows the evolution of vehicle density as a function of position on the road at different moments in time. Each colored line represents a specific time instant (as indicated in the legend) and describes how vehicle density changes across the various segments of the simulated road.

For this vehicular flow simulation, conducted using the SUMO tool, the Continuity Equation was applied to ensure the conservation of vehicle density over time and space. The equation was implemented using a finite difference scheme, allowing for the approximation of partial derivatives and analysis of traffic behavior by dividing the road into segments.

The graph illustrates how the initial density in certain segments creates areas of high congestion, while other segments show lower densities. Observing the evolution of density over time reveals how congestion forms and dissipates in response to simulated traffic conditions. For instance, certain peaks reflect a concentration of vehicles in specific areas, while a decrease in density in other segments suggests traffic dispersion or a reduction in congestion at those points.

The analysis using the Continuity Equation allows us to visualize how density varies in each segment and at each moment. As time progresses, changes in density reflect the movement and redistribution of vehicles, which may be influenced by friction and resistance forces (such as aerodynamic resistance and interaction with the road surface), also modeled in the system. The relationship between density and speed is inverse; that is, higher density in a segment generally implies a reduction in the average speed of vehicles at that point, which aligns with the behavior observed in the graph.

(b) Implementation of the Adapted Navier–Stokes Equations: The implementation of the Adapted Navier–Stokes Equations follows a similar approach to that used in the simulation of the Continuity Equation, but incorporates terms that represent the dynamics of traffic, such as velocity, pressure, and external forces. The implementation of the adapted Navier–Stokes equations in a basic simulation involves the following steps:

- Initial parameters: In this step, the length of the road, spatial step, time step, maximum speed, maximum density, and traffic viscosity are defined.
- Vehicle density and initial speed: In this step, an initial density is established in a segment of the road, and the initial speed is calculated based on the density.
- External force parameters: In this step, intersections, traffic lights, and road incidents are established as external forces that impact the performance of vehicular flow.

The two simulations of vehicle density and speed as a function of road position, in which the adapted Navier–Stokes Equations were implemented to model traffic flow, incorporate parameters and external forces that influence vehicular dynamics, such as speed, pressure, and external elements. In the upper graphs of Fig. 28.2, which represent vehicle density, a variable distribution along the road is observed. In the first simulation (on the left), a central density peak indicates a high congestion area, while in the second simulation (on the right), an abrupt increase in density is observed, which subsequently stabilizes. The lower graphs, showing vehicle speed, reflect the expected inverse relationship between density and speed in traffic: in high-density areas, speed decreases. In the first simulation, speed drops in the center of the road, coinciding with the density peak, while in the second simulation, speed drops sharply at the point of highest density.

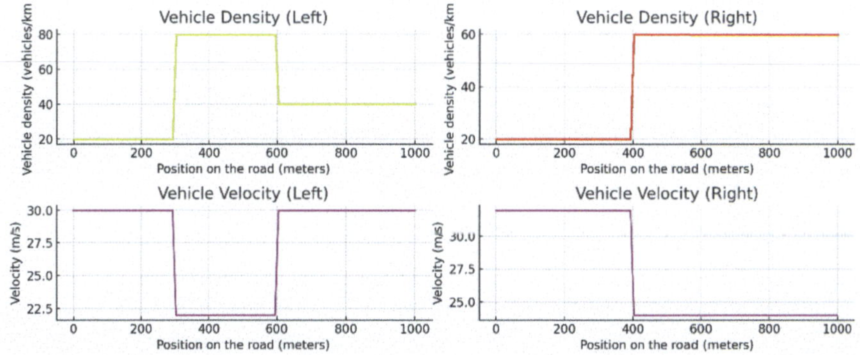

Fig. 28.2 Vehicle density over time

For these simulations, initial parameters were defined, such as road length, spatial and time steps, maximum speed, maximum density, and traffic viscosity. Additionally, an initial density was established in a specific segment of the road, and the initial speed was calculated based on this density. Finally, external factors such as intersections, traffic lights, and possible incidents on the road were considered as external forces impacting vehicular flow. Together, the image illustrates how density and speed vary in response to these initial conditions and simulated external forces, providing a dynamic representation of traffic behavior according to the adapted Navier–Stokes Equations.

(c) The implementation of Friction and Resistance is carried out through the following steps:
- Aerodynamic resistance: Refers to the assignment of resistance values that represent the friction of air against the surface of the vehicle as it moves.
- Friction resistance: Consists of the resistance values generated by the contact between the vehicle's tires and the road surface, influenced by factors such as the type of pavement and weather conditions.
- Gradient resistance: Refers to the assignment of values that are applied when vehicles move over inclined terrain.

In scenarios with a high drag coefficient (Cd), as represented in (b), (c), and (d) shown in Fig. 28.3, aerodynamic resistance increases significantly at all speeds, as a higher Cd intensifies the force exerted by the air on the moving vehicle. On the other hand, frictional resistance remains constant in all cases, as it depends solely on the friction coefficient (μ) and the slope angle, and is not influenced by the increase in speed.

The resistance due to the slope is also constant in each scenario, although it varies between scenarios depending on the angle of inclination; for example, in scenario (d), with a slope angle of 15°, this resistance is considerably greater compared to scenario (a), which has a slope of only 5°, and scenario (c), with an angle of 3°, where the resistance is even lower.

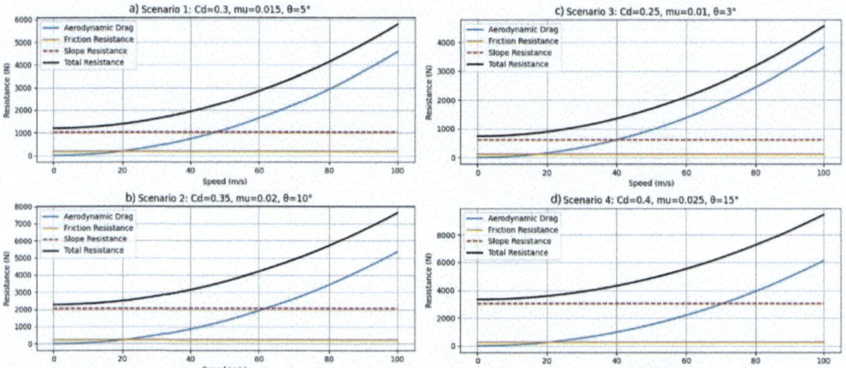

Fig. 28.3 Resistances as a function of speed in different scenarios

As for total resistance, at low speeds, friction and slope have a more significant impact in all scenarios. However, as speed increases, aerodynamic resistance becomes the predominant factor, especially in scenarios with a higher Cd, as observed in scenarios (b), (c), and (d). This trend is evident in the graphs, where aerodynamic resistance progressively surpasses other resistances as speed increases, becoming the main component of total resistance at high speeds.

28.3.6 Model Validation and Adjustment

For the validation and adjustment of the model presented in Eq. 28.8, sensitivity tests were conducted to identify how small changes in parameters such as the initial density.

ρ, traffic pressure $p_{traffic}$, traffic viscosity $\mu_{traffic}$, and the resistance factor.

k affect the model's behavior. Specific values were assigned to these variables to reflect different traffic scenarios: initial density ranged from 10 to 100 vehicles per kilometer; traffic pressure was set between 0.8 and 1.5 Pa, increasing with density; traffic viscosity was assigned values between 0.3 and 0.7 Pa·s to represent resistance between vehicles; and the resistance factor was set to 0.05 N for low densities and 0.15 N for high densities. In Fig. 28.4, several curves illustrate the evolution of average traffic speed over time for different densities. The X-axis shows time in hours over a simulation period of 1 h, and the Y-axis represents average speed in km/h.

The results show that higher density leads to a faster decrease in speed, as vehicle interactions increase resistance and amplify the effect of traffic viscosity. In low-density conditions, vehicles experience less resistance, resulting in a more gradual speed decrease. The model validation with simulated data indicates that the speed

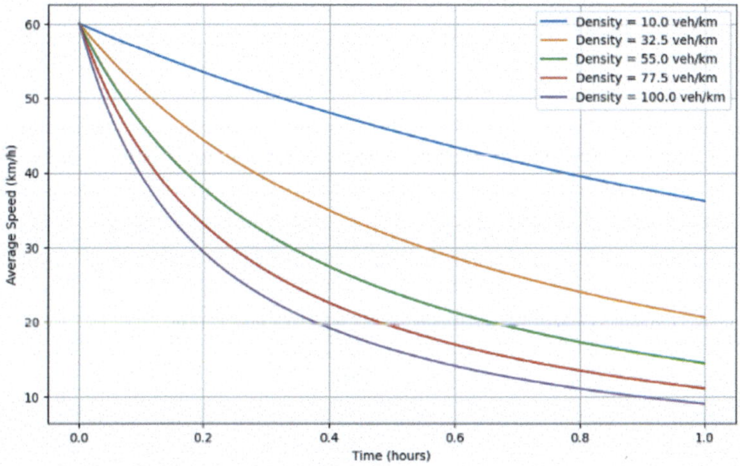

Fig. 28.4 Sensitivity test: average speed versus time

reduction pattern aligns with experimental data: for a density of 80 vehicles per kilometer, the initial speed of 35 km/h decreased to 10 km/h within 30 min, reflecting progressive congestion; whereas, with 20 vehicles per kilometer, the initial speed of 45 km/h dropped to 30 km/h in the same period. These results demonstrate that the model effectively captures traffic behavior at different density levels, offering a useful predictive tool for urban congestion analysis.

28.4 Results

This research introduces a new perspective on vehicular traffic analysis by applying advanced fluid dynamics concepts to traffic management problems. The results show that the Navier–Stokes equations, traditionally used to model the behavior of fluids such as air and water, can be effectively adapted to create a model that simulates the flow of vehicles on a road network. Through detailed simulations and comparisons with real traffic data, the model accurately captures dynamic traffic characteristics, such as congestion formation and variations in vehicular flow.

One of the key findings of this study is that the Navier–Stokes-based approach allows for the development of a mathematical model that provides a faithful representation of traffic and serves as an effective tool for characterizing and analyzing the evolution of congestion under different conditions. The model has proven useful for simulating traffic patterns, which is essential for optimizing traffic management and designing strategies to mitigate congestion. However, some limitations have been identified, such as the need for accurate real-world data and the associated computational complexity, which must be considered when applying this model in practical scenarios.

Compared to traditional methods, such as cell models and those based on queueing theory, the Navier–Stokes approach offers significant advantages. While conventional techniques tend to simplify vehicular flow, the use of the Navier–Stokes equations allows traffic to be modeled as a continuous flow, capturing both rapid fluctuations in traffic volume and the gradual formation and dispersion of congestion. This precision is particularly valuable in cities with complex and nonlinear traffic variations, where traditional models may prove insufficient in their predictions.

Another relevant advantage is the adaptability of the model to predict congestion patterns in different traffic scenarios. Unlike traditional models, which often require specific adjustments for each situation, the Navier–Stokes approach enables generalizable simulations and allows for parameter adjustments to adapt to diverse urban environments, weather conditions, and schedules. This flexibility facilitates more effective preparation and response to critical situations, such as peak-hour congestion or the effects of unexpected events.

28.5 Conclusions and Future Work

The application of the Navier–Stokes equations, commonly used to model moving fluids, has provided an effective approach for analyzing vehicular flow in simulations. The results obtained show that this method allows for a precise simulation of traffic dynamics, effectively capturing both the formation of congestion and variations in vehicle flow. The proposed model has demonstrated a high degree of accuracy in representing traffic patterns. However, it is acknowledged that the model's success depends on the availability of accurate data and the computational complexity required for its implementation, which may limit its application in scenarios with large traffic volumes or in real-time.

For the analysis of the Navier–Stokes equations, a flow was assumed that can be either laminar or turbulent, depending on the Reynolds number; incompressible under low-velocity conditions and compressible at higher velocities. Similarly, the flow may be stationary or transient, depending on the temporal evolution of system properties. Finite differences were used for their simplicity in structured meshes, although the Finite Volume method is preferred in fluid dynamics for its accuracy in property conservation. The control volume size was adjusted according to gradient variations: it is smaller in areas with high variation to improve precision and larger in homogeneous zones, thereby optimizing computation time.

In modeling vehicular traffic using the Navier–Stokes equations, the article suggests several directions for future work. First, it is necessary to develop methods to simplify the model's computational complexity, making it more accessible for real-time implementation and large-scale scenarios. This could involve optimizing algorithms or using data reduction techniques.

Second, additional studies are recommended that integrate more detailed and diverse data, as well as real-time data, to improve the model's accuracy and its ability to handle different traffic conditions. It would be particularly interesting to incorporate temporal factors, such as signal timing and traffic light cycles, which significantly influence urban traffic dynamics. Including these temporal factors would allow the model to capture traffic flow interruptions and fluctuations more realistically, further enhancing its ability to predict patterns in different traffic contexts and under varying load conditions.

Moreover, it would be beneficial to explore combining the Navier–Stokes-based approach with other modeling methodologies, such as machine learning and network theory, to develop hybrid models that can capture both fluid dynamics and complex interactions between vehicles. The incorporation of artificial intelligence techniques could help optimize the model's real-time responsiveness and improve its adaptability in complex scenarios, integrating temporal and environmental factors for more efficient traffic management.

References

1. E.R. Wong Leon, M.A. Coral Ygnacio, A systematic literature review of traffic control system implementations. Interfaces **19**, 157–178 (2024). https://doi.org/10.26439/interfases2024.n19.6779
2. H. Dorantes-Benavidez, M.Á. Martínez-Cruz, M. Chávez-Pichardo, T. Arenas-Reséndiz, A. Trejo-Martínez, Evolutionary simulation system of vehicular traffic at two intersections. Amazonia Invest. **13**(74), 22–34 (2021). https://doi.org/10.34069/ai/2024.74.02.2
3. J. Zhao, V.L. Knoop, M. Wang, Two-dimensional vehicular movement modelling at intersections based on optimal control. Transp. Res. Part B **138**, 1–22 (2020). https://doi.org/10.1016/j.trb.2020.04.001
4. J.V. Gutiérrez-Santacreu, M.A. Rojas-Medar, On the approximation of turbulent fluid flows by the Navier–Stokes-α equations on bounded domains. Phys. D **448**, 133724 (2023). https://doi.org/10.1016/j.physd.2023.133724
5. B. Hu, D. McDaniel, Applying physics-informed neural networks to solve Navier-stokes equations for laminar flow around a particle. Math. Comput. Appl. **28**(5), 102 (2003). https://doi.org/10.3390/mca28050102
6. A. Nayak, K. Goyal, Traffic modeling and accidental data analysis using GIS: a review. IOP Conference Series Earth and Environmental Science. 1327(1), 0–10. (2024). https://doi.org/10.1088/1755-1315/1327/1/012028
7. M. Modi, G. Agarwal, V. Patil, A. Khare, S. Shukla, A. Sankhala, Minimization of traffic congestion by using queuing theory. Int. J. Sci. Technol. Res. **8**(10), 532–541 (2019)
8. A. Oblakova, A. Al Hanbali, R.J. Boucherie, J.C.W. van Ommeren, W.H.M. Zijm, An analytical model for a tandem of two traffic-light intersections under semi-actuated and fixed control. Transp. Res. Interdisc. Perspect. **16**, 100715 (2022). https://doi.org/10.1016/j.trip.2022.100715
9. M.G.R. Alam, Queueing theory based vehicular traffic management system through jackson network model and optimization. IEEE Access **9**, 136018–136031 (2021). https://doi.org/10.1109/ACCESS.2021.3116503
10. E. Harahap, D. Darmawan, Y. Fajar, R. Ceha, A. Rachmiatie, Modeling and simulation of queue waiting time at traffic light intersection. J. Phys. Conf. Series **1188**(1), 012001 (2019). https://doi.org/10.1088/1742-6596/1188/1/012001
11. L. An, X. Yang, J. Hu, Modeling system dynamics of mixed traffic with partial connected and automated vehicles. Intell. Transp. Syst. Conf. **23**(9), 15755–15764 (2022). https://doi.org/10.1109/TITS.2022.3145395
12. P. Ross, Traffic dynamics. Transp. Res. Part B Methodol. **22**(6), 421–435 (1988). https://doi.org/10.1016/0191-2615(88)90023-9
13. M.J. Lighthill, G.B. Whitham, On kinematic waves II. A theory of traffic flow on long crowded roads. Proceed. R. Soc. Lond. **229**(1178), 317–345 (1955). https://doi.org/10.1098/rspa.1955.0089
14. S. Kim, M. Kleiber, S. Weber, Microscopic traffic models, accidents, and insurance losses. ASTIN Bull. **54**(1), 1–24 (2024). https://doi.org/10.1017/asb.2023.36
15. E.R. Maiorov, I.R. Ludan, J.D. Motta, O.N. Saprykin, Developing a microscopic city model in SUMO simulation system. J. Phys. Conf. Series **1368**(4), 042081 (2019). https://doi.org/10.1088/1742-6596/1368/4/042081
16. N. Bellomo, M. Delitala, V. Coscia, On the mathematical theory of vehicular traffic flow I. Fluid dynamic and kinetic modelling. Math. Models Methods Appl. Sci. **12**(12), 1801–1843 (2002). https://doi.org/10.1142/S0218202502002343
17. M.H. Zawawi et al., A review: fundamentals of computational fluid dynamics (CFD). AIP Conf. Proc. **2030**, 020252 (2018). https://doi.org/10.1063/1.5066893
18. J. Monreal, General Navier-Stokes-like momentum and mass-energy equations. Ann. Phys. **354**, 565–569 (2015). https://doi.org/10.1016/j.aop.2015.01.026
19. A.A. Kashyap, S. Raviraj, A. Devarakonda, K.S.R. Nayak, S. Kv, S.J. Bhat, Traffic flow prediction models–a review of deep learning techniques. Cogent Eng. **9**(1), 1–24 (2022). https://doi.org/10.1080/23311916.2021.2010510

20. F. Harrou, A. Zeroual, F. Kadri, Y. Sun, Enhancing road traffic flow prediction with improved deep learning using wavelet transforms. Res. Eng. **23**, 102342 (2021). https://doi.org/10.1016/j.rineng.2024.102342
21. H. Ulvi, M.A. Yerlikaya, K. Yildiz, Urban traffic mobility optimization model: a novel mathematical approach for predictive urban traffic analysis. Appl. Sci. **14**(13), 5873 (2024). https://doi.org/10.3390/app14135873
22. R. Qu, Research on optimization algorithm for urban traffic flow based on computer simulation. Acad. J. Sci. Technol. **7**(3), 65–68 (2023). https://doi.org/10.54097/ajst.v7i3.12719
23. S. Afandizadeh, S. Abdolahi, H. Mirzahossein, Deep learning algorithms for traffic forecasting : a comprehensive review and comparison with classical ones. J. Adv. Transp. (1), (2024). https://doi.org/10.1155/2024/9981657
24. S.A. Sayed, Y. Abdel-Hamid, H.A. Hefny, Artificial intelligence-based traffic flow prediction: a comprehensive review. J. Elect. Syst. Inf Technol. 10(1), (2023). https://doi.org/10.1186/s43067-023-00081-6
25. L. Jiang, Traffic flow prediction method based on deep learning. J. Phys. Conf. Series **1646**(1), 012050 (2020). https://doi.org/10.1088/1742-6596/1646/1/012050
26. X. Wang, A survey on deep learning for cellular traffic prediction. Intell. Comput. **3**, 1–17 (2024). https://doi.org/10.34133/icomputing.0054

Chapter 29
From Modules to Communities: Fostering Flexibility and Aggregation in School Infrastructure for Sustainable Smart Territories

Paolo Piantanida, Claudia Pilar, and Antonio Vottari

Abstract Developing smart cities implies developing networks with the territory: not only communication and data networks, but also service networks that avoid unsustainable depopulation or impoverishment of rural and peripheral territories. With this purpose, the outcome of a research on how to respond to the need for new school infrastructure in northern Argentina is presented. In this way, the school "network" can be sustainability-oriented and offer a certain level of education to the entire population, regardless of the location of residence. For this purpose, a prefabricated wooden module is proposed that is designed according to the very hot climatic conditions found in northern Argentina (bioenvironmental zone I according to IRAM 11,603) and that can be the basic cell for the formation of new schools, buildings for public services and even residences or guest houses. The dimensions of the module and the stratigraphies and materials that make up the walls, floor and roof of the prefab are described. The various possibilities of aggregation and the HVAC system are also explained. The energy needs for the entire year are estimated, which, thanks to the installation of rooftop PV, can be fully met even when the location is "off grid," if adequate batteries are provided. Connected to the power grid, which can be used as a virtual accumulator, the module is still energy self-sufficient and therefore suitable for carbon-free operation.

Keywords Wood construction · Sustainability · Energy efficiency

---

P. Piantanida · A. Vottari (✉)
DISEG-Politecnico di Torino, Torino, Italy
e-mail: antonio.vottari@polito.it

C. Pilar
FAU-Universidad Nacional del Nordeste, Resistencia, Argentina

D. Rossit et al. (eds.), *Proceedings of the VII Ibero-American Congress of Smart Cities, ICSC-Cities 2024, 12-14 November, San Carlos, Costa Rica*, Springer Proceedings in Energy, https://doi.org/10.1007/978-981-96-4301-1_29

29.1 Foreword

This paper aims to propose a contribution about the transition and integration of school infrastructure into the inclusive and sustainable system of smart cities. The results are the outcome of an inter-university research activity involving researchers from UNNE Universidad Nacional del Nordeste (Chaco, ARG) and the Politecnico di Torino (Italy) about the response to the needs of new schools in the Argentine North.

First of all, underlying the research work is the distinction between 'smartization' and digitization: the latter must eventually be ancillary to the former, otherwise the goal of making stably diffuse, smart and sustainable the social, infrastructural and building system that constitutes the city would be missed, because it would be limited to management strategies (algorithms) that are often feedback-driven and that improve the contingency of the present i.e., the effects on the inhabitants, but do not govern the evolution of the city as an inhabited territory. The spin-off would thus be more about citizens than about the city, whereas this study is focused on the "physical" infrastructure represented by buildings.

The objectives of this study address the need for a more widespread and inclusive school infrastructure through the proposal of building modules for sustainable and energy efficient buildings.

29.2 Context

29.2.1 The Reason Why

Why this research? In a context of transition to smart cities, the deficit of classrooms in Argentina identified in the report *Infrastructura Escolar 2023–2033* by CAMARCO's Office of Studies [1] raises the need for new construction, but leaves the identification of typological modes and characters to local initiative. This may induce inertia and inefficiencies that result from the prototypical character of each construction site and may favor the choice of the big city for new settlements, due to better infrastructure and lower supply costs, exacerbating the imbalance between rural or peripheral areas and the city, contrary to SDGs Nos. 4 & 10 (Quality education & Reduce inequalities).

The Argentine government has developed several projects to improve school infrastructure and the quality of education in Chaco province, including through the *Plan Nacional de Infraestructura Educativa.* In recent years, a significant number of new school constructions and renovations have been completed, along with ongoing interventions throughout the country, with 27 projects already completed, 38 in execution, and another 120 funded in 2024 alone.

These investments are part of a broader effort that includes logistical support, distribution of educational materials, and enhancement of essential services for schools in disadvantaged areas. You can find additional details on the official website of the Argentine government or in documentation published by the *Ministerio de Educación.*

29.2.2 Sustainable and Smart

In this frame, a design proposal for a reversible (i.e., deconstructible) building system that has versatility of aggregation, flexibility of use and technological integration, and ease in adapting to environmental, even rural, contexts is designed. It allows to conjecture a built environment that itself behaves smartly, preventing depopulation of rural and peripheral areas through the deployment of low-cost but not low-quality school infrastructure. The school very close to residences can contain costs to the State through metabolization of digital resources and network integration, this facilitating the daily presence or co-presence in teleducation. That reduces the resources used for faculty and pupil travel, thus the number of decentralized schools can be increased without disproportionately multiplying personnel costs. The idea is for aggregations, varying according to the context, of modular building units that can offer, depending on local needs, teaching units (classrooms), services (bathrooms, kitchen, school canteen etc.), guest quarters for teachers, social gathering and co-working spaces, etc., even growing or decreasing rapidly (dry construction with disassemblable prefabricated elements) according to changing needs, with a "smart" and self(?)-adaptive strategy.

A smart school can then provide for the digitization of the infrastructure and is not merely "digitizing" its students (e.g., equipping them with interconnected tools), but is organized for active sharing of networked teaching, where, for example, teleducation is bidirectional and not throttled by the limitations of a tablet screen. For this to happen the building must be sufficiently flexible, given that the lifespan of the building container is much longer than that of the network and the technical components that constitute its digitization, and maintenance and expansions of these services should be able to be carried out while containing interference with the activities taking place there.

29.3 Methodology

First of all, the researchers considered the situation of public school buildings in the different Argentine provinces (as described in institutional documentation available online) and identified the greatest building needs in relation to school-age population and lower average income in the northern provinces, which greatly limits the access to private schools. The provinces of Misiones, Chaco and Corrientes are among the poorest in Argentina, with high unemployment rates and low fiscal capacity. These provinces are often unable to finance school construction due to limited economic resources. According to a report by the *Ministerio de Educación de la Nación* [2], schools in these regions suffer from a chronic lack of resources for construction, maintenance, and expansion of school buildings.

For this reason, Chaco, Corrientes and Misiones were selected as the setting for this study. Although the Argentine government has launched several construction and renovation programs to address the deficit of school buildings, the effectiveness of these programs has been limited by economic and administrative difficulties. The *Plan Nacional de Infraestructura Educativa* (PNIE), launched in 2016, included investment in the construction of new schools and renovation of existing ones, but according to the aforementioned report [2], the provinces of Misiones, Chaco, and Corrientes received only a fraction of the funds allocated to the wealthier provinces [2].

For the province of Chaco, Argentina, substantial investment in school construction has been undertaken, aimed at improving the quality and accessibility of educational facilities. The provincial government, in collaboration with national and international funds, has planned 180 school construction and renovation projects. This plan includes maintenance works for more than 1,200 schools, with special efforts in remote areas such as "El Impenetrable". Some specific projects are supported by organizations such as FONPLATA, which has funded educational facilities to ensure an inclusive and safe educational environment for students with special needs.

In particular, the school in Fuerte Esperanza, for example, received funding for a new 230 million pesos' facility, including classrooms, laboratories and specialized services for students with disabilities. Other interventions include building technical schools and improving infrastructure for rural schools through specific programs such as PROMER II [3], in partnership with UNICEF.

The research group then focused on climate and its relationship to building sustainability and resilience to climate change. The goal is to devise a module that can be as energy self-sufficient as possible by harnessing solar energy: e.g., the sunnier the day, the more energy is obtained for cooling needs, with a self-adaptive effect with respect to heat waves, considering the daytime use of the school. To evaluate this behavior, an energy simulation software (compliant with ISO 52016–1:2017 and ISO 52022–1:2017) was used to compare the photovoltaic production feeding a heat pump system with the needs for seasonal air conditioning, adopting climate and irradiance data from Resistencia (Chaco). The building envelope was characterized through its transimittance and thermal wave phase shift and attenuation (the latter only for opaque parts). All envelope calculations were performed in accordance with the standards ISO 6946:2017 and ISO 13786:2017.

29.4 Results

First, the module has been conceptualized and designed by reflecting on Matías Taborda, Gerardo Esteche and Carlos Marcial's design for the Tamandua school in Misiones [4]. This research's architectural result can meet almost any school building requirement with several prefabricated modules joined together. For example, the CAMARCO report for the period 2021–2033 estimates the full-time elementary school building demand in Chaco, Misiones and Corrientes as follows: 139 new schools in Chaco (corresponding to 834 classrooms), 129 new schools (772 classrooms) for Corrientes Province, and 148 new schools (889 classrooms) for Misiones Province. Even conjecturing that full-time is not extensively adopted (halving the need for space because each school can operate on two shifts) and that resources can be found to meet half of the demand that thus remains, building demand remains high with more than thirty new elementary school per Province (6 classrooms each), expected to be built at the rate of about 3 per year.

Each building can be organized through the replication of the prefabricated module in multiple floor plan configurations, all with a single floor elevated from the ground (with an accessible cavity, good to protect the building from flooding and soil moisture) so as to promote accessibility and safety of displacement. The adoption of large openable transparent surfaces increases natural light and the relationship with the surrounding environment (each room accesses the outside) that are stimulating learning, as indicated, e.g., by the studies of Peter Barrett et al. [5].

29.4.1 Sustainable Goals

The purposes are the containment of construction costs and time combined together with SDG No. 4 "Ensure inclusive and equitable quality education and promote lifelong learning opportunities for all": the choice of a prefabricated wooden module with the same structure for different uses in fact allows to control production costs (thanks to industrial replication of identical units) and to ensure the same level of construction quality in each school building, regardless of its location.

The design phase reasoned about the entire lifespan of the building: management and maintenance costs are reduced because the module lends itself to different destinations, interior distributions and furnishings, as well as the possibility of being aggregated in multiple configurations and, at the end of its service life, disassembled and reused. Moreover, the main wiring and piping run in the ceiling of the porch and are therefore serviceable, reconfigurable and expandable while working without interfering with classrooms. The air conditioning system is suspended from the underside of the module floor and therefore maintainable from the gap formed between the natural ground and the building floor. In addition, the possibility of equipping the roof with photovoltaic panels that power the electrical services and the heat pump air conditioning system makes the building almost self-sufficient.

Table 29.1 List of components of a prefabricated module

Material	Unit	Quantity	Thickness (mm)	Mass (kg)
Plasterboard	m^2	97.50	12.50	853
Wooden planking	m^2	207.53	20	2283
Aluminium roofing sheet	m^2	36.94	1	103
Waterproof PVC sheet	m^2	36.94	1	51
Vinyl flooring	m^2	36.76	5	221
Wood fiber panel	m^2	134.01	100	3015
Windows & entrance door	m^2	21.6	70	756
			Total	7282

29.4.2 Structure

The entire module, which measures 9 × 5 m in plan and has a usable internal area of 4.76 × 6.70 m, is built using prefabricated wooden elements.

The load-bearing structure consists of two vertical frames that support the lower floor and the roof, on which any partitions are anchored. On the decks are placed the side walls and those toward the loggias. The former consist of opaque wafer panels that can be omitted in the union between modules; the latter consist entirely of window and door frames, opaque only for the entrance door and for the upper part to close the different height of the envelope for the two glazed façades.

The large windows protected by the loggia roof-overhang allows a very favorable visual relationship with the outside and a high daylight factor with good control of direct solar radiation. The choice of glazing type (single or double, or triple glazing) can be reconsidered depending on climate and financial availability, so as to contain the resources allocated to the most expensive component of the vertical envelope.

Each module is composed of the elements and layers shown in the previous Table 29.1. All the components can be moved by a single medium-duty truck (US class 6).

29.4.3 Thermal Specifications of the Envelope

In the specific case (northern Argentina), a median solution was decided upon, with double normal glass and a wooden frame. The layers of the opaque parts, including floor and roof, are summarized in Table 29.2. Thermal transmittances are calculated according to ISO 6946:2017 [6]; thermal wave phase shift and attenuation are calculated according to ISO 13786:2017 [7].

In view of the current global climatic evolution, the research team decided to equip the modules with air-to-air cooling system, so as to improve indoor comfort even on days when the relative humidity and/or outdoor temperature do not allow

Table 29.2 Thermal transmittances and layers specs of the opaque envelope

Component	Thermal transmittance (W/m^2·K)	Layers (from inside to outside)	Thickness (mm)	Conductivity (W/m·K)
Floor	0.408	Vinyl flooring	5	0.17
		Wooden planking	20	0.14
		Wood fiber panel	100	0.056
		Wooden planking	20	0.14
		Wooden planking	20	0.14
Wall	0.428	Plasterboard	12.5	0.21
		Air gap	40	0.222
		Wood fiber panel	100	0.056
		Wooden covering	20	0.14
Roof	0.409	Plasterboard	12.5	0.21
		Air gap	100	0.625
		Wood fiber panel	100	0.056
		Wooden planking	20	0.14
		Waterproof PVC sheet	1	0.17
		Weakly ventilated air gap	30	–
		Aluminium multilayer roofing	3	160

for comfort conditions with cross ventilation alone, also in view of the foreseeable crowding of the rooms. The thermal insulation of the vertical opaque envelope, which might seem overabundant (K = 0.428 W/m^2·K compared with the limit of 1 W/m^2·K for level B compliance), is in fact aimed at containing heat re-entry in summer. Its periodic transmittance is 0. 265 W/m^2·K and get a thermal wave attenuation of 0.631 and a phase shift of 6.258 h: in this way, the outdoor temperature peak reaches the classrooms when the evening hours allow a better performance of the cooling machine and, especially if classes are over, can be compensated with natural night ventilation.

The designed glazed envelope has an overall thermal transmittance K around 2 W/m^2·K, including the wooden frame, opaque and transparent parts. The latter are made with normal double glazing.

The roof was the subject of special consideration. It is designed with 10-cm thick wood wool thermal insulation and results in a thermal transmittance K of 0.409 W/m^2·K. The periodic transmittance is worth 0.248 W/m^2·K, the attenuation of the thermal wave is 0.614 and its time lag is 6.76 h, with thermal wave management similar to that of the opaque wall. The K parameter is already thus half of the maximum allowed by the B level (0.83 W/m^2·K) of IRAM 11,605, and in this way the effect of heat radiation from the ceiling is greatly reduced, benefiting summer comfort.

In order to further improve indoor comfort, the thickness of the wood wool insulation layer of the roof can be increased to 18 cm: in this way the K is reduced to 0.258 $W/m^2 \cdot K$ and the roof not only complies with Level A of IRAM 11,605 for design temperatures down to – 6 °C, but also greatly improves summer behavior. In fact, with the 1.8-fold increase in thermal insulation thickness, periodic transmittance is reduced by 4 times (0.061 $W/m^2 \cdot K$), thermal wave attenuation by 2.5 times (0.238), and phase shift by 1.8 times (12.13 h: the thermal lag is directly proportional to insulation thickness). The improved behavior of the roof reduces the heat peak from the ceiling by 1.6 times (according to the summer thermo-physical model with the heat-storage factors by the "Carrier-Pizzetti" method [8]). At 6 PM on a typical summer day, the heat arriving from the ceiling drops from 335 to 206 W: these are not major numbers, but a significant reduction in radiation, and thus a consequent improved well-being for the occupants. The higher thermal resistance of the roof causes a negligible worsening of the overall summer behavior, due to the lower nighttime cooling by transmission: this is about 3 kWh/year, which is irrelevant anyway because it is provided by the photovoltaic system. In contrast, the winter situation, where PV may be less effective due to bad weather days, improves by 86 kWh, or about 14.9%.

29.4.4 HVAC Needs

In estimating the winter needs, the presence of people was not taken into account: this is a conservative assessment, which allows the empty rooms to be prepared for the comfort temperature even on the coldest days and also allows the real electricity demand to be reduced, which benefits the behavior of the system on cloudy or rainy weather days. With similar caution, the prefabricated module, for the sake of safety, is considered fully exposed on all sides to the outdoor climate, even in the case of the summer power calculation in which, however, full room crowding was considered.

The thermal power required for each module is about 3 kW in winter and 4.3 kW in summer and is delivered by an air-to-air packaged conditioner, e.g., flat type, hanging from the soffit of the slab, in the cavity under the floor. Due to the conformation of the module, raised above the excavation of the foundations, the space is ventilated from multiple sides and accessible for maintenance (height of at least 1.5–1.6 m), and this placement keeps the system permanently shaded, with better efficiency due to the absence of overheating caused by solar radiation, and protects it from rain.

Eight photovoltaic panels with a total peak power of 3280 W are installed on the roof, which can be used off-grid with suitable batteries or connected to the power grid as virtual accumulator. In this way, each module is self-sufficient and carbon free throughout the year, as shown in Fig. 29.1.

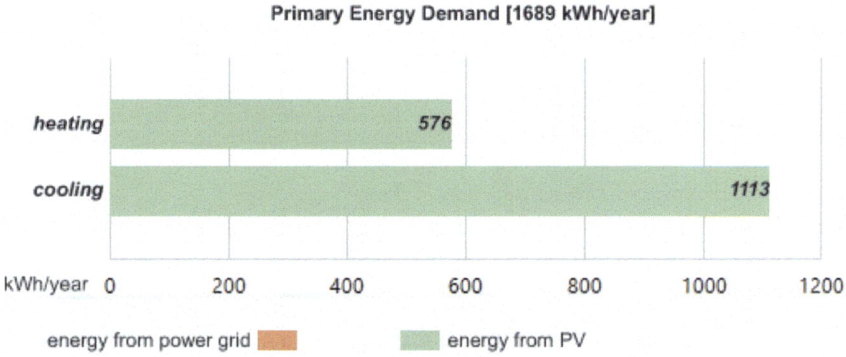

Fig. 29.1 Primary energy demand for a single module in Resistencia (Chaco) with 3,28 kW PV on the roof: no emission of carbon dioxide all year round

29.5 Discussion

The study of the prefabricated module resulted in maximum flexibility of use, with the possibility of aggregating and re-aggregating modules according to changes that may occur in their uses (Fig. 29.2), and also of disassembling and reassembling them elsewhere if necessary.

The stratigraphies proposed in this study meet the Level B of IRAM 11,605:1996 [9] for any outdoor temperature stipulated in the standard, for bioclimatic zones I "very hot" (and possibly II "hot") according to IRAM 11,603:2012 [10].

The research team focused on adapting to the climatic context and meeting the demand for new schools with an easy, fast and reliable construction system to meet the requirements of quantity, quality and speed. The climatic data suggest the use of a floor plan organization with open connectives (e.g., porch or loggia corridors), which is beneficial for construction cost containment and natural cross ventilation of spaces, which, precisely because of the open distribution of porches, is effective and easily achieved in every room. The adoption of perimeter loggias, which echoes recurring construction patterns in Argentine rural schools up to the 1950s and in some of Soto & Rivarola's wooden schools [11, 12], also allows external sunscreens to be placed away from the windows and doors and to protect the two room faces with mosquito nets as well.

In the future, the research group aims to work on refining the adaptation of the energy calculation software used in this project (Edilclima® compliant with ISO 52016–1:2017 and ISO 52022–1:2017). This software also includes the estimation of production from photovoltaics, which it is intended to make assessable in the future at other locations in the area of interest (besides Resistencia) for which the available database needs to be adapted and implemented.

Although the survey locations, building morphology, and climatic conditions are far from Europe and the EU Directive 2018/844 on the Energy Performance of Buildings (EPBD), it is interesting to count how many of the aspects included in

Fig. 29.2 3D perspective views of possible aggregations of the proposed prefabricated module

the elements to be evaluated for the determination of the European Smart Readiness Indicator (SRI) are addressed by the module design: the SRI consists of 72 checks originating from 9 technical domains (heating, cooling, domestic hot water, ventilation, lighting, dynamic building envelope, electricity, electric vehicle charging, and monitoring & control), each of which must be managed to achieve 7 objectives (energy efficiency, maintenance and fault prediction, comfort, convenience, health, wellness and accessibility, occupant information, energy flexibility, and storage). The design of the building consisting of the proposed prefabricated modules appears to address about 40% of the 72 design aspects from the European Smart Readiness Indicator (SRI) (27 directly and 2 via add-on software), as in Table 29.3.

Of course, it is not possible to calculate SRI, since the reference smart building has not yet been defined for the climate zones of Argentina or Latin America.

Aggregating multiple modules can yield 26-pupil classrooms (2 modules side by side on the long side), *comedor* (canteen) or *biblioteca* (library), *salón de usos*

Table 29.3 EU smart readiness indicator (SRI): topics addressed in the module design

Goals	Energy efficiency	Maintenance & fault prediction	Comfort	Convenience	Health	Well-being & accessibility	Information to occupants	Energy flexibility & storage
Technical domains								
Heating	✓		✓	✓	✓	✓	(✓)	✓
Cooling	✓		✓	✓	✓	✓	(✓)	
Domestic hot water	✓			✓	✓	✓		✓
Ventilation			✓	✓	✓			
Lighting	✓		✓	✓				
Dynamic building env								
Electricity	✓			✓	✓		✓	✓
Electric vehicle charging								
Monitor & control			✓	✓			(✓)	✓

multiples (multiple modules side by side on the short side and on the long side, as needed), as in Figs. 29.4 and 29.5, to which coordinate modules intended for kitchen, warehouse, toilets, office, etc.

Further development of this research is aimed at building a prototype without furnishings and facilities to test the real issues of construction and transportation.

Any individual module is suitable for and can be equipped as a collective toilet unit, as a locker room and shower unit (e.g., near sports facilities), as an office with secretariat, medical clinic with waiting room, etc., etc.

Having similar spaces to other proposals for prefabricated houses [13], it can also be intended as a minimal living unit, as in Fig. 29.3, suitable for four people and equipped with a kitchenette over a living room with a sofa bed, so that its capacity can occasionally be increased to six people. This unit, replicated if necessary, can be useful as well as a residence for the school's janitors, or permanent teachers, as guest quarters for traveling teachers or for students living far away from the school.

The examples are offered to demonstrate the flexibility and modifiability of the uses of prefabricated modules, which, remaining energy self-sufficient, can grow in number without negatively impacting the environment and energy burdens. Because they are structurally identical, they can change their intended use and type of aggregation over time without much of the initial investment or materials being wasted, benefiting long-term sustainability.

29.6 Conclusions

The proposed prefabricated module is made of wood to improve its sustainability requirements and is also energy self-sufficient. It can be moved by a medium-sized truck and thus can be the answer to local school and community building requirements that can be easily adapted to local needs.

By adopting these practices, schools can become more responsive, efficient, and effective in serving and aggregating students and communities, ultimately improving the overall educational experience and outcomes. The multiple mode of module aggregation is the key to having public buildings that can evolve as the population for which they are intended evolves. It is the pivot to have a mix of public services (school, community center, medical services, and so on) that is web connected and also maximizes the value for money of construction.

This that can become an "innovative practice" satisfies the initial instance of responding intelligently to the need for a more widespread and inclusive school infrastructure through the proposal of construction modules for sustainable and energy efficient buildings (SDG #4) and through the possibility of sending them where they are actually required, freely configuring them to meet the needs of the inhabitants: the research group trusts that in this way the smart city will spread and can 'smartize' the territory without functioning as an engine of depopulation of suburban and rural areas.

Fig. 29.3 3D axonometric views of the minimal residential unit

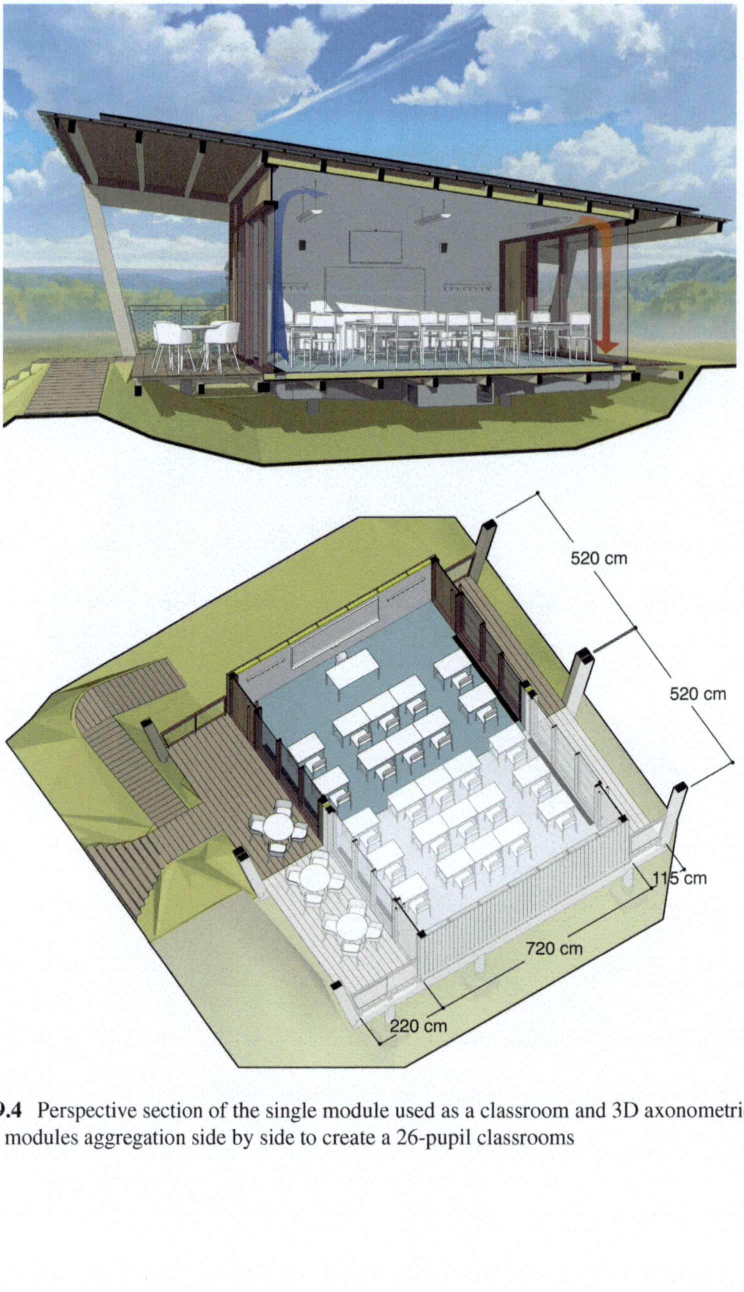

Fig. 29.4 Perspective section of the single module used as a classroom and 3D axonometric view of two modules aggregation side by side to create a 26-pupil classrooms

Fig. 29.5 3D perspective view of a co-working space created by the aggregation of 4 modules side by side along the long side

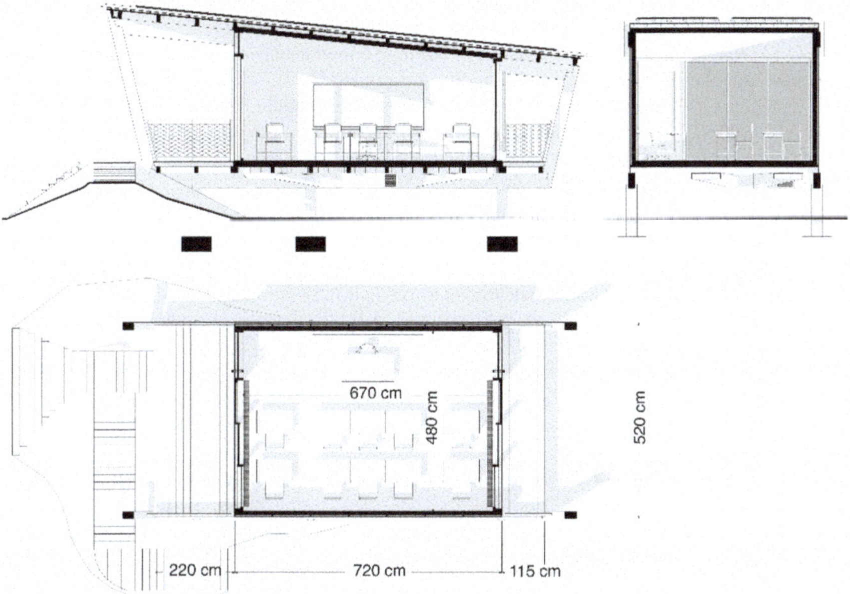

Fig. 29.6 Cross-sections and floor plan of the designed module

Disclosure of Interests The authors have no competing interests to declare that are relevant to the content of this article. This study is part of the institutional research activity of the Politecnico di Torino and the Universidad Nacional del Nordeste.

References

1. Cámara Argentina de la Construcción (CAMARCO), Área de Pensamiento Estratégico: Construir 2034—Infraestructura Escolar 2023–2033, Buenos Aires (2023)
2. Ministerio de Educación de la Nación, Plan Nacional de Infraestructura Educativa al 2025. SITEAL, Buenos Aires (2021)
3. D.C. Diz, Argentina/Latin America and Caribbean—P133195—Argentina Second Rural Education Improvement Project 'PROMER-II'—Procurement Plan (English). World Bank Group Washington D.C., (2018)
4. ARQA website (social, cultural, technological and commercial network of Architecture, Design and Construction in Latin America), https://arqa.com/arquitectura/sustentable/escuela-tamandua-en-misiones.html. Last accessed 2024/09/11
5. P. Barrett, F. Davies, Y. Zhang, L. Barret, The impact of classroom design on pupils' learning. Final results of a holistic, multi-level analysis. Build. Environ. **89**, 118–133 (2015)
6. ISO 6946:2017, Building components and building elements—thermal resistance and thermal transmittance—calculation methods. Int. Organ. Stand. Vernier (2017)
7. ISO 13786:2017, Thermal performance of building components—dynamic thermal characteristics—calculation methods. Int. Organ. Stand. Vernier (2017)
8. C. Pizzetti, *Condizionamento dell'aria e refrigerazione* (Masson, Milano, 1989)
9. IRAM 11605:1996, Acondicionamiento térmico de edificios—condiciones de habitabilidad en edificios—valores máximos de transmitancia térmica en cerramientos opacos. Inst. Argent. de Normalización y Certificación Buenos Aires. (1996)
10. IRAM 11603:2012, Acondicionamiento térmico de edificios—clasificación bioambiental de la República Argentina. Inst. Argent. de Normalización y Certificación Buenos Aires. (2012)
11. C. Noetzly, Arquitecturas cívicas y propuestas urbanas a partir de la provincialización de los territorios nacionales. El caso de Misiones a través de la obra de Mario Soto y Raúl Rivarola. In: Secretaría de Ciencia y Tecnología, Proceedings of VII Encuentro de Docentes e Investigadores del Diseño, la Arquitectura y la Ciudad. FAPyD-UNR, Rosario. (2017)
12. G.S. Varela Freire, La arquitectura escolar rural en su devenir. Cuatro escuelas de montaña en Tucumán, Argentina. AREA, 29(2), ISSN 2591–5312, Universidad de Buenos Aires Facultad de Arquitectura, Diseño y Urbanismo Secretaría de Investigaciones, Buoneso Aires. pp. 1–17. (2023)
13. P. Piantanida, C. Pilar, V. Villa, A. Vottari, Sustainable house manufacturing for smart matching cities. In: Moreno-Bernal, P., Hernández-Callejo, L., Nesmachnow, S., Rossit, D., Ochoa-Correa, D. (eds.) ICSC-CITIES 2022, Proceedings of the V Ibero-American Congress of Smart Cities, pp. 502–516, UCuenca Press, Cuenca. (2022)

Chapter 30
WGAN to Augment Tabular Energy Consumption Data of High Performance Computing Nodes

Jonathan Muraña, Sergio Nesmachnow, and Juan J. Durillo

Abstract This article examines the potential of Wasserstein Generative Adversarial Networks (WGANs) for generating tabular energy data for high performance computing (HPC) nodes. Generating data that retains statistical properties similar to real data is especially valuable in scenarios where data access is costly. In this regard, information on the usage of HPC nodes in data centers is expensive to collect, mainly due to the high security levels maintained in these infrastructures. Four HPC nodes are studied through three different types of estimation models. Results indicate that the generated data preserve the statistical properties of the real data when evaluated using relevant techniques for distribution similarity analysis. Moreover, the model estimation quality remains at acceptable levels when replacing real data with generated data. In fact, improvements are observed when using generated data in combination with real data.

Keywords WGAN · High performance computing · Energy models

30.1 Introduction

Smart grids offer novel opportunities for data centers and supercomputing facilities to leverage information flow and communication technologies via intelligent energy management, leading to cost reduction and environmental preservation. This leverage is possible due to the nature of the computing infrastructure sector, where large

J. Muraña (✉) · S. Nesmachnow
Universidad de la República, Montevideo, Uruguay
e-mail: jmurana@fing.edu.uy

S. Nesmachnow
e-mail: sergion@fing.edu.uy

J. J. Durillo
Leibniz-Rechenzentrum, Munich, Germany
e-mail: juan.durillobarrionuevo@lrz.de

D. Rossit et al. (eds.), *Proceedings of the VII Ibero-American Congress of Smart Cities, ICSC-Cities 2024, 12-14 November, San Carlos, Costa Rica*, Springer Proceedings in Energy, https://doi.org/10.1007/978-981-96-4301-1_30

consumers have the operational flexibility to defer or slow down tasks, or scale up workloads according to situational demands.

Intelligent management necessitates predicting task execution behaviors concerning energy consumption, which requires reliable energy consumption models. Energy consumption models of computing nodes have been extensively studied in the literature, built using several computational intelligence techniques, achieving satisfactory results [10]. However, the bottleneck of these studies lies in the difficulty of accessing realistic measurements of computing nodes data, given the cost of high-end computing nodes and the restrictions that infrastructures impose on access to them. In this context, synthetic data generation techniques emerge as a straightforward and cost-effective alternative, with the potential to enhance the predictive quality of models.

Synthetic data is data that has not been collected from real experiments but retains statistical properties similar to real data. Among its main uses are enhancing the quality of data available for training and testing models, and protecting sensitive data [4]. In recent years, techniques such as Variational Autoencoders and Generative Adversarial Networks (GANs) have proven to be promising for the generation of synthetic data across a wide range of phenomena [7]. GANs [8] are a generative models based on neural networks, whose competitive training, grounded in game theory, converges to an equilibrium point.

This work utilizes a variant known as the Wasserstein GANs (WGAN), to generate synthetic data from measurements of energy consumption and computing resource usage statistics. The objective is to study the viability of synthetic data generation by analyzing the similarity between real and generated distributions, and by comparing the accuracy of model estimations using datasets containing real, generated, or a combination of both types of data for training and evaluation.

The main results indicate that the generated data, when compared to the real data, retain the statistical properties. Additionally, the model estimation quality remains within acceptable limits when substituting real data with generated data. Moreover, for three of the four HPC nodes studied, some neural network-based models improve their estimation when using generated data in combination with real data, compared to using only real data.

The article is organized as follows. Section 30.2 describes the background, including the energy models used, data description, and WGAN theory. Related work is presented in Sect. 30.3. Section 30.4 outlines the methodology followed for experiment design and results evaluation. Results and discussion are presented in Sect. 30.5. Finally, the conclusions and main directions for future work are formulated in Sect. 30.6.

30.2 Background

This section presents the related background: energy models used, energy data description, and WGAN theory.

30.2.1 Energy Models of Computing Nodes

The energy models of computing nodes studied in this work use the utilization of various computing resources, collected via performance counters, as independent variables, and the energy consumption as the dependent variable. In our previous work [11], energy consumption models were built based on three different forecasting techniques: linear regression (LRG), a fully connected neural network (FCN), and a convolutional neural network (CNN). Unlike the previous work, in this study, the LRG model adjusts the independent term. The quality of the models estimation was measured using the Normalized Root Mean Squared Error (NRMSE), a widely used metric for comparing regression models performance.

30.2.2 Data Description

The data, containing measurements of energy consumption and performance counters, was gathered during executions of the Apex-MAP benchmark [12] on high-performance computing nodes. Several parameter combinations were considered, resulting in 6000 different experiments, this is, 6000 data points. The Performance Application Programming Interface [13] was used for collecting performance counter values. Gathered data contains values of the following performance counters: the number of CPU instructions, the number of L2 cache misses, the number of L3 cache misses, the number of double precision float operations (DPS), the ratio of vectorization per cycle of double precision floating point operations using 128-bit, 256-bit and 512-bit vectorization per CPU cycle (DPV), and the number of CPU cycles. The energy consumption was recorded from two sources: the PKG package, which represents the energy consumption of the CPU, and the DRAM package, which represents the energy consumption of the memory. Four HPC nodes are considered in the experiments:

- CN node: two Intel Xeon Golden 6146 processors, Skylake microarchitecture, 12 cores (24 threads) at 3.2 GHz and 192 GB RAM.
- ESB node: one Intel Xeon Silver 4215 processor, Cascade Lake microarchitecture, 8 cores at 2.50 GHz and 48 GB RAM.
- DAM node: two Intel Xeon Platinum 8260M processors, Cascade Lake microarchitecture, 24 cores at 2.4 GHz and 384 GB RAM.

- SKN node: two Intel Xeon Phi CPU 7210F processors, Knights Landing microarchitecture, 64 cores at 1.30 GHz and 92 GB RAM

30.2.3 Wasserstein GANs

A WGAN is a type of GAN, so traditional GANs (only GANs from this point forward) are described first. A GAN [8] consists of two neural networks: a generator (G) and a discriminator (D). The generator takes random vectors z (with z, for example, in a normal distribution p_z) as input, commonly referred to as noise. The generator output consists of vectors that attempt to approximate real data. The discriminator receives inputs from both the generator outputs and real data (a vector $x \sim p_{\text{data}}$, where p_{data} is the real distribution), aiming to learn to distinguish which data comes from the generator (i.e., generated) and which is real data. Equation 30.1 shows the mathematical model of GAN training. The objective of the discriminator is to maximize the probability that any real data input x is classified as real, denoted by $D(x)$, and to minimize the probability that any generated sample is classified as real, denoted by $D(G(z))$, where $G(z)$ represents the output of the generator given input z. In turn, the objective of the generator is to maximize the probability that the discriminator classifies its generated samples as real, aiming to increase $D(G(z))$. In practice, the logarithm of probability is commonly used for objective functions (called loss functions in the context of neural networks) and the Sample Mean Estimator is used as the estimator of the expected value $\mathbb{E}$.

$$\min_G \max_D \left[\mathbb{E}_{x\sim p_{\text{data}}}[\log D(x)] + \mathbb{E}_{z\sim p_z}[\log(1 - D(G(z)))]\right] \tag{30.1}$$

As is common in neural networks, training in GANs consists of progressively adjusting the weights, guided by the loss function. Note that the weight adjustment by the generator affects only the second term of Eq. 30.1, while the weight adjustment by the discriminator affects both terms. In the case of GANs, the training process is a joint training process that determines a min-max game, which ideally converges to an equilibrium point where $D(x)$ equals to 0.5. This convergence means that the discriminator cannot distinguish between real and generated data. The min-max game that determines the Eq. 30.1 can be interpreted as a minimization of the Jensen-Shannon divergence (a metric for quantifying the similarity between two probability distributions) between the real and generated distributions [8].

In WGANs [1], modifications are introduced to the architecture of GANs. Specifically, the Wasserstein distance is used between real and generated distributions, instead of the Jensen-Shannon divergence, which helps to address stability issues and allows for more efficient training through the introduction of a Critic rather than a traditional Discriminator.

The Wasserstein distance (or Earth-Mover distance) intuitively expresses the minimum cost required to move mass from one probability distribution to another in order to transform the first distribution into the second one. Given two probability distributions μ and ν, the Wasserstein distance is given by Eq. 30.2, where $\Gamma(\mu, \nu)$ denotes the set of all joint distributions γ with marginals μ and ν respectively.

$$W(\mu, \nu) = \inf_{\gamma \in \Gamma(\mu,\nu)} \mathbb{E}_{(X,Y)\sim\gamma} [\|X - Y\|] \tag{30.2}$$

The Kantorovich-Rubinstein Duality [14] provides an alternative formulation of the Wasserstein distance that is computationally more feasible. Equation 30.3 shows the duality, where the supremum is taken over all 1-Lipschitz functions $f : M \to \mathbb{R}$, meaning that f satisfies the condition $|f(X) - f(Y)| \leq \|X - Y\|$ for all $X, Y \in M$. M is the space where X and Y are defined.

$$W(\mu, \nu) = \sup_{\|f\|_{\mathrm{Lip}} \leq 1} \left(\mathbb{E}_{X\sim\mu}[f(X)] - \mathbb{E}_{Y\sim\nu}[f(Y)]\right) \tag{30.3}$$

Rewriting Eq. 30.3 in terms of neural networks defined for GANs, and considering the discriminator D as the critic C, we obtain Eq. 30.4.

$$W(p_{\mathrm{data}}, p_{\mathrm{G(z)}}) = \sup_{\|C\|_{\mathrm{Lip}} \leq 1} \left(\mathbb{E}_{x\sim p_{\mathrm{data}}}[C(x)] - \mathbb{E}_{z\sim p_z}[C(G(z))]\right) \tag{30.4}$$

The min-max game in WGANs takes the following form: on the one hand, the Critic C aims to maximize the expression in Eq. 30.4, which means it tries to estimate the Wasserstein distance between the real and generated distributions. On the other hand, the generator aims to minimize the expression in Eq. 30.4, i.e., to minimize the Wasserstein distance so that the distributions are as similar as possible. To achieve this, the generator receives feedback from the critic. Since the function determined by the Critic must be Lipschitz continuous, it is necessary to ensure this condition, for example, by clipping the weights, or applying gradient penalty after each update during training [9]. Figure 30.1 presents a schema of WGAN training.

The blue and red data flows correspond to the Critic training, while the green data flow corresponds to the Generator training. The blue data flow starts from the real data, which consists of batches of vectors. These vectors are processed by the Critic to produce outputs and the average of this outputs is calculated. The Critic attempts, by adjusting its weights during training, to assign the highest possible value to the outputs of the real data, thereby making the average as high as possible. In turn, the red data flow starts from noise, i.e., vectors whose values are generated following a Gaussian distribution. These noise vectors pass through the Generator, producing vectors of the same dimension as the real data vectors. The generated vectors are then processed by the Critic, which aims to achieve the lowest possible average for each batch by adjusting its weights. This average can be interpreted as a score indicating how realistic the data appears. As the Critic seeks to maximize the expression in Eq. 30.4, the optimization involves increasing the value of the first term in the sum, decreasing the value of the second term, or both. Finally, the green flow corresponds to the Generator training, which also starts from noise, generating data and aiming

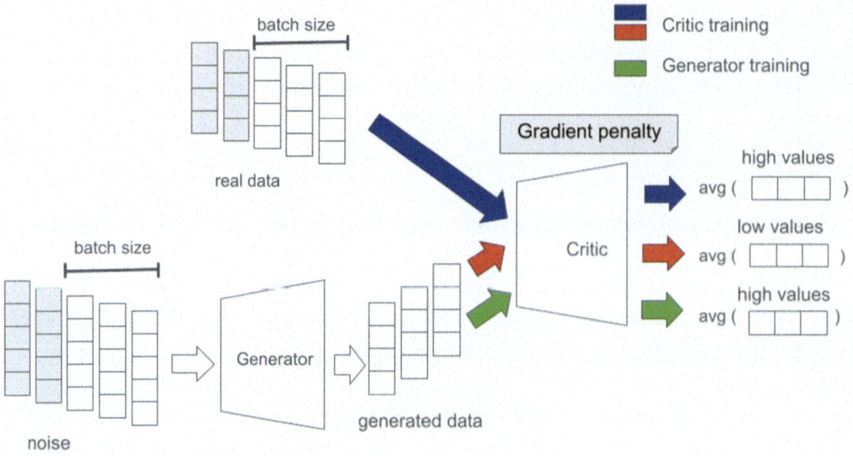

Fig. 30.1 Training scheme of critic and generator in WGAN

to achieve the highest possible score when the generated vectors pass through the Critic. It is important to note that the red and green data flows use the same batches of generated vectors in each iteration of the training.

Gradient penalty is the method used to ensure the Lipschitz continuous condition of the Critic. The gradient penalty has achieved better experimental results compared to the weight clipping method [9]. Since Lipschitz continuous functions are those whose gradient has a norm less than or equal to one at every point in their domain, the gradient penalty method involves penalizing network configurations where the gradient norm deviates from one in sample points by adding a term to the loss function. This term is called regularizer. Equation 30.5 shows the final expression of the Critic loss function ($\mathcal{L}_C$), considering the gradient penalty.

$$\mathcal{L}_C = \mathbb{E}_{z\sim p_z}[C(G(z))] - \mathbb{E}_{x\sim p_{data}}[C(x)] + \lambda\mathbb{E}_{\hat{x}\sim\mathbb{P}_{\hat{x}}}\left[(\|\nabla_{\hat{x}} D(\hat{x})\|_2 - 1)^2\right] \quad (30.5)$$

The first and second terms of Eq. 30.5 are the same as those included in Eq. 30.4, but with inverted signs since the loss function is minimized. The third term includes two factors: a coefficient λ, which represents the importance given to the regularizer (i.e., the importance of the penalty) within the loss function, and the second factor, which is the regularizer itself. The regularizer is calculated as the expected value of the squared difference between the norm of the gradient of the Critic function ($C(X)$) and one. That is, the regularizer measures how far the Critic gradient is from one, but it does so only for computational reasons, for certain points $\hat{x}$ from a distribution $\mathbb{P}_{\hat{x}}$. This distribution $\mathbb{P}_{\hat{x}}$ is constructed using points taken from a convex combination of the batch of real points and the batch of generated points. The epsilon used in the combination is randomly generated from a uniform distribution.

30.3 Related Work

Data generation using GANs and WGANs has become a significant focus in recent years. A review of the most relevant related work is presented below.

Fekri et al. [6] propose a GAN for generating temporal energy data of non-residential buildings. The dataset contains one year of hourly records of energy and other features such as humidity and temperature. The design of the GAN combines Long Short-Term Memory (LSTM) networks and WGANs for the discriminator and generator, respectively. A preprocessing method based on autoRegressive integrated moving average and fourier transform was applied to incorporate new features into the original data, aiming to improve the quality of data generation. The results indicate that the model trained with synthetic data achieves an accuracy comparable to that of the model trained with real data. Integrating features generated through preprocessing enhances the quality of the generated data.

Baasch et al. [2] proposed a GANs based in Temporal Convolutional Networks for augmenting the energy data of commercial and residential buildings. Metrics were used to evaluate both the predictive capabilities of models trained with generated data and validated with real data, compared to training and validation with real data, as well as to compare the similarity of real and generated distributions. These similarity metrics include Principal Component Analysis (PCA) and Standard Variance and Mean Load. The authors claim that despite considering a reduced amount of data, the presented GANs achieve competitive results compared to other GANs in the literature.

Zhao et al. [16] proposed a conditional GAN (CGAN) for generating tabular data with both continuous and categorical columns. The authors aim to enhance the quality of the generated data by adding a estimation or classification model to the GAN architecture that operates in parallel with the discriminator. This model provides feedback to the generative models based on the discrepancy between the estimation and the generated data, thereby ensuring semantic consistency. Additionally, the proposal includes incorporating Wasserstein loss with gradient penalty to improve the convergence of the training process. The comparison between the proposed CGAN and other generators focuses on evaluating improvements in estimations made by five classification models. The estimation of the classification models is compared by using training data generated by each model, and testing is conducted with real data. By this method, the quality of the generated data is assessed. The models trained with data generated by the proposed CGAN achieve a 21% higher F1 score compared to other generating approaches.

Engelmann et al. [5] introduced a CGAN for generating financial data to calculate credit scores. In this application, GANs are utilized to manage imbalanced data, as high-risk cases are generally underrepresented, leading to reduced precision in classifier models when dealing with real-world datasets. The proposal utilizes Wasserstein distance and an auxiliary classification during training to provide feedback to both the generator and the discriminator. The comparisons are made against imbalanced data and other balancing techniques. The evaluation involved seven real-world credit scoring datasets and utilized five different classification algorithms, assessing performance with three classification metrics. The proposed CGAN outperforms other oversampling methods, but it does not achieve the performance observed with training on imbalanced data.

Hazra et al. [3], proposed a GAN for generating synthetic time series tabular data of energy consumption. The predictive models utilizing data aim to forecast energy consumption based on factors such as humidity, temperature, and wind, among others. Electric companies rely on these models to plan efficiently for energy management. The authors proposed a GAN model that integrates the architecture of both Tabular GAN [15] and Wasserstein GAN with gradient penalty. The results indicate that the generated data from the proposed model, when integrated with original data, leads to a reduction in estimation error rates compared to using solely real data for predicting energy consumption.

According to the reviewed literature, the generation of tabular data using GANs, particularly Wasserstein GANs, has been widely employed to address issues such as data privacy, scarcity, and imbalance. However, to the best of our knowledge, there have been no studies on synthetic data generation for energy consumption models of computing nodes.

30.4 Methodology

This section describes the workflow, the techniques for assessing the quality of the generated data, and the approach for analyzing the energy models estimation.

30.4.1 Workflow for Energy Data Generation

The workflow is as follows: first, augment the data using WGANs. Second, the quality of the generated data is assessed using relevant techniques. Third, the generated data, combined with real data, is used for training and evaluating the energy models from previous work to analyze and compare the results.

30.4.2 Evaluation of Distribution Similarity

To compare the real and generated data, the following techniques were used. The methodology consisted of applying the techniques to both the real and generated data, and then analyzing if they yield similar results.

- *Boxplots*: Boxplots provide a visual summary of the distribution of data, highlighting key statistics such as the median, quartiles, and potential outliers. This technique allows for analyzing the features independently.
- *Principal Component Analysis (PCA)*: PCA is a dimensionality reduction technique that transforms high-dimensional data into a lower-dimensional space while preserving the most significant variance.
- *t-distributed Stochastic Neighbor Embedding (t-SNE)*: t-SNE is another dimensionality reduction method, particularly effective for visualizing high-dimensional data by mapping it into a lower-dimensional space. It is used to reveal clusters and relationships between data points.
- *Correlation matrix*: The correlation matrix, derived from Pearson correlation coefficients, displays the relationships between pairs of features. The absolute values closer to one indicate a stronger correlation, while the sign of each coefficient reveals whether the correlation is direct or inverse.

30.4.3 Evaluation of Models Estimation

Once the data was generated, it was used to train and evaluate the models. The experiments were designed to assess the quality of the estimations and analyze the quality of data generation, based on the source of the training and evaluation data. In this context, all possible combinations were considered, involving computing nodes, model types, training and evaluation phases, and the use of real, generated, or mixed data. Figure 30.2 summarizes the experimental design, showing the possible values that each dimension of the experiments can take. In summary, 27 experiments were conducted for each of the four computing nodes, resulting in a total of 108 different experiments. To refer to specific experiments, the following nomenclature is used: Computing Node—Model Type—Training Data—Evaluation Data, for example, CN-LRG-Real-Gen

30.5 Experimental Results

This section analyses the experimental results.

30.5.1 Experiment Overview

The WGAN was used to generate synthetic data of CN, ESB, DAM, and SKN computing nodes. A total of 5000 epochs were executed in the training. The generator architecture consisted of 4 layers with 64 neurons and ReLU activation, followed by

Computing node	Model type	Training data	Evaluation data
CN | ESB | DAM | SKN	LRG | FCN |CNN	Real | Gen | Both	Real | Gen | Both

Fig. 30.2 Design of experiments and nomenclature

an output layer with 11 neurons and sigmoid activation. The discriminator architecture consisted of one layer with 64 neurons and ReLU activation, 5 layers with 128 neurons and linear activation, and a final layer with one neuron and linear activation.

30.5.2 Analysis of the Similarity of Distributions

Figure 30.3 presents the boxplot for CN. The visual representation enables the comparison of the median, Q1 and Q3 quartiles, and the interquartile range of the real distribution (blue) and the generated distribution (orange). The boxplot also highlights the outlier points. As shown, the generated data has similar values for the mean, Q1, and Q3 quartiles. The maximum differences in the mean, Q1, and Q3 across all features are 0.02 , 0.03, and 0.03, respectively. A key difference in the distributions is that the generated data covers a broader range, particularly in cases where the real data has less representation. For example, in the Energy and Cycles features, lower whiskers are higher for the real data but reach zero for the generated data.

Figure 30.4 presents the boxplot for ESB computing node.

The maximum differences in the mean, Q1, and Q3 across all features are 0.05, 0.03, and 0.09 corresponding to Instructions. For DPV, the generated data shows outlier points distributed throughout the entire interval outside the outside the 1.5 interquartile range, which is not observed in the real data.

Figure 30.5 presents the boxplot for DAM. The maximum differences in the mean, Q1, and Q3 across all features are 0.04, 0.04, and 0.08, respectively.

Figure 30.6 presents the boxplot for SKN. The maximum differences in the mean, Q1, and Q3 across all features are 0.01, 0.01, and 0.02, respectively.

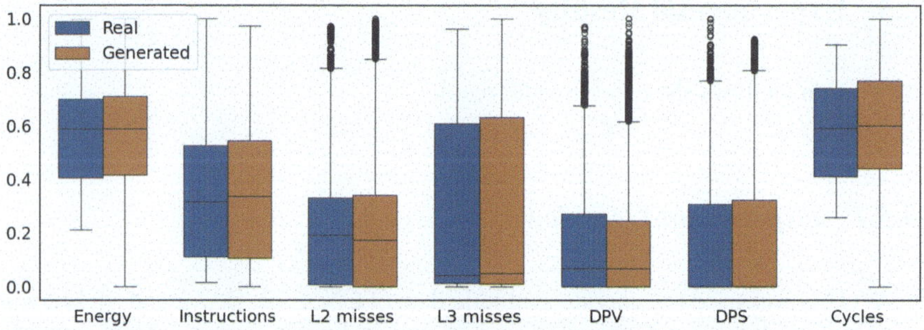

Fig. 30.3 Statistics comparison of real and generated data for CN

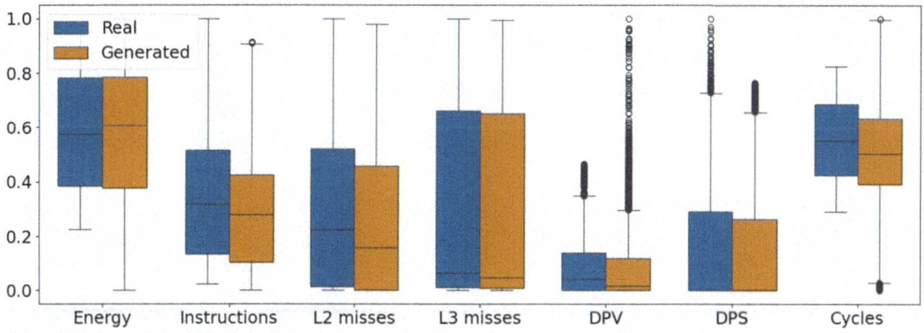

Fig. 30.4 Statistics comparison of real and generated data for ESB

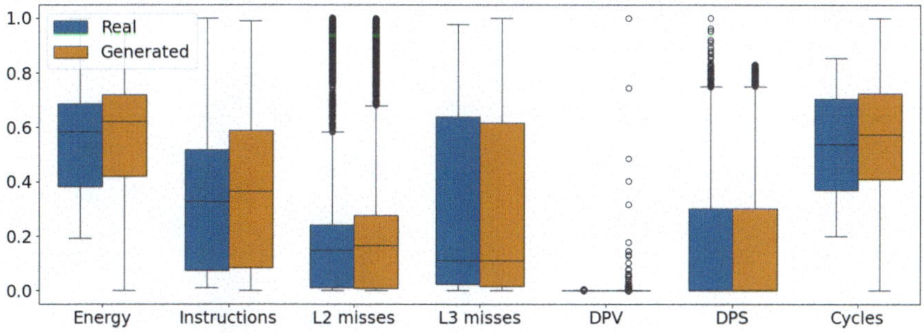

Fig. 30.5 Statistics comparison of real and generated data for DAM

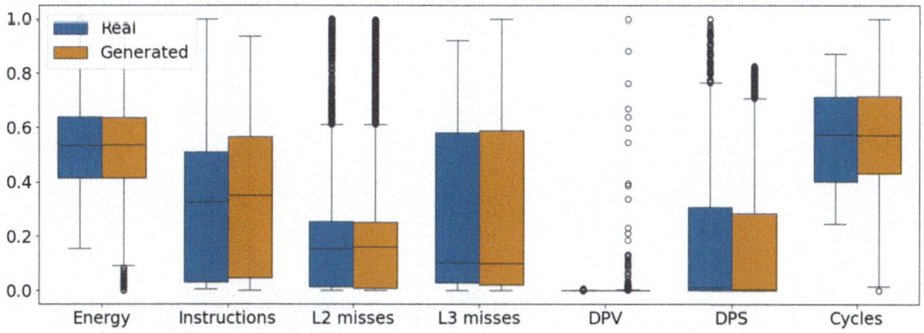

Fig. 30.6 Statistics comparison of real and generated data for SKN

Figures 30.7 and 30.8 show the PCA and t-SNE analisis for CN. The data presented in the PCA and t-SNE analyses are scaled according to quartiles Q1 and Q3 (Robust scaler). It can be observed that, on both PCA and t-SNE, the generated data largely preserves the structure of the real data, while also presenting points in other areas, expanding the representation of points in the domain.

Figures 30.9 and 30.10 show the PCA and t-SNE analisis for ESB. In the case of ESB, the t-SNE technique shows similar regions in the real and generated data. The

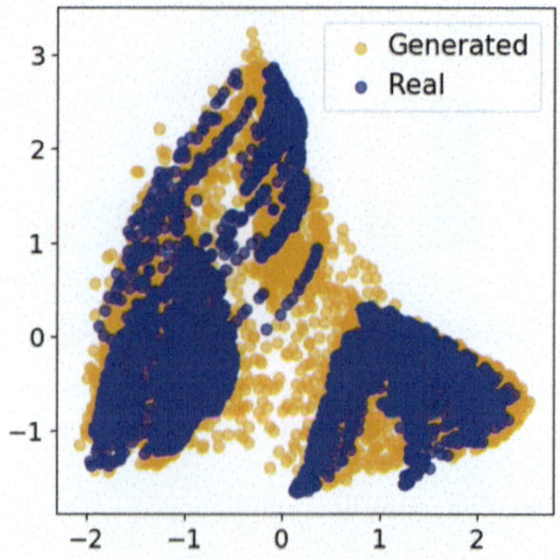

Fig. 30.7 PCA analysis for CN

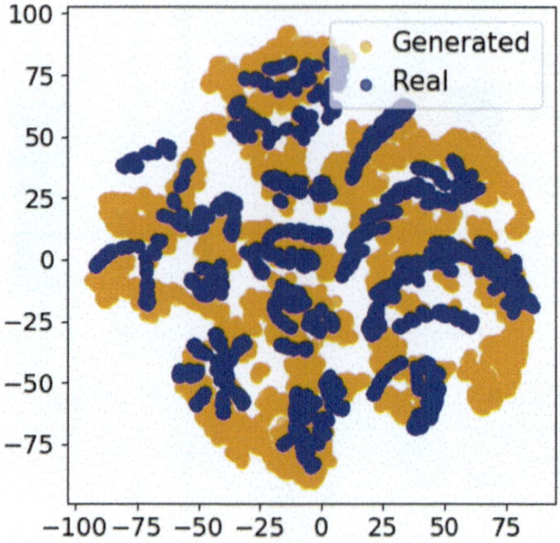

Fig. 30.8 t-SNE analysis for CN

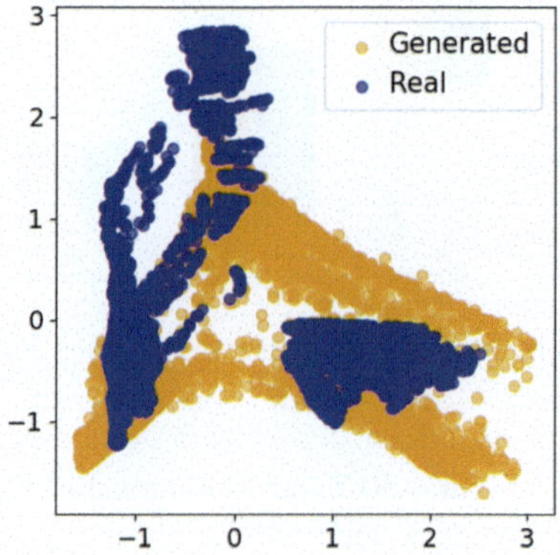

Fig. 30.9 PCA analysis for ESB

PCA technique, although the generated data exhibit similar behavior in the lower-left region of the quadrant, does not seem to capture the variation of the real data corresponding to the upper-left region.

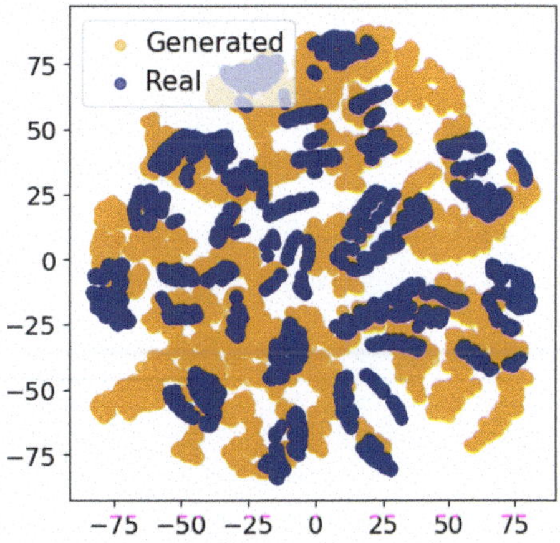

Fig. 30.10 t-SNE analysis for ESB

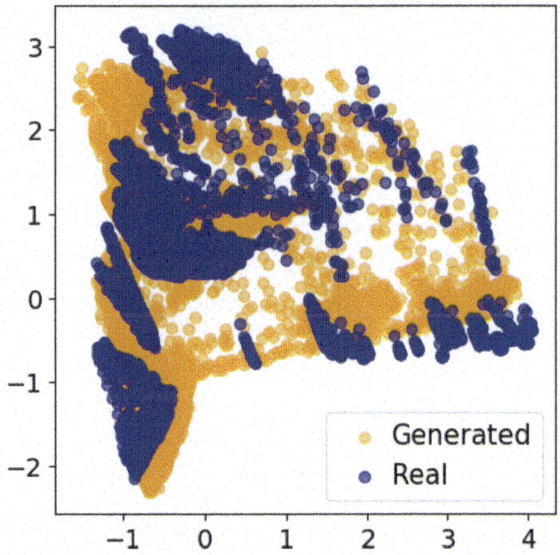

Fig. 30.11 PCA analysis for DAM

Figures 30.11 and 30.12 show the PCA and t-SNE analysis for DAM. In the case of DAM, the overall data points present in the same regions as the real data, as well as in other regions not covered by the real data.

Figures 30.13 and 30.14 show the PCA and t-SNE for SKN. The generated data shows similar behavior in the left region of the PCA, but the behavior on the right differs from the real data. In t-SNE, clustering regions are largely identifiable.

Figure 30.15 shows the correlation matrix for CN. The relationship between pairs of features is significantly captured by the generated data. The cell with the greatest difference between the correlation matrices corresponds to Cyc-Energy, with a difference of only 0.03.

Figure 30.16 shows the correlation matrix for ESB. The cells with the greatest difference (0.11) are Intr-Energy and Cyc-Energy.

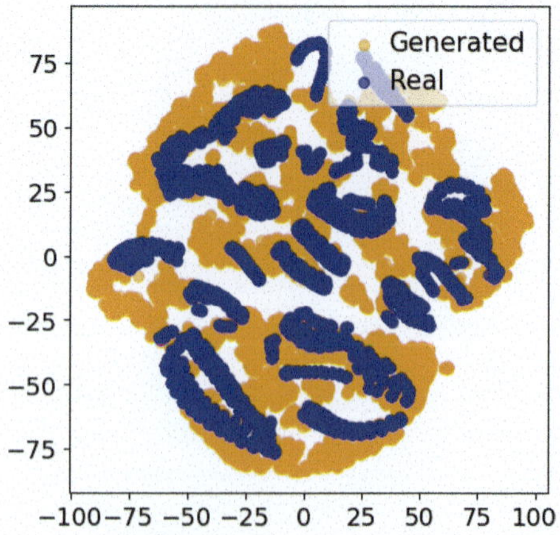

Fig. 30.12 t-SNE analysis for DAM

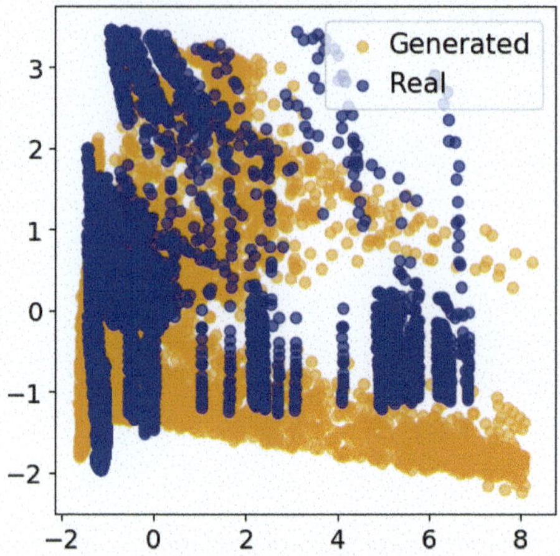

Fig. 30.13 PCA analysis for SKN

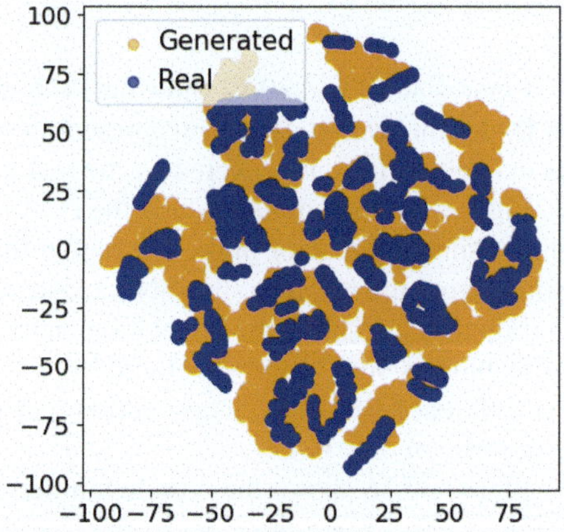

Fig. 30.14 t-SNE analysis for SKN

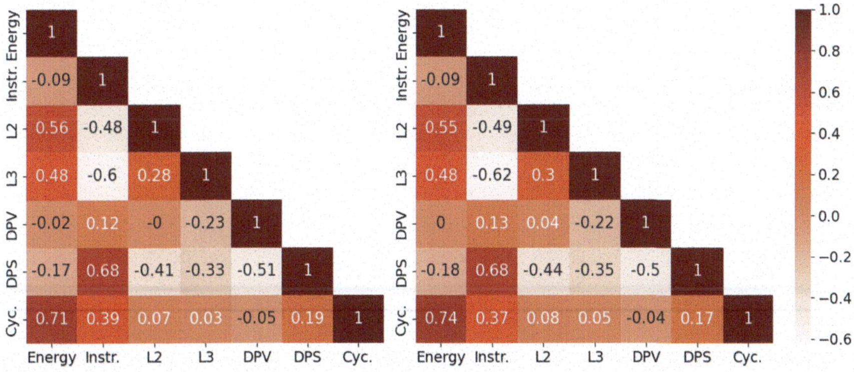

Fig. 30.15 Correlation matrix of real (left) and generated (right) data for CN

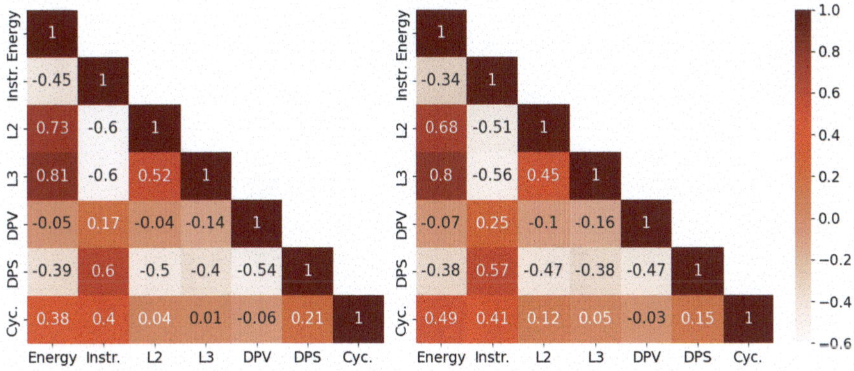

Fig. 30.16 Correlation matrix of real (left) and generated (right) data for ESB

Figure 30.17 shows the correlation matrix for DAM. The cells with the greatest difference (0.11) is L3-L2.

Figure 30.18 shows the correlation matrix for SKN. The cells with the greatest difference (0.16) is DPV-L2.

30.5.3 Analysis of Energy Model Estimation

Table 30.1 shows the results of model estimation for CN, measured using NRSME. The rows correspond to different training data. In the first row, only real data is used for training; in the second row, only generated data is used; and in the third row, both are utilized. The columns correspond to each model type, and within each model type, to the data used for evaluation in each case. Training with generated data, whether alone or in combination, manages to maintain the quality of the model when only real

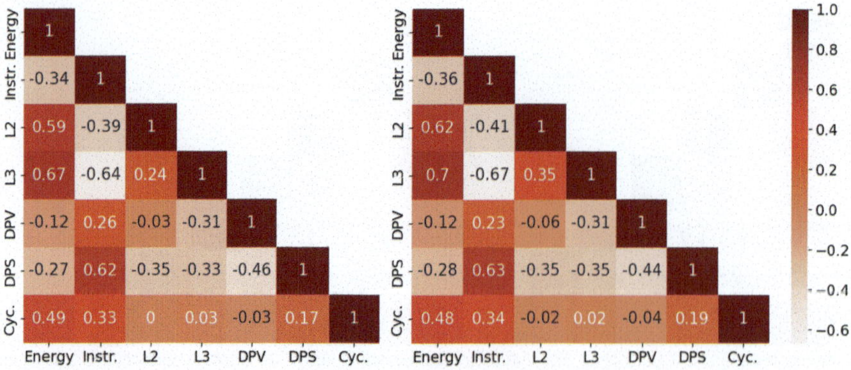

Fig. 30.17 Correlation matrix of real (left) and generated (right) data for DAM

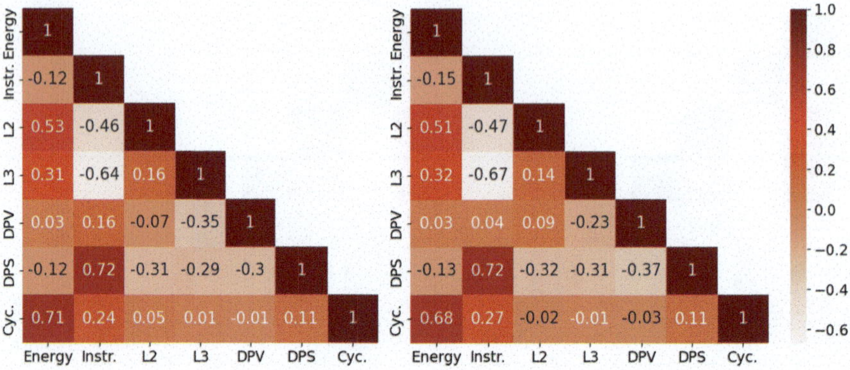

Fig. 30.18 Correlation matrix of real (left) and generated (right) data for SKN

data is used for evaluation. For example, comparing the value of CN-LRG-Real-Real with CN-LRG-Gen-Real there is only a difference of 0.09. Additionally, in the case of the CNN model and training with combined data (CN-CNN-Both-Real), it surpasses the NRMSE reported when trained solely with real data (CN-CNN-Real-Real) by 0.6.

Table 30.2 shows NRSME for ESB.

In this case, the quality of the estimation using generated data for training is also maintained at an acceptable level, although to a lesser extent than in the CN case. For example, the difference between ESB-LRG-Real-Real and ESB-LRG-Gen-Real is 1.26. It is noted that the NRSE increases to 9.50 in the ESB-LRG-Gen-Real case, possibly due to the differences in the range of the values reported in the section on distribution similarity analysis.

Table 30.3 shows the NRSME for DAM. In this case, the difference between DAM-LRG-Real-Real and DAM-LRG-Gen-Real is 2.38, resulting in worse results than the previous computing nodes. On the other hand, training with combined data

Table 30.1 NRSME of energy estimation for CN computing node

Training	Estimation								
	LRG			FCN			CNN		
	Real	Gen	Both	Real	Gen	Both	Real	Gen	Both
Real	5.79	5.54	5.59	5.99	5.90	5.86	3.33	4.93	4.47
Gen	5.88	5.39	5.56	6.17	5.64	5.79	4.50	4.18	4.29
Both	5.82	5.39	5.54	6.11	5.70	5.80	2.73	2.96	2.88

Table 30.2 NRSME of energy estimation for ESB computing node

Training	Estimation								
	LRG			FCN			CNN		
	Real	Gen	Both	Real	Gen	Both	Real	Gen	Both
Real	5.31	9.50	5.77	5.47	8.08	6.07	2.65	3.52	3.05
Gen	6.57	5.25	5.55	5.53	6.43	5.65	3.24	2.55	2.75
Both	5.88	5.47	5.36	5.49	6.36	5.49	3.53	2.38	2.74

Table 30.3 NRSME of energy estimation for DAM computing node

Training	Estimation								
	LRG			FCN			CNN		
	Real	Gen	Both	Real	Gen	Both	Real	Gen	Both
Real	5.16	5.37	5.19	5.50	5.73	5.58	2.41	3.08	2.75
Gen	7.54	4.85	6.22	7.35	5.42	6.36	5.18	2.18	3.99
Both	5.23	5.14	5.13	5.39	5.16	5.27	2.71	2.71	2.72

achieves good results when comparing the experiments DAM-FCN-Real-Real and DAM-FCN-Both-Real, where the second surpasses the first by 0.11.

Table 30.4 shows the NRSME for SKN. the difference between SKN-LRG-Real-Real and SKN-LRG-Gen-Real is 1.96. Similar to the CN and DAM cases, the experiment that trains with both data using the CNN model, SKN-CNN-Both-Real, improves upon the model that trains only with real data (SKN-CNN-Real-Real). In this case, the difference is 0.48.

30.6 Conclusion and Future Work

This work proposed to use Wasserstein Generative Adversarial Networks to augment the data of energy consumption of HPC nodes.

Relevant techniques such as boxplots, PCA and t-SNE analysis, and correlation matrices were used to describe and compare real and generated data. The results show

Table 30.4 NRSME of energy estimation for SKN computing node

Training	Estimation								
	LRG			FCN			CNN		
	Real	Gen	Both	Real	Gen	Both	Real	Gen	Both
Real	6.07	6.30	5.90	7.13	6.81	6.71	5.06	4.78	4.81
Gen	8.03	6.10	6.80	8.84	6.71	7.46	6.28	3.73	5.04
Both	6.10	6.19	5.89	6.53	6.26	6.17	4.58	3.24	3.87

that the generated data maintain statistical properties similar to the real data, such as mean, interquartile range, and variable correlation coefficients. In the PCA analysis, the generated data exhibit similar variation in most of the computing nodes studied. The t-SNE analysis also shows that the generated and real data occupy similar regions in the domain.

The generated data, in combination with real data, were used to train and evaluate energy estimation models in HPC nodes. A total of 108 experiments were conducted, combining computing node, model type, training data, and evaluation data. The results show that training with generated data, whether alone or in combination, manages to maintain the quality of the model estimates when only real data is used for evaluation. Moreover, in the case of the model using convolutional networks, the estimation improves when using a combination of real and generated data for training.

Future work includes increasing the number of computing nodes studied and comparing the results with different generation techniques.

References

1. M. Arjovsky, S. Chintala, L. Bottou, in *Proceedings of the 34th International Conference on Machine Learning*, ed. by D. Precup, Y.W. Teh. Proceedings of Machine Learning Research, vol 70. Wasserstein generative adversarial networks (PMLR, 2017), pp. 214–223
2. G. Baasch, G. Rousseau, R. Evins, A conditional generative adversarial network for energy use in multiple buildings using scarce data. Energy AI **5**, 100087 (2021)
3. H. Debapriya, Y.C.B. Wafa Shafqat, Generating synthetic data to reduce prediction error of energy consumption. Comput. Mater. Continua **70**(2), 3151–3167 (2022)
4. K. El Emam, L. Mosquera, R. Hoptroff, *Practical Synthetic Data Generation: Balancing Privacy and the Broad Availability of Data* (O'Reilly Media, 2020)
5. J. Engelmann, S. Lessmann, Conditional wasserstein gan-based oversampling of tabular data for imbalanced learning. Exp. Syst. Appl. **174**, 114582 (2021)
6. M.N. Fekri, A.M. Ghosh, K. Grolinger, Generating energy data for machine learning with recurrent generative adversarial networks. Energies **13**(1) (2020)
7. A. Figueira, B. Vaz, Survey on synthetic data generation, evaluation methods and gans. Mathematics **10**(15), 2733 (2022)
8. I. Goodfellow, J. Pouget-Abadie, M. Mirza, B. Xu, D. Warde-Farley, S. Ozair, A. Courville, Y. Bengio, in *Advances in Neural Information Processing Systems*, vol 27, ed. by Z. Ghahra-

mani, M. Welling, C. Cortes, N. Lawrence, K. Weinberger. Generative adversarial nets (Curran Associates, Inc., 2014)
9. I. Gulrajani, F. Ahmed, M. Arjovsky, V. Dumoulin, A.C. Courville, Improved training of wasserstein gans. Adv. Neural Inf. Process. Syst. **30** (2017)
10. C. Jin, X. Bai, C. Yang, W. Mao, X. Xu, A review of power consumption models of servers in data centers. Appl. Energy **265**, 114806 (2020)
11. J. Muraña, C. Navarrete, S. Nesmachnow, in *High Performance Computing*. ed. by H. Jagode, H. Anzt, H. Ltaief, P. Luszczek. Machine Learning for Generic Energy Models of High Performance Computing Resources (Springer International Publishing, Cham, 2021), pp. 314–330
12. E. Strohmaier, H. Shan, in *2005 ACM/IEEE Conference on Supercomputing*. Apex-Map: A Global Data Access Benchmark to Analyze HPC Systems and Parallel Programming Paradigms (2005), pp. 49–49
13. D. Terpstra, H. Jagode, H. You, J. Dongarra, in *Tools for High Performance Computing*. Collecting Performance Data with PAPI-C (2010), pp. 157–173
14. C. Villani, in *Optimal Transport: Old and New*, vol 338. Chapter 5: Kantorovich-Rubinstein Duality (Springer, 2009)
15. L. Xu, K. Veeramachaneni, Synthesizing tabular data using generative adversarial networks (2018). arXiv preprint arXiv:1811.11264
16. Z. Zhao, A. Kunar, R. Birke, H. Van der Scheer, L.Y. Chen, Ctab-gan+: enhancing tabular data synthesis. Front. Big Data **6** (2024)

The manufacturer's authorised representative in the EU is Springer Nature Customer Service Centre GmbH, Europaplatz 3, 69115 Heidelberg, Germany. If you have any concerns regarding our products, please contact ProductSafety@springernature.com

Printed and bound by CPI Group (UK) Ltd, Croydon, CR0 4YY
01/10/2026
02225935-0017